T0201764

INTRODUCTION TO DIGITAL COMMUNICATION SYSTEMS

INTRODUCTION TO DIGITAL COMMUNICATION SYSTEMS

Krzysztof Wesołowski
Poznań University of Technology
Poland

A John Wiley and Sons, Ltd., Publication

Library of Congress Cataloging-in-Publication Data:
Wesołowski, Krzysztof.
   Introduction to digital communication systems/Krzysztof Wesołowski.
       p. cm.
   Includes bibliographical references and index.
   ISBN 978-0-470-98629-5 (cloth)
   1. Digital communications. I. Title.
   TK5103.7.W48 2009
   004.6–dc22
                          2009015950
A catalogue record for this book is available from the British Library.

ISBN   978-0-470-98629-5 (H/B)

Typeset in 10/12 Times by Laserwords Private Limited, Chennai, India.
Printed and bound in Singapore by Markono Print Media Pte Ltd.

To my wife Maria

Contents

Preface

Knowledge of basic rules of operation of digital communication systems is a crucial factor in understanding contemporary communications. Digital communication systems can be treated as a medium for many different systems and services. Digital TV, cellular telephony or Internet access are only three prominent examples of such services. Basically, each kind of communication between human beings and between computers requires a certain kind of transmission of digitally represented messages from one location to another, or, alternatively, from one time instant to another, as it is in the case of digital storage. It often happens in technology that its current state is a result of a long engineering experience and numerous experiments. However, most of the developments in digital communications are the result of deep theoretical studies. Thus, theoretical knowledge is needed to understand the operation of many functional blocks of digital communication systems.

There are numerous books devoted to digital communication systems and they are written for different readers; simpler books are directed to undergraduate students specializing in communication engineering, whereas more advanced ones should be a source of knowledge for graduate or doctoral students. The number of topics to be described and the details to be explained grow very quickly, so some of these books are very thick indeed. As a result, there is a problem of appropriate selection of the most important topics, leaving the rest to be studied in more specialized books.

The author of this textbook has tried to balance the number of interesting topics against the moderate size of the book by showing the rules of operation of several communication systems and their functional blocks rather than deriving deep analytical results. Whether this aim has been achieved can be evaluated by the reader. This textbook is the result of many years of lectures read to students of Electronics and Telecommunications at Poznań University of Technology. One-semester courses were devoted to separate topics reflected in the book chapters, such as information theory, channel coding and digital modulations. The textbook was first published in Polish. The current English version is an updated and extended translation of the Polish original. To make this textbook more attractive and closer to the telecommunication practice, almost each chapter has been enriched with a case study that shows practical applications of the material explained in this chapter.

Unlike many other textbooks devoted to digital communication systems, we start from the basic course on information theory in Chapter 1. This approach gives us some knowledge on basic rules and performance limitations and ideas that are applied later in the following chapters. Such an approach allows us to consider a digital communication system in a top-to-bottom direction, i.e. starting from very general rules and models and going deeper into particular solutions and details.

Chapter 2 is devoted to protection of digital messages against errors. The basic rules of this protection are derived from information theory. We start from very simple error correction codes and end up with basic information on turbo codes and LDPC codes. Error detection codes and several automatic request-to-repeat strategies are also tackled.

The subject of Chapter 3 is the baseband transmission. We show how to shape baseband pulses and how to form the statistical properties of data symbols in order to achieve the desired spectral properties of the transmitted signal. We derive the structure of the optimum synchronous receiver and we analyze basic methods of digital signaling.

In Chapter 4 we use our results derived in Chapter 3 for analysis of passband transmission and digital modulations of a sinusoidal carrier. We consider simple one- and more dimensional modulations, continuous phase modulations, trellis-coded modulations and present respective receivers. In most cases we derive the probability of erroneous detection in selected types of receivers.

In Chapters 3 and 4 we consider baseband and passband digital signaling assuming an additive Gaussian noise and limited channel bandwidth as the only impairments. In turn, Chapter 5 is devoted to the description of representative physical channel properties. Such considerations allow us to evaluate the physical limitation that can be encountered in practice.

One such limitation occurring in band-limited digital communication systems is intersymbol interference. This phenomenon is present in many practical cases and many digital communication systems have to cope with it. The methods of eliminating intersymbol interference or decreasing its influence on the system performance are presented in Chapter 6.

Chapter 7 overviews basic types of digital communication systems based on the spread spectrum principle. Many contemporary communication systems, in particular wireless ones, use spectrum spreading for reliable communications.

Synchronization is another important topic that must be understood by a communication engineer. Basic synchronization types and configurations are explained in Chapter 8.

Finally, Chapter 9 concentrates on the overview of multiple access methods, including new methods based on multicarrier modulations.

Most of the chapters are appended with the problems that could be solved in the problem sessions accompanying the lecture.

This book would not be in its present form if it had not been given attention and time by many people. First of all, I would like to direct my thanks to the anonymous reviewers of the English book proposal, who encouraged me to enrich the book with some additional problems and slides that could be useful for potential lecturers using this book as a basic source of material. I am also grateful to Mark Hammond, the Editorial Director of John Wiley & Sons Ltd, and Sarah Tilley, the Project Editor, who showed their patience and help. Someone who substantially influenced the final form of the book is Mrs Krystyna Ciesielska (MA, MSc) who was the language consultant and as an electrical engineer was a particularly critical reader of the English translation. I would like to thank Mr Włodzimierz Mankiewicz who helped in the preparation of some drawings. Finally, the book would not have appeared if I did not have the warm support of my family, in particular my wife Maria.

KRZYSZTOF WESOŁOWSKI

About the Author

 Krzysztof Wesołowski has been employed at Poznan University of Technology (PUT), Poznan, Poland, since 1976. He received PhD and *Doctor Habilitus* degrees in communications from PUT in 1982 and 1989, respectively. Since 1999 he has held the position of Full Professor in telecommunications. Currently he is Head of the Department of Wireless Communications at the Faculty of Electronics and Telecommunications at PUT. In his scientific activity he specializes in digital wireline and wireless communication systems, information and coding theory and DSP applications in digital communications. He is the author or co-author of more than 100 scientific publications, including the following books: "Systemy radiokomunikacji ruchomej" (in Polish, WKL, Warsaw, 1998, 1999, 2003), translated into English as "Mobile Communication Systems", John Wiley & Sons, Chichester, 2003, and into Russian as "Sistiemy podvizhnoy radiosvyazi", Hotline Telecom, Moscow, 2006, and "Podstawy cyfrowych systemow telekomunikacyjnych" (in Polish, WKL, Warsaw, 2003). The current book is an extended and updated translation of the latter publication. He published his results, among others, in *IEEE Transactions on Communications*, *IEEE Journal on Selected Areas in Communications*, *IEEE Transactions on Vehicular Technology*, *IEE Proceedings*, *European Transactions on Telecommunications*, *Electronics Letters* and *EURASIP Journal on Wireless Communications and Networking*.

Professor Wesołowski was a Postdoctoral Fulbright Scholar at Northeastern University, Boston, in 1982–1983 and a Postdoctoral Alexander von Humboldt Scholar at the University of Kaiserslautern, Germany, in 1989–1900. He also worked at the University of Kaiserslautern as a Visiting Professor. His team participates in several international research projects funded by the European Union within the Sixth and Seventh Framework Programs.

1

Elements of Information Theory

In this chapter we introduce basic concepts helpful in learning the rules of operation of digital communication systems that have their origin in information theory. We present basic theorems of information theory that establish the limits on effective representation of messages using symbol sequences, i.e. we consider the limits of *source coding*. We analyse the conditions for ensuring reliable transmission over distorting channels with the maximum data rate. Sometimes we encounter complaints that information theory sets the limits on the communication system parameters without giving recipes on how to reach them. As modern communication systems are becoming more and more sophisticated, the information theory hints are more and more valuable in optimization of these systems. Therefore, knowing its basic results seems to be necessary for better understanding of modern communication systems.

1.1 Introduction

As already mentioned, only basic concepts and the most important results of information theory are presented in this chapter. The reader who is interested in more detailed knowledge on information theory can find a number of books devoted to this interesting discipline, such as the classical book by Abramson (1963) and others by Gallager (1968), Cover and Thomas (1991), Mansuripur (1987), Heise and Quatrocchi (1989), Roman (1992), Blahut (1987) or MacKay (2003). Their contents and level of presentation are different and in some cases the reader should have a solid theoretical background to profit from them. Some other books feature special chapters devoted to information theory, e.g. Proakis' classics (Proakis 2000) and the popular handbook by Haykin (2000).

The contents of the current chapter are as follows. First, we introduce the concept of an amount of information, and we present various message source models and their properties. Then we introduce and discuss the concept of source entropy. We proceed to the methods of source coding and we end this part of the chapter with Shannon's theorem on source coding. We also give some examples showing source coding in practical applications such as data compression algorithms.

The next section is devoted to discrete memoryless channel models. The concepts of mutual information and channel capacity are introduced in the context of message transmission over memoryless channels. Then, the notion of a decision rule is defined

Introduction to Digital Communication Systems   Krzysztof Wesołowski
© 2009 John Wiley & Sons, Ltd

and a few decision rules are derived. Subsequently, we present the basic Shannon's theorem showing conditions that have to be fulfilled to ensure reliable transmission over distorting channels. These conditions motivate the application of channel coding. Next, we extend our considerations on mutual information and related issues onto continuous random variables. The concept of differential entropy is introduced. The achieved results are applied to derive the formula describing the capacity of a band-limited channel with additive white Gaussian noise. Some practical examples illustrating the meaning of this formula are given. Then, the channel capacity formula is extended onto channels with a specified transfer function and distorted by Gaussian noise with a given power spectral density. Channel capacity and signaling strategy are also considered for time varying, flat fading channels. Finally, channel capacity is considered for cases when transmission takes place over more than one transmit and/or more than one receive antenna, i.e., capacity of multiple-input multiple-output channels is derived.

1.2   Basic Concepts

However amazing it may seem, the foundations for information theory were laid in a single forty-page-long paper written by a then young scientist, Claude Shannon (1948). From that moment this area developed very quickly, providing the theoretical background for rapidly developing telecommunications. Information theory was also treated as a tool for the description of phenomena that were far from the technical world, with varying success.

Although Shannon founded the whole discipline, the first elements of information theory can already be found a quarter of a century earlier. H. Nyquist in his paper entitled "Certain Factors Affecting Telegraph Speed" (Nyquist 1924) formulated a theorem on the required sampling frequency of a band-limited signal. He showed indirectly that time in a communication system has a discrete character because in order to acquire full knowledge of an analog signal it is sufficient to know the signal values in sufficiently densely located time instants.

The next essential contribution to information theory was given by R. V. L. Hartley, who in his work entitled "Information Transmission" (Hartley 1928) associated the information content of a message with the logarithm of the number of all possible messages that can be observed on the output of a given source.

However, the crucial contribution to information theory came from Claude Shannon who in 1948 presented his famous paper entitled "A Mathematical Theory of Communication" (Shannon 1948). The contents of this paper are considered to be so significant that many works written since that time have only supplemented the knowledge contained in Shannon's original paper.

So what indeed is information theory? And what is the subject of its sister discipline – coding theory?

Information theory formulates performance limits and states conditions that have to be fulfilled by basic functional blocks of a communication system in order for a certain amount of information to be transferred from its source (sender) to the sink (recipient). Coding theory in turn gives the rules of protecting the digital signals representing sequences of messages from errors, which ensure sufficiently low probability of erroneous reception at the receiver.

1.3   Communication System Model

Before we formulate basic theorems of information theory let us introduce a model of a communication system. As we know, a model is a certain abstraction or simplification of reality; however, it contains essential features allowing the description of basic phenomena occurring in reality, neglecting at the same time those features that are insignificant or rare.

Let us first consider a model of a discrete communication system. It is conceptually simpler than a model of a continuous system and reflects many real cases of transmission in digital communication systems in which a source generates discrete messages. The case of a continuous system will be considered later on.

A model of a discrete communication system is shown in Figure 1.1.

Its first block is a *message source*. We assume that it generates messages selected from a given finite set of *elementary messages* at a certain clock rate. We further assume that the source is stationary, i.e. its statistical properties do not depend on time. In particular, messages are generated with specified probabilities that do not change in time. In other words, the probability distribution of the message set does not depend on a specific time instant.[1] The properties of message sources will be discussed later.

The *source encoder* is a functional block that transforms the message received from the message source into a sequence of elementary symbols. This sequence in turn can be further processed in the next blocks of the communication system. The main task of the source encoder is to represent messages using the shortest possible sequences of elementary symbols, because the most frequent limitation occurring in real communication systems is the maximum number of symbols that can be transmitted per time unit.

The *channel encoder* processes the symbols received from the source encoder in a manner that guarantees reliable transmission of these symbols to the receiver. The channel encoder usually divides the input sequence into disjoint blocks and intentionally augments each input block with certain additional, redundant symbols. These symbols allow

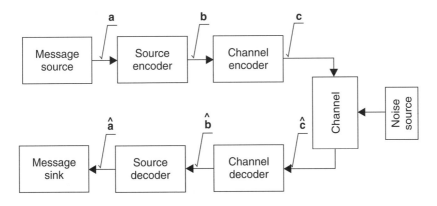

Figure 1.1   Basic model of a discrete communication system

---

[1] As we remember from probability theory, this feature is called *stationarity in a narrow sense*.

the decoder to make a decision about the transmitted block with a high probability of correctness despite errors made on some block symbols during their transmission.

The *channel* is the element of a communication system that is independent of other system blocks. In the scope of information theory a channel is understood as a serial connection of a certain number of physical blocks whose inclusion and structure depend on the construction of the specific, considered system. In this sense, the channel block can represent for example a mapper of the channel encoder output symbols into data symbols, a block shaping the waves representing the data symbols and matching them to the channel bandwidth, and a modulator that shifts the signal into the passband of the physical channel. The subsequent important block of the channel is the physical transmission channel, which reflects the properties of the transmission medium. It is probably obvious to each reader that, for example, a pair of copper wires operating as a subscriber loop has different transmission properties than a mobile communication channel. On the receiver side the channel block can contain an amplifier, a demodulator, a receive filter, and a decision device producing the estimates of the signals acceptable by the channel decoder. These estimates sometimes can be supplemented by additional data informing the following receiver blocks about the reliability of the supplied symbols. Figure 1.2 presents a possible scheme of part of a communication system that can be integrated in the form of a channel block.

A channel can have spacial or time character. A spacial channel is established between a sender and recipient of messages who are located in different geographical places. Communication systems that perform such message transfer are called *telecommunication* (or *communication*) *systems*. We speak about time channels, on the other hand, with reference to computer systems, in which signals are stored in memory devices such as tape, magnetic or optical disk, and after some time are read out and sent to the recipient. The properties of a memory device result from its construction and the physical medium on which the memory is implemented.

Estimates of signal sequences received on the channel output are subsequently processed in a functional block called a *channel decoder*. Its task is to recover the transmitted signal block on the basis of the signal block received on the channel output. The channel decoder applies the rule according to which the channel encoder produces its output signal blocks. Typically, a channel decoder memorizes the signals received from the channel in the form

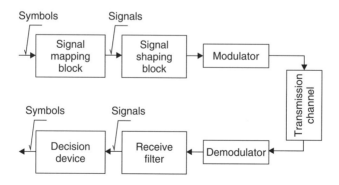

Figure 1.2 Example of the internal structure of the channel block

of n-element blocks, and on this basis attempts to recover such a k-element block, which uniquely indicates a particular n-element block that is "the most similar" to the received n-element block. Three cases are possible:

- On the basis of the channel output block, the channel decoder reconstructs the signal block that was really transmitted.
- The channel decoder is not able to reconstruct the transmitted block, however it detects the errors in the received block and informs the receiver about this event.
- The channel decoder selects the signal block; however it is different from the block that was actually transmitted. Although the decision is false, the block is sent for further processing.

If the communication system has been correctly designed the latter case occurs with an extremely low probability.

The task of a *source decoder* is to process the symbol blocks produced by the channel decoder to obtain a form that is understandable to the recipient (*message sink*).

Example 1.3.1 *As an example of a communication system, let us consider transmission of human voice over the radio. There are many ways to assign the particular elements of such a system to the functional blocks from Figure 1.1. One of them is presented below. Let the human brain be the source of messages. Then the vocal tract can be treated as a source encoder, which turns the messages generated by the human brain into acoustic waves. The channel encoder is the microphone, which changes the acoustic wave into electrical signals. The channel is a whole sequence of blocks, the most important of which are the amplifier, radio transmitter with transmit antenna, physical radio channel, receive antenna and receiver. The loudspeaker plays the role of a channel decoder, which converts the received radio signal into an acoustic signal. This signal hits the human ear, which can be considered as a source decoder. Through the elements of the nervous system the "decoded" messages arrive in the human brain – the message sink.*

Let us now consider a more technical example.

Example 1.3.2 *Let the message source be a computer terminal. Alphanumeric characters (at most 256 if the ASCII code is applied) are considered as elementary messages. The source encoder is the block that assigns an 8-bit binary block (byte) to each alphanumeric character according to the ASCII code. Subsequent bytes representing alphanumeric characters are grouped into blocks of length k, which is a multiple of eight. Each k-bit block is supplemented with r appropriately selected additional bits. The above operation is in fact channel coding. Its aim is to protect the information block against errors. The resulting binary stream is fed to the modem input. The latter device turns the binary stream into a form that can be efficiently transmitted over a telephone channel. On the receive side the signal is received by the modem connected to a computer server. The cascade of functional elements consisting of a modem transmitter, a telephone channel and a modem receiver is included in the channel block in the sense of the considered communication system model. On the receive side, based on the reception of the k-bit block, r additional bits are derived and compared with the additional received bits. This operation constitutes channel decoding. Next, the transmitter of the modem on the server side sends a short feedback signal to* *the modem on the remote terminal side informing the latter about the required operation, depending on the result of comparison of the calculated and received redundant bits; it can be the transmission of the next binary block if both bit blocks are identical, or block repetition if the blocks are not identical. The division of the accepted k-bit block into bytes and assigning them appropriate alphanumeric blocks displayed on the screen or printed by the printer connected to the server is a source decoding process. Thus, a printer or a display monitor can be considered as a message sink.*

The above example describes a very simple case of a digital transmission with an automatic request to repeat erroneous blocks. The details of such an operation will be given in the next chapter.

1.4   Concept of Information and Measure of Amount of Information

The question "what is information?" is almost philosophical in nature. In the literature one can find different answers to this question. Generally, information can be described in the following manner.

Definition 1.4.1 *Information is a piece of knowledge gained on the reception of messages that allows the recipient to undertake or improve his/her activity* (Seidler 1983).

This general definition implies two features of information:

- potential character – it can, but need not, be utilized in the recipient's current activity;
- relative character – what can be valuable knowledge for one particular recipient can be disturbance for another recipient.

Let us note that we have not defined the notion of *message*. We will treat it as a *primary idea*, as with a *point* or a *straight line* in geometry, which are not definable in it.

A crucial feature associated with information transfer is energy transfer. A well constructed system transmitting messages transfers a minimum amount of energy required to ensure an appropriate quality of received signal.

The definition of information given above has a descriptive character. In science it is often required to define a measure of quantity of a given value. Such a measure is the amount of information and should result from the following intuitive observations:

- If we are certain about the message that occurs on the source output, there is no information gained by observing this message.
- The occurrence of a message either provides some or no information, but never brings about a loss of information.
- The more unexpected the received message is, the more it can influence the recipient's activity; the amount of information contained in a message should be associated with the message probability of appearance – the lower the probability of message occurrence, the higher the amount of information contained in it.
- Observation of two statistically independent messages should be associated with the amount of information, which is the sum of amounts of information gained by observation of each message separately.

The above requirements for measure of information are reflected in the definition given by Hartley.

Definition 1.4.2 *Let a be a message that is emitted by the source with a probability $P(a)$. We say that on observing message a, its recipient acquires*

$$I(a) = \log_r \frac{1}{P(a)} \tag{1.1}$$

units of amount of information.

In information theory the logarithm base r is usually equal to 2 and then the unit of amount of information is called a *bit*.[2] The logarithm base $r = e$ implies denoting the unit of amount of information as a *nat*, whereas taking $r = 10$ results in a unit of amount of information described as *Hartley*. Unless stated otherwise, in the current chapter the logarithm symbol will denote the logarithm of base 2.

From the above definition we can draw the following conclusion: Gaining a certain amount of information due to observation of the specified message on the source output is associated with a stochastic nature of the message source.

1.5   Message Sources and Source Coding

In this section we will focus our attention on the description of message sources. We will present basic source models and describe their typical parameters. We will define the concepts of entropy and conditional entropy. We will also consider basic rules and limits of source coding. We will quote Shannon's theorem about source coding. We will also present some important source coding algorithms applied in communication and computer practice.

1.5.1   Models of Discrete Memory Sources

As we have already mentioned, a message source has a stochastic nature. Thus, its specification should be made using the tools of description of random signals or sequences. In consequence, a sequence of messages observed on the source output can be treated as a sample function of a stochastic process or of a random sequence. A source generates messages by selecting them from the *set of elementary messages*, called the *source alphabet*. The source alphabet can be continuous or discrete. In the first case, in an arbitrarily close neighborhood of an elementary message another elementary message can be found. In the case of a discrete message source the messages are countable, although their number can be infinitely high. A source is discrete and finite if its elementary messages are countable and their number is finite. In the following sections we will concentrate on the models of discrete sources, leaving the problems of continuous sources for later consideration.

---

[2] We should not confuse "bit" denoting a measure of amount of information with a "bit", which is a binary symbol taking two possible values, "0" or "1".

1.5.2 Discrete Memoryless Source

The simplest source model is the model of a discrete memoryless source. Source memory is considered as a statistical dependence of subsequently generated messages. A source is memoryless if generated messages are statistically independent. It implies that the probability of generation of a specific message at a given moment does not depend on what messages have been generated before. Let us give a formal definition of a discrete memoryless source.

Definition 1.5.1 *Let $X = \{a_1, \ldots, a_K\}$ be a discrete and finite set of elementary messages generated by source* **X**. *We assume that this set is time invariant. Source* **X** *is discrete and memoryless if elementary messages are selected mutually independently from set X in conformity with the time-invariant probability distribution $\{P(a_1), \ldots, P(a_K)\}$.*

In order to better characterize the properties of a discrete memoryless source we will introduce the notion of average amount of information, which is acquired by observation of a single message on the source output. An average amount of information is a weighted sum of the amount of information acquired by observing subsequently all elementary messages from the source with the alphabet X, where the weights of particular messages are the probabilities of occurrence of these messages. In the mathematical sense, this value is an ensemble average (*expectation*) of the amount of information $I(a_i)$. It is denoted by the symbol $H(X)$ and called the entropy of source **X**. Formalizing the above considerations, we will give the definition of the entropy of the source **X**.

Definition 1.5.2 *The entropy of a memoryless source* **X**, *characterized by the alphabet $X = \{a_1, \ldots, a_K\}$ and the probability distribution $\{P(a_1), \ldots, P(a_K)\}$, is the average amount of information acquired by observation of a single message on the source output, given by the formula*

$$H(X) = E[I(a_i)] = \sum_{i=1}^{K} P(a_i) \log \frac{1}{P(a_i)} \tag{1.2}$$

Since the source entropy is the average amount of information acquired by observation of a single message, its unit is also a bit. The source entropy characterizes our uncertainty in guessing which message will be generated by the source in the next moment (or generally in the future). The value of entropy results from the probability distribution of elementary messages, therefore the following properties hold.

Property 1.5.1 *Entropy $H(X)$ of a memoryless source* **X** *is non-negative.*

Proof. Since for each elementary message of the source **X** the following inequality holds

$$1 \geq P(a_i) > 0, \quad (i = 1, \ldots, K)$$

then for each message a_i

$$\log \frac{1}{P(a_i)} \geq 0$$

which implies that the weighted sum of the above logarithms is non-negative as well, i.e.

$$\sum_{i=1}^{K} P(a_i) \log \frac{1}{P(a_i)} \geq 0$$

It can be easily checked that the entropy is equal to zero, i.e. it achieves its minimum if and only if a certain message a_j from the source alphabet X is sure (i.e. $P(a_j) = 1$). This implies the fact that the alphabet reduces to a single message. The amount of information acquired by observing this message is zero, in other words, our uncertainty associated with forthcoming messages is null.

Property 1.5.2 *The entropy of a memoryless source does not exceed the logarithm of the number of elementary messages constituting its alphabet, i.e.*

$$H(X) \leq \log K \tag{1.3}$$

Proof. We will show that $H(X) - \log K \leq 0$, using the formula allowing calculation of the logarithm to the selected base, given the value of the logarithm to a different base

$$\log_r x = \frac{\log_a x}{\log_a r} \tag{1.4}$$

Knowing that $\sum_{i=1}^{K} P(a_i) = 1$, we have

$$H(X) - \log K = \sum_{i=1}^{K} P(a_i) \log \frac{1}{P(a_i)} - \sum_{i=1}^{K} P(a_i) \log K$$

$$= \sum_{i=1}^{K} P(a_i) \log \frac{1}{K P(a_i)}$$

Recall that the logarithm base $r = 2$. In the proof we will apply the inequality $\ln x \leq x - 1$ (cf. Figure 1.3) and the formula

$$\log x = \ln x \log e$$

We have

$$H(X) - \log K = \log e \sum_{i=1}^{K} P(a_i) \ln \frac{1}{K P(a_i)}$$

$$\leq \log e \sum_{i=1}^{K} P(a_i) \left(\frac{1}{K P(a_i)} - 1 \right) = \log e \left(\sum_{i=1}^{K} \frac{1}{K} - \sum_{i=1}^{K} P(a_i) \right) = 0$$

so indeed

$$H(X) - \log K \leq 0$$

which concludes the proof.

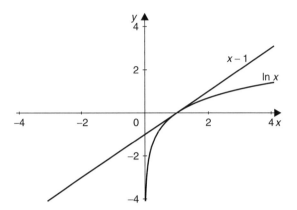

Figure 1.3   Plots of the functions $\ln x$ and $x - 1$ (Goldsmith and Varaiya (1997)) © 1997 IEEE

In this context a question arises when the entropy is maximum, i.e. what conditions have to be fulfilled to have $H(X) = \log K$. In the proof of Property 1.2 we applied the boundary $\ln x \leq x - 1$ separately for each element $1/K P(a_i)$. One can conclude from Figure 1.3 that the function $\ln x$ is bounded by the line $x - 1$ and the boundary is exact, i.e. $\ln x = x - 1$ if $x = 1$. In our case, in order for the entropy to be maximum and equal to $\log K$, for each elementary message a_i the following equality must hold

$$\frac{1}{K P(a_i)} = 1, \quad \text{i.e. } P(a_i) = \frac{1}{K} \ (i = 1, \ldots, K) \tag{1.5}$$

It means that the entropy of the memoryless source is maximum if the probabilities of occurrence of each message are the same. It also means that uncertainty with respect to our observation of the source messages is the highest – none of the messages is more probable than the others.

Consider now a particular example – a memoryless source with a two-element alphabet $X = \{a_1, a_2\}$. Let the probability of message a_1 be $P(a_1) = p$. The sum of probabilities of generation of all the messages is equal to 1, so $P(a_2) = 1 - p = \overline{p}$. Therefore, the entropy of this two-element memoryless source is

$$H(X) = p \log \frac{1}{p} + \overline{p} \log \frac{1}{\overline{p}} \tag{1.6}$$

As we see, the entropy $H(X)$ is a function of probability p. Therefore let us introduce the so-called *entropy function* given by the formula

$$H(p) = p \log \frac{1}{p} + \overline{p} \log \frac{1}{\overline{p}} \tag{1.7}$$

The plot of the entropy function, which will be useful in our future considerations, is shown in Figure 1.4. For obvious reasons (its argument has a sense of probability) the function has the argument in the range $(0, 1)$. The values of the entropy function are contained in the range $(0, 1]$, achieving maximum for $p = 0.5$, which agrees with formula (1.5).

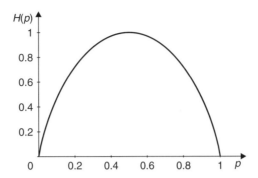

Figure 1.4   Plot of the entropy function versus probability p

1.5.3   Extension of a Memoryless Source

A discrete memoryless source is the simplest source model. A slightly more sophisticated model is created if an n-element block of messages subsequently generated by a memoryless source $\mathbf{X}$ is treated jointly as a single message from a new message source, called the nth *extension of source* $\mathbf{X}$. We will now present a formal definition of an nth extension of source $\mathbf{X}$.

Definition 1.5.3 *Let a memoryless source* $\mathbf{X}$ *be described by an alphabet* $X = \{a_1, \ldots, a_K\}$ *and associated probability distribution of the elementary messages* $\{P(a_1), \ldots, P(a_K)\}$. *The nth extension of the source* $\mathbf{X}$ *is a memoryless source* $\mathbf{X}^n$, *which is characterized by a set of elementary messages* $\{b_1, \ldots, b_{K^n}\}$ *and the associated probability distribution* $\{P(b_1), \ldots, P(b_{K^n})\}$, *where message* b_j $(j = 1, \ldots, K^n)$ *is defined by a block of messages from source* $\mathbf{X}$

$$b_j = (a_{j_1}, a_{j_2}, \ldots, a_{j_n}) \tag{1.8}$$

Index j_i $(i = 1, \ldots, n)$ *may take the values from the interval* $(1, \ldots, K)$, *and the probability of occurrence of message* b_j *is equal to*

$$P(b_j) = P(a_{j_1}) \cdot P(a_{j_2}) \cdot \ldots \cdot P(a_{j_n}) \tag{1.9}$$

The number of messages of the nth source extension $\mathbf{X}^n$ is equal to K^n. Messages of $\mathbf{X}^n$ are all n-element combinations of the messages of the primary source $\mathbf{X}$.

Let us calculate the entropy of the source extension described above. The entropy value can be derived from the following theorem.

Theorem 1.5.1 *The entropy of the nth extension* $\mathbf{X}^n$ *of a memoryless source* $\mathbf{X}$ *is equal to the nth multiple of the entropy* $H(X)$ *of source* $\mathbf{X}$.

Proof. The entropy of source $\mathbf{X}^n$ is given by the formula

$$H(X^n) = \sum_{j=1}^{K^n} P(b_j) \log \frac{1}{P(b_j)}$$

However, message b_j is a message block described by expression (1.8), with probability given by formula (1.9). Therefore enumerating all subsequent messages by selection of the whole index block $(j_1, j_2, \ldots, j_n)$, $j_i = 1, 2, \ldots K$ $(i = 1, 2, \ldots n)$, we obtain the n-fold sum

$$H(X^n) = \sum_{j_1=1}^{K} \sum_{j_2=1}^{K} \cdots \sum_{j_n=1}^{K} P(a_{j_1}) \cdot \ldots \cdot P(a_{j_n}) \log \frac{1}{P(a_{j_1}) \cdot \ldots \cdot P(a_{j_n})} \tag{1.10}$$

Knowing that the logarithm of the product of factors is equal to the sum of logarithms of those factors, we can write formula (1.10) in the form

$$H(X^n) = \sum_{j_1=1}^{K} \sum_{j_2=1}^{K} \cdots \sum_{j_n=1}^{K} P(a_{j_1}) \cdot \ldots \cdot P(a_{j_n}) \log \frac{1}{P(a_{j_1})} + \cdots$$

$$+ \sum_{j_1=1}^{K} \sum_{j_2=1}^{K} \cdots \sum_{j_n=1}^{K} P(a_{j_1}) \cdot \ldots \cdot P(a_{j_n}) \log \frac{1}{P(a_{j_n})} \tag{1.11}$$

Consider a single component of formula (1.11), in which the argument of the logarithm is $1/P(a_{j_1})$. Exclude in front of the appropriate sums the factors that do not depend on the index with respect to which the sum is performed. Then we obtain

$$\sum_{j_1=1}^{K} \sum_{j_2=1}^{K} \cdots \sum_{j_n=1}^{K} P(a_{j_1}) \cdot \ldots \cdot P(a_{j_n}) \log \frac{1}{P(a_{j_1})}$$

$$= \sum_{j_1=1}^{K} P(a_{j_1}) \log \frac{1}{P(a_{j_1})} \sum_{j_2=1}^{K} P(a_{j_2}) \cdots \sum_{j_n=1}^{K} P(a_{j_n})$$

In turn, knowing that the sum of probabilities of all elementary messages of source **X** is equal to 1, we receive the following expression describing the above component

$$\sum_{j_1=1}^{K} \sum_{j_2=1}^{K} \cdots \sum_{j_n=1}^{K} P(a_{j_1}) \cdot \ldots \cdot P(a_{j_n}) \log \frac{1}{P(a_{j_1})}$$

$$= \sum_{j_1=1}^{K} P(a_{j_1}) \log \frac{1}{P(a_{j_1})} = H(X) \tag{1.12}$$

Performing similar steps for all remaining $n - 1$ components, we obtain the same result, i.e. each component is equal to entropy $H(X)$. Adding these results together, we obtain the thesis of the theorem, i.e. the formula

$$H(X^n) = nH(X) \tag{1.13}$$

Example 1.5.1 *Consider a memoryless source* **X** *with the alphabet* $X = \{a_1, a_2, a_3\}$ *and associated probability distribution* $\{P(a_1), P(a_2), P(a_3)\} = \{\frac{1}{2}, \frac{1}{4}, \frac{1}{4}\}$. *In the table below we describe the second extension* X^2 *of source* **X** *by giving its elementary messages and associated probability distribution. We also calculate the source entropy and compare it with the entropy of source* **X**.

Messages of X^2	b_1	b_2	b_3	b_4	b_5	b_6	b_7	b_8	b_9
Messages of X	a_1a_1	a_1a_2	a_1a_3	a_2a_1	a_2a_2	a_2a_3	a_3a_1	a_3a_2	a_3a_3
$P(b_j)$	$\frac{1}{4}$	$\frac{1}{8}$	$\frac{1}{8}$	$\frac{1}{8}$	$\frac{1}{16}$	$\frac{1}{16}$	$\frac{1}{8}$	$\frac{1}{16}$	$\frac{1}{16}$

The entropy of source X^2 *can be calculated on the basis of this table. The entropy of source* **X** *is*

$$H(X) = \frac{1}{2}\log 2 + \frac{1}{4}\log 4 + \frac{1}{4}\log 4 = \frac{3}{2}$$

whereas the entropy of the second extension, X^2, *of source* **X** *is*

$$H(X^2) = \frac{1}{4}\log 4 + 4 \cdot \frac{1}{8}\log 8 + 4 \cdot \frac{1}{16}\log 16 = 3$$

so in fact $H(X^2) = 2H(X)$.

1.5.4   Markov Sources

A discrete memoryless source is a very simple model and it does not reflect sufficiently precisely how the messages or their sequences are generated by the source. A simple example such as a text written in a specified language, in which alphanumerical characters are treated as elementary messages, shows us that subsequent messages are statistically dependent on each other. There exist typical combinations of characters constituting words in a given language while some other combinations do not occur. Thus, messages are statistically dependent. A model that takes statistical dependence of generated messages into account is called a *model of Markov sequences*. Below we give its formal definition.

Definition 1.5.4 *Let* **X** *be a source with the message alphabet* $X = \{a_1, \ldots, a_K\}$. *We say that source* **X** *is a Markov source of the mth order, if the probability of gener- ation of a message* $x_i \in \{a_1, \ldots, a_K\}$ *in the ith time instant depends on the sequence of m messages generated by the source in the previous moments. This means that the Markov source is described by the alphabet* X *and the set of conditional probabilities* $\{P(x_i|x_{i-1}, \ldots, x_{i-m})\}$, *where* $x_{i-j} \in X$, $(j = 0, \ldots, m)$.

The message block $(x_{i-1}, \ldots, x_{i-m})$ describes the current state of a Markov source. Since in the $(i-j)$th moment the source can generate one of K messages from its alphabet, the number of possible states is equal to K^m. As message x_i is generated at the

ith timing instant, the source evolves from the state $(x_{i-1}, \ldots, x_{i-m})$ in the ith moment to the state $(x_i, \ldots, x_{i-m+1})$ in the next moment.

A Markov source can be efficiently described by its state diagram, as it is done when describing automata. The *state diagram* presents all K^m source states with appropriate connections reflecting possible transitions from the state in the ith moment to the state in the $(i+1)$st moment and their probabilities.

Example 1.5.2 *Figure 1.5 presents the state diagram of a second-order Markov source with a two-element alphabet $X = \{0, 1\}$ and with the following conditional probabilities $\{P(x_i | x_{i-1}, x_{i-2})\}$ given below*

$$P(0|00) = P(1|11) = 0.6$$

$$P(1|00) = P(0|11) = 0.4$$

$$P(0|01) = P(0|10) = P(1|01) = P(1|10) = 0.5$$

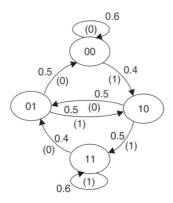

Figure 1.5   Example of the state diagram of a second-order Markov source

In a typical situation, we consider *ergodic* Markov sources. Let us recall that a random process is ergodic if time averages of any of its sample functions are equal (with probability equal to 1) to the adequate ensemble average calculated in any time instant. One can also describe a Markov source as ergodic (Abramson 1963) if it generates a "typical" sequence of messages with a unit probability. Below we show an example of a source that does not fulfill this condition, i.e. that is not ergodic (Abramson 1963).

Example 1.5.3 *Consider a Markov source of second order with the binary alphabet $X = \{0, 1\}$. Let its probability distribution have the form*

$$P(0|00) = P(1|11) = 1$$

$$P(1|00) = P(0|11) = 0$$

$$P(0|01) = P(0|10) = P(1|01) = P(1|10) = 0.5$$

The state diagram of this source is given in Figure 1.6. If following the generation of a specific message sequence the message source achieves state 00 or 11, it will stay in it

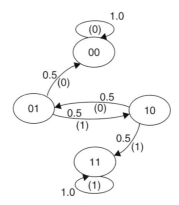

Figure 1.6 State diagram of the Markov source considered in Example 1.5.3

forever. Let us assume that each initial state of the source is equiprobable. If the source generates a sufficiently large number of messages with the probability equal to 0.5, it will reach either state 00 or state 11 with the same probability. After reaching state 00 the following sequence of messages will have the form 000 . . . . Similarly, from the moment of achieving state 11 the source will emit an infinite sequence 111 . . . . We see that none of the sequences is typical and the time averages calculated for both sample functions of the process of message generation are different. On the basis of a single sample function one cannot estimate the probability that the source is in a given state. Thus, the source is not ergodic.

From now on we will consider the ergodic Markov source. Since it generates "typical" message sequences, after selection of the initial source state in a long time span we observe generated messages, and in consequence we observe the sequence of states that the source subsequently reaches. On the basis of long-term observation of subsequent states one can estimate the values of probabilities of each state. Moreover, the obtained state probability distribution does not depend on the choice of the initial state (this is understandable as the source is ergodic). The obtained probability distribution is called a *stationary distribution* and is one of the characteristics of the Markov source.

This distribution can be found on the basis of probabilities of state transitions, which characterize the Markov source. We will show how to find the stationary distribution for the source considered in Example 1.5.2.

Example 1.5.4 *Let us return to the state diagram shown in Figure 1.5. Since the source is stationary, the probability of reaching a given state can be found on the basis of the probability that the source is in one of the previous states and the probability of transition from that state to the state in the next moment. So in order for the source to be in state 00, at the previous moment it must have been in state 00 or 01. Taking into account the probabilities of transitions between the states, we receive the following equation*

$$P(00) = P(0|00) \cdot P(00) + P(0|01) \cdot P(01)$$

Similar equations can be formulated for the remaining states

$$P(01) = P(0|10) \cdot P(10) + P(0|11) \cdot P(11)$$

$$P(10) = P(1|00) \cdot P(00) + P(1|01) \cdot P(01)$$

$$P(11) = P(1|11) \cdot P(11) + P(1|10) \cdot P(10)$$

It is easy to see that the above equation system is not independent. However, there is one more equation that can be applied, namely

$$P(00) + P(01) + P(10) + P(11) = 1$$

because the sum of probabilities of the event that the source is in a given state is equal to unity. Replacing any of the system equations with the last one, we obtain a new equation system that can be uniquely solved. Simple calculations lead us to the following result

$$P(00) = P(11) = \frac{5}{18}, \qquad P(01) = P(10) = \frac{2}{9}$$

1.5.5   Entropy of the Markov Source

We will now introduce the concept of entropy of the Markov source. As a result, we will be able to compare this entropy with the average amount of information obtained by observing a single message on the output of the memoryless source.

Recall that the state of an mth-order Markov source at the ith moment can be denoted as $(x_{i-1}, x_{i-2}, \ldots, x_{i-m})$. If at this moment the source emits a message $x_i \in \{a_1, \ldots, a_K\}$, then the amount of information we receive is equal to

$$I(x_i|x_{i-1}, \ldots, x_{i-m}) = \log \frac{1}{P(x_i|x_{i-1}, \ldots, x_{i-m})}$$

By averaging this result with respect to all possible messages, and assuming that the source is in the state $(x_{i-1}, x_{i-2}, \ldots, x_{i-m})$, we receive the entropy of the source in this state

$$H(X|x_{i-1}, \ldots, x_{i-m})$$

$$= \sum_{j=1}^{K} P(x_i = a_j|x_{i-1}, \ldots, x_{i-m}) \cdot I(x_i = a_j|x_{i-1}, \ldots, x_{i-m}) \qquad (1.14)$$

In turn, calculating the average amount of information with the assumption that the source is in any possible state, we obtain the ensemble average of expression (1.14), i.e.

$$H(X) = E\left[H(X|x_{i-1}, \ldots, x_{i-m})\right]$$

$$= \sum_{j_1=1}^{K} \cdots \sum_{j_m=1}^{K} P(x_{i-1} = a_{j_1}, \ldots, x_{j-m} = a_{j_m}) H(X|x_{i-1}, \ldots, x_{i-m}) \qquad (1.15)$$

Using expression (1.14) in (1.15), we receive

$$H(X) = \sum_{x_i} \ldots \sum_{x_{i-m}} P(x_{i-1}, \ldots, x_{i-m}) P(x_i | x_{i-1}, \ldots, x_{i-m})$$

$$\cdot \log \frac{1}{P(x_i | x_{i-1}, \ldots, x_{i-m})}$$

Applying Bayes' formula to the probability products in the above expression,[3] we obtain the final formula for the entropy of a Markov source

$$H(X) = \sum_{x_i} \ldots \sum_{x_{i-m}} P(x_i, x_{i-1}, \ldots, x_{i-m}) \log \frac{1}{P(x_i | x_{i-1}, \ldots, x_{i-m})} \qquad (1.16)$$

As we see, the entropy of a Markov source is an amount of information averaged over all possible states and all messages that can be generated by the source remaining in each of these states.

Example 1.5.5 *Let us calculate the entropy of the source from Example 1.5.2. For this source we can build the following table of probabilities*

x_i, x_{i-1}, x_{i-2}	$P(x_i | x_{i-1}, x_{i-2})$	$P(x_{i-1}, x_{i-2})$	$P(x_i, x_{i-1}, x_{i-2})$
000	0.6	5/18	3/18
001	0.5	2/9	1/9
010	0.5	2/9	1/9
011	0.4	5/18	1/9
100	0.4	5/18	1/9
101	0.5	2/9	1/9
110	0.5	2/9	1/9
111	0.6	5/18	3/18

On the basis of this table we can calculate the entropy of the Markov source as

$$H(X) = \sum_{x_i} \ldots \sum_{x_{i-m}} P(x_i, x_{i-1}, \ldots, x_{i-m}) \log \frac{1}{P(x_i | x_{i-1}, \ldots, x_{i-m})}$$

$$= 2 \cdot \frac{3}{18} \log \frac{10}{6} + 4 \cdot \frac{1}{9} \log \frac{10}{5} + 2 \cdot \frac{1}{9} \log \frac{10}{4} = 0.9839 \text{ [bit/message]}$$

1.5.6 Source Associated with the Markov Source

Knowing already the stationary distribution of the Markov source, it would be interesting to calculate the probability of generation of specific messages by the source. For the mth-order Markov source these probabilities can be derived from the stationary

---

[3] $P(A, B) = P(B|A)P(A)$.

distribution and the conditional probabilities describing the probability of generation of a given message on condition that the source is in a given state.

Example 1.5.6 *For the source considered in Example 1.5.2 we have*

$$P(0) = P(0|00)P(00) + P(0|01)P(01) + P(0|10)P(10) + P(0|11)P(11)$$

$$P(1) = P(1|00)P(00) + P(1|01)P(01) + P(1|10)P(10) + P(1|11)P(11)$$

The substitution of the probabilities calculated in the previous example leads us to $P(0) = P(1) = 0.5$.

In the above example the probabilities of generation of particular messages by the source are identical. If the source were memoryless it would have the highest possible entropy, $H(X) = 1$. Calculation of the entropy of the Markov source, performed in Example 1.5.5, indicates that its entropy is lower. Let us compare the value of this entropy with the entropy of the memoryless source characterized by the same probabilities of generation of particular messages. For that purpose the definition of the source associated with the Markov source is introduced.

Definition 1.5.5 *Let $X = \{a_1, \ldots, a_K\}$ be the alphabet of an mth-order Markov source. Let $P(a_1), \ldots, P(a_K)$ be the probabilities of occurrence of respective messages on the source output. Source $\overline{X}$ associated with the Markov source X is a memoryless source with the same alphabet X and identical probability distribution of elementary messages.*

Below we will show that the entropy of a Markov source is lower than or equal to the entropy of the source associated with it. First we will prove a useful inequality that will be used subsequently in the course of this chapter.

Let p_i and q_i $(i = 1, \ldots, N)$ be interpreted as probabilities, so the following property holds for them

$$\sum_{i=1}^{N} p_i = \sum_{i=1}^{N} q_i = 1 \quad \text{and} \quad p_i \geq 0, \; q_i \geq 0 \;\text{ for } i = 1, \ldots, N$$

We will show that

$$\sum_{i=1}^{N} p_i \log \frac{q_i}{p_i} \leq 0 \tag{1.17}$$

For this purpose let us use the inequality $\ln x \leq x - 1$ again. We have

$$\sum_{n=1}^{N} p_n \log \frac{q_n}{p_n} = \frac{1}{\ln 2} \sum_{n=1}^{N} p_n \ln \frac{q_n}{p_n} \leq \frac{1}{\ln 2} \sum_{n=1}^{N} p_n \left(\frac{q_n}{p_n} - 1 \right)$$

$$= \frac{1}{\ln 2} \left(\sum_{n=1}^{N} q_n - \sum_{n=1}^{N} p_n \right) = 0$$

We will use inequality (1.17) to find the relationship of the entropy of the first-order Markov source to the entropy of the memoryless source associated with it. For this purpose we apply the following substitution

$$p_n = P(x_i, x_{i-1}) = \Pr\{x_i = a_k, x_{i-1} = a_j\}$$

$$q_n = P(x_i)P(x_{i-1}) = \Pr\{x_i = a_k\}\Pr\{x_{i-1} = a_j\}$$

$$k, j = 1, \ldots, K$$

Using inequality (1.17) and the expression written above, we obtain the following inequality, expressed in simplified notation as

$$\sum_{x_i}\sum_{x_{i-1}} P(x_i, x_{i-1}) \log \frac{P(x_i)P(x_{i-1})}{P(x_i, x_{i-1})} \leq 0 \qquad (1.18)$$

Knowing from Bayes' formula that

$$P(x_i, x_{i-1}) = P(x_i|x_{i-1})P(x_{i-1})$$

on the basis of (1.18) we can write the following inequality

$$\sum_{x_i}\sum_{x_{i-1}} P(x_i, x_{i-1}) \log \frac{P(x_i)}{P(x_i|x_{i-1})} \leq 0$$

so

$$\sum_{x_i}\sum_{x_{i-1}} P(x_i, x_{i-1}) \left(\log \frac{1}{P(x_i|x_{i-1})} - \log \frac{1}{P(x_i)} \right) \leq 0$$

From the latter we conclude that

$$\sum_{x_i}\sum_{x_{i-1}} P(x_i, x_{i-1}) \log \frac{1}{P(x_i|x_{i-1})} \leq \sum_{x_i}\sum_{x_{i-1}} P(x_i, x_{i-1}) \log \frac{1}{P(x_i)} \qquad (1.19)$$

Therefore, the left-hand side, which is, effectively, the entropy of the Markov source, is bounded from above by the expression

$$\sum_{x_i}\sum_{x_{i-1}} P(x_i, x_{i-1}) \log \frac{1}{P(x_i)} = \sum_{x_i} \log \frac{1}{P(x_i)} \sum_{x_{i-1}} P(x_i, x_{i-1})$$

$$= \sum_{x_i} P(x_i) \log \frac{1}{P(x_i)} \qquad (1.20)$$

We used the fact that

$$\sum_{x_{i-1}} P(x_i, x_{i-1}) = P(x_i)$$

Finally, on the basis of (1.19) and (1.20) we receive the dependence

$$H(X) \leq H(\overline{X}) \qquad (1.21)$$

Let us try to establish when the Markov source entropy achieves its maximum, equal to $H(\overline{X})$. We should consider the situation in which expression (1.18) is fulfilled with the equality sign. We can easily notice that it occurs when

$$P(x_i, x_{i-1}) = P(x_i)P(x_{i-1})$$

However, it means that the messages generated at particular moments are statistically independent, so the Markov source loses its memory, i.e. it becomes a memoryless source.

Our considerations can be easily extended on mth-order Markov sources. It is sufficient to replace a single message x_{i-1} from the $(i-1)$st timing instant by their whole block $(x_{i-1}, \ldots, x_{i-m})$.

1.6 Discrete Source Coding

As we remember from the introductory section, the process of assignment of symbol sequences to the source messages is called source coding. The level of efficiency of the source coding process determines the size of the symbol stream that has to be transmitted to the receiver. In the case of a typical computer system, the memory size needed to memorize a particular message sequence depends on the efficiency of the source coding. Similar dependence also occurs for the continuous sources whose messages, with acceptable loss of information, are represented by streams of discrete symbols.

Example 1.6.1 *Consider a binary representation of a color picture on the color monitor. Knowing that a single pixel has a 24-bit representation, a picture of the size 800×600 pixels would require 11.52 milion binary symbols (bits). However, thanks to the currently used methods of picture coding (known as picture compression) it is posible to represent such a picture in a much more effective manner. Usually, typical properties of such pictures are taken into account, e.g. the fact that part of the picture plane is a uniform surface or that neighboring points do not differ much from each other. Methods of picture compression are currently an important branch of digital signal processing.*

Our considerations on source coding will start from a formal definition of code (Steinbuch and Rupprecht 1982).

Definition 1.6.1 *Let $X = \{a_1, a_2, \ldots, a_K\}$ denote the source alphabet (the set of messages that is the subject of coding), and let $Y = \{y_1, y_2, \ldots, y_N\}$ be a set of code symbols. A code is a relation in which each message of the source alphabet is mutually uniquely assigned a sequence of symbols selected from the set of code symbols. The code sequence representing a given message is called a* codeword *(or a code sequence).*

Example 1.6.2 *Let a message source have the alphabet $X = \{a_1, a_2, a_3, a_4\}$. Assume that the set of code symbols is binary, i.e. $Y = \{0, 1\}$. An example of the relation between source messages and code sequences is shown in the table below.*

Messages	Codewords
a_1	0
a_2	11
a_3	00
a_4	01

This relation is not a code in the sense of the above definition because it is not mutually unequivocal. The code sequence 00 *can be a representation both of the message a_3 and a message sequence $a_1 a_1$.*

Let us now consider some examples of codes. The source is the same as in the last example.

Example 1.6.3 *Denote by $\mathcal{A}$, $\mathcal{B}$ and $\mathcal{C}$ three codes presented in the table below* (Abramson 1963).

Messages	$\mathcal{A}$	$\mathcal{B}$	$\mathcal{C}$
a_1	00	0	0
a_2	01	10	01
a_3	10	110	011
a_4	11	1110	0111

Each of the above codes has different features. Code $\mathcal{A}$, as opposed to the other two codes, has codewords of equal length. The characteristic feature of code $\mathcal{B}$ is a construction of codewords relying on the application of a specific number of "1"s followed by a single zero symbol. In the case of code $\mathcal{C}$ a zero symbol starts each codeword, whereas the codewords differ by the number of "1"s following the zero symbol.

This simple example makes us aware of the multitude of possible codes. Thus, a question arises as to how we should evaluate them and which of them should be selected. The answer to this question is not easy. In the selection and evaluation process we should consider:

- coding efficiency – we aim at possibly the smallest number of coding symbols representing a given sequence of messages; in the statistical sense we would like to minimize the average number of symbols needed to represent a single message;
- simplicity of the coding and decoding process – the software and hardware complexity is the consequence of both processes;
- allowable delay introduced by the coding and, in particular, the decoding processes.

As we conclude from the definition of code, encoding is an operation of mutually unique assignment. In consequence, the sequence of code symbols observed in the receiver can be unambiguously divided into codewords. This is obviously the necessary condition of correct functioning of the whole coding/decoding process and clearly results from the code difinition. The case is straightforward if the codewords have equal lengths, as in the case of code $\mathcal{A}$ in Example 1.6.3. Only the knowledge of the initial moment is necessary for correct decoding. Comparing codes $\mathcal{B}$ and $\mathcal{C}$ we see that in the case of code $\mathcal{B}$ the decoder should detect the occurrence of a zero symbol, which indicates the end of a codeword. In code $\mathcal{C}$ a zero symbol signals the begining of a codeword. In order to decompose the whole symbol sequence into particular codewords and to extract the current codeword, one has to observe the first symbol of the next codeword. In this sense it is not possible to decode the codewords of code $\mathcal{C}$ without delay. On the contrary, code $\mathcal{B}$ enables decoding without delay. For decoding codewords of a given code without delay, none of the codewords may be a *prefix* of another codeword. Therefore, code $\mathcal{B}$ is often called a *prefix code*. The prefix is defined in the following way.

Definition 1.6.2 *Let $\mathbf{c}_i = (c_{i_1}, c_{i_2}, \ldots, c_{i_m})$ be a codeword of a given code. Any sequence of symbols $(c_{i_1}, c_{i_2}, \ldots, c_{i_j})$, where $j \leq m$, is a prefix of codeword $\mathbf{c}_i$.*

Note that in code $\mathcal{C}$ each codeword listed on a higher position in the code table is a prefix of codewords appearing below it.

An essential task is a construction of a prefix code. In the next example (Abramson 1963) a heuristic approach to this task is presented.

Example 1.6.4 *Assume that a memoryless source is characterized by a five-element alphabet $\{a_1, a_2, \ldots, a_5\}$. We construct a prefix code in the following manner. Assign message a_1 the symbol "0". Thus, it is the first selected codeword. If this symbol is not to be a prefix of another codeword, all remaining codewords should start with "1" in their first position. Therefore, let message a_2 be assigned the symbol sequence "10". All remaining codewords will have to start with the sequence "11". So message a_3 can be assigned the codeword "110". The remaining two messages can be assigned codewords starting with the sequence "111" supplemented with "0" and "1", respectively. The result of our code design is presented in the table below as code $\mathcal{A}$.*

Message	Code $\mathcal{A}$	Code $\mathcal{B}$
a_1	0	00
a_2	10	01
a_3	110	10
a_4	1110	110
a_5	1111	111

Is this the only way of assigning the codewords to the source messages? For sure not! Let us inspect the column containing the codewords of code $\mathcal{B}$. In creation of this code the same basic rule is applied as in the construction of code $\mathcal{A}$, i.e. none of the codewords is a prefix of another codeword. However, we start from assigning message a_1 the sequence "00". As a result we obtain a different code! Therefore, the following question arises: How to evaluate these codes? Generally, we can say that the smaller number of symbols required, on average, for representation of a single message, the better code. It is intuitively clear that in order to achieve a high degree of efficiency of using the coding symbols, messages that occur frequently should be assigned short codewords, whereas messages with low probability of occurrence should be assigned longer codewords.

In order to assess the quality of the source coding process we introduce the concept of the *average codeword length*.

Definition 1.6.3 *Consider a memoryless source* **X** *with the alphabet* $\{a_1, \ldots, a_K\}$ *and let* $\{P_1, \ldots, P_K\}$ *be the probability distribution of occurrence of elementary messages. Let codewords of length* $\{l_1, \ldots, l_K\}$ *be assigned to these messages, respectively. The average length L of codeword is an ensemble average of a codeword length described by the formula*

$$L = \sum_{i=1}^{K} l_i P_i \qquad (1.22)$$

In order to evaluate the quality of source coding we have to be sure that there exists a prefix code with the selected set of codeword lengths $\{l_1, \ldots, l_K\}$. In order to check this we apply the *Kraft-McMillan inequality*.

Theorem 1.6.1 *The necessary and sufficient condition of the existence of an r-nary prefix code characterized by a set of codeword lengths* $\{l_1, \ldots, l_K\}$ *is the fulfilling of inequality*

$$\sum_{i=1}^{K} r^{-l_i} \leq 1 \qquad (1.23)$$

For a binary code ($r = 2$) this inequality obtains the form

$$\sum_{i=1}^{K} 2^{-l_i} \leq 1 \qquad (1.24)$$

As we said, the Kraft-McMillan inequality is useful for checking if there exists a prefix code with a given set of codeword lengths. Unfortunately, this inequality does not facilitate the process of finding such a code. We illustrate this statement by the next example quoted after Abramson (1963).

Example 1.6.5 *Consider five different source codes proposed for a memoryless source* **X** *with four elementary messages* $\{a_1, a_2, a_3, a_4\}$. *The codes are presented in the table below.*

Message	$\mathcal{A}$	$\mathcal{B}$	$\mathcal{C}$	$\mathcal{D}$	$\mathcal{E}$
a_1	00	0	0	0	0
a_2	01	100	10	100	10
a_3	10	110	110	110	110
a_4	11	111	111	11	11

If we denote the left-hand side of formula (1.24) as W, then simple calculations show that for codes $\mathcal{A}$, $\mathcal{C}$ *and* $\mathcal{D}$ *we have* $W = 1$, *for code* $\mathcal{B}$ *we have* $W = 7/8$, *whereas for code*

$\mathcal{E}$ we obtain $W = 9/8$. Code $\mathcal{A}$ is uniquely decodable without delay because its codewords have constant length and each of them is unique. The codeword lengths of code $\mathcal{B}$ fulfill the Kraft-McMillan inequality and none of the codewords is a prefix of another codeword. Using one codeword shorter than those in code $\mathcal{B}$ results in the allowable set of codeword lengths as well; however, not all codes from the set of codes characterized by the same codeword length are decodable without delay. Code $\mathcal{C}$ belongs to the prefix codes, whereas code $\mathcal{D}$ does not. The reason for that is that the last codeword of code $\mathcal{D}$ is a prefix of the codeword assigned to message a_3. Finally, code $\mathcal{E}$ does not satisfy the Kraft-McMillan inequality, so not only this code but also any other code with the same set of codeword lengths will not be a prefix code.

There are many possible source codes characterized by a single set of codeword lengths satisfying the Kraft-McMillan inequality. However, not all of them are the prefix codes. Therefore, there is a problem of how to find the code that will have the shortest average codeword length among all r-nary prefix codes used to represent the messages of a given memoryless source **X**. Such a code will be called a *compact code*.

In the first step of our search for compact codes we will find a minimum average codeword length for a prefix code.

As previously, consider a memoryless message source **X** described by the message alphabet $X = \{a_1, a_2, \ldots, a_K\}$ and the set of respective probabilities $P(a_i) = P_i$ ($i = 1, 2, \ldots, K$). The code symbols that make up codewords are selected from an r-nary alphabet Y. Denote the length of the codeword assigned to message a_i as l_i. As we remember, the entropy of such a memoryless source is given by formula (1.2). Recall again that we have already proven the inequality

$$\sum_{i=1}^{K} P_i \log \frac{Q_i}{P_i} \leq 0 \qquad (1.25)$$

where both P_i and Q_i can be interpreted as probabilities, i.e. $P_i \geq 0$, $Q_i \geq 0$ ($i = 1, 2, \ldots, K$), $\sum_{i=1}^{K} P_i = 1$, $\sum_{i=1}^{K} Q_i = 1$. In consequence of (1.25) we have

$$\sum_{i=1}^{K} P_i \log \frac{1}{P_i} \leq \sum_{i=1}^{K} P_i \log \frac{1}{Q_i} \qquad (1.26)$$

Let us note that (1.25) becomes an equity if $P_i = Q_i$ ($i = 1, 2, \ldots, K$). So, recalling the definition of the entropy of a memoryless source, we have

$$H(X) \leq \sum_{i=1}^{K} P_i \log \frac{1}{Q_i} \qquad (1.27)$$

Let us assume now that

$$Q_i = \frac{r^{-l_i}}{\sum_{j=1}^{K} r^{-l_j}} \qquad (1.28)$$

Let us note that the variables Q_i defined in this way satisfy the conditions $Q_i \geq 0$ and $\sum_{i=1}^{K} Q_i = 1$. Substituting (1.28) into (1.27), we obtain

$$H(X) \leq \sum_{i=1}^{K} P_i \log \frac{\sum_{j=1}^{K} r^{-l_j}}{r^{-l_i}} = \sum_{i=1}^{K} P_i \log \frac{1}{r^{-l_i}} + \sum_{i=1}^{K} P_i \log \left(\sum_{j=1}^{K} r^{-l_j} \right)$$

$$= \log r \sum_{i=1}^{K} P_i l_i + \log \left(\sum_{j=1}^{K} r^{-l_j} \right) \sum_{i=1}^{K} P_i = L \log r + \log \left(\sum_{j=1}^{K} r^{-l_j} \right)$$

So finally

$$H(X) \leq L \log r \qquad (1.29)$$

The last inequality is a consequence of the fact that for a prefix code the Kraft-McMillan inequality is satisfied so the second term of formula (1.29) is non-positive. Finally we obtain

$$L \geq \frac{H(X)}{\log r} = H_r(X) \qquad (1.30)$$

The following theorem is a consequence of inequality (1.30).

Theorem 1.6.2 *(The first Shannon theorem) The average codeword length in a prefix code used for representation of messages generated by a memoryless source* **X** *is not lower than the source entropy (calculated in r-nary units).*

Theorem 1.6.2 formulates an important limit related to source coding. Let us now consider the requirements for achieving this boundary. The analysis of subsequent derivation steps in (1.29) indicates that in order for the boundary to be reached the following requirements have to be fulfilled

$$1° \qquad H(X) = \sum_{i=1}^{K} P_i \log \frac{\sum_{j=1}^{K} r^{-l_j}}{r^{-l_i}} \qquad (1.31)$$

$$2° \qquad \sum_{j=1}^{K} r^{-l_j} = 1 \qquad (1.32)$$

We conclude from requirements 1° and 2° that the boundary of source coding efficiency can be reached if the probability of occurrence P_i for each message of the memoryless source is expressed by the formula

$$\bigwedge_i \quad P_i = Q_i = \frac{r^{-l_i}}{\sum\limits_{j=1}^{K} r^{-l_j}} = r^{-l_i} \tag{1.33}$$

Thus, for each message a_i whose probability is given by (1.33) the codeword representing it should be of length

$$l_i = \log_r \frac{1}{P_i} \tag{1.34}$$

Obviously, the codeword length has to be an integer number. It is possible only if the probabilities of message occurrence have the form

$$P_i = \left(\frac{1}{r}\right)^{\alpha_i} \tag{1.35}$$

where α_i is an integer. Then for each i the codeword length would be $l_i = \alpha_i$. Obviously it is a special case that can occur in practice very rarely. The message probabilities are in fact properties of the message source and we are not able to shape them according to our needs. Thus, let us think how to proceed if condition (1.35) does not hold. The solution presented below approches the limit of source coding efficiency asymptotically.

If the logarithm of the reciprocal of a particular message probability (1.35) is not an integer number, then the codeword length, as an integer, is definitely contained in the interval

$$\log_r \frac{1}{P_i} \leq l_i < \log_r \frac{1}{P_i} + 1 \tag{1.36}$$

Choosing the codeword lengths according to rule (1.36) ensures the result in the form of the prefix code, because if for each i the inequality $l_i \geq \log \frac{1}{P_i}$ is fulfilled, the Kraft-McMillan inequality holds as well. We conclude that

$$\bigwedge_i \quad P_i \geq r^{-l_i}$$

so we have in consequence

$$\sum_{i=1}^{K} r^{-l_i} \leq \sum_{i=1}^{K} P_i = 1 \tag{1.37}$$

One can easily ascertain that the choice of the codeword length using (1.36) does not lead to a very effective code because the average codeword length for such a code is contained in the interval between the value of the source entropy and that value increased by 1. We receive this result after multiplying, for each i, both sides of inequality (1.36) by the respective message probability P_i and subsequently summing the sides of all the inequalities. We namely have

$$\bigwedge_i P_i \log_r \frac{1}{P_i} \leq P_i l_i < P_i \log_r \frac{1}{P_i} + P_i$$

Summing all sides together we obtain

$$\sum_{i=1}^{K} P_i \log_r \frac{1}{P_i} \leq \sum_{i=1}^{K} P_i l_i < \sum_{i=1}^{K} P_i \log_r \frac{1}{P_i} + \sum_{i=1}^{K} P_i$$

and finally

$$H_r(X) \leq L < H_r(X) + 1 \tag{1.38}$$

For small values of entropy such an interval is relatively wide and coding according to (1.36) is ineffective. Thus, it is more effective to code n-element message blocks instead of single messages separately. In other words, the efficiency of source coding increases if the subjects of coding are the messages of the nth extension of source $\mathbf{X}$. Denoting L_n as the average length of the codeword representing a message of the nth source extension $\mathbf{X}^n$, on the basis of (1.38) we receive the inequality

$$H_r(X^n) \leq L_n < H_r(X^n) + 1 \tag{1.39}$$

Recalling that $H_r(X^n) = n H_r(X)$ we conclude that

$$n H_r(X) \leq L_n < n H_r(X) + 1$$

As a result

$$H_r(X) \leq \frac{L_n}{n} < H_r(X) + \frac{1}{n}$$

Finally, we conclude from the last inequality that

$$\lim_{n \to \infty} \frac{L_n}{n} = H_r(X) \tag{1.40}$$

Note that L_n/n is the average number of code symbols needed to encode a single message in a block of n messages. The limit (1.40) is in agreement with the Shannon theorem, once more indicating that the average codeword length of a decodable code applied in representation of memoryless source messages cannot be lower than the entropy of this source (given in the r-nary units).

Similar considerations to those shown above can be performed for a Markov source (Abramson 1963). The conclusions derived from them are the same and lead to equality (1.40).

Example 1.6.6 *Consider a memoryless source* $\mathbf{X}$ *characterized by the elementary messages "0" and "1" appearing with the probabilities* $P_0 = 0.1$ *and* $P_1 = 0.9$, *respectively.*

It can be easily calculated that the entropy of this source is $H(X) = 0.4690$. Assuming application of a binary source code ($r = 2$) and direct encoding of messages "0" and "1" using single-element codewords 0 and 1, we conclude that in this case the average codeword length is $L = 1$. Thus, the coding efficiency is

$$\eta = \frac{H_r(X)}{L} = \frac{0.4690}{1} = 46.9\%$$

Consider now the second extension $\mathbf{X}^2$ of source $\mathbf{X}$. The table below presents the messages of source $\mathbf{X}^2$, their probabilities and the codeword lengths associated with them, selected according to rule (1.36).

Message	P_i	l_i
11	0.81	1
10	0.09	4
01	0.09	4
00	0.01	7

It turns out that in this case the average codeword length is $L_2 = 1.6$, so the coding efficiency is

$$\eta = \frac{2H_r(X)}{L_2} = 58,6\%$$

One can easily find that the application of rule (1.36) has not led to finding a compact code. Already the set of codeword lengths such as $(1, 2, 3, 3)$ assigned to the messages 11, 10, 01 and 00, respectively, makes construction of a prefix code possible. The average codeword length is then equal to $L_2 = 1.29$, so the coding efficiency is $\eta = 72\%$. The codeword length selected according to rule (1.36) becomes closer and closer to the optimum when we encode messages of higher and higher source extensions of source $\mathbf{X}$.

The example shown above indicates that it could be interesting to determine a coding method resulting in a compact code not only in the asymptotic sense by increasing source extensions of message source $\mathbf{X}$, but also for any source. It turns out that there are some coding methods resulting in a compact code. In the following sections we will consider the most important among them.

1.6.1 Huffman Coding

In 1952 Huffman presented a procedure that determines a r-nary compact code for any memoryless message source. Below we describe it by the example of a binary code synthesis.

Consider a memoryless message source characterized by the message alphabet $X = \{a_1, a_2, \ldots, a_K\}$ with related elementary message probabilities equal to $P_1, P_2, \ldots, P_K$, respectively. Assume without loss of generality that the message probabilities are ordered in decreasing order, i.e. $P_1 \geq P_2 \geq \ldots \geq P_K$. In the first part of the procedure we construct the sequence of reduced sources $\mathbf{X} = \mathbf{X}_0, \mathbf{X}_1, \ldots, \mathbf{X}_{K-2}$ in the following way.

Consider the ith step of the reduced source construction. Let us represent two of the least probable messages of source $\mathbf{X}_i$, denoted as $x_{K-i}^{(i)}$ and $x_{K-i-1}^{(i)}$, in the message set of source $\mathbf{X}_{i+1}$ by a new message $x_j^{(i+1)}$, whose probability of occurrence is $P(x_j^{(i+1)}) = P(x_{K-i}^{(i)}) + P(x_{K-i-1}^{(i)})$. The other messages of the newly created reduced source $\mathbf{X}_{i+1}$ remain unchanged; however, all messages are again ordered in decreasing order of probabilities. Let us note that source $\mathbf{X}_{i+1}$ has one message less when compared with message source $\mathbf{X}_i$. In this sense it is reduced with respect to source $\mathbf{X}_i$. In the following steps analogous to the previous one, the number of messages is gradually reduced until we obtain a two-message source $\mathbf{X}_{K-2}$. Deriving a compact code for such a source is straightforward: one message is represented by the codeword consisting of the zero symbol, whereas the second message is represented by the codeword consisting of a single "1".

In the following steps we find compact codes for each of the sources $\mathbf{X}_{K-3}$, $\mathbf{X}_{K-4}, \ldots, \mathbf{X}_1, \mathbf{X}_0 = \mathbf{X}$. Assume that a compact code for source $\mathbf{X}_{i+1}$ is already known. We intend to find a compact code for source $\mathbf{X}_i$. In order to perform this task we assign codewords to all messages of source $\mathbf{X}_i$, which are the same as those that have been assigned to these messages in the source $\mathbf{X}_{i+1}$ except messages $x_{K-i}^{(i)}$ and $x_{K-i-1}^{(i)}$. The codewords assigned to messages $x_{K-i}^{(i)}$ and $x_{K-i-1}^{(i)}$ consist of a codeword associated with message $x_j^{(i+1)}$ of source $\mathbf{X}_{i+1}$, extended by the symbols "0" and "1", respectively. Recall that message $x_j^{(i+1)}$ was created in the source reduction process by merging messages $x_{K-i}^{(i)}$ and $x_{K-i-1}^{(i)}$. Figure 1.7 presents an example of deriving a compact code for a six-message memoryless source with respective message probabilities. As we see, the Huffman algorithm renders a code in which the less probable the message the longer its codeword.

The compact code achieved as a result of the Huffman algorithm can also be represented by a tree in which the branches leaving a certain node in the left direction are associated with the code symbol "0", whereas those diverging in the right direction are associated with the symbol "1". The ending of such a branch is associated with a certain message

Message	$X = X_0$ P_i	Codeword	X_1 P_i	Codeword	X_2 P_i	Codeword	X_3 P_i	Codeword	X_4 P_i	Codeword
a_1	0.5	0	0.5	0	0.5	0	0.5	0	0.5	0
a_2	0.3	10	0.3	10	0.3	10	0.3	1<u>0</u>	0.5	1
a_3	0.1	110	0.1	110	0.1	11<u>0</u>	0.2	1<u>1</u>		
a_4	0.05	1110	0.05	111<u>0</u>	0.1	11<u>1</u>				
a_5	0.03	1111<u>0</u>	0.05	111<u>1</u>						
a_6	0.02	1111<u>1</u>								

Figure 1.7   Synthesis of the compact code using the Huffman algorithm

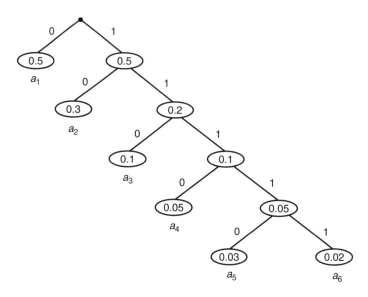

Figure 1.8 Code tree determined for the code from Figure 1.7 derived using the Huffman procedure

and the codeword assigned to it is a sequence of symbols associated with the path from the tree root to the ending of this branch. The example of a tree for the code shown in Figure 1.7 is presented in Figure 1.8.

The Huffman algorithm can be easily generalized for nonbinary source codes. If the applied source code is intended to be r-nary, then in each source reduction step r least probable messages are replaced by a single message of the reduced source. As a result of the last reduction step we obtain a source with r messages, which can be assigned one-symbol codewords each. However, in order for the reduction process to yield this result, K (the number of messages of source $\mathbf{X}$) has to fulfill the condition

$$K = (r - 1)m + r \qquad (1.41)$$

where m is an integer, because in each reduction step the number of messages decreases by $r - 1$, and the final number of messages is r. If the number of source messages does not fulfill condition (1.41), then the source can be supplemented by a number of messages occurring with zero probability. Thus, the resulting number of source messages does fulfill condition (1.41).

The proof of the Huffman algorithm can be found, among others, in (Abramson 1963).

1.6.2 Shannon-Fano Coding

Huffman coding is obviously not the only method of derivation of a compact code. Another method known as Shannon-Fano coding is presented in the form of the following algorithm.

1. For a given list of messages construct the list of probabilities of message occurrence.

2. Reorder the list of messages in decreasing order of message probabilities.
3. Divide the list of messages into two sets, W_0 and W_1, in such a way that the sums of probabilities are either identical or as close to each other as possible.
4. Create one-bit prefixes of codewords for messages, setting them to "0" for the codewords related to the messages contained in set W_0 and to "1" for the codewords related to the messages contained in set W_1.
5. Apply the above algorithm recursively for each set of messages by dividing it into subsets and adding subsequent bits to the codewords created in this way until all the subsets contain single messages.

Figure 1.9 illustrates the creation of the source code using the above method. The respective code tree is also presented.

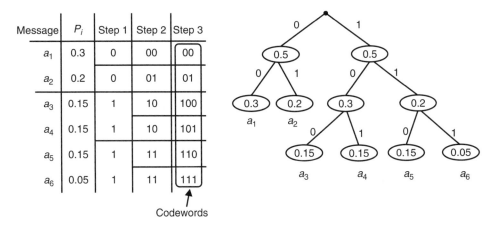

Message	P_i	Step 1	Step 2	Step 3
a_1	0.3	0	00	00
a_2	0.2	0	01	01
a_3	0.15	1	10	100
a_4	0.15	1	10	101
a_5	0.15	1	11	110
a_6	0.05	1	11	111

Codewords

Figure 1.9   A compact code and code tree obtained using the Shannon-Fano algorithm

1.6.3   Dynamic Huffman Coding

In Huffman or Shannon-Fano source encoding algorithms the statistics of source messages are taken into account. In practice, blocks of messages such as data, text, single pictures or video sequences featuring different statistics are the objects of coding. So far, for the selected source code, the knowledge of these statistics in the form of message probabilities is assumed. In consequence, these probabilities have to be estimated by initial analysis of the message sequence to be encoded. On the basis of these estimates, during the second round of analysis the message block can be a subject of encoding. The necessity of double analysis of the message block to be encoded is disadvantageous because the processing requirements increase. An additional problem is the necessity to supply the decoder with the message source statistics or, equivalently, the obtained mapping of the messages onto the codewords, i.e. the codebook. This was the motivation for developing a coding method that, although optimum in the asymptotic sense for encoding of very long message blocks, does not require double browsing of the message block and allows independent estimation of the source statistics both at the source encoder and at the decoder. Such requirements are fulfilled by the dynamic Huffman encoding. Its operation will be explained by inspecting the example of encoding of a given text, which is represented in the form of an ASCII-encoded character sequence. Each character is considered to be a single message. As an example, let us encode the sequence "*This_is_mom*". For simplicity, assume that the character "_" denotes space.

Both encoder and decoder construct a *code tree*. Initially, the code tree contains a single branch with a "zero leaf" at its end and denoted by the symbol $e\phi$. In general, leaves terminating each branch are marked by the symbol xN, where x denotes a message (a single character) and N is the number of its appearances from the start of the encoding process. As previously, a branch growing in the left direction is assigned code symbol "0", whereas a branch directed to the right is assigned "1". The process of encoding of the analyzed text together with construction of the code tree is presented in Figures 1.10–1.13.

The encoder starts from encoding the message "*T*". As this message occurs for the first time, it is sent to the receiver in the open form, i.e. as an ASCII codeword. We denote its occurrence on the code tree by the symbol $T1$ at the end of the branch accompanied by digit "1". In each step, besides tree modification, the list of nodes and leaves is created, picking them up from the left of the lowest tree layer to the right and repeating the same move shifting to the higher tree layers. In the first step the list consists only of two nodes, $e\phi$ and $T1$. The list is arranged from the lowest to the highest weight of the nodes and leaves. If, after a simple tree modification resulting from the occurrence of the next message, the list is not ordered according to the increasing weights, we have to perform its ordering by appropriate modification of the code tree. For the modified tree we create the list again.

In the second step a new message, "*h*", appears. This fact is signaled using the current codeword assigned to an empty leaf, followed by the ASCII code of the character "*h*". In our case sequence $0'h'$ is sent to the receiver. The created list $e\phi$, $h1$, 1, $T1$ still remains appropriately ordered.

In the third step the message "*i*" is encoded. The codeword currently assigned to an empty leaf $e\phi$ is 00, therefore the sequence $00'i'$ is sent to the receiver. From the branch terminated by an empty leaf a new empty leaf grows in the left direction and a new leaf denoted by $i1$ grows in the right direction. The list of nodes and leaves of the newly created tree is $e\phi$, $i1$, 1, $h1$, 2, $T1$ and it requires reordering. The position of symbol "2" has to be exchanged with the symbol $T1$. In consequence, appropriate tree modification is needed and a new list of nodes and leaves is created.

In the fourth and fifth steps the messages "*s*" and "*_*" are encoded in a similar way. In each step, the code tree is expanded and modified and the sequences sent to the receiver are $100's'$ and $000'\_'$, respectively. Eventually, in the sixth step (Figure 1.11), message "*i*" appears, which has already occurred before. In this case the encoder emits the codeword 01 and the number of occurrences of "*i*" in the leaf $i1$ is increased by one ($i2$). This results in the need for the next tree modification. We recommend that the motivated reader creates the tree on his/her own and generates the codeword sequence for the whole analyzed text. The whole process of encoding and creation of the code tree is shown in Figures 1.10–1.13.

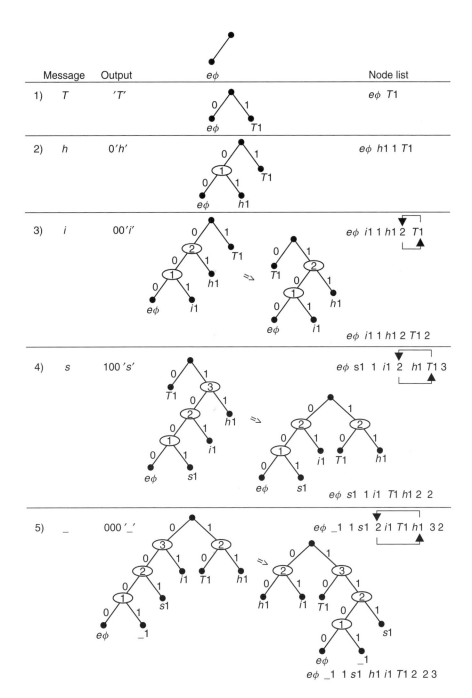

Message	Output
1) T	'T'
2) h	0'h'
3) i	00'i'
4) s	100's'
5) _	000'_'

Figure 1.10   Operation of the dynamic Huffman algorithm (Part 1)

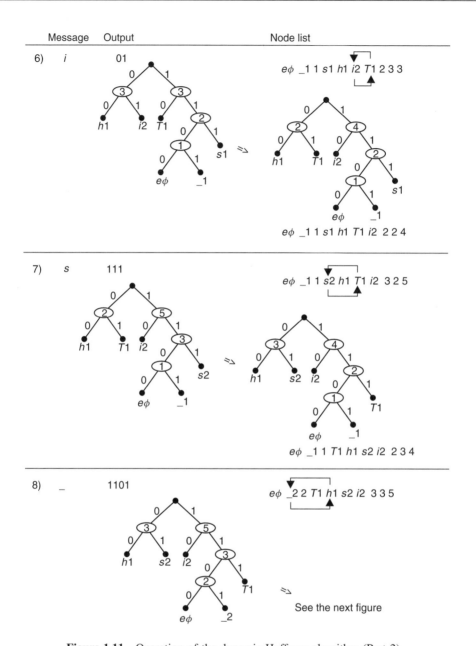

Figure 1.11 Operation of the dynamic Huffman algorithm (Part 2)

Let us note that on the basis of the codewords sent to the receiver the same code tree can be created in it. The signal indicating the introduction of a new message (character) to the code tree is the current codeword of the empty leaf, which is followed by a new ASCII-encoded character. The occurrence of a character that has already appeared before is signaled by the codeword currently assigned to it. The efficiency of dynamic Huffman coding increases with the increase in encoded message sequence length.

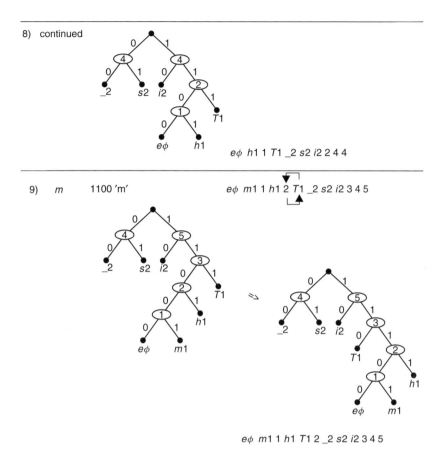

Figure 1.12   Operation of the dynamic Huffman algorithm (Part 3)

1.6.4 Arithmetic Coding

The method of source coding described in this paragraph was invented by Peter Elias around 1960; however, it was not implementable until the 1970s. The description of arithmetic coding is based on the work of Kieffer (2003).

Denote the message sequence that is the subject of arithmetic coding as the vector $(x_1, x_2, \ldots, x_n)$. Subinterval I_i of a unit interval $[0, 1]$ is assigned to each message x_i in such a way that the following expression holds

$$I_1 \supset I_2 \supset \ldots I_n \tag{1.42}$$

Subinterval I_i is determined recursively on the basis of I_{i-1} and message x_i ($i \geq 2$). When subinterval I_n is finally specified, the binary sequence $(b_1, b_2, \ldots, b_k)$ is selected in such a way that the number

$$\frac{b_1}{2} + \frac{b_2}{4} + \cdots + \frac{b_k}{2^k} \tag{1.43}$$

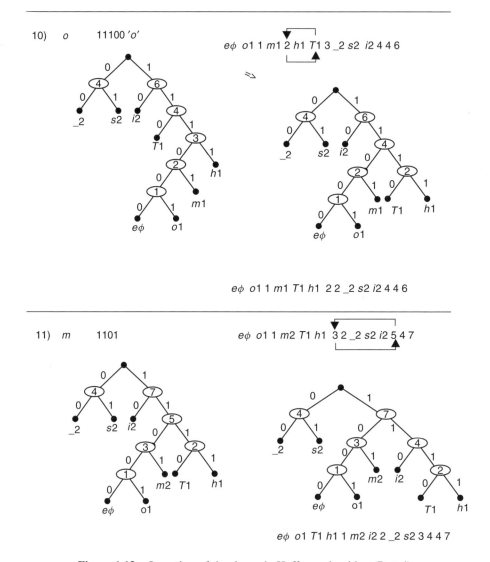

Figure 1.13 Operation of the dynamic Huffman algorithm (Part 4)

is contained inside subinterval I_n. The length k of the codeword is approximately equal to $-\log P(x_1, x_2, \ldots, x_n)$.

Let us now present a formal description of arithmetic coding and decoding. Consider the nth extension of the memoryless source $X = \{0, 1, \ldots, j-1\}$. As we see, the elementary messages of source X are simply denoted as subsequent numbers. The probability of the message sequence $(x_1, x_2, \ldots, x_n)$ is equal to

$$P_n(x_1, x_2, \ldots, x_n) = \prod_{i=1}^{n} P(x_i) \tag{1.44}$$

where $P(0), P(1), \ldots, P(j-1)$ are the probabilities of occurrence of individual elementary messages. Denote the lower and upper limit of the subinterval I_i as a_i and b_i, respectively. Let $a_0 = 0$ and $a_1 = 1$. The coding algorithm can be formally presented in the following three steps.

1. For each i $(i = 2, 3, \ldots, n)$, recursively determine the subinterval $I_i = [a_i, b_i]$ according to the formula

$$I_i = \begin{cases} [a_{i-1}, (a_{i-1} + (b_{i-1} - a_{i-1})P(0))] & \text{for } x_i = 0 \\[2mm] [a_i, b_i] & \text{for } x_i > 0 \end{cases} \tag{1.45}$$

where

$$a_i = a_{i-1} + \big(P(0) + \cdots + P(x_{i-1})\big)(b_{i-1} - a_{i-1})$$
$$b_i = a_{i-1} + \big(P(0) + \cdots + P(x_i)\big)(b_{i-1} - a_{i-1})$$

After applying the subinterval construction rule formulated above, the last subinterval I_n will have the length equal to $P_n(x_1, x_2, \ldots, x_n)$.

2. Determine the number of bits, k, used to encode sequence $(x_1, x_2, \ldots, x_n)$

$$k = \left\lceil \log \frac{1}{P_n(x_1, x_2, \ldots, x_n)} \right\rceil + 1 \tag{1.46}$$

where $\lceil x \rceil$ is the lowest integer greater than or equal to x.

3. Determine the number that is the middle of subinterval I_n. Perform k-bit expansion (1.43) of this number. Digits of this expansion, $(b_1, b_2, \ldots, b_k)$, constitute a codeword assigned to the message sequence $(x_1, x_2, \ldots, x_n)$.

The decoding of the received sequence $(b_1, b_2, \ldots, b_k)$ can be formally described in the following steps.

1. Determine number $\widehat{M}$ described by formula (1.43). Owing to preceding selection of the number of bits k used in the binary expansion of number $\widehat{M}$, the latter is located inside the subinterval I_n.
2. There are j possible positions of subinterval I_1 that depend on message x_1, however the number $\widehat{M}$ contained in this subinterval is situated in one position only. Determine the subinterval I_1 in this way and, based on that, decide upon the message x_1.
3. For each i $(i = 2, \ldots, n)$ and knowing subinterval I_{i-1}, determine that location of subinterval I_i out of j possible locations of this subinterval in which the number $\widehat{M}$ is contained. Determine message x_i based on I_i.

Let us explain the operation of coding and decoding with an example (Kieffer 2003).

Example 1.6.7 *Let the subject of arithmetic coding be a sequence of messages of a binary memoryless source with message alphabet $X = \{0, 1\}$. The probabilities of particular messages are $P(0) = 2/5$, $P(1) = 3/5$. Consider encoding the messages of the 5th source* *extension X^5 of source X. Let the binary message sequence be $(1, 0, 1, 1, 0)$. First, we determine subintervals I_1, I_2, I_3, I_4 and I_5.*

$I_1 = \frac{3}{5}$ *of the length from the right end of the interval* $[0, 1]$, *i.e.* $I_1 = \left[\frac{2}{5}, 1\right]$

$I_2 = \frac{2}{5}$ *of the length from the left end of the subinterval* I_1, *i.e.* $I_2 = \left[\frac{2}{5}, \frac{16}{25}\right]$

$I_3 = \frac{3}{5}$ *of the length from the right end of the subinterval* I_2, *i.e.* $I_3 = \left[\frac{62}{125}, \frac{16}{25}\right]$

$I_4 = \frac{3}{5}$ *of the length from the right end of the interval* I_3, *i.e.* $I_4 = \left[\frac{346}{625}, \frac{16}{25}\right]$

$I_5 = \frac{2}{5}$ *of the length from the left end of the interval* I_4, *i.e.* $I_5 = \left[\frac{346}{625}, \frac{1838}{3125}\right]$

The width of subinterval I_5 is equal to $108/3125$. Let us note that it equals the probability of generation of the considered encoded binary message sequence. Based on the width of the subinterval, the length of the codeword k is determined from formula (1.46). Thus

$$k = \left\lceil \log_2 \frac{3125}{108} \right\rceil + 1 = 6$$

The number $M = 1784/3125$ is the middle of subinterval I_5. The binary expansion of the number M is in turn equal to

$$\frac{1784}{3125} = .100100\ldots$$

so, the binary sequence $(b_1, b_2, \ldots, b_k)$ representing the encoded messsage sequence $(1, 0, 1, 1, 0)$ is (100100). Let us note that the number of bits in the codeword is higher than the length of the binary message sequence itself. However, the arithmetic coding becomes more effective when the message sequence gets longer.

Consider now the process of decoding the codeword (100100). The decoder knows that the message sequence that is the subject of coding has the form $(x_1, x_2, x_3, x_4, x_5)$. The decoder has received a 6-bit sequence, therefore $k = 6$. On that basis an approximate value of the number M equal to $\widehat{M} = 1/2 + 1/16 = 9/16$ is determined. In subsequent steps the subintervals I_i $(i = 1, 2, \ldots, 5)$ are found in such a way that the number $\widehat{M}$ is contained in them. Depending on whether a subsequent subinterval is the lower or upper part of the preceding subinterval, the decoded message is equal to 0 or 1, respectively. So the decoder subsequently determines:

- *subinterval I_1: out of two alternatives $\left[0, \frac{2}{5}\right]$ or $\left[\frac{2}{5}, 1\right]$ the decoder selects the subinterval $\left[\frac{2}{5}, 1\right]$ because $\widehat{M}$ belongs to it; therefore $x_1 = 1$;*
- *subinterval I_2: out of two alternatives $\left[\frac{2}{5}, \frac{16}{25}\right]$ or $\left[\frac{16}{25}, 1\right]$ the decoder selects the subinterval $\left[\frac{2}{5}, \frac{16}{25}\right]$ because $\widehat{M}$ belongs to it; therefore $x_2 = 0$;*
- *subinterval I_3: out of two alternatives $\left[\frac{2}{5}, \frac{62}{125}\right]$ or $\left[\frac{62}{125}, \frac{16}{25}\right]$ the decoder selects the subinterval $\left[\frac{62}{125}, \frac{16}{25}\right]$ because $\widehat{M}$ is contained in it; therefore $x_3 = 1$;*
- *subinterval I_4: out of two alternatives $\left[\frac{62}{125}, \frac{346}{625}\right]$ or $\left[\frac{346}{625}, \frac{16}{25}\right]$ the decoder selects the subinterval $\left[\frac{346}{625}, \frac{16}{25}\right]$ because $\widehat{M}$ is contained in it; therefore $x_4 = 1$;*

- *subinterval I_5: out of two alternatives $\left[\frac{346}{625}, \frac{1838}{3125}\right]$ or $\left[\frac{1838}{3125}, \frac{16}{25}\right]$ the decoder selects the subinterval $\left[\frac{346}{625}, \frac{1838}{3125}\right]$ because $\widehat{M}$ is contained in it; therefore $x_5 = 0$.*

The example of arithmetic coding and decoding considered above has not brought spectacular results – the codeword is longer than the encoded binary message sequence itself. As we noted before, the arithmetic coding becomes effective in the case of much longer message sequences, i.e. when the messages to be coded are emitted by the extensions of the memoryless source with the extension level much higher than 5. One can prove that if the source extension level n is sufficiently high, the average number of code symbols per single encoded message, denoted as L_n/n, is contained within the limits

$$H(X) < \frac{L_n}{n} < H(X) + \frac{2}{n} \tag{1.47}$$

Figure 1.14 illustrates the process of arithmetic coding and decoding of a binary message sequence considered in the above example.

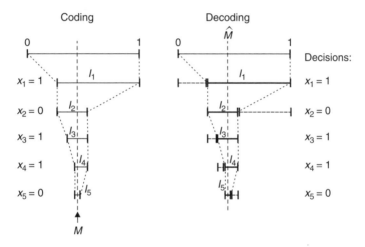

Figure 1.14   Illustration of arithmetic coding and decoding (subintervals rejected in the decoding process are denoted by dashed lines)

1.6.5 Lempel-Ziv Algorithm

The source coding algorithm developed by Lempel and Ziv (Ziv and Lempel 1977, 1978) belongs to the category of source coding methods resulting in a so-called *dictionary code*. Lempel-Ziv encoding became the basis for data compression algorithms applied in the UNIX operating system. Below we describe the operation of this algorithm in the version presented in (Ziv and Lempel 1978).

Assume that the subject of coding is a certain message sequence. Source encoding of such a sequence consists of a few steps. The first step is *parsing* of the message sequence into *phrases*. In the second step each phrase is assigned an address in the phrase dictionary built on the basis of messages encoded so far. The phrase dictionary changes dynamically along with the progress of sequence message coding. In the next step a codeword is assigned to each already determined address. Finally, the codeword assigned to the message sequence $\{x_1, x_2, \ldots, x_n\}$ is a serial concatenation of the codewords assigned to the subsequent addresses. Below we describe in detail the successive steps of the algorithm and illustrate them with a simple example. The steps are:

1. Let the message sequence $\{x_1, x_2, \ldots, x_n\}$ selected from the source alphabet $X = \{0, 1, \ldots, K - 1\}$ be the subject of coding. This sequence is parsed into subsequent phrases. The first phrase obtained in this way from the sequence $\{x_1, x_2, \ldots, x_n\}$ is x_1. The second phrase is the shortest prefix of the sequence $\{x_2, \ldots, x_n\}$ that is not equal to x_1. Let it be the sequence $\{x_2, \ldots, x_j\}$. The subsequent phrase is then the shortest prefix of the sequence $\{x_{j+1}, x_{j+2}, \ldots, x_n\}$ that is different from the previously obtained phrases. Assume that l initial phrases $B_1, B_2, \ldots, B_l$ of the sequence $\{x_1, x_2, \ldots, x_n\}$ have already been determined. Denote the remaining part of the message sequence as $x^{(l)}$. The next phrase B_{l+1} received in the process of parsing the sequence $x^{(l)}$ is the shortest prefix of sequence $x^{(l)}$ that is different from any of the previously derived phrases $B_1, B_2, \ldots, B_l$. In the case when there is no such prefix, $B_{l+1} = x^{(l)}$ and the process of parsing is finished.
2. Denote the sequence of phrases obtained in the process of parsing of the sequence of messages $\{x_1, x_2, \ldots, x_n\}$ as $B_1, B_2, \ldots, B_m$. A pair of integers is assigned to each phrase according to the following rule. Each phrase B_j of length equal to 1 is represented by a pair of numbers $(0, B_j)$. If the phrase length is higher than one, then the phrase is represented by a pair (i, s), in which s is the last symbol in phrase B_j, while i is the index of that phrase, which is identical in each symbol position of phrase B_j except the last symbol of that phrase. Next, the indices $I_j = Ki + s$ $(j = 1, 2, \ldots, m)$ are generated based on the so-created pairs (i, s). Recall that K is the cardinality of the alphabet of the message source, whereas m is the number of phrases obtained in the previous step of the algorithm in which $\{x_1, x_2, \ldots, x_n\}$ was parsed into phrases.
3. Assume that the previous step of the algorithm resulted in the indices $I_1, I_2, \ldots, I_m$. In the current step of the algorithm these indices are changed into binary sequences, which are concatenated and yield the codeword representing the message sequence $\{x_1, x_2, \ldots, x_n\}$. Each integer I_j $(j = 1, 2, \ldots, m)$ receives a binary representation that is preceded by a sequence of zeros of such length that the number of bits representing index I_j is equal to $\lceil \log_2(Kj) \rceil$. Denote the received binary sequence as C_j. The number of bits of this sequence results from determination of the maximum possible value of the index I_j. Assume that sequence C_j has been created on the basis of the integer pair (i, s). Thus, the maximum value of i is $j - 1$, whereas s can take the maximum value equal to K. Then the maximum value of the index I_j is equal to $K(j - 1) + (K - 1) = Kj - 1$ and the number of bits necessary for binary representation of that number is exactly $\lceil \log_2(Kj) \rceil$. In the end, the codeword assigned to message sequence $\{x_1, x_2, \ldots, x_n\}$ achieves the form $(C_1, C_2, \ldots, C_m)$.

Let us illustrate the formal description of the algorithm with the following example.

Example 1.6.8 *Let the subject of coding be the message sequence of the form*

$$1010110100100$$

so the source has a binary alphabet $X = \{0, 1\}$ and its cardinality is $K = 2$. The subsequent phrases achieved in the first step of the algorithm have the form

$$B_1 = (1), \qquad B_2 = (0), \qquad B_3 = (1, 0), \qquad B_4 = (1, 1),$$
$$B_5 = (0, 1), \qquad B_6 = (0, 0), \qquad B_7 = (1, 0, 0)$$

Integer pairs (i, s) corresponding to the received phrases are

$$(0, 1), \ (0, 0), \ (1, 0), \ (1, 1), \ (2, 1), \ (2, 0), \ (3, 0)$$

In turn, the indices corresponding to the received phrases are in this case determined by the formula $I_j = 2i + s$ and they are equal to

$$I_1 = 1, \ I_2 = 0, \ I_3 = 2, \ I_4 = 3, \ I_5 = 5, \ I_6 = 4, \ I_7 = 6$$

The lengths of the codewords C_j received from the binary representation of indices I_j supplemented from the left side by an appropriate number of zeros are calculated from the formula $\lceil \log_2(2j) \rceil$ and they are respectively equal to

$$1, \ 2, \ 3, \ 3, \ 4, \ 4, \ 4$$

In consequence, knowing the lengths of the codewords C_j and the values of indices I_j the codewords can be determined as

$$C_1 = (1), \qquad C_2 = (0, 0), \qquad C_3 = (0, 1, 0), \qquad C_4 = (0, 1, 1),$$
$$C_5 = (0, 1, 0, 1), \qquad C_6 = (0, 1, 0, 0), \qquad C_7 = (0, 1, 1, 0)$$

Summarizing, the binary message sequence 1010110100100 is assigned the following code sequence

$$(C_1, C_2, C_3, C_4, C_5, C_6, C_7) = (1, 0, 0, 0, 1, 0, 0, 1, 1, 0, 1, 0, 1, 0, 1, 0, 0, 0, 1, 1, 0)$$

As we see, also in this example the length of the codewords exceeds the length of the binary message sequence that is the subject of coding. However, this is the case for short message sequences only. The algorithm becoms effective for long mesage sequences. It can be proved that the following condition is fulfilled for the Lempel-Ziv encoding

$$H(X) \leq \frac{L_n}{n} \leq H(X) + \frac{\rho_K}{\log_2 n} \tag{1.48}$$

where ρ_K is a positive constant depending on the cardinality K of the source alphabet.

Generally, the Lempel-Ziv algorithm is a suboptimal procedure. However, it has a meaningful advantage – it does not depend on the properties of the message source. Because of this feature the Lempel-Ziv code is called a *universal code*.

1.6.6   Case study: Source Coding in Facsimile Transmission

A facsimile (fax) machine is a device that converts black and white pictures into binary sequences, performs their compression and, finally, using a transmission device called a *modem*, sends them to the receiver. In the receiver, the received binary sequences are decompressed and the printing device maps them back into a black and white picture. Note that compression of the binary sequence is virtually the same as a source coding.

Fax transmission is standardized by the recommendations issued by the International Telecommunication Union (ITU). Two fax standards denoted as T4 and T6 are currently applied. They set the rules of operation of faxes from Group 3 and Group 4 related to the Public Switched Telephone Network (PSTN) and Integrated Services Digital Network (ISDN), respectively.

In a typical fax machine, a picture to be processed is divided into primary elements called pixels with a resolution of 100 or 200 lines per inch (3.85 or 7.7 lines per millimeter). Let a white pixel be represented by the digit "0" and a black pixel by the digit "1". In consequence, a single A4 page is equivalent to about two million binary digits. During the analysis of subsequent lines of a picture with the above-mentioned resolution one can easily see that the binary sequence representing one line is either dominated by long sequences of zeros or by zeros or ones grouped in blocks. These observations refer to a typical form of scanned document featuring a black text or black drawings with geometrical lines against a white background. These features are taken into account in the choice of coding method used to compress the binary sequence received in the process of scanning (processing of the sampled picture). The source code applied in the transmitter belongs to the so-called *run-length codes* in which the subject of encoding is the length of sequence of identical digits.

The data compressing block applies two code tables: the first one containing the so-called *termination code sequences* (Figure 1.15) and the second one the *make-up code sequences* (Figure 1.16). Both tables have been derived on the basis of *Modified Huffman* (MH) coding. In general, encoding relies on representing the length of bit sequence of the same kind in an effective manner by decomposing this length into two components: the largest possible natural number from the list of lengths in the make-up code sequences and the complementing number from the terminating code sequences, which is in the range 0–64. For example, the sequence of 140 black pixels occurring in a single line can be written in the form of two components, $128 + 12$. From the table of make-up code sequences we find a code sequence associated with the number 128, which is 000011001000, whereas in the table of termination code sequences we find a code sequence associated with the number 12, which is a sequence 0000111. Summarizing, the block of 140 black pixels, which without compression would be represented by 140 "1"s, is compressed to the form 0000110010000000111. Therefore the number of bits that have to be sent to the receiver decreases from 140 to 19.

Let us note that the last element of the make-up code sequence table is a codeword associated with the control message denoting the end of line – EOL (*End-of-Line*). In Group 3 fax error protection is not applied and each scanned line is independently encoded. Therefore, if one or more bits in the code sequences representing a given line are received erroneously, then the decoder starts to parse the received binary sequence into code sequences in an erroneous manner. The EOL codeword is applied, among others, to

Number of white pixels	Code sequence	Number of black pixels	Code sequence	Number of white pixels	Code sequence	Number of black pixels	Code sequence
0	00110101	0	0000110111	32	00011011	32	000001101010
1	000111	1	010	33	0010010	33	000001101011
2	0111	2	11	34	00010011	34	000011010010
3	1000	3	10	35	00010100	35	000011010011
4	1011	4	011	36	00010101	36	000011010100
5	1100	5	0011	37	00010110	37	000011010101
6	1110	6	0010	38	00010111	38	000011010110
7	1111	7	00011	39	00101000	39	000011010111
8	10011	8	000101	40	00101001	40	000001101100
9	10100	9	000100	41	00101011	41	000001101101
10	00111	10	0000100	42	00101011	42	000011011010
11	01000	11	0000101	43	00101100	43	000011011011
12	001000	12	0000111	44	00101101	44	000001010100
13	000011	13	00000100	45	00000100	45	000001010101
14	110100	14	00000111	46	00000101	46	000001010110
15	110101	15	000011000	47	00001010	47	000001010111
16	101010	16	0000010111	48	00001011	48	000001100100
17	101011	17	0000011000	49	01010010	49	000001100101
18	0100111	18	0000001000	50	01010011	50	000001010010
19	0001100	19	00001100111	51	01010100	51	000001010011
20	0001000	20	00001101000	52	01010101	52	000000100100
21	0010111	21	00001101100	53	00100100	53	000000110111
22	0000011	22	00000110111	54	00100101	54	000000111000
23	0000100	23	00000101000	55	01011000	55	000000100111
24	0101000	24	00000010111	56	01011001	56	000000101000
25	0101011	25	00000011000	57	01011010	57	000001011000
26	0010011	26	000011001010	58	01011011	58	000001011001
27	0100100	27	000011001011	59	01001010	59	000000101011
28	0011000	28	000011001100	60	01001011	60	000000101100
29	00000010	29	000011001101	61	00110010	61	000001011010
30	00000011	30	000001101000	62	00110011	62	000001100110
31	00011010	31	000001101001	63	00110100	63	000001100111

Figure 1.15 Table of termination code sequences for white and black pixel sequences

recover synchronism. If the decoder is not able to identify a valid codeword after a sequence of bits of length equal to the maximum code length, it starts to look for the EOL sequence. In turn, if after an assumed number of decodable lines the decoder is not able to find the EOL sequence, the decoding process is interrupted and the transmitter is notified about this event. A single EOL codeword precedes the codewords of each scanned page, whereas a sequence of six EOL codewords indicates the end of a page.

Besides the Modified Huffman coding described above, some more sophisticated fax machines additionally use Modified Read coding. Line processing is performed in blocks of a few lines. The first line in the block is encoded in conformance with MH coding, described above. The next line in the processed block is compared with the previous one and their difference is the subject of encoding. This is an effective solution as most lines differ little from their predecessor. Such an approach is continued until the last line in the block is processed. The whole procedure is repeated in the next blocks of lines. The number of lines in a block is two for standard resolution and four in higher resolution fax machines.

Number of white pixels	Code sequence	Number of black pixels	Code sequence	Number of black pixels	Code sequence	Number of black pixels	Code sequence
64	11011	64	0000001111	1344	011011010	1344	0000001010011
128	10010	128	000011001000	1408	011011011	1408	0000001010100
192	010111	192	000011001001	1472	010011000	1472	0000001010101
256	0110111	256	000001011011	1536	010011001	1536	0000001011010
320	00110110	320	000000110011	1600	010011010	1600	0000001011011
384	00110111	384	000000110100	1664	011000	1664	0000001100100
448	01100100	448	000000110101	1728	010011011	1728	0000001100101
512	01100101	512	0000001101100	1792	00000001000	1792	00000001000
576	01101000	576	0000001101101	1856	00000001100	1856	00000001100
640	01100111	640	0000001001010	1920	00000001101	1920	00000001101
704	011001100	704	0000001001011	1984	000000010010	1984	000000010010
768	011001101	768	0000001001100	2048	000000010011	2048	000000010011
832	011010010	832	0000001001101	2112	000000010100	2112	000000010100
896	011010011	896	0000001110010	2176	000000010101	2176	000000010101
960	011010100	960	0000001110011	2240	000000010110	2240	000000010110
1024	011010101	1024	0000001110100	2304	000000010111	2304	000000010111
1088	011010110	1088	0000001110101	2368	000000011100	2368	000000011100
1152	011010111	1152	0000001110110	2432	000000011101	2432	000000011101
1216	011011000	1216	0000001110111	2496	000000011110	2496	000000011110
1280	011011001	1280	0000001010010	2560	000000011111	2560	000000011111
				EOL	0000000001	EOL	0000000001

Figure 1.16   Table of make-up code sequences for white and black pixel sequences

This kind of source coding algorithm is effective for scanning and coding documents featuring the above-mentioned properties. However, if the subject of source coding were a black and white photograph in which grey levels are achieved by the intensity of black pixels, the length of the code sequence received due to compression would be higher than that of the uncoded binary sequence. This is caused by the fact that very short binary sequences dominate in the sampled lines and such sequences are assigned long codewords.

The source coding algorithms presented in the above sections cover a few basic coding methods only. All of them belong to the methods of *lossless encoding*. This means that on the basis of correctly received codewords the receiver is able to recover a message sequence identical to that produced on the transmitting side. In turn, coding of still and moving pictures or sound coding is often matched to the human eye or ear perception properties. In such cases the amount of information needed to recover the picture or sound at the receiver can be substantially lower without subjective quality deterioration declared by the recipient. The coding methods that use this fact are called *lossy coding*. This is an interesting domain of digital signal processing that is outside the scope of this book.

1.7   Channel Models from the Information Theory Point of View

So far in the course of information theory we have considered message source models and source encoding. The next block of a communication system that requires our study is a channel encoder, however the topics related to channel coding are so broad that most of them are presented in the next chapter. Instead we will concentrate our attention on the next block in the communication system block chain: a channel and its models. The channel models reflect the influence of physical properties of the transmission channel and all the elements of the communication system that are hidden in the joint block, often called a *channel*. Below we present the simplest but most important channel models.

1.7.1 Discrete Memoryless Channel

A *discrete memoryless channel* is a channel model describing the statistical dependence of the input symbols X on the channel output symbols Y. From a statistical point of view X and Y are random variables. In each time instant the channel accepts and subsequently transmits a single symbol X selected from the input symbol alphabet $\mathcal{X}$. As a result, a single output symbol Y from alphabet $\mathcal{Y}$ appears on the channel output. If the alphabets $\mathcal{X}$ and $\mathcal{Y}$ are finite, then the channel is *discrete*. The channel is *memoryless* if the current output symbol Y depends exclusively on the current input symbol X, and does not depend on the previous input symbols. Formalizing the above statements, let us present the following definition of a discrete memoryless channel model.

Definition 1.7.1 *A discrete memoryless channel model is a statistical model of a channel determined by the following elements:*

- *the input symbol alphabet* $\mathcal{X} = \{x_1, x_2, \ldots, x_J\}$;
- *the output symbol alphabet* $\mathcal{Y} = \{y_1, y_2, \ldots, y_K\}$;
- *the set of transition probabilities* $P(y_k|x_j) = \Pr\{Y = y_k | X = x_j\}$, *for* $k = 1, \ldots, K$, $j = 1, \ldots, J$.

The cardinality of input and output alphabets does not need to be the same. In practice, we usually have $K \geq J$. The transition probabilities $P(y_k|x_j)$ describe statistical properties of the channel and are often presented in matrix form, as in formula (1.49)

$$
P = \begin{bmatrix}
P(y_1|x_1) & P(y_2|x_1) & \cdots & P(y_K|x_1) \\
P(y_1|x_2) & P(y_2|x_2) & \cdots & P(y_K|x_2) \\
\vdots & \vdots & \vdots & \vdots \\
P(y_1|x_J) & P(y_2|x_J) & \cdots & P(y_K|x_J)
\end{bmatrix}
\tag{1.49}
$$

Matrix P is called a *channel matrix* or *transition matrix*. Each matrix row is associated with a single input symbol. In turn, each column is related to a single output symbol. Because for each generated channel input symbol a single output symbol is received, the following equality holds

$$
\sum_{k=1}^{K} P(y_k|x_j) = 1 \quad \text{for} \quad j = 1, 2, \ldots, J
\tag{1.50}
$$

As we see, the sum of probabilities in a single row is equal to unity.

For simplicity, denote $P(x_j) = \Pr\{X = x_j\}$, $j = 1, 2, \ldots, J$. Then the probability of occurrence of a particular output symbol can be calculated on the basis of the formula

$$
\Pr\{Y = y_k\} = P(y_k) = \sum_{j=1}^{J} P(y_k|x_j) P(x_j)
\tag{1.51}
$$

Let us temporarily assume that the numbers of channel input and output symbols are the same, i.e. $K = J$. Assume that if the index of the received channel output symbol is the same as the index of the transmitted channel input symbol, then the reception is correct. If the reception correctness were described by another relation between the input and output symbols, then we would be able to renumerate the output symbols in such a way that the condition $j = k$ would reflect the correct reception. The reception is incorrect if $k \neq j$. In consequence of the above assumptions, the probability of incorrect reception, i.e. of an error, can be determined from the formula

$$P(\mathcal{E}) = \sum_{k=1, k \neq j}^{K} \sum_{j=1}^{J} \Pr\{Y = y_k, X = x_j\} = \sum_{k=1, k \neq j}^{K} \sum_{j=1}^{J} P(y_k|x_j)P(x_j) \qquad (1.52)$$

It is often easier to calculate this probability by deriving the probability $P(C)$ of the correct reception first. Namely, we have

$$P(\mathcal{E}) = 1 - P(C) = 1 - \sum_{k=1}^{K} \Pr\{Y = y_k, X = x_k\}$$

$$= 1 - \sum_{k=1}^{K} P(y_k|x_k)P(x_k) = \sum_{k=1}^{K} P(x_k) - \sum_{k=1}^{K} P(y_k|x_k)P(x_k)$$

$$= \sum_{k=1}^{K} \left[1 - P(y_k|x_k)\right] P(x_k) = \sum_{k=1}^{K} P(\mathcal{E}|x_k)P(x_k) \qquad (1.53)$$

The probability $P(x_k)$ of generation of a given input symbol x_k is often called *a priori probability*.[4]

1.7.2 Examples of Discrete Memoryless Channel Models

Below we present a few basic discrete memoryless channel models. Despite their simplicity they are often used as a tool in selection of a channel code and its decoding method.

1.7.2.1 Binary Symmetric Memoryless Channel

A binary symmetric memoryless channel is the most common channel model. In this model we assume that both the input and output symbol alphabets are binary (often described by symbols "0" and "1"), and subsequent output symbols are dependent only on single input symbols. As we remember, this is a statistical description of the absence of channel memory. Adoption of model symmetry results in assuming the same statistical channel behavior in the case of generation of symbols "0" and "1". The binary symmetric memoryless channel is presented in Figure 1.17a. The arrows illustrate the occurrence of

---

[4] The term *a priori* originates from the Latin language and denotes something given in advance, before experiencing it.

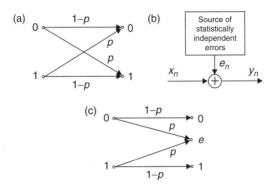

Figure 1.17 Models of binary memoryless channels: (a) binary symmetric channel; (b) another form of binary symmetric channel with binary error source; (c) binary erasure channel

the output symbol depending on the generated input symbol, with appropriate transition probabilities placed above them. For our channel model we have

$$\Pr\{Y = 0|X = 0\} = P(0|0) = 1 - p \quad \Pr\{Y = 0|X = 1\} = P(0|1) = p$$
$$\Pr\{Y = 1|X = 0\} = P(1|0) = p \quad \Pr\{Y = 1|X = 1\} = P(1|1) = 1 - p$$

Let us now calculate the probability of error. Denoting the *a priori* probabilities of the input symbols as $\Pr\{X = 0\} = \alpha$ and $\Pr\{X = 1\} = 1 - \alpha$, respectively (as we know, the sum of probabilities of occurrence of all the input symbols is equal to unity), we obtain

$$P(\mathcal{E}) = \Pr\{Y = 1|X = 0\}\Pr\{X = 0\} + \Pr\{Y = 0|X = 1\}\Pr\{X = 1\}$$
$$= p\alpha + p(1 - \alpha) = p \tag{1.54}$$

The model shown in Figure 1.17a presents the channel operation in a single moment. In Figure 1.17b another form of binary symmetric memoryless channel is presented. In subsequent moments, indexed by n, the input symbols x_n take on the value "0" or "1". The occurrence of errors in the channel is modeled by the exclusive-or addition of the input symbol with the binary error symbol e_n. If an error occurs in the channel, the error source generates the symbol $e_n = 1$, otherwise $e_n = 0$. As we know from formula (1.54), the probability of error is equal to p, so the error source is a binary digit generator that emits "1"s statistically independently of other symbols, with the probability p.

The model seems to be highly abstract. However, in practice many channel codes, and in particular decoding algorithms, are constructed by taking into account such an error source model. Very often errors in real communication channels are not uniformly distributed in time, but they are grouped in *error bursts*. Thus, there are time intervals for which the error probability is high and intervals in which error bursts do not happen. A remedy for this disadvantageous situation is the application of a so-called *interleaver* at the transmitter and *deinterleaver* at the receiver. They are blocks that perform mutually dual operations. The interleaver changes the order of the sequence of transmitted channel input symbols, whereas the deinterleaver recovers the initial order of the sequence of

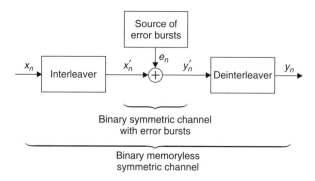

Figure 1.18   Application of interleaver and deinterleaver in spreading of error bursts

input symbols operating on the channel output symbols. Both operations compensate each other with respect to the data symbols; however, the error sequence is the subject of deinterleaving only. Thus the error bursts occurring in the channel are spread in time and become almost independent statistically. Figure 1.18 illustrates a general rule of interleaving and deinterleaving. Owing to this idea the binary symmetric memoryless channel model remains valuable, despite the fact that the errors occurring in the channel are bursty and therefore statistically dependent.

1.7.2.2   Binary Symmetric Erasure Channel

Figure 1.17c presents a binary symmetric erasure channel. As we see, the number of different output symbols is increased to three. Besides the symbols "0" and "1" there is a third symbol denoted by "e" , called *erasure*. This symbol reflects the situation in which the receiver is not able to perform detection and decide if the received symbol is "0" or "1". This can occur if there is a transmission outage or if another transmitter placed in the receiver's vicinity temporarily saturates this receiver. This model does not take into account the possibility of conversion of "0" into "1" or vice versa. Derivation of error probability is very simple and leads to the same result as in the case of a binary symmetric memoryless channel. Namely, assuming again that $\Pr\{X = 0\} = \alpha$ and $\Pr\{X = 1\} = 1 - \alpha$ we have

$$P(\mathcal{E}) = \Pr\{Y = e | X = 0\}\alpha + \Pr\{Y = e | X = 1\}(1 - \alpha)$$

$$= p\alpha + p(1 - \alpha) = p \qquad\qquad (1.55)$$

1.7.2.3   Memoryless Channel with Binary Input and m-ary Output

The memoryless binary input m-ary output channel model is more and more often used in the analysis of digital communication systems with channel coding. As we remember, in the binary memoryless channel model the output symbols are binary. This model reflects the cases in which the demodulator, supplemented with a decision device, produces binary decisions. Thus, the channel decoder loses a part of the knowledge that otherwise could be used in the channel decoding. An important step towards avoiding this drawback is the

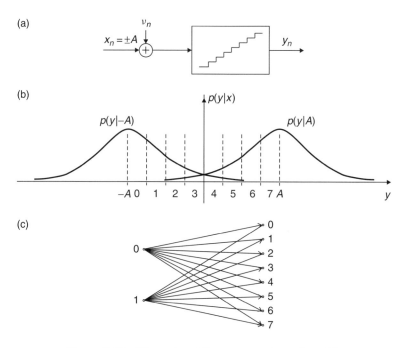

Figure 1.19 Binary input 8-ary output channel model

application of a decision device that not only generates binary decisions but also gives an additional measure of decision quality. An example of such an approach is the application of an m-level quantizer replacing the binary one. The simplest model of such a system is shown in Figure 1.19a. The binary symbols are transmitted in the form of pulses of amplitude $\pm A$, which are distorted by an additive noise. A sample of a signal that undergoes m-ary quantization is taken once per single pulse duration. In Figure 1.19b the quantization thresholds of an 8-level quantizer are shown against the background of conditional probability density functions of the channel output samples, whereas in Figure 1.19c a corresponding channel model is drawn. The transition probabilities for a particular input and output signal pair (not shown above the arrows showing appropriate transitions in Figure 1.19c) are given by the area limited by the appropriate conditional probability curve and neighboring quantization thresholds. The selection of the optimum quantization thresholds will not be considered here. The described model is an example of the channel model applied in situations in which the decision device generates *soft decisions* when our knowledge about received symbols is larger than the decided binary values only.

1.7.3 Example of a Binary Channel Model with Memory

As we said before, in some communication channels errors tend to appear in bursts. It happens in particular on radio channels featuring fading. The channel in the fading phase is characterized by a low signal-to-noise power ratio, which leads, as we will learn later in this book, to a high error probability. The channel that is not suffering from fading in

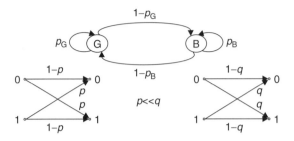

Figure 1.20   Gilbert-Elliot model

a given moment features a higher signal-to-noise power ratio, so the error probability at the receiver is lower. Evolution from one state to the other has a random and dynamic nature, therefore the channel model has to take into account the memory of the channel state. The simplest statistical channel model that reflects this situation is the *Gilbert-Elliot model* (Bossert 1999) shown in Figure 1.20.

The channel can be in one of two states: a good one (G) or a bad one (B). Transition from good to bad state occurs with the probability $1 - p_G$, whereas transition in the opposite direction occurs with the probability $1 - p_B$. The channel remains in a good state with the probability p_D, and in a bad state with the probability p_B. When the channel remains in a good state the binary memoryless symmetric channel with a low error probability p is applied. Transition to bad state B implies a change of the properties of the binary memoryless symmetric channel model, resulting in an increase of the error probability to $q \gg p$. As we see in Figure 1.20, the transition state diagram of the channel model resembles the state diagram of the Markov source.

The Gilbert-Elliot model is one of the simplest channel models with memory. Determination of the probabilities p, q, p_B and p_G is often performed on the basis of long-lasting observation of occurrence of errors and their statistical grouping in bursts. However, we have to stress that if the delay introduced by the communication system is not of the first importance, the best solution is application of the interleaver in the transmitter and the deinterleaver in the receiver in order to turn these blocks, jointly with the channel with memory, into the memoryless channel with statistically independent errors. Most of the channel codes and associated decoding algorithms are designed for the latter channel model.

1.8   Mutual Information

Below we introduce the term *mutual information*, which will be applied in the course of this chapter to determine the limits of channel parameters. This concept has a very general meaning; however, for practical reasons we will relate it to the characteristics of input and output channel symbols X and Y, respectively.

Since each input symbol x_j ($j = 1, 2, \ldots, J$) is given to the channel input with a specified probability, it is possible to determine the entropy $H(\mathcal{X})$ of the input symbol alphabet $\mathcal{X}$. Intuitively, the entropy $H(\mathcal{X})$ is a measure of the observer's uncertainty with respect to the occurrence of a channel input symbol. It is obviously maximum if all the input symbols are equiprobable. In relation to this fact the question arises as to how the observation of a channel output symbol influences the uncertainty (or knowledge, respectively) of the observer about the symbol given to the channel input. In order to answer this question we introduce the notion of *conditional entropy*.

Assume that a specific symbol y_k has been observed at the channel output. The input symbol entropy conditioned on the observation of that output symbol is determined by the formula

$$H(\mathcal{X}|Y = y_k) = \sum_{j=1}^{J} P(x_j|y_k) \log \frac{1}{P(x_j|y_k)} \tag{1.56}$$

If we average this formula with respect to all possible output symbols, we end up with the following expression describing the conditional entropy

$$H(\mathcal{X}|\mathcal{Y}) = \sum_{k=1}^{K} H(\mathcal{X}|Y = y_k) P(y_k)$$

$$= \sum_{k=1}^{K} \sum_{j=1}^{J} P(x_j|y_k) P(y_k) \log \frac{1}{P(x_j|y_k)}$$

$$= \sum_{k=1}^{K} \sum_{j=1}^{J} P(x_j, y_k) \log \frac{1}{P(x_j|y_k)} \tag{1.57}$$

We can state once more that $H(\mathcal{X}|\mathcal{Y})$ represents the observer's uncertainty with respect to the channel input symbols X that remains after observation of the output symbols Y. Since, as we remember, $H(\mathcal{X})$ represents the observer's uncertainty before his/her observation of the channel output (*a priori*), the difference $H(\mathcal{X}) - H(\mathcal{X}|\mathcal{Y})$ represents the value by which the observer's uncertainty with respect to an input symbol has decreased after observation of the appropriate channel output symbol. Let us note that this value has a meaning of an average amount of information about the channel input symbol gained by the observer on the basis of the channel output symbol. Due to the above-presented properties, the value $H(\mathcal{X}) - H(\mathcal{X}|\mathcal{Y})$ is called an *average amout of mutual information*, or in short *mutual information*, and is denoted by $I(\mathcal{X}; \mathcal{Y})$. It can be also interpreted in the following way: the observer needs, on average, $H(\mathcal{X})$ bits of information to determine a channel input symbol. However, after observing a channel output symbol he/she needs on average only $H(\mathcal{X}|\mathcal{Y})$ bits of information to determine which channel input symbol caused the appearance of the observed channel output symbol. Thus, one can say that observation of a single channel output symbol results in gaining, on average, $H(\mathcal{X}) - H(\mathcal{X}|\mathcal{Y})$ bits of information. Applying the expressions describing the *a priori* and conditional entropies in the expression $I(\mathcal{X}; \mathcal{Y}) = H(\mathcal{X}) - H(\mathcal{X}|\mathcal{Y})$ results in the following formula for mutual information

$$I(\mathcal{X}; \mathcal{Y}) = H(\mathcal{X}) - H(\mathcal{X}|\mathcal{Y})$$

$$= \sum_{j=1}^{J} P(x_j) \log \frac{1}{P(x_j)} - \sum_{k=1}^{K} \sum_{j=1}^{J} P(x_j, y_k) \log \frac{1}{P(x_j|y_k)}$$

Knowing that $P(x_j) = \sum_{k=1}^{K} P(x_j, y_k)$ we obtain

$$I(X; Y) = \sum_{k=1}^{K} \sum_{j=1}^{J} P(x_j, y_k) \left(\log \frac{1}{P(x_j)} - \log \frac{1}{P(x_j|y_k)} \right)$$

$$= \sum_{k=1}^{K} \sum_{j=1}^{J} P(x_j, y_k) \log \frac{P(x_j|y_k)}{P(x_j)} \qquad (1.58)$$

1.9 Properties of Mutual Information

Mutual information is characterized by the following properties.

Property 1.9.1 *Mutual information is symmetric, i.e.*

$$I(\mathcal{X}; \mathcal{Y}) = I(\mathcal{Y}; \mathcal{X}) \qquad (1.59)$$

Proof. Knowing that mutual information $I(\mathcal{X}; \mathcal{Y})$ is determined by formula (1.58) and using Bayes' formula in the following form

$$P(x_j|y_k) = \frac{P(x_j, y_k)}{P(y_k)} = \frac{P(y_k|x_j) P(x_j)}{P(y_k)}$$

we obtain

$$I(\mathcal{X}; \mathcal{Y}) = \sum_{k=1}^{K} \sum_{j=1}^{J} P(x_j, y_k) \log \frac{P(y_k|x_j) P(x_j)}{P(y_k) P(x_j)}$$

$$= \sum_{k=1}^{K} \sum_{j=1}^{J} P(x_j, y_k) \log \frac{P(y_k|x_j)}{P(y_k)} = I(\mathcal{Y}; \mathcal{X}) \qquad (1.60)$$

Property 1.9.2 *Mutual information is non-negative, i.e.* $I(\mathcal{X}; \mathcal{Y}) \geq 0$.

Proof. In order to prove this property we will apply the following property, previously used in the course of our lecture. It states that if P_i and Q_i denote probabilities (i.e. $P_i \geq 0$, $Q_i \geq 0$, $\sum_i P_i = 1$, $\sum_i Q_i = 1$), then the following inequality holds

$$\sum_i P_i \log \frac{Q_i}{P_i} \leq 0 \quad \text{i.e.} \quad \sum_i P_i \log \frac{P_i}{Q_i} \geq 0 \qquad (1.61)$$

As we remember, the equality occurs when $\bigwedge_i P_i = Q_i$. Applying the formula for conditional probability in (1.58), we obtain

$$I(\mathcal{X}; \mathcal{Y}) = \sum_{k=1}^{K} \sum_{j=1}^{J} P(x_j, y_k) \log \frac{P(x_j|y_k)}{P(x_j)}$$

$$= \sum_{k=1}^{K} \sum_{j=1}^{J} P(x_j, y_k) \log \frac{P(x_j, y_k)}{P(x_j) P(y_k)} \tag{1.62}$$

Comparing (1.62) with (1.61) we note that $P(x_j, y_k)$ can be interpreted as P_i in (1.61) whereas the product of probabilities $P(x_j) P(y_k)$ can be interpreted as Q_i. Summation with respect to both indices, j and k, exhausts the whole set of probabilities. Then, on the basis of (1.61), it becomes obvious that $I(X; Y) \geq 0$.

Let us note that mutual information is equal to zero if and only if for each index k and j the following equality holds: $P(x_j, y_k) = P(x_j) P(y_k)$. This means that channel input and output symbols are statistically independent. In practice it means that the channel acts in such a way that there is no dependence of channel output symbols on those given to its input. From the point of view of information transfer, such a channel is useless.

From Property 1.9.1 and the general formula describing mutual information, the following property can be deduced:

Property 1.9.3 *Mutual information can be determined from the formula*

$$I(X; Y) = H(Y) - H(Y|X) \tag{1.63}$$

Let us illustrate the above-mentioned properties of mutual information with the following example.

Example 1.9.1 *Let us calculate mutual information for input and output of a binary symmetric memoryless channel. Recall that a binary symmetric memoryless channel is depicted in Figure 1.17a. In calculations of mutual information we will use Property 1.9.3. Assume that $\Pr\{X = 0\} = \alpha$, which results in $\Pr\{X = 1\} = 1 - \alpha = \overline{\alpha}$. Let us start with the calculation of the conditional entropy $H(Y|X)$. Using formula (1.57) we have*

$$H(Y|X) = \sum_{k=1}^{2} \sum_{j=1}^{2} P(y_k|x_j) P(x_j) \log \frac{1}{P(y_k|x_j)}$$

$$= \alpha \left(\overline{p} \log \frac{1}{\overline{p}} + p \log \frac{1}{p} \right) + \overline{\alpha} \left(\overline{p} \log \frac{1}{\overline{p}} + p \log \frac{1}{p} \right)$$

$$= (\alpha + \overline{\alpha}) \left(\overline{p} \log \frac{1}{\overline{p}} + p \log \frac{1}{p} \right) = H(p) \tag{1.64}$$

As we see, the conditional entropy depends only on the channel properties and it does not depend on the channel input symbol statistics. Let us calculate now the entropy of the channel output symbols. For this purpose we have to determine the channel output probabilities, which are equal to (see Figure 1.17a)

$$\Pr\{Y = 0\} = \Pr\{Y = 0|X = 0\} \Pr\{X = 0\} + \Pr\{Y = 0|X = 1\} \Pr\{X = 1\}$$

$$= \overline{p}\alpha + p\overline{\alpha}$$

$$\Pr\{Y = 1\} = \Pr\{Y = 1 | X = 0\} \Pr\{X = 0\} + \Pr\{Y = 1 | X = 1\} \Pr\{X = 1\}$$

$$= p\alpha + \overline{p}\,\overline{\alpha} = 1 - \Pr\{Y = 0\} \qquad (1.65)$$

Since in our channel model there are only two input symbols, the probability of one of them is the complement to unity of the probability of the other one. As a result, the entropy of the channel output symbols $H(\mathcal{Y})$ is the entropy function of the argument $\Pr\{Y = 0\}$ calculated from formula (1.65). Therefore, $H(\mathcal{Y}) = H(\overline{p}\alpha + p\overline{\alpha})$. Finally, mutual information of the input and output channel symbols for the binary symmetric memoryless channel is

$$I(\mathcal{X}; \mathcal{Y}) = H(\overline{p}\alpha + p\overline{\alpha}) - H(p) \qquad (1.66)$$

Mutual information is shown graphically in Figure 1.21 with the entropy function in the background. Let us note that if the error probability in a binary symmetric memoryless channel is lower than $1/2$, then $\overline{p}\alpha + p\overline{\alpha} > p$, so, taking into account the shape of the entropy function, we find that the difference of entropies in formula (1.66) is positive. The case in which $p = 1/2$ is the only one for which the arguments of both entropy functions are identical, therefore their difference is zero. In conclusion, if the error probability in a binary symmetric memoryless channel is $p = 1/2$, then the average amount of information transferred by the channel is zero.

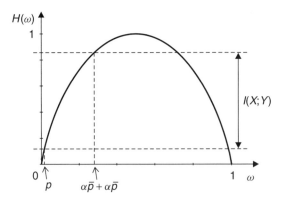

Figure 1.21 Illustration of calculations of average amount of mutual information in the case of transmission over a binary symmetric memoryless channel

1.10 Channel Capacity

Consider once more a discrete memoryless channel model characterized by the input symbol alphabet $\mathcal{X}$, the output symbol alphabet $\mathcal{Y}$ and the set of transition probabilities $P(y_k | x_j)$ ($k = 1, 2, \ldots, K$, $j = 1, 2, \ldots, J$). The conclusion that can be drawn from equation (1.60) leads us to the statement that mutual information $I(\mathcal{X}; \mathcal{Y})$ can be expressed as a function of the probabilities of input and output symbols as well as transition probabilities. Thus, it can be calculated using the formula

$$I(\mathcal{X}; \mathcal{Y}) = \sum_{k=1}^{K} \sum_{j=1}^{J} P(x_j, y_k) \log \frac{P(y_k|x_j)}{P(y_k)}$$

Knowing that $P(x_j, y_k) = P(y_k|x_j)P(x_j)$ and $P(y_k) = \sum_{j=1}^{J} P(y_k|x_j)P(x_j)$ we are able to express $I(\mathcal{X}; \mathcal{Y})$ using exclusively the input symbol probabilities and transition probabilities. Namely we have

$$I(\mathcal{X}; \mathcal{Y}) = \sum_{k=1}^{K} \sum_{j=1}^{J} P(y_k|x_j)P(x_j) \log \frac{P(y_k|x_j)}{\sum_{l=1}^{J} P(y_k|x_l)P(x_l)} \qquad (1.67)$$

The transition probabilities are known in advance because they characterize the channel. Thus, mutual information is influenced only by the channel input symbol probabilities, which in turn can result from particular procedures of source or channel coding. Therefore, the following question arises: What are the values of input symbol probabilities for which the average amount of mutual information is maximum? This maximum simultaneously determines the maximum amount of information that can be transmitted on average through the channel by sending a single input symbol and receiving a single output symbol. This value is called *channel capacity*. The channel capacity C of a discrete memoryless channel is given by the formula

$$C = \max_{\{P(x_j)\}} I(\mathcal{X}; \mathcal{Y}) \qquad (1.68)$$

Formally the channel capacity C is a maximum of mutual information calculated with respect to the input symbol probability distribution for a single channel use. From a mathematical point of view, calculation of the channel capacity is an optimization problem with constraints. These constraints result from the fact that the set of arguments for which the maximum is searched is a set of probabilities, so it fulfills the conditions $P(x_j) \geq 0$ $(j = 1, 2, \ldots, J)$ and $\sum_{j=1}^{J} P(x_j) = 1$. In a general case, finding the maximum can be a very complicated problem.

Example 1.10.1 *Let us calculate the capacity of a binary symmetric memoryless channel. Let us denote, as previously, the probability of the zero input channel symbol as α. Thus, the probability of channel input symbol 1 is $1 - \alpha$. As we see, the independent variable with respect to which the optimization is performed is exclusively α. Taking into account Example 1.9.1 and formula (1.66) we conclude that the capacity of this channel is given by the formula*

$$C = \max_{\alpha} \left[H(\overline{p}\alpha + p\overline{\alpha}) - H(p) \right] \qquad (1.69)$$

The second term of expression (1.69) does not depend on α; therefore, maximization of (1.69) reduces to the derivation of such a value α for which $H(\overline{p}\alpha + p\overline{\alpha})$ approaches the maximum. As we remember, the entropy function reaches its maximum if its argument is equal to 1/2. A simple inspection of the expression $\overline{p}\alpha + p\overline{\alpha}$ allows us to conclude that for α = 1/2, independently of the value of p, the expression $\overline{p}\alpha + p\overline{\alpha}$ takes the value exactly equal to 1/2. Eventually, we obtain a formula for the capacity of a binary symmetric memoryless channel

$$C = 1 - H(p) \tag{1.70}$$

Figure 1.22 presents the plot of capacity of this channel versus channel error probability p. As we see, the maximum capacity of such a channel would be equal to one bit per symbol and would be reached if the channel error probability were equal to zero. Since an appropriate system design results in a decrease of the error probability, the channel capacity is maximized at the same time.

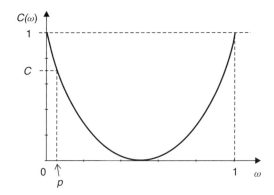

Figure 1.22   Capacity of the binary symmetric memoryless channel versus error probability *p*

1.11   Decision Process and its Rules

1.11.1   Idea of Decision Rule

When receiving a sequence of symbols at the channel output we usually wish to get to know the sequence of symbols given to the channel input in order to determine which messages have been generated by the message source. For this purpose we have to apply a projection in which all the channel output symbols are assigned the channel input symbols. Therefore, we introduce the notion of a *decision rule*.

Definition 1.11.1 *Consider a discrete memoryless channel model with the input alphabet containing J symbols and the output alphabet consisting of K symbols. A function $d(y_k)$ that unambiguously assigns an input symbol to each output symbol y_k is called the decision rule.*

There are J^K possible unambiguous assignments of K output symbols to J input symbols, therefore there are J^K different decision rules. In this context the following question arises: Which rule should be selected? In order to answer this question we must propose a selection criterion.

The basic quality measure of a digital communication system is the probability of an erroneous decision about the transmitted symbol performed in the receiver on the basis of the received symbol. Therefore the decision rule should minimize the probability of an erroneous decision. Let the channel input symbol $x^* = d(y_k)$ be selected according to the established decision rule when the received channel output symbol is y_k. There is an error event if, on reception of symbol y_k, a symbol x_j different from $x^* = d(y_k)$ has been fed to the channel input. Thus, the error probability can be, as in expression (1.53), determined from the following formula

$$P(\mathcal{E}) = 1 - P(C) = 1 - \sum_{k=1}^{K} \Pr\{X = d(y_k), Y = y_k\}$$

Therefore

$$P(\mathcal{E}) = 1 - \sum_{k=1}^{K} P\big(d(y_k)|y_k\big)P(y_k) = \sum_{k=1}^{K} P(y_k) - \sum_{k=1}^{K} P\big(d(y_k)|y_k\big)P(y_k)$$

$$= \sum_{k=1}^{K} \Big[1 - P\big(d(y_k)|y_k\big)\Big]P(y_k) = \sum_{k=1}^{K} P(\mathcal{E}|y_k)P(y_k) \qquad (1.71)$$

where the probability of the decision error conditioned on reception of the output symbol y_k is $P(\mathcal{E}|y_k) = 1 - P\big(d(y_k)|y_k\big)$. Analysis of formula (1.71) indicates that because all terms of the form $P(\mathcal{E}|y_k)$ in the sum are non-negative, the error probability is minimum if the decision rule is selected in such a way that for each k the conditional probability $P(\mathcal{E}|y_k)$ is minimized. It is a very important observation, which leads us to the *Maximum a Posteriori Probability* (MAP) decision rule.[5]

1.11.2   Maximum a Posteriori Probability (MAP) Decision Rule

Minimization of the conditional probability $P(\mathcal{E}|y_k)$ for each k is equivalent to maximization of the conditional probability $P(d(y_k)|y_k)$. Thus, the MAP – *Maximum a Posteriori Probability* – rule relies on the choice of such a channel input symbol $x^* = d(y_k)$, for which the following inequality holds

$$P(x^*|y_k) \geq P(x_j|y_k) \quad \text{for } j = 1, 2, \ldots, J \qquad (1.72)$$

However, we conclude from Bayes' formula that

$$P(x_j|y_k) = \frac{P(y_k|x_j)P(x_j)}{P(y_k)}$$

---

[5] Expression *a posteriori* denotes "after experience", which in our particular case means after observation of the channel output symbol.

so we can write

$$\frac{P(y_k|x^*)P(x^*)}{P(y_k)} \geq \frac{P(y_k|x_j)P(x_j)}{P(y_k)} \tag{1.73}$$

Since $P(y_k)$ occurs in the denominator on both sides of inequality (1.73), this value does not influence the result of the comparison. Therefore the equivalent form of the MAP rule is described by the inequality

$$P(y_k|x^*)P(x^*) \geq P(y_k|x_j)P(x_j) \quad \text{for } j = 1, 2, \ldots, J \tag{1.74}$$

Let us note that channel transition probabilities and *a priori* probabilities of the input symbols appear in expression (1.74), as opposed to formula (1.72). Therefore, the MAP rule in the version shown by (1.74) is much more useful than that described by formula (1.72).

1.11.3 Maximum Likelihood Decision Rule

If the channel input symbols are equiprobable, the MAP rule expressed by formula (1.74) can be further simplified and it takes the form

$$P(y_k|x^*) \geq P(y_k|x_j) \tag{1.75}$$

Sometimes the decision rule given by (1.75) is applied even though the channel input symbol probabilities are unknown to the receiver. In that case (1.75) is a suboptimum procedure and it is called the *Maximum Likelihood* (ML) decision rule. This rule is applied in data detection in many receivers. It is often used as a base of decoding algorithms for channel code decoding.

Let us consider a simple example of the ML decision rule setting.

Example 1.11.1 *Let us consider a discrete memoryless channel model with input and output symbol alphabets of equal size. Let $J = K = 3$. Let the channel transition matrix for the considered channel have the form*

$$P = \begin{bmatrix} 0.7 & 0.2 & 0.1 \\ 0.3 & 0.6 & 0.1 \\ 0.1 & 0.4 & 0.5 \end{bmatrix}$$

As we remember, the columns of the channel transition matrix $P = [P(y_k|x_j)]$ ($j = 1, \ldots, J$, $k = 1, \ldots, K$) are associated with the same output symbol, whereas the rows are associated with the same input symbol. Therefore, applying decision rule (1.75) in which a given output symbol has to be assigned the input symbol for which the transition probability is maximum, we obtain

$$d(y_1) = x_1, \quad d(y_2) = x_2, \quad d(y_3) = x_3$$

From first glance it seems that both MAP and ML rules are quite abstract from the implementation point of view; however, as we will learn, application of the ML decision rule leads to highly practical solutions.

So far we have considered a discrete memoryless channel model. Let us focus for a moment on one particular case, i.e. on the channel model with binary input and continuous output. This channel model additionally supplemented with the quantizer is shown in Figure 1.19a. Let y_n be the unquantized channel output at the nth timing instant and let the input symbol $x_n = \pm A$. Since the channel output can take continuous values, the ML decision rule changes its form to

$$p(y_n|x_n = A) \lessgtr p(y_n|x_n = -A) \tag{1.76}$$

As we see, conditional probabilities have been replaced by appropriate conditional probability density functions. The receiver selects the value, $+A$ or $-A$, for which the conditional probability density function is higher. The decision rule will obviously remain the same if both sides of (1.76) are replaced by their natural logarithms, i.e.

$$\ln p(y_n|x_n = A) \lessgtr \ln p(y_n|x_n = -A) \tag{1.77}$$

or, equivalently, if we calculate the expression

$$\Lambda(y_n) = \ln \frac{p(y_n|x_n = A)}{p(y_n|x_n = -A)} \tag{1.78}$$

and check if it is higher or lower than zero. The function $\Lambda(y_n)$ given by (1.78) is called the *Log-Likelihood Ratio* (LLR) function and is an alternative tool in performing the ML decision rule. For common probability density functions the general expression (1.78) can be significantly simplified. For example, consider the channel model with an additive Gaussian noise source, which is shown in Figure 1.19a. The conditional probability density functions of a Gaussian shape are given in Figure 1.19b and are described by the formula

$$p(y_n|x_n) = \frac{1}{\sqrt{2\pi}\sigma} \exp\left[-\frac{(y_n - x_n)^2}{2\sigma^2}\right] \tag{1.79}$$

For this example the LLR function reduces to

$$\Lambda(y_n) = \ln \frac{\dfrac{1}{\sqrt{2\pi}\sigma} \exp\left[-\dfrac{(y_n - A)^2}{2\sigma^2}\right]}{\dfrac{1}{\sqrt{2\pi}\sigma} \exp\left[-\dfrac{(y_n + A)^2}{2\sigma^2}\right]} = \frac{2A}{\sigma^2} y_n \tag{1.80}$$

The reader is asked to perform a simple derivation leading to the right-hand side of (1.80). The additive noise variance is denoted by σ^2. As we see, in the particular case of bipolar transmission[6] the ML criterion using the LLR function is reduced to checking if y_n is positive or negative.

---

[6] We call that transmission *bipolar* if data symbols take the form $\pm A$.

Now let us consider a binary symmetric memoryless channel with the error probability $p < 1/2$. Let the channel transmit a block of n subsequent binary symbols. Let us treat this block as a whole. Thus, we can state that we deal with a discrete memoryless channel for which the input and output symbols are n-element binary blocks. Let the input symbol alphabet $\mathcal{X}$ consist of $J = 2^k$ ($k < n$) blocks selected from 2^n possible binary combinations, and let the output symbol alphabet $\mathcal{Y}$ consist of $K = 2^n$ blocks (all possible n-element binary blocks). During transmission of subsequent bits of the block x_j a bit is received in error with the probability p and it is received correctly with the probability $1 - p$. As a result of feeding the symbol x_j to the channel input, we receive a single symbol y_k at its output. On the basis of the received symbol y_k in the form of an n-element block, the ML decision rule should ensure the selection of such an input symbol x^* for which condition (1.75) is fulfilled.

In order to find the ML rule for the considered case, let us introduce the idea of *the Hamming distance* between two binary blocks of the same length.

Definition 1.11.2 *The Hamming distance between binary blocks x_j and y_k of the same length, denoted by $d(x_j, y_k)$, is the number of positions at which both blocks differ from each other.*

Let the Hamming distance between the input block x_j and the output block y_k be $D = d(x_j, y_k)$. Knowing that transmission of binary symbols consitituting an n-element block is a sequence of statistically independent events, the probability of reception of y_k conditioned on transmission of x_j is given by the expression

$$P(y_k|x_j) = p^D(1 - p)^{n-D} \tag{1.81}$$

In a typical situation the error probability p is lower than $1/2$. Thus, the following sequence of inequalities holds true

$$(1 - p)^n > p(1 - p)^{n-1} > p^2(1 - p)^{n-2} > \ldots \tag{1.82}$$

We conclude from (1.82) that the ML rule is reduced in this case to the selection of block $x^* = d(y_k)$ from all possible blocks x_j, for which the Hamming distance to the received block y_k is the lowest. This situation is symbolically shown in Figure 1.23. The received block y_k is denoted by a cross. Block x_5 is the closest one in the Hamming distance sense to y_k among the input symbols $x_1, x_2, \ldots, x_9$.

Consider now a particular case of the above example. Let the input symbol alphabet $\mathcal{X}$ consist of two symbols $x_1 = (000\ldots0)$ and $x_2 = (111\ldots1)$ of length n. Let n be an odd number. Symbol x_1 can be assigned a message "0" and x_2 the message "1", respectively. Theoretically, these messages could be represented by 0 or 1; however, instead of that they are represented by whole sequences of these symbols of length n. During transmission of subsequent binary symbols of block x_1 or x_2 over the binary symmetric memoryless channel with the error probability p, the received channel output block y_k can take one of 2^n possible forms. The ML rule allows selection of the input sequence that is closest to the received block in the Hamming sense. The decision will be erroneous if the number of binary errors committed during transmission of an n-element block of "0"s or "1"s

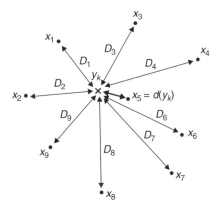

Figure 1.23 Process of finding the sequence x^* featuring the minimum Hamming distance from the received sequence y_k

exceeds $n/2$. Assuming the independence of binary error events, we can easily deduce that the probability of i errors occurring in the n-element block is $p^i(1-p)^{n-i}$. The number of possible combinations of i errors in the block of length n is $\binom{n}{i}$. Therefore the probability of an erroneous decision on the transmitted message is given by the formula

$$P(\mathcal{E}) = \sum_{i=(n+1)/2}^{n} \binom{n}{i} p^i (1-p)^{n-i} \tag{1.83}$$

Assuming, for example, the value $p = 0.01$ and calculating the values of the probability $P(\mathcal{E})$ for subsequent odd block lengths n we obtain $P(\mathcal{E}) = 10^{-2}, 3 \times 10^{-4}, 10^{-5}, 4 \times 10^{-7}, \ldots$ for n equal to $1, 3, 5, 7, \ldots$, respectively. From the above we conclude that if we want to achieve a very low decision error probability related to a single message, we should increase the size of the "0" and "1" blocks appropriately. Since each binary message is represented by an n-bit block, the efficiency of this representation, and therefore of the coding, is $R = 1/n$. As we see, the price paid for increasing the transmission quality is the n-fold lowering of its rate. Need this price really be paid?

The answer to this question was given by Claude Shannon, who formulated the famous theorem on the reliable transmission of messages over unreliable channels. The form of this theorem for the case of binary symmetric memoryless channel is as follows.

Theorem 1.11.1 *Consider a binary symmetric memoryless channel with the error probability p and capacity $C = 1 - H(p)$. Let ε be an arbitrarily small positive constant and let $M = 2^{n(C-\varepsilon)}$. For a sufficiently large number n, from 2^n possible binary blocks of length n one can select a subset of M blocks in such a way that the probability of erroneous decoding of the received block will be arbitrarily small.*

The proof of this theorem can be found in the original paper by Shannon (1948) and in more advanced books on information theory. The above-quoted theorem, called the *second Shannon theorem* for a binary symmetric memoryless channel, states that in order to ensure transmission with arbitrarily low probability of error, the coding rate R cannot be higher than the channel capacity C. Since the number of allowed transmitted blocks is $M = 2^{n(C-\varepsilon)}$, each allowed block of n bits represents in fact one of $n(C-\varepsilon)$ binary messages. Therefore the coding rate is equal to

$$R = \frac{n(C-\varepsilon)}{n} = C - \varepsilon \qquad (1.84)$$

As we see, $R < C$. The coding rate R gets closer to channel capacity C as ε decreases. Thus, we conclude that the application of repetition coding is not a necessary solution, because the coding rate is in reality limited by the channel capacity only. However, the condition for achieving an arbitrarily small error probability is the application of the appropriately long symbol block.

The theorem states that there exists a set of M blocks that ensure arbitrarily low probability of erroneous decoding. However, it does not propose how to select them. In this sense the above theorem is not constructive. However, in the 1990s some good codes with performance very close to the limit stated by Shanon's theorem have been constructed. They will be presented in the next chapter.

In a more general case of the discrete memoryless channel the Shannon theorem has the following formulation.

Theorem 1.11.2 *Consider a memoryless source characterized by the alphabet X and entropy $H(X)$. Let the source emit a message every T_s seconds. Let there be given a discrete memoryless channel with capacity C, through which the symbols representing the messages of source X are sent every T_c seconds. Then, if the following inequality holds*

$$\frac{H(X)}{T_s} \leq \frac{C}{T_c} \qquad (1.85)$$

there exists a code for which encoded messages of source X can be decoded at the channel output with an arbitrarily low error probability. However, if

$$\frac{H(X)}{T_s} > \frac{C}{T_c}$$

there is no code that ensures reception of the transmitted message sequence with an arbitrarily low error probability.

Both theorems establish a basic limit on the rate of reliable message transmission over unreliable channels.

1.12 Differential Entropy and Average Amount of Information for Continuous Variables

So far we have concentrated on a description of discrete sources and have considered reliable transmission over unreliable channels. Now we will analyze the case of continuous variables because it is obvious that many physical communication systems have continuous character.

Consider a continuous random variable X that is characterized by the probability density function $p_X(x)$. By analogy between discrete and continuous random variables we define the so-called *differential entropy*, using the following formula

$$h(X) = \int_{-\infty}^{\infty} p_X(x) \log \frac{1}{p_X(x)} \, dx \qquad (1.86)$$

Theoretically, a continuous random variable has an infinitely high entropy because it can take an infinite number of values. Let us explain the sense of differential entropy definition by considering a continuous random variable as the boundary of a discrete random variable with values in the form $x_k = k\Delta x$ ($k = 0, \pm 1, \pm 2, \ldots$). If the incremental value Δx tends to zero, we can define the entropy of the random variable X using the expression

$$H(X) = \lim_{\Delta x \to 0} \sum_{k=-\infty}^{\infty} p_X(x_k) \Delta x \log \frac{1}{p_X(x_k)\Delta x} \qquad (1.87)$$

Let us note that $p_X(x_k)\Delta x$ is the approximate probability of the value of the random variable X being contained in the interval $[x_k, x_k + \Delta x]$. Expanding formula (1.87) further, we receive

$$H(X) = \lim_{\Delta x \to 0} \left[\sum_{k=-\infty}^{\infty} p_X(x_k) \log \frac{1}{p_X(x_k)} \Delta x - \log \Delta x \sum_{k=-\infty}^{\infty} p_X(x_k) \Delta x \right]$$

$$= \int_{-\infty}^{\infty} p_X(x) \log \frac{1}{p_X(x)} dx - \lim_{\Delta x \to 0} (\log \Delta x) \int_{-\infty}^{\infty} p_X(x) dx$$

$$= h(X) - \lim_{\Delta x \to 0} (\log \Delta x) \qquad (1.88)$$

It turns out from formula (1.88) that the entropy of a continuous random variable is indeed infinitely high, because an infinite value has the second term of this formula if $\Delta x \to 0$. However, the entropy $H(X)$ can be expressed as the sum of the continuous term $h(X)$ and the component tending to infinity, which can be treated as a reference term. Let us note that in particular during derivation of channel capacity the difference of two entropies is usually calculated and then the reference terms reduce each other. As a result the channel capacity depends only on differential entropies. From formula (1.88) we also conclude that the differential entropy $h(X)$ is essentially the difference between the exact entropy of the random variable X and the reference term. This is then probably the source of its name.

As in the case of discrete random variables, we want to find a probability density function of the random variable X that maximizes the value of the differential entropy $h(X)$. Formally, we search for a probability density function $p_X(x)$ that maximizes expression (1.86) for the following constraints:

1. Function $p_X(x)$ is a probability density function, therefore

$$\int_{-\infty}^{\infty} p_X(x)dx = 1 \qquad (1.89)$$

2. Variance σ^2 of the random variable X is finite, i.e.

$$\int_{-\infty}^{\infty} (x - \mu)^2 p_X(x)dx = \sigma^2 \qquad (1.90)$$

The probability density function that maximizes expression (1.86) will be derived by applying the following theorem of variation calculus.

Theorem 1.12.1 *Let us have the integral*

$$I = \int_a^b F(x, p)dx \qquad (1.91)$$

If we search for parameter p that maximizes (1.91) for the following constraints:

$$\int_a^b \varphi_1(x, p)dx = \alpha_1, \quad \int_a^b \varphi_2(x, p)dx = \alpha_2, \quad \ldots, \quad \int_a^b \varphi_k(x, p)dx = \alpha_k \qquad (1.92)$$

then p can be derived as a solution of the equation

$$\frac{\partial F(x, p)}{\partial p} + \lambda_1 \frac{\partial \varphi_1}{\partial p} + \cdots + \lambda_k \frac{\partial \varphi_k}{\partial p} = 0 \qquad (1.93)$$

In our case $F(x, p) = -p \log p$, $\varphi_1(x, p) = p$, $\varphi_2(x, p) = (x - \mu)^2 p$. Therefore, applying formula (1.93) we receive

$$\frac{\partial}{\partial p}\left(p \log \frac{1}{p}\right) + \lambda_1 + \lambda_2 \frac{\partial}{\partial p}\left((x - \mu)^2 p\right) = 0 \qquad (1.94)$$

Deriving p from the above equation, we get

$$p = e^{(\lambda_1 - 1)} e^{\lambda_2 (x - \mu)^2} \qquad (1.95)$$

Substituting this value in the formulas of constraints (1.89) and (1.90), after short calculations we find that the probability density function that maximizes the differential entropy is given by the formula

$$p_X(x) = \frac{1}{\sqrt{2\pi}\sigma} \exp\left(-\frac{(x - \mu)^2}{2\sigma^2}\right) \qquad (1.96)$$

As we see, the random variable X featuring the maximum differential entropy has a Gaussian distribution with mean μ and variance σ^2.

Calculate now the maximum value of the differential entropy. Substituting expression (1.96) in formula (1.86) we obtain

$$h(X) = \int_{-\infty}^{\infty} p_X(x) \log \left[\sqrt{2\pi}\sigma \exp\left(\frac{(x-\mu)^2}{2\sigma^2}\right) \right] dx$$

$$= \int_{-\infty}^{\infty} p_X(x) \log \left(\sqrt{2\pi\sigma^2}\right) dx + \int_{-\infty}^{\infty} p_X(x)(x-\mu)^2 \frac{\log e}{2\sigma^2} dx$$

$$= \frac{1}{2}\log\left(2\pi\sigma^2\right) + \frac{\log e}{2\sigma^2}\sigma^2 = \frac{1}{2}\log\left(2\pi e \sigma^2\right) \qquad (1.97)$$

The achieved result indicates that the maximum differential entropy calculated for a Gaussian random variable depends only on the variance and does not depend on the mean of the random variable.

By analogy between discrete and continuous random variables, we define the mutual information between continuous random variables X and Y as

$$I(X;Y) = \int_{-\infty}^{\infty}\int_{-\infty}^{\infty} p_{X,Y}(x,y) \log \frac{p_X(x|y)}{p_X(x)} dx dy \qquad (1.98)$$

One can also prove that this function can be expressed as

$$I(X;Y) = h(X) - h(X|Y) \qquad (1.99)$$

The following property also holds true

$$I(X;Y) = I(Y;X) = h(Y) - h(Y|X) \qquad (1.100)$$

with

$$h(X|Y) = \int_{-\infty}^{\infty}\int_{-\infty}^{\infty} p_{X,Y}(x,y) \log \frac{1}{p_X(x|y)} dx dy \qquad (1.101)$$

We will use the above dependencies in the derivation of a band-limited channel capacity.

1.13   Capacity of Band-Limited Channel with Additive White Gaussian Noise

Consider a zero-mean random signal $X(t)$ whose band is limited to B Hz. As it is a band-limited signal, it can be represented by a sequence of samples collected with the frequency of $2B$ Hz. On the basis of this sequence of samples the original random signal

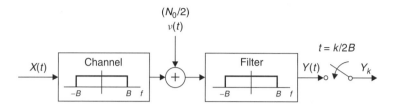

Figure 1.24   Scheme of the system with the band-limited channel and additive noise

$X(t)$ can be recovered with a probability of 1. Assume that we analyze the transmission of the signal $X(t)$ over the channel limited to B Hz in the time period of T seconds. The number of analyzed samples is then equal to $n = 2BT$. During transmission through this channel the signal is disturbed by Additive White Gaussian Noise (AWGN) with the power density of $N_0/2$. Figure 1.24 shows a scheme of such a system. Because the signal on the output of the filter is $Y(t) = X(t) + v'(t)$, where $v'(t)$ is a result of filtration of the noise $v(t)$ by the lowpass filter of bandwidth B Hz, at the output of the sampler at timing instants $t = k/2B$ we receive

$$Y_k = X_k + N_k \tag{1.102}$$

In Chapter 3 we will prove that the noise sample N_k is zero-mean and has variance $\sigma^2 = N_0 B$. Subsequent samples are mutually uncorrelated, and as they are Gaussian they are also statistically independent. Let statistical independence also be a feature of samples X_k of the input signal $X(t)$. Thus, transmission of signal $X(t)$ through the channel limited to B Hz during the period of T seconds can be treated as $n = 2BT$ independent transmissions of samples X_k through the discrete memoryless channel described by expression (1.102), often called the discrete time *Gaussian memoryless channel*. Let us note that all the variables appearing in formula (1.102) are continuous random variables: N_k is a Gaussian variable and X_k has the probability distribution $p_X(x)$. A natural assumption is that the mean power of the input signal is finite, i.e.

$$E[X_k^2] = P, \quad k = 1, 2, \ldots, n \tag{1.103}$$

Let us define the capacity of a discrete memoryless Gaussian channel as

$$C_s = \max_{p_X(x)} \left\{ I(X_k; Y_k) \colon E[X_k^2] = P \right\} \tag{1.104}$$

where $I(X_k; Y_k) = h(Y_k) - h(Y_k|X_k)$. In order to derive the capacity C_s let us first calculate $h(Y_k|X_k)$:

$$h(Y_k|X_k) = \int_{-\infty}^{\infty} \int_{-\infty}^{\infty} p_{X,Y}(x, y) \log \frac{1}{p_Y(y|x)} dx dy$$

$$= \int_{-\infty}^{\infty} p_X(x) \left(\int_{-\infty}^{\infty} p_Y(y|x) \log \frac{1}{p_Y(y|x)} dy \right) dx \tag{1.105}$$

As we remember, the noise sample N_k has a Gaussian distribution with zero-mean and variance σ^2. Then the probability density function of the channel output sample Y_k conditioned on the occurrence of a specific value of the input sample X_k is Gaussian with the same variance, but with the mean equal to the value of the sample X_k. The conditional probability density function $p_Y(y|x)$ is therefore described by the formula

$$p_Y(y|x) = \frac{1}{\sqrt{2\pi}\sigma} \exp\left[-\frac{(y-x)^2}{2\sigma^2}\right] \tag{1.106}$$

Using the latter expression in formula (1.105), we obtain

$$h(Y_k|X_k) = \int_{-\infty}^{\infty} p_X(x)\left[\int_{-\infty}^{\infty} p_Y(y|x)\left(\log(\sqrt{2\pi}\sigma) + \frac{(y-x)^2}{2\sigma^2}\log e\right)dy\right]dx$$

$$= \int_{-\infty}^{\infty} p_X(x)\left(\log\sqrt{2\pi\sigma^2}\cdot 1 + \frac{\log e}{2\sigma^2}\sigma^2\right)dx = \frac{1}{2}\log(2\pi e\sigma^2) \tag{1.107}$$

As we see, the conditional differential entropy depends exclusively on the noise variance and it does not depend on the distribution of the samples X_k of the input signal. Thus, in order to maximize the value of mutual information $I(X_k; Y_k)$ it is necessary to maximize the entropy of samples Y_k of the channel output signal. As we have already proven, the differential entropy of a continuous random variable achieves its maximum if the variable is Gaussian. The sample Y_k, which is the sum of two random variables X_k and N_k, has a probability density function that is a convolution of the Gaussian distribution of the noise sample N_k and the probability distribution of the input signal sample X_k. Fortunately, if the probability density function of the sample X_k is Gaussian, then the probability density function of the sample Y_k is also Gaussian, because a convolution of two Gaussian curves is also Gaussian. Concluding, the differential entropy $h(Y_k)$ achieves its maximum if the random signal $X(t)$ and its samples are Gaussian. Since the samples X_k and N_k are statistically independent, the mean power of their sum is equal to the sum of mean powers of both components. So on the basis of formula (1.97) for the maximum differential entropy we have

$$h(Y_k) = \frac{1}{2}\log\left[2\pi e(P + \sigma^2)\right] \tag{1.108}$$

As a result, the capacity of a time-discrete Gaussian memoryless channel is

$$C_s = \frac{1}{2}\log\left[2\pi e(P+\sigma^2)\right] - \frac{1}{2}\log\left(2\pi e\sigma^2\right)$$

$$= \frac{1}{2}\log\left(1 + \frac{P}{\sigma^2}\right) \tag{1.109}$$

So far we have calculated the capacity for a single sample only. Since in the time period T there are $n = 2BT$ samples, the capacity of the channel band-limited to B Hz and disturbed by additive Gaussian noise with the power density $N_0/2$ can be calculated from the formula

$$C = \frac{n}{T} C_s = \frac{2BT}{T} \cdot \frac{1}{2} \log\left(1 + \frac{P}{\sigma^2}\right)$$

$$= B \log\left(1 + \frac{P}{N_0 B}\right) \quad \text{[bit/s]} \tag{1.110}$$

The above calculations allow us to formulate the following theorem.

Theorem 1.13.1 *The capacity of the channel band-limited to B Hz, in which the signal is disturbed by the additive Gaussian noise of the power density equal to $N_0/2$, is described by the formula*

$$C = B \log\left(1 + \frac{P}{N_0 B}\right) \quad \text{[bit/s]} \tag{1.111}$$

where P is the mean power of the transmitted signal.

The above theorem defines an essential limit for the data rate of errorless transmission in a Gaussian band-limited channel with the input signal of a limited power. Let us note that in order to approach the limit established by (1.111) as closely as possible, the transmitted signal should be Gaussian. At first glance this requirement seems to be difficult to fulfill, however probability distributions of some digital signals are well approximated by the Gaussian distribution. Figure 1.25a plots the channel capacity per Hz (measured in bit/s/Hz), versus signal-to-noise ratio (SNR, in dB). We see that the capacity per Hz

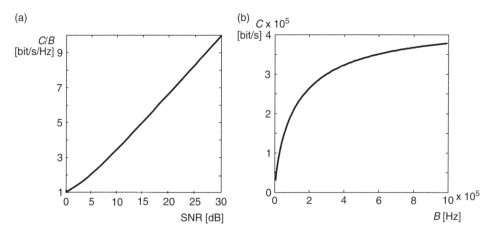

Figure 1.25 Channel capacity plot: (a) per spectrum unit versus the signal-to-noise ratio (SNR); (b) for constant power of the input signal at $P/N_0 = 3 \times 10^5$ versus bandwidth of the signal

increases almost linearly for high SNR on the decibel scale. At the assumption of a constant bandwidth, the capacity increases linearly along with the increase of transmitted signal power. Assume now that the mean power of the transmitted signal is constant but the channel bandwidth B changes. Figure 1.25b shows the channel capacity as a function of the channel bandwidth when $P/N_0 = 3 \times 10^5$. As we see, the capacity increases along with the increase of the channel and transmitted signal bandwidth, however the rise has a shape similar to the curve of $1 - e^{-\alpha x}$ type. It turns out that for the constant mean power of the transmitted signal and the increasing bandwidth the channel capacity tends to the asymptotic value equal to $\frac{P}{N_0} \log_2 e$. A channel type in which a very large band is occupied at the constant value of P/N_0 is applied in the so-called *spread spectrum systems*. They will be presented in one of the later chapters. In such systems, if the signal bandwidth is very large, the SNR can be established even below $0\,\mathrm{dB}$.

Example 1.13.1 *Let us calculate the theoretical capacity of the acoustic telephone channel with the passband in the range 300–3400 Hz for the following values of SNR: 10, 15, 20, 25, 30, 35 dB. Assume that the amplitude channel characteristic is flat in its passband, which is in fact far from reality. The selected SNR levels are related to the following values of $P/(N_0 B)$ on the linear scale: 10, 31.62, 100, 316.23, 1000, 3162.28. Using these values successively in formula (1.111) for the bandwidth of $B = 3100$ Hz we receive the following approximate values of the channel capacity, given in kbit/s: 10.7, 15.6, 20.6, 25.8, 30.9, 36.0. Currently, data transmission methods applied on acoustic telephone channels allow a 28.8 kbit/s data rate to be achieved if highly sophisticated transmission and reception algorithms are used. Therefore, such a value is close to the limit at the SNR of 30 dB. Let us note that the telephone modems offered on the market, which conform to ITU-T V.90 Recommendation, use in fact the subscriber loop channel in a different manner than a typical acoustic modem, which uses only the bandwidth of 3100 Hz; therefore, they achieve higher data rates.*

1.14 Implication of AWGN Channel Capacity for Digital Transmission

Let us analyze formula (1.111) once more. Let us express the mean signal power as a product of two terms: energy per transmitted bit E_b and the binary data rate R_b given in bit/s. Thus, the AWGN channel capacity normalized with respect to the channel bandwidth is described by the formula

$$\frac{C}{B} = \log_2\left(1 + \frac{E_b}{N_0}\frac{R_b}{B}\right) \qquad (1.112)$$

If we were able to design an *ideal system* that achieves the data rate R_b equal to the channel capacity C, then formula (1.112) would evolve to the following

$$\frac{C}{B} = \log_2\left(1 + \frac{E_b}{N_0}\frac{C}{B}\right) \qquad (1.113)$$

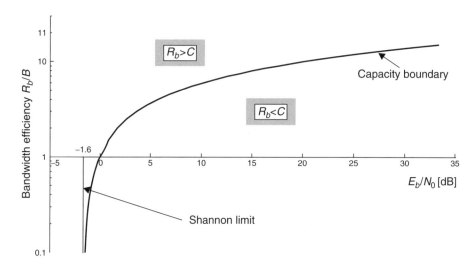

Figure 1.26 Bandwidth efficiency diagram

It follows from (1.113) that the energy per bit to noise power density spectrum required to achieve the channel capacity is

$$\frac{E_b}{N_0} = \frac{2^{\frac{C}{B}} - 1}{\frac{C}{B}} \qquad (1.114)$$

The curve visualizing formula (1.114) in reverse form, i.e. as $C/B = f(E_b/N_0)$, is shown in Figure 1.26. The curve is in fact a capacity boundary and it is related to the ideal system, for which $R_b = C$. Let us note that the curve divides the surface determined by the variables E_b/N_0 and R_b/B into two fields. In the area above the curve $R_b > C$, so for combinations of E_b/N_0 and R_b/B located in it it is not possible to construct the system that can achieve sufficiently low probability of error (cf. Shannon's theorem). However, for the combinations of E_b/N_0 and R_b/B located below the capacity boundary curve it is possible to construct a system that, owing to sufficiently strong coding and other transmission and reception procedures applied in it, can achieve an arbitrarily low error probability. Thus, all the real systems are characterized by the required E_b/N_0, the data rate R_b and the channel bandwidth B used. These parameters determine a certain operation point on the surface below the capacity boundary. As we will learn in one of the future chapters, the error probability is a direct function of E_b/N_0. We can improve the system by decreasing the E_b/N_0 required to achieve a given value of the error probability at a constant R_b/B. This operation is equivalent to moving the system operation point to the left along the horizontal axis. On the other hand, owing to other possible improvements in the system design, for a given E_b/N_0 and a required error probability the ratio R_b/B can be increased, resulting in spectrum savings or an increase in the data rate.

Let us consider one more aspect of the capacity boundary curve. Let us determine the value of E_b/N_0 required to achieve the data rate equal to the channel capacity if the channel bandwidth tends to infinity

$$\lim_{B \to \infty} \frac{E_b}{N_0} = \lim_{B \to \infty} \frac{2^{\frac{C}{B}} - 1}{\frac{C}{B}} = \lim_{B \to \infty} \frac{\ln 2 \cdot 2^{\frac{C}{B}} \cdot \left(-\frac{C}{B^2}\right)}{\left(-\frac{C}{B^2}\right)} = \ln 2 \qquad (1.115)$$

Thus, in the decibel scale, the asymptotic value of E_b/N_0 required to achieve the data rate equal to the channel capacity when the channel bandwidth tends to infinity is equal to $-1.6\,\mathrm{dB}$. This value is called the *Shannon limit* (see Figure 1.26).

1.15 Capacity of a Gaussian Channel with a Given Channel Characteristic

So far we have considered a band-limited channel that does not introduce any signal distortions in its passband. Now we will generalize our considerations to the channel band-limited to B Hz, which has a given transfer function $H(f)$. Assume that the input signal is Gaussian, it has a power density spectrum $P(f)$ and its power is limited to P. This means that $\int_B P(f)\mathrm{d}f = P$. The signal is disturbed by additive Gaussian noise characterized by the power density spectrum $G_n(f)$. An example of such a characteristic is shown in Figure 1.27. In order to determine the channel capacity we divide the channel band into N frequency intervals of width Δf so that $B = N\Delta f$. If N is sufficiently high, the width of a single component channel is so small that it has approximately flat characteristics and we can use (1.111) to derive its capacity. Thus, transmission through the channel with the transfer function $H(f)$ can be treated as a parallel transmission through N ideal passband channels of bandwidth Δf. Using formula (1.111) we can obtain the following formula for the capacity of the ith component channel

$$C_i = \Delta f \log\left[1 + \frac{\Delta f P(f_i)|H(f_i)|^2}{\Delta f G_n(f_i)}\right] \qquad (1.116)$$

If the channel input signal is characterized by the power density spectrum $P(f)$ then the power density spectrum at the output of the channel with the transfer function $H(f)$

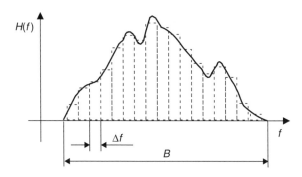

Figure 1.27 Example of the channel transfer function of the channel with bandwidth B Hz

is given by the expression $P(f)|H(f)|^2$. Therefore, the power of the signal seen at the output of the ith component channel of bandwidth Δf with the center frequency f_i is $\Delta f P(f_i)|H(f_i)|^2$. The capacity of the whole channel approximated by N ideal passband channels is equal to

$$C = \sum_{i=1}^{N} C_i = \Delta f \sum_{i=1}^{N} \log \left[1 + \frac{P(f_i)|H(f_i)|^2}{G_n(f_i)} \right] \qquad (1.117)$$

If the bandwidth Δf of component channels tends to an infinitely low value df, then the sum evolves into the integral and discrete frequency values f_i change into a continuous variable f. Finally, we obtain the following formula for the channel capacity

$$C = \int_B \log \left[1 + \frac{P(f)|H(f)|^2}{G_n(f)} \right] df \qquad (1.118)$$

The capacity of the channel with transfer function $H(f)$ depends both on its characteristics and the power density spectra of the input signal and noise. The properties of physical channel and noise are often difficult to change; however, it is possible to change the power density spectrum $P(f)$ of the input signal. Recall that the capacity calculations require the input signal to be Gaussian. Thus, we would like to determine the power density spectrum of the input signal for which the channel capacity is maximum, i.e. the highest number of bits in a time unit that can be transmitted over the channel $H(f)$. Searching for the best shape of the power density spectrum $P(f)$ that maximizes capacity (1.118) with the assumption that the signal power is constant and is equal to P is an optimization problem with a constraint. The solution method is similar to the method that was applied in derivation of the probability density function for which the differential entropy is maximized.

Let us apply Theorem 1.12.1 again. This time we deal with maximization of function of form

$$C = \int_B F(f, P(f)) df = \int_B \log \left[1 + \frac{P(f)|H(f)|^2}{G_n(f)} \right] df \qquad (1.119)$$

for the constraint

$$\int_B \varphi(f, P(f)) df = \int_B P(f) df = P \qquad (1.120)$$

As we remember from Theorem 1.12.1, we find the best $P(f)$ by solving equation (1.93). In our case this equation has the form

$$\frac{\partial}{\partial P(f)} \left\{ \log_2 \left[1 + \frac{P(f)|H(f)|^2}{G_n(f)} \right] \right\} + \lambda \frac{\partial}{\partial P(f)} [P(f)] = 0 \qquad (1.121)$$

In equation (1.121) function $P(f)$ is treated as a variable. The calculation of the derivative with respect to $P(f)$ leads to the following equation

$$\log_2 e \cdot \frac{G_n(f)}{G_n(f) + P(f)|H(f)|^2} \cdot \frac{|H(f)|^2}{G_n(f)} + \lambda = 0 \qquad (1.122)$$

Substituting $1/K = -\lambda/\log_2 e$ we receive

$$\frac{|H(f)|^2}{G_n(f) + P(f)|H(f)|^2} = \frac{1}{K} \qquad (1.123)$$

which after simple calculations leads to the formula

$$P(f) = K - \frac{G_n(f)}{|H(f)|^2} \qquad (1.124)$$

If the additive noise is white, i.e. $G_n(f) = N_0/2$, then

$$P(f) = K - \frac{N_0/2}{|H(f)|^2} \qquad (1.125)$$

Substituting (1.124) into the equation describing the constraint $\int_B P(f)\mathrm{d}f = P$, we end up with

$$\int\limits_B \left[K - \frac{G_n(f)}{|H(f)|^2} \right] \mathrm{d}f = P \qquad (1.126)$$

From this formula the following expression arises (see Figure 1.28)

$$KB = P + \int\limits_B \frac{G_n(f)}{|H(f)|^2} \mathrm{d}f \qquad (1.127)$$

Let us note that KB is the area of a rectangle of width B, which is limited by the horizontal axis and the horizontal straight line located at the height K. The second term of (1.127) is the area under the curve $G_n(f)/|H(f)|^2$. Thus, the input signal power P is the area denoted in grey color above the mentioned curve, which fills out the area above the curve to the level K (Figure 1.28).

The analysis of the optimized input signal power density shape, which leads to the maximum capacity of the channel with a given characteristic $H(f)$ and noise power density $G_n(f)$, leads us to interesting conclusions. It turns out that the channel capacity is maximized if we assign the highest input signal power to the channel sub-bands with the lowest attenuation of the input signal. Less power should be placed in those frequency intervals in which the signal is heavily attenuated. It is against our intuition, because at first glance it seems that we apparently should amplify the transmitted signal in those frequency ranges in which the channel attenuates it heavier. The process of shaping of the input signal power density is called *power loading*. The rule of power loading reminds

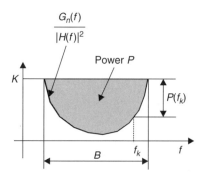

Figure 1.28 Illustration of the choice of input signal power density maximizing the capacity of the channel with a given transfer function

us of pouring water into a basin, therefore this rule is often known as the *water pouring principle*. Power loading is performed in the frequency domain. However, we will learn in the next section that it is also possible in the time domain.

Let us note that power loading requires a feedback channel from the receiver back to the transmitter. In order to assign the input signal power optimally, the receiver has to derive the channel characteristic (we say that it performs *channel estimation*) and then it has to transmit it back to the transmitter via a feedback channel. Thus, it is a case in which *Channel State Information* (CSI) is known both to the transmitter and receiver. In suboptimum systems the channel state information is known only at the receiver and it can be applied only in the signal detection. In this case power loading is not possible and the feedback channel is not required.

1.16   Capacity of a Flat Fading Channel

Consider now the capacity of the channel whose model is shown in Figure 1.29. In our considerations we follow the work of Goldsmith and Varaiya (1997). The channel input signal is band-limited to B Hz. The channel is modeled by a multiplier that performs signal amplification by $\sqrt{g(i)}$, where i is the current timing instant. Let us assume that $g(i)$ $(g(i) \geq 0)$ is a sample function of the stationary and ergodic random process characterized

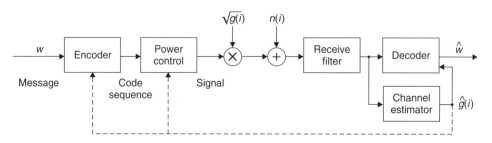

Figure 1.29 System model with flat fading channel, channel estimation and feedback channel (dashed line) (Goldsmith and Varaiya 1997 © IEEE 1997)

by a unit mean and a given probability density function. Additive white Gaussian noise with the power density $N_0/2$ is added to the signal that is modified by the channel coefficient $\sqrt{g(i)}$. The time varying coefficient $\sqrt{g(i)}$ models a situation often appearing on radio channels, in which the received signal level is varying in time and the whole signal spectrum is basically attenuated in the same way. We say that it is a *flat fading channel* and the transmitted signal is the subject of *flat fading*.[7] At the receiver, the input filter limits the bandwidth of the received signal to B Hz, the channel estimator determines the current value of the channel coefficient $g(i)$, and the decoder decodes the received codeword. The dashed line in Figure 1.29 denotes the feedback channel from the receiver to the transmitter, which allows for selection of the appropriate transmitted power level and the particular coding scheme. Let us denote the mean power of the transmitted signal as P. Thus, the SNR at the output of the receive filter is

$$\gamma(i) = Pg(i)/(N_0 B) \tag{1.128}$$

At the ith moment the channel is practically flat with the bandwidth limited to B Hz. In this case its capacity is given by formula (1.111), i.e. for a given value of γ it is equal to

$$C_\gamma = B \log(1 + \gamma) \tag{1.129}$$

Let the probability distribution of the SNR γ be $p_\Gamma(\gamma)$. In practice, it refers to the process $g(i)$. In this case the channel capacity can be understood as an ensemble average of the capacity C_γ, i.e.

$$C = \int_\gamma C_\gamma p_\Gamma(\gamma)\mathrm{d}\gamma = \int_\gamma B \log(1 + \gamma) p_\Gamma(\gamma)\mathrm{d}\gamma \tag{1.130}$$

One can show that the capacity defined by formula (1.130) is lower than the capacity of a flat channel band-limited to B Hz with the SNR equal to the average SNR of $P/(N_0 B)$.

So far we have presented the formula for the capacity of a flat fading channel when the input signal has a constant power equal to P. One can state the following problem: How should we select the transmitted signal power with respect to the current value of the SNR, γ, at the given mean signal power P, in order to maximize the capacity given by the formula

$$C(P) = \max_{P(\gamma)} \int_\gamma B \log\left[1 + \frac{P(\gamma)\gamma}{P}\right] p_\Gamma(\gamma)\mathrm{d}\gamma \tag{1.131}$$

The constraint for the choice of the transmitted signal power is its mean power, which is expressed by the formula

$$\int_\gamma P(\gamma) p_\Gamma(\gamma)\mathrm{d}\gamma = P \tag{1.132}$$

---

[7] The channel model is called *selective fading* if in some parts of the passband substantial attenuation is introduced.

Let us note that if the ratio γ is given by formula (1.128), then the expression $P(\gamma)\gamma/P = P(\gamma)g(i)/(N_0 B)$ determines the current value of the SNR. As shown by Goldsmith and Varaiya (1997), there exists a channel coding scheme that achieves efficiency $R < C(P)$ with a sufficiently small codeword detection error probability when the mean input signal power P is applied. In contrast, the probability of erroneous codeword decoding of the channel code applied in the considered channel with the efficiency $R > C(P)$ is higher than zero.

Let us find the rule that should govern the selection of the power level $P(\gamma)$ of the channel input signal depending on parameter γ, so that the channel capacity $C(P)$ described by formula (1.131) for the constraint (1.132) is maximized. For this purpose we apply once more a similar procedure to that applied in derivation of the optimum input signal power density spectrum that maximizes the capacity of the channel with transfer function $H(f)$. As previously, let us apply Theorem 1.12.1. In the current case, the integrated function of the maximized integral and the integrated function of the constraint are of the form

$$F(\gamma, P(\gamma)) = B \log_2 \left[1 + \frac{P(\gamma)\gamma}{P} \right] p_\Gamma(\gamma)$$

$$\varphi(\gamma, P(\gamma)) = P(\gamma)p_\Gamma(\gamma) \tag{1.133}$$

The optimum value of the applied input power $P(\gamma)$ results from solution of the following equation

$$\frac{\partial F(\gamma, P(\gamma))}{\partial P(\gamma)} + \lambda \frac{\partial \varphi(\gamma, P(\gamma))}{\partial P(\gamma)} = 0 \tag{1.134}$$

Calculation of equation (1.134) by applying (1.133) gives the following dependence

$$B p_\Gamma(\gamma) \log_2 e \cdot \frac{P}{P + P(\gamma)\gamma} \cdot \frac{\gamma}{P} + \lambda p_\Gamma(\gamma) = 0 \tag{1.135}$$

The coefficient γ is selected from the range in which $p_\Gamma(\gamma) > 0$, therefore the equation from which we derive $P(\gamma)$ has a simpler form

$$B \log_2 e \cdot \frac{P}{P + P(\gamma)\gamma} \cdot \frac{\gamma}{P} + \lambda = 0 \tag{1.136}$$

Applying the following substitution

$$\gamma_0 = -\frac{\lambda P}{B \log_2 e} \tag{1.137}$$

after simple calculations we achieve the following result

$$P(\gamma) = \begin{cases} P\left(\dfrac{1}{\gamma_0} - \dfrac{1}{\gamma}\right) & \text{for } \gamma \geq \gamma_0 \\[2mm] 0 & \text{for } \gamma < \gamma_0 \end{cases} \tag{1.138}$$

Let us analyze the meaning of (1.138). For $\gamma \geq \gamma_0$, the transmitted power should increase with increase of the mean SNR γ. However, if this ratio falls below a certain threshold value γ_0, we should abandon transmission of the signal. The value of γ_0 results from the established limit for the mean power (1.132) and is a solution of the equation

$$\int_{\gamma_0}^{\infty} P \left(\frac{1}{\gamma_0} - \frac{1}{\gamma} \right) p_{\Gamma}(\gamma) d\gamma = P \qquad (1.139)$$

In turn, after simple calculations, applying formula (1.138) in expression (1.131) we obtain the following result

$$C(P) = \int_{\gamma_0}^{\infty} B \log_2 \left(\frac{\gamma}{\gamma_0} \right) p_{\Gamma}(\gamma) d\gamma \qquad (1.140)$$

For comparison let us consider a situation in which the transmitter applies the knowledge about the channel attenuation in a nonoptimal way, namely it transmits the signal with a higher power if the channel attenuates the signal more. This means that the transmitter power is selected according to the rule

$$P(\gamma) = P \frac{\sigma}{\gamma} \qquad (1.141)$$

where σ is the mean value of the SNR and constraint (1.132) holds. Thus, the constant σ results from this constraint, which has the form

$$\int_{\gamma} P \frac{\sigma}{\gamma} d\gamma = P \quad \text{i.e.} \quad \sigma = \frac{1}{E[1/\gamma]} \qquad (1.142)$$

and the channel capacity is

$$C(P) = B \log_2(1 + \sigma) = B \log_2 \left(1 + \frac{1}{E[1/\gamma]} \right) \qquad (1.143)$$

At the end of this section consider the case in which there is no feedback channel that could be used to transmit data related to the channel estimated at the receiver to the transmitter. This time the knowledge about the channel can be used only by the receiver. The knowledge of the channel coefficient $\sqrt{g(i)}$ allows us to equalize the level of the received signal, i.e. multiply it by $1/\sqrt{g(i)}$. Thus, the received signal power is constant and equal to P, whereas the instantaneous noise power is $BN_0/g(i)$. Therefore, the SNR is $\gamma = Pg(i)/(BN_0)$ and it is the same as in (1.128). We conclude that in this case the channel capacity is also described by formula (1.130).

Figure 1.30 cited after Goldsmith and Varaiya (1997) presents examples of the capacity curves, normalized with respect to the channel bandwidth B, as a function of the mean SNR in a dB scale for the log-normal probability density function of the channel coefficient

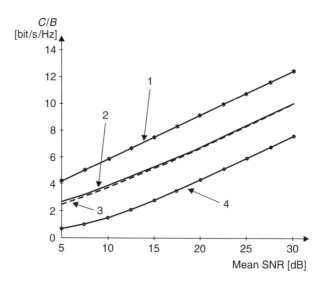

Figure 1.30 Capacity per spectum unit of the channel with log-normal fading ($\sigma_\gamma = 8$ dB): (1) system with AWGN flat channel, (2) system with the optimum use of the channel state information at the transmitter and receiver, (3) system with the optimum use of the channel state information at the receiver only, (4) system with inversion of the power level at the transmitter. Reproduced by permission of IEEE (Goldsmith and Varaiya 1997 © IEEE 1997)

$\sqrt{g(i)} = q$. This probability density function is described by the formula

$$
p(q) = \begin{cases} \dfrac{1}{\sqrt{2\pi}\,\sigma_q q} \exp\left(-\dfrac{\left(\ln q - m_q\right)^2}{2\sigma_q^2}\right) & \text{for } q \geq 0 \\[2em] 0 & \text{for } q < 0 \end{cases} \tag{1.144}
$$

Knowing the rules of transformation of random variables, on the basis of probability density function $p(q)$ one can easily receive the probability density function of the variable $\gamma = Pq^2/(BN_0)$. The normalized capacity of the flat AWGN channel provides a reference curve in Figure 1.30.

From analysis of the curves shown in Figure 1.30 we can see that, in general, fading decreases channel capacity because the flat AWGN channel has the highest capacity. The next channel, as far as quality is concerned, is the one in which optimum power control is performed at the transmitter and the channel state information is used both by the transmitter and receiver. The capacity of the channel for which the channel state information is applied exclusively at the receiver by compensating the channel attenuation is only slightly lower. Finally, the lowest capacity is achieved when the transmitted signal power is increased if the channel attenuation increases. Goldsmith and Varaiya (1997) present similar results for other probability density functions $p(q)$, however the described tendencies are basically preserved.

1.17   Capacity of a Multiple-Input Multiple-Output Channel

Multiple-input multiple-output (MIMO) systems are a relatively new invention in communications. Their particular value, expecially for the development of wireless commnica-
tions, have been proven on the basis of information theory. Let us consider such systems
and show their capacity. In addition to our analysis we will also consider the capacity of
some other system configurations typical for wireless communications.

So far we have analyzed the systems in which a single transmitter sends symbols
representing the source messages and a single receiver transfers them to the message sink.
We call them *Single-Input Single-Output* (SISO) systems. Let us extend our considerations
onto the systems that have n_T transmitters and n_R receivers applied to transmit the
messages from a single message source to a single message sink. We will show how the
capacity of such a system with a MIMO channel depends on the number of transmitters
and receivers. Our considerations will lead us to very important conclusions showing a
potentially considerable improvement in capacity as compared with a SISO system. Our
derivations are quoted after Vucetic and Yuan (2003).

Consider the MIMO system shown in Figure 1.31. The messages from the message
source are source encoded and the resulting code symbols are subsequently assigned
to n_T transmitters. The assignment scheme depends on the system designer. It can
be a simple demultiplexer that forms the input symbols into n_T-element blocks. Each
such element is subsequently emitted in parallel by an appropriate transmitter. Another
possibility is using a code with a given coding rate and generating a certain number
of n_T-element blocks that are transmitted through n_T transmitters in subsequent timing
instants. Since n_T transmitters are typically distributed in space, such a coding scheme
is known as a *Space-Time coding*. Signals emitted by n_T transmitters are received
by n_R receivers. On the way from the jth transmitter ($j = 1, \ldots, n_T$) to the ith
receiver ($i = 1, \ldots, n_R$) the signal undergoes attenuation, which is symbolized by
the channel gain coefficient h_{ij}. As we see in Figure 1.31 each composite channel
is characterized by the gain coefficient, so if it is time varying then the channels
are flat fading. Besides channel attenuation, the transmitted signals are the subject
of disturbance by additive Gaussian noise. Denote the block of transmitted signals

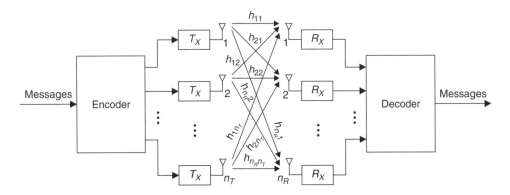

Figure 1.31   General scheme of the MIMO system

at a given time instant as $\mathbf{x}$ and the block of received signals at this time instant as $\mathbf{r}$, i.e.

$$\mathbf{x} = \begin{bmatrix} x_1 \\ x_2 \\ \vdots \\ x_{n_T} \end{bmatrix}, \qquad \mathbf{r} = \begin{bmatrix} r_1 \\ r_2 \\ \vdots \\ r_{n_R} \end{bmatrix} \qquad (1.145)$$

As we have already learnt, in order to calculate the capacity of such a system we assume that the input signals have to be Gaussian distributed. Thus, we assume that each element of vector $\mathbf{x}$ is a zero-mean Gaussian variable. The distributions of all the vector elements are identical and statistically independent of each other. The operation of the whole system can be described by the following matrix equation

$$\mathbf{r} = H\mathbf{x} + \mathbf{n} \qquad (1.146)$$

where $\mathbf{n}$ is the n_R-long sample noise vector and H is a channel matrix of the form

$$H = \begin{bmatrix} h_{11} & h_{12} & \cdots & h_{1n_T} \\ h_{21} & h_{22} & \cdots & h_{2n_T} \\ \cdots & \cdots & \cdots & \cdots \\ h_{n_R 1} & h_{n_R 2} & \cdots & h_{n_R n_T} \end{bmatrix} \qquad (1.147)$$

We assume that in general the input signals and noise vectors are complex random variables. As we will learn in the course of this book, this assumption about complex signal representation allows us to consider most types of modulations applicable in digital communication systems. We further assume that the elements of the noise vector $\mathbf{n}$ are mutually uncorrelated, i.e.

$$R_{\mathbf{nn}} = E\left[\mathbf{n}\mathbf{n}^H\right] = \sigma^2 I_{n_R} \qquad (1.148)$$

where σ^2 is the noise variance and I_{n_R} is the identity matrix of size $[n_R \times n_R]$. The symbol $(.)^H$ denotes Hermitian transposition, which is equivalent, as we know, to a regular vector transposition and complex conjugation of its elements. Similarly, let us define the autocorrelation matrix of the input signal $\mathbf{x}$ as

$$R_{\mathbf{xx}} = E[\mathbf{x}\mathbf{x}^H] \qquad (1.149)$$

The power of the signals transmitted by n_T transmitters is then equal to

$$P = \sum_{j=1}^{n_T} E\left[|x_j|^2\right] = \mathrm{tr}\left(R_{\mathbf{xx}}\right) \qquad (1.150)$$

where $\mathrm{tr}(.)$ is a matrix trace, i.e. the sum of the main diagonal matrix entries. If the channel matrix is unknown at the transmitter, then we assume that the powers of signals generated by each transmitter are identical, i.e. equal to P/n_T. Moreover, we assume that the transmitted signals are mutually uncorrelated. Thus

$$R_{xx} = \frac{P}{n_T} I_{n_T} \tag{1.151}$$

Our next assumption is related to the receive side. Namely, we assume that the power of the signals received by each of the n_R receivers is equal to the total power P. This means that we assume the normalized attenuation in the transmission chain and for each receiver the following equation holds true

$$\sum_{j=1}^{n_T} |h_{ij}|^2 = n_T, \quad i = 1, 2, \ldots, n_R \tag{1.152}$$

In the case of random channel coefficients the above equation becomes

$$\sum_{j=1}^{n_T} E\left[|h_{ij}|^2\right] = n_T, \quad i = 1, 2, \ldots, n_R \tag{1.153}$$

Similarly to the transmit side, the autocorrelation matrix can be determined for the receive side. For known channel coefficients, this is given by the expression

$$R_{rr} = E\left[rr^H\right] = E\left[(Hx + n)(Hx + n)^H\right]$$
$$= HE[xx^H]H^H + \sigma^2 I_{n_R} = H R_{xx} H^H + \sigma^2 I_{n_R} \tag{1.154}$$

After the above introductory considerations let us derive the general formula for MIMO channel capacity. Let us assume that the channel matrix H is perfectly known at the receivers and unknown at the transmitters. Inspecting the form of the channel matrix H we see that at each receiver there is mutual interaction of all signals generated by all transmitters. In order to present the nature of MIMO transmission in a more clear way let us replace equation (1.146), characterizing basic channel behavior by another one in which mutual interaction of the transmitted signal at the receivers is avoided. In order to perform this task let us decompose the channel matrix using the procedure known as *Singular Value Decomposition* (SVD), according to which the channel matrix H of size $[n_R \times n_T]$ can be replaced by a product of three matrices

$$H = UDV^H \tag{1.155}$$

in which D is a non-negative diagonal matrix of size $[n_R \times n_T]$ and U and V are unitary matrices[8] of size $[n_R \times n_R]$ and $[n_T \times n_T]$, respectively, i.e.

$$UU^{-1} = I_{n_R}, \quad VV^{-1} = I_{n_T} \quad \text{and} \quad UU^H = I_{n_R}, \quad VV^H = I_{n_T} \tag{1.156}$$

In SVD decomposition, the elements of the main diagonal of matrix D are non-negative square roots of eigenvalues λ of the matrix HH^H, i.e. they are the singular values of

---

[8] Matrix U is called unitary if the product of U with its own Hermitian transpose is a unity matrix.

matrix H. Thus, the following eigenvalue equation holds

$$HH^H\mathbf{y} = \lambda\mathbf{y}, \quad \mathbf{y} \neq \mathbf{0} \tag{1.157}$$

where $\mathbf{y}$ is the eigenvector associated with the eigenvalue λ. Applying SVD decomposition (1.155) in the system equation (1.146) we obtain

$$\mathbf{r} = UDV^H\mathbf{x} + \mathbf{n} \tag{1.158}$$

Let us introduce the following transformations:

$$\mathbf{r'} = U^H\mathbf{r}, \quad \mathbf{x'} = V^H\mathbf{x}, \quad \mathbf{n'} = U^H\mathbf{n} \tag{1.159}$$

Therefore multiplying both sides of equation (1.158) on their left side by U^H we have

$$\mathbf{r'} = U^H\mathbf{r} = U^H UDV^H\mathbf{x} + U^H\mathbf{n}$$
$$= D\mathbf{x'} + \mathbf{n'} \tag{1.160}$$

The number of nonzero values $\sqrt{\lambda_i}$ in the main diagonal of the matrix D is equal to the rank r of the matrix HH^H. If the size of the matrix H is, as previously assumed, $[n_R \times n_T]$, then the rank r is at most equal to

$$m = \min(n_R, n_T) \tag{1.161}$$

Thus, the vector equation (1.160) can be equivalently expressed by a set of individual equations of the form

$$r'_i = \sqrt{\lambda_i}x'_i + n'_i, \quad i = 1, 2, \ldots, r$$
$$r'_i = n'_i \qquad\qquad i = r + 1, r + 2, \ldots, n_R \tag{1.162}$$

This means that the elements r'_i for $i > r$ do not depend on the transmitted signal, i.e. the channel coefficients are equal to zero. For $i = 1, \ldots, r$ the signal r'_i depends only on the single signal x'_i. Thus, owing to the SVD decomposition we have represented MIMO transmission in the form of r parallel transmissions over independent subchannels. Each subchannel is associated with a singular value of the channel matrix H. The power gain in a given subchannel is equal to the appropriate eigenvalue of the matrix HH^H. The above considerations are visualized in Figure 1.32.

Based on the definition of the autocorrelation matrix we have

$$R_{\mathbf{r'r'}} = E[\mathbf{r'r'}^H] = E[U^H\mathbf{rr}^H U] = U^H E[\mathbf{rr}^H]U = U^H R_{\mathbf{rr}}U \tag{1.163}$$

Similarly

$$R_{\mathbf{x'x'}} = V^H R_{\mathbf{xx}}V \quad \text{and} \quad R_{\mathbf{n'n'}} = U^H R_{\mathbf{nn}}U \tag{1.164}$$

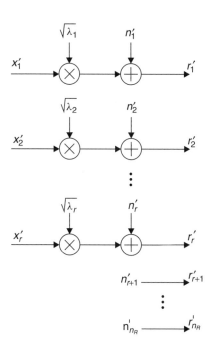

Figure 1.32 Equivalent form of the MIMO system in the form of parallel transmission over independent subchannels

From the matrix properties one can conclude that

$$\text{tr}(R_{\mathbf{r'r'}}) = \text{tr}(R_{\mathbf{rr}}), \quad \text{tr}(R_{\mathbf{x'x'}}) = \text{tr}(R_{\mathbf{xx}}), \quad \text{tr}(R_{\mathbf{n'n'}}) = \text{tr}(R_{\mathbf{nn}}) \quad (1.165)$$

where $\text{tr}(R)$ denotes a trace of matrix R, i.e. the sum of its main diagonal entries. The latter equations indicate that vectors $\mathbf{r'}$, $\mathbf{x'}$ and $\mathbf{n'}$ have the same mean square value (i.e. the power) as the vectors $\mathbf{r}$, $\mathbf{x}$ and $\mathbf{n}$.

As we have represented the MIMO system in the form of $r = \text{rank}(HH^H)$ parallel independent transmission systems, their capacities add together, resulting in the joint capacity

$$C = W \sum_{i=1}^{r} \log\left(1 + \frac{P_{r_i}}{\sigma^2}\right) \quad (1.166)$$

where $P_{r_i} = \frac{\lambda_i P}{n_T}$. In consequence

$$C = W \sum_{i=1}^{r} \log\left(1 + \frac{\lambda_i P}{n_T \sigma^2}\right) = W \log\left[\prod_{i=1}^{r}\left(1 + \frac{\lambda_i P}{n_T \sigma^2}\right)\right] \quad (1.167)$$

Let us show now how the channel capacity depends on the channel matrix H. Again, let $m = \min(n_R, n_T)$. From the equation for eigenvalues and eigenvectors of matrix Q

we have

$$(\lambda I_m - Q)\,\mathbf{y} = \mathbf{0}, \quad \mathbf{y} \neq \mathbf{0} \tag{1.168}$$

or equivalently

$$Q\mathbf{y} = \lambda\mathbf{y} \tag{1.169}$$

where

$$Q = \begin{cases} HH^H & \text{for } n_R < n_T \\ H^H H & \text{for } n_R \geq n_T \end{cases} \tag{1.170}$$

The eigenvector $\mathbf{y}$ is different from zero if $\det(\lambda I_m - Q) = 0$, i.e. if matrix Q is singular. Thus λ is the eigenvalue of matrix Q. As a result

$$\det(\lambda I_m - Q) = \prod_{i=1}^{m}(\lambda - \lambda_i) \tag{1.171}$$

Let us substitute λ in (1.171) by the expression

$$\lambda = -\frac{n_T\sigma^2}{P}$$

Thus, equation (1.171) receives the form

$$\det\left(-\frac{n_T\sigma^2}{P}I_m - Q\right) = \prod_{i=1}^{m}\left(-\frac{n_T\sigma^2}{P} - \lambda_i\right)$$

Equivalently

$$\det\left[-\frac{n_T\sigma^2}{P}\left(I_m + \frac{P}{n_T\sigma^2}Q\right)\right] = \left(-\frac{n_T\sigma^2}{P}\right)^m\prod_{i=1}^{m}\left(1 + \frac{P}{n_T\sigma^2}\lambda_i\right)$$

or

$$\left(-\frac{n_T\sigma^2}{P}\right)^m\det\left(I_m + \frac{P}{n_T\sigma^2}Q\right) = \left(-\frac{n_T\sigma^2}{P}\right)^m\prod_{i=1}^{m}\left(1 + \frac{P}{n_T\sigma^2}\lambda_i\right) \tag{1.172}$$

Comparing (1.172) with (1.167) we conclude that the MIMO channel capacity can be expressed using the formula

$$C = W\log\left[\det\left(I_m + \frac{P}{n_T\sigma^2}Q\right)\right] \tag{1.173}$$

where, as previously

$$
Q = \begin{cases} H H^H & \text{for } n_R < n_T \\[2mm] H^H H & \text{for } n_R \geq n_T \end{cases}
$$

Based on the above formula, let us consider a few particular examples that allow us to illustrate the practical meaning of MIMO systems with respect to the previously known system configurations. First consider the simplest case we already know, i.e. the SISO (*Single-Input Single-Output*) system. In this system there is a single transmitter and receiver, i.e. $n_T = n_R = 1$. Furthermore, let the channel be normalized, i.e. let the channel matrix be $H = h = 1$. As a result, matrix $Q = h = 1$ and $I_m = 1$ ($m = 1$). For this case the channel capacity is given by the well-known formula

$$
C = W \log \det \left(1 + \frac{P |h|^2}{\sigma^2} \right) = W \log \left(1 + \frac{P}{\sigma^2} \right) \tag{1.174}
$$

Let the SNR be $10 \log_{10}(P/\sigma^2) = 15\,\text{dB}$. This means that $P/\sigma^2 = 31.62$. Using this value in formula (1.174) we receive the channel capacity per spectrum unit: $C/W = 5.02$ b/s/Hz.

Consider now the case with a single transmitter and multiple receivers, i.e. $n_T = 1$ and $n_R > 1$. Here, the channel matrix H has the form

$$
H = \left(h_1, h_2, \ldots, h_{n_R} \right)^T
$$

and the channel capacity is described by the expression

$$
C = W \log \left[\det \left(I_{n_T} + \frac{P}{n_T \sigma^2} H^H H \right) \right] \tag{1.175}
$$

However,

$$
H^H H = \sum_{i=1}^{n_R} |h_i|^2, \quad n_T = 1 \quad \text{and} \quad I_{n_T} = 1
$$

so

$$
C = W \log \left[\det \left(1 + \frac{P}{\sigma^2} \sum_{i=1}^{n_R} |h_i|^2 \right) \right] \tag{1.176}
$$

If the channel coefficients are normalized, i.e. $|h_i|^2 = 1$, then

$$
C = W \log \left(1 + \frac{P}{\sigma^2} n_R \right) \tag{1.177}
$$

As we see, the channel capacity grows logarithmically with the number of receivers.

We can draw another important conclusion from formula (1.176). This formula indicates how the signals from component receivers should be combined to create a single output

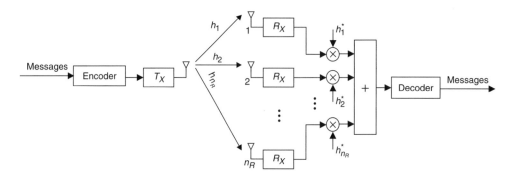

Figure 1.33   SIMO system configuration with optimum combining

signal. As the channel from the transmitter to the ith receiver has the channel coefficient h_i, the ith receiver output signal should be weighted by the factor h_i^* before summing with other receiver outputs. This scheme is shown in Figure 1.33. Such a system configuration is called SIMO (*Single-Input Multiple-Output*) and this type of reception is called *receive diversity*. The above-mentioned method of signal combining is called *Maximum Ratio Combining* (MRC). It can be proved that it maximizes the SNR at the combiner's output. Let us note that due to the fact that each received signal is multiplied by the complex conjugate of its own channel coefficient, the strong signals (for which channel coefficients are higher) are amplified, whereas weeker signals are summed with lower weights. There are a few other receive diversity methods that are suboptimum with respect to the MRC method but they will not be considered here.

Let us illustrate the achievable capacity with an example, as for the previous system. Consider the receiver consisting of the $n_R = 4$ or 8 component receivers. Let the SNR be 15 dB, as before. Using formula (1.177) we receive $C/W = 6.99$ bit/s/Hz for $n_R = 4$ and $C/W = 7.99$ bit/s/Hz for $n_R = 8$, so we observe increases in channel capacity by 37 and 59 percent, respectively.

The next particular case is the so-called *transmit diversity*, in which there are $n_T > 1$ transmitters and a single receiver ($n_R = 1$). This configuration is often called MISO (*Multiple-Input Single-Output*). This time the channel matrix is

$$H = \left(h_1, h_2, \ldots, h_{n_T} \right)$$

and

$$H H^H = \sum_{j=1}^{n_T} \left| h_j \right|^2$$

As a result

$$C = W \log \left[\det \left(1 + \sum_{j=1}^{n_T} \left| h_j \right|^2 \frac{P}{n_T \sigma^2} \right) \right]$$

$$= W \log \left(1 + \sum_{j=1}^{n_T} |h_j|^2 \, \frac{P}{n_T \sigma^2} \right) \tag{1.178}$$

Assuming $|h_j|^2 = 1$, we have

$$C = W \log \left(1 + \frac{P}{\sigma^2} \right)$$

As we see, in this case the channel capacity is the same as in the SISO system.

Finally, consider the MIMO (*Multiple-Input Multiple-Output*) system in which the number of transmitters and receivers is the same, i.e. $n_T = n_R = n$. In calculating the capacity let us take into account the idealized case in which the channels are mutually orthogonal, so there is no interference between different channels. This can be performed practically using spread spectrum techniques, explained in Chapter 7. Channel orthogonality also means that the channel matrix H is diagonal. Assuming that $\sum_{j=1}^{n_T} |h_{ij}|^2 = n_T = n$, the entries of the channel matrix are

$$h_{ij} = \begin{cases} \sqrt{n} & \text{for } i = j \\ 0 & \text{for } i \neq j \end{cases}$$

Thus

$$\sum_{j=1}^{n} |h_{ij}|^2 = n \quad \text{and} \quad H H^H = n I_n \tag{1.179}$$

As a result, the capacity is given by the formula

$$C = W \log \left[\det \left(I_n + \frac{P}{n\sigma^2} n I_n \right) \right]$$

As the matrices for which the determinant is calculated are diagonal, this determinant is

$$\det \left(I_n + \frac{P}{\sigma^2} I_n \right) = \left(1 + \frac{P}{\sigma^2} \right)^n$$

therefore

$$C = W \log \left[\left(1 + \frac{P}{\sigma^2} \right)^n \right] = n W \log \left(1 + \frac{P}{\sigma^2} \right) \tag{1.180}$$

The most important conclusion from formula (1.180) is that the capacity linearly depends on the number n of transmiters and receivers. For an SNR of $15\,\text{dB}$ and

$n_T = n_R = 4$ or 8, respectively, the capacity per herz is equal to 20.08 and 40.16 bit/s/Hz, respectively. This is an enormous increase in capacity, which amounts to 400% and 800%! As we see, in order to design a system with high capacity it is advised to apply both transmit and receive diversities and to orthogonalize channels as much as possible.

The above relatively simple capacity calculations gave a significant impulse in the design of high capacity radio systems for which, due to spectrum scarsity, the high spectral efficiency is a crucial feature. However, we have to be aware that the above example illustrates an idealized case. In practice there is dependence between particular channels and they are not fully orthogonal. Despite that, the increase in data rates achievable in MIMO systems is very significant compared with SISO systems.

In this chapter we have presented only the most important and simplest elements of information theory and they will allow us to analyze digital communication systems with deeper understanding. Having in mind the theoretical performance limits related both to the source coding as well as to transmit and receive strategies, it makes the evaluation of the possible margin that still remains to be reduced through appropriate system design much easier. For this reason, information theory, although a relatively theoretical discipline, brings more and more to the development of modern digital communication systems.

Problems

Problem 1.1 *Calculate the entropy of a discrete memoryless source featuring the message alphabet* $\mathbf{X} = \{a_1, a_2, \ldots, a_6\}$. *The probability of appearance of each message at the source output is equal to* $1/6$.

Problem 1.2 *Let the message a have the probability of occurrence at the source output equal to p, i.e.,* $P(a) = p$. *Draw a plot of amount of information obtained by observing the message a, as a function of its probability of occurrence p.*

Problem 1.3 *Consider a discrete memoryless source with the message alphabet* $\mathbf{X} = \{a_1, a_2, a_3, a_4\}$ *and respective probabilities* $P(a_1) = 0.5$, $P(a_2) = 0.25$, $P(a_3) = 0.15$, $P(a_4) = 0.1$. *Find the entropy of source* $\mathbf{X}$ *and its second extension.*

Problem 1.4 *A typical TFT screen of a mobile phone has the size of 240 × 320 pixels. The color of each pixel is encoded in 18 bits. Assuming that each color of the pixel is equally probable and all pixels are statistically independent, calculate the entropy of a single picture shown on this screen.*

Problem 1.5 *A random signal x(t) of zero mean is sampled every T seconds. The received samples are converted into digital form by the analog-to-digital converter. The probability distribution of the signal samples and the characteristics of the analog-to-digital converter are shown in Figure 1.34. Calculate the entropy of the samples observed at the output of the converter.*

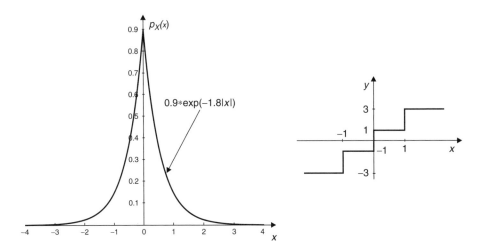

Figure 1.34 Probability distribution function of the samples of signal $x(t)$ and the input–output characteristics of the analog-to-digital converter

Problem 1.6 *Solve Problem 1.5 for the signal at the input of the analog-to-digital converter that has the uniform distribution shown in Figure 1.35.*

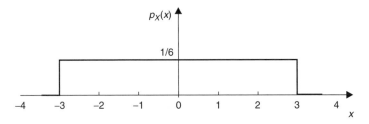

Figure 1.35 Probability distribution function of signal samples at the input of the analog-to-digital converter

Problem 1.7 *Consider a discrete memoryless source with an infinite number of messages $\{a_1, a_2, \ldots\}$ whose distribution is given by the formula*

$$P(a_i) = \alpha p^i, \quad i = 1, 2, \ldots$$

What is the correct value of α? Calculate the entropy of this source and plot it as a function of probability p.

Problem 1.8 *Consider a second-order Markov source $\mathbf{X}$ whose state diagram is shown in Figure 1.36. Is this source ergodic? Calculate the entropy of this source. Calculate the stationary probabilities and the entropy of the memoryless source $\overline{\mathbf{X}}$ associated with source $\mathbf{X}$. Compare the entropies of source $\mathbf{X}$ and source $\overline{\mathbf{X}}$.*

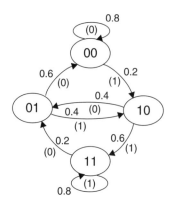

Figure 1.36   State diagram of the second-order Markov source from Problem 1.8

Problem 1.9   *A discrete memoryless source has eight messages* $X = \{a_1, a_2, \ldots, a_8\}$ *that appear on its output with probabilities shown in Table 1.1. Six different mappings denoted as $\mathcal{A}, \mathcal{B}, \ldots, \mathcal{F}$ are considered as potential source codes. Check which mappings consitute a source code and which are prefix codes. Calculate the average code length for each code and the respective coding efficiency. Which code is the best from the coding efficiency point of view?*

Table 1.1   Mapping of the source messages onto symbol sequences

Message	$P(a_i)$	$\mathcal{A}$	$\mathcal{B}$	$\mathcal{C}$	$\mathcal{D}$	$\mathcal{E}$	F
a_1	1/4	000	0	0	0	00	0
a_2	1/4	001	01	10	10	01	100
a_3	1/8	010	011	110	110	100	101
a_4	1/8	011	0111	1110	1110	101	110
a_5	1/16	100	01111	11110	111100	1100	111
a_6	1/16	101	011111	111110	111101	1101	1110
a_7	1/16	110	0111111	1111110	111110	1110	1000
a_8	1/16	111	01111111	11111110	111111	1111	11110

Problem 1.10   *Construct a compact code for the message source from Problem 1.9 using the Huffman algorithm. Repeat the problem for the Shannon-Fano algorithm.*

Problem 1.11   *For a given message source, two source codes are called nontrivially different if they have different distributions of codeword lengths. For the message source described by Table 1.2 construct two different compact codes using the Huffman algorithm. Compare their average lengths and coding efficiencies.*

Table 1.2   Table of source messages and their probabilities

Message	a_1	a_2	a_3	a_4	a_5
Probability	0.4	0.2	0.2	0.1	0.1

Problem 1.12 *Find the coding efficiency of the compact code constructed for the discrete memoryless source* **X** *with the alphabet* $\{a_1, a_2, a_3\}$ *for which* $P(a_1) = 0.5$, $P(a_2) = 0.3$ *and* $P(a_3) = 0.2$. *Construct a compact code for the second extension of the source* **X**. *Compare the coding efficiencies of the constructed compact codes.*

Problem 1.13 *Use the dynamic Huffman coding procedure to encode the text "It is science".*

Problem 1.14 *Perform dynamic Huffman code decoding of the sequence obtained in Problem 1.13.*

Problem 1.15 *Let us treat the binary sequence* 00101100001011011000011 *as an output sequence of messages from the memoryless message source* **X**. *The probabilities of particular messages are* $P(0) = 0.1$ *and* $P(1) = 0.9$, *respectively. Encode the sequence of the first six messages using the arithmetic coding algorithm. Then decode the received codeword. Calculate the entropy of the memoryless source and compare it with the average number of source symbols per single message achieved in the encoding process.*

Problem 1.16 *Apply the Lempel-Ziv algorithm to encode the sequence of messages from Problem 1.15. Recall that the Lempel-Ziv algorithm does not require the knowledge of probabilities of messages generated by the message source. Compare the length of codewords achieved in both encoding methods. Calculate the number of source symbols per single message achieved owing to the encoding process.*

Problem 1.17 *Let us consider the communication link transmitting binary symbols that consists of a cascade of component segments. This is a typical situation in transmission systems built of optical fiber links or terrestrial radio links (see Chapter 5). On the output of each communication segment the received signals are detected and the decided symbols are subsequently transmitted through the next communication segment. The scheme of such a link is shown in Figure 1.37. The communication block that detects the received symbols and transmits them in the regenerated form through the next segment is sometimes called a* regenerative repeater. *Let us assume that each segment can be represented by a binary symmetric memoryless channel model characterized by the error probability p. Assume that binary symbols fed to the link input are equally probable. What is the error probability on the output of a cascade connection of: (a) two segments, (b) three segments? Knowing the error probability at the output of the* $(n-1)$*st segment, derive the error probability at the output of the nth segment.*

Figure 1.37 Communication link consisting of the link segments with regenerative repeaters

Problem 1.18 *Using the results of Problem 1.17 write a program, e.g. in Matlab, C, Pascal or any other computer language you know, that iteratively calculates the probability*

of error on the output of n segments assuming that binary input symbols are equally probable. Draw the results on a single plot for p = 0.01, 0.001 and 0.0001 as a function of the number of segments n (let n be in the range between 1 and 10). Draw the conclusions from the plotted curves.

Problem 1.19 *Calculate the capacity of the binary symmetric erasure channel model shown in Figure 1.17.*

Problem 1.20 *Calculate the capacity of a cascade connection of (a) two and (b) three binary symmetric memoryless channels. Each component channel is characterized by the error probability p. On the same plot draw the channel capacity as a function of p for one, two and three binary symmetric memoryless channels connected in cascade. Find the channel capacity for an increasing number n of composite channels if the error probability for a single channel is p = 0.001.*

Problem 1.21 *Determine the capacity of the channel shown in Figure 1.38.*

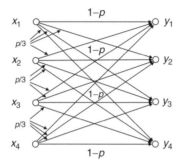

Figure 1.38 Model of a 4-ary input–4-ary output memoryless symmetric channel

Problem 1.22 *Let us consider the discrete memoryless channel described by the channel transition matrix P*

$$P = \begin{bmatrix} 0.6 & 0.2 & 0.1 & 0.1 \\ 0.4 & 0.5 & 0.03 & 0.07 \\ 0.1 & 0.1 & 0.1 & 0.7 \\ 0.1 & 0.2 & 0.5 & 0.2 \end{bmatrix}$$

Determine the maximum likelihood decision rule for this channel.

Problem 1.23 *Let us consider the channel model with bipolar input $x_n = \pm A$ and continuous output determined by the formula*

$$y_n = x_n + v_n$$

where v_n is a noise sample characterized by the probability density function $p_N(v_n)$. Assume that both subsequent data input symbols x_n and noise samples v_n are statistically independent. The probabilities of input signals are respectively equal to $\Pr\{x_n = A\} = P_0$

and $\Pr\{x_n = -A\} = P_1$. *Show how the MAP decision can be implemented for this case. Derive the Log-Likelihood Ratio (LLR) in which the input symbol probabilities* (a priori *probabilities*) *are taken into account. Repeat the derivation for the particular case of the Gaussian probability density function of noise* v_n. *Assume that the noise* v_n *has a zero mean and variance* σ^2.

Problem 1.24   *Calculate the differential entropy for a uniformly distributed random variable with the probability density function shown in Figure 1.35. Calculate the variance of this random variable and then calculate the differential entropy of the Gaussian random variable with the same variance. Compare both results.*

Problem 1.25   *Consider the ideally flat band-limited channel. Let the SNR in this channel be equal to* $10\,dB$. *Let the input signal power be uniformly distributed over the whole channel band and the additive noise be Gaussian and white. What is the value of the channel bandwidth B required to achieve the channel capacity C equal to 10 kbit/s? Now let us assume that the SNR is equal to* $0\,dB$. *How much wider does the channel bandwidth have to be in order for the channel to achieve the same capacity? We further assume that the input signal power is uniformly spread over the whole channel band.*

Problem 1.26   *Consider the communication system with diversity reception shown in Figure 1.39. Data symbols* d_n *of the mean power P are transmitted through two channels characterized by the channel coefficients* h_1 *and* h_2, *respectively. The signals transmitted through each channel are disturbed by the additive Gaussian noise samples of zero mean and variance* σ_1^2 *and* σ_2^2, *respectively. In the receiver the signals* $x_{n,i}$ $(i = 1, 2)$ *received from each channel are appropriately weighted by the coefficients* a_1 *and* a_2 *and subsequently combined, resulting in the signal* $y_n = a_1 x_{n,1} + a_2 x_{n,2}$. *Calculate the optimal weighting coefficients* a_1 *and* a_2 *that ensure maximization of the SNR on the output of the combiner, assuming that the power of the useful signal on the combiner output remains constant and equal to P. Calculate the SNR on the combiner output for the optimal weighting coeffcients and the resulting channel capacity. Assume that* $\sigma_1^2 = \sigma_2^2 = \sigma^2$. *Calculate the SNR and channel capacity for this case and compare it with formulas (1.176) and (1.177). Comment on the results.*

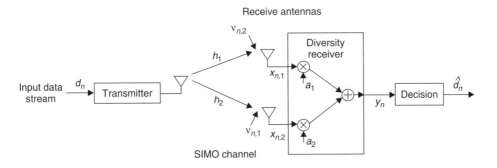

Figure 1.39   System with diversity reception

Problem 1.27   *Let us consider the* 2×2 *MIMO system shown in Figure 1.40. In order to achieve a more reliable transmission as compared with the respective SISO system,*

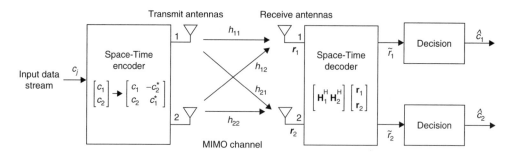

Figure 1.40 General block diagram of the MIMO system with the Alamouti ST code

the so-called Alamouti Space-Time (ST) code has been applied. The input symbols are grouped into blocks of two symbols $[c_1, c_2]$ and transmitted by two parallel transmitters in two consecutive time instants in the form of blocks $[c_1, c_2]$ and $[-c_2^*, c_1^*]$, where $(.)^*$ denotes a complex conjugate. Let us denote r_{kn} as the signal sample received at the kth receiver at the nth moment. Thus, we have

$$r_{11} = h_{11}c_1 + h_{12}c_2 + v_{11}$$
$$r_{12} = h_{11}(-c_2^*) + h_{12}c_1^* + v_{12}$$
$$r_{21} = h_{21}c_1 + h_{22}c_2 + v_{21}$$
$$r_{22} = h_{21}(-c_2^*) + h_{22}c_1^* + v_{22} \qquad (1.181)$$

Denoting

$$\mathbf{r}_1 = \begin{bmatrix} r_{11} \\ r_{12}^* \end{bmatrix}, \quad \mathbf{r}_2 = \begin{bmatrix} r_{21} \\ r_{22}^* \end{bmatrix}, \quad v_1 = \begin{bmatrix} v_{11} \\ v_{12}^* \end{bmatrix}, \quad v_2 = \begin{bmatrix} v_{21} \\ v_{22}^* \end{bmatrix}, \quad \mathbf{c} = \begin{bmatrix} c_1 \\ c_2 \end{bmatrix}$$

we receive

$$\mathbf{r}_1 = \mathbf{H}_1\mathbf{c} + v_1$$
$$\mathbf{r}_2 = \mathbf{H}_2\mathbf{c} + v_2 \qquad (1.182)$$

where

$$\mathbf{H}_1 = \begin{bmatrix} h_{11} & h_{12} \\ h_{12}^* & -h_{11}^* \end{bmatrix}, \quad \mathbf{H}_2 = \begin{bmatrix} h_{21} & h_{22} \\ h_{22}^* & -h_{21}^* \end{bmatrix} \qquad (1.183)$$

The ST code decoder performs the operation shown in Figure 1.40, where $(.)^H$ denotes matrix transposition and conjugation (Hermitian transposition).

1. *Prove that matrices $\mathbf{H}_1$ and $\mathbf{H}_2$ are orthogonal.*
2. *Calculate the signals at both decoder outputs.*
3. *Find the formula for the maximum likelihood criterion for finding decisions $\widehat{c}_1$ and $\widehat{c}_2$ on the basis of the decoder output, assuming that all noise samples are Gaussian and white.*

4. Assume that $E[|h_{ij}|^2] = 1$ $(i, j = 1, 2)$ and the transmitted signal power P is equally divided between two antennas. Calculate the SNR at the output of the decoder, assuming the mean noise power at each receiver input is equal to σ^2, and compare it with the output of the regular SISO system transmitting data symbols c_i over a flat channel with the channel coefficient h, where $E[|h|^2] = 1$, and additive noise with zero mean and variance σ^2.

5. Calculate the capacity of the system with the Alamouti code and compare it with the capacity of a regular SISO system featuring the same P/σ^2.

2

Channel Coding

Information transfer over channels introducing distortions often requires protection against errors. Due to distortions and noise, physical channels hardly ever ensure satisfactory transmission quality. User applications often require binary error rates of the order of 10^{-5}–10^{-6}, so application of channel coding that protects binary sequences against errors is necessary. Sometimes the system requirements are much higher. For example, due to a very high compression rate of the video signal, in Digital Video Broadcasting (DVB) correct video signal decompression requires *Quasi Error-Free* (QEF) reception, i.e. the error rate should be of the order of 10^{-10}–10^{-12}. Ensuring such quality is certainly a demanding task. Fortunately, it is achievable owing to the progress in coding theory and communication technology.

Channel coding applied in a given digital transmission system is strictly associated with its structure, required transmission quality and limitations resulting from the applications of the system. In some systems the data sequence needs to be transmitted at a constant rate and rate fluctuations cannot be tolerated. In some others an allowable transmission delay is a system limitation. In certain systems a feedback channel from the data receiver to the transmitter can be established in order to send messages about data blocks reaching the receiver. This enables repetition of the erroneously received blocks. Such a feedback channel has a crucial influence on the choice of the channel code. Generally, channel coding is applied to ensure error detection and/or correction. The latter task is usually much more costly than the former one. In modern digital transmission systems, in particular those applying radio channels, both tasks are usually performed.

2.1 Idea of Channel Coding

The essence of channel coding is based on two rules: introduction of information redundancy and averaging the noise influence. Introduction of information redundancy is realized by attaching an additional symbol sequence to the information block representing a given message. This sequence is selected in such a way that the transmitted message could be easily distinguished from other messages that could potentially be transmitted. Messages are represented by the symbol sequences in such a way that it is very unlikely that channel perturbations distort so high a number of symbols in the sequence that these erroneous symbols would destroy the possibility of a unique association of the received

Introduction to Digital Communication Systems   Krzysztof Wesołowski
© 2009 John Wiley & Sons, Ltd

symbol sequence with the transmitted message. The effect of noise averaging, in turn, is achieved by association of the redundant symbols with a few different information symbols representing a given message.

Recall for a moment our considerations from Chapter 1. We observed that the longer the coded sequence, the easier it is to be closer to the asymptotic value of the code rate determined by the channel capacity C. This statement is also illustrated by the following argumentation (Clark and Cain 1981).

Consider the problem of binary error events occurring during transmission over a binary memoryless symmetric channel. In this channel errors are mutually independent. Let the error probability of a single symbol be $p = 0.01$. Figure 2.1 presents the plot of probability $\Pr\{e/n > \rho\}$ of the event that the ratio of the error number e in an n-element block to the block length n exceeds a certain threshold value ρ. For any assumed value of n the plot has a staircase shape, because the ratio e/n has a discrete nature. Calculation of the probability $\Pr\{e/n > \rho\}$ is easy when we use the fact that $\Pr\{e/n > \rho\} = 1 - \Pr\{e/n \le \rho\}$. It is easy to show that

$$\Pr\{e/n \le \rho\} = \sum_{i=0}^{\lfloor \rho n \rfloor} \binom{n}{i} p^i (1-p)^{n-i} \tag{2.1}$$

where $\lfloor x \rfloor$ denotes the highest natural number not higher than x.

Conclusions resulting from Figure 2.1 are the following. Let the receiver make the decision upon the whole received binary data block instead of individual decisions upon successive symbols. In order to achieve acceptably low block error probability, it is necessary to apply a sufficiently strong coding system in which the decoder is able to

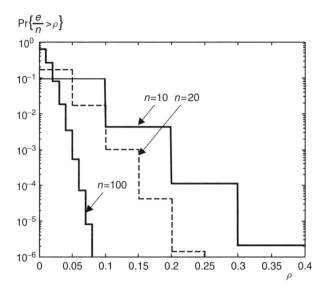

Figure 2.1 Probability of the event that the ratio of the number of errors to the block length exceeds the given threshold ρ

correct a certain number of erroneous symbols in a block. Unless the number of errors exceeds a certain threshold, the decoder is able to interpret the received binary block correctly. Let us note that the longer the binary block, the more rapid the decrease of the probability of the event that the relative number of errors e/n exceeds a given value ρ. In turn, assuming a certain value of $\Pr\{e/n > \rho\}$, e.g. equal to 10^{-4}, we observe that the longer the block, the smaller the relative number of errors that need to be corrected. Therefore a longer code has to be able to correct a relatively smaller number of errors.

The next problem related to the reliable message transfer with channel coding involves the choice of the method of error detection and correction. Let us again consider a binary memoryless channel model in which errors occur independently, with probability p. As we remember from the previous chapter, probability of the event that in n-element block a particular sequence of d errors has occurred is equal to $p^d(1 - p)^{n-d}$. According to the maximum likelihood rule, having received a particular symbol sequence, from possible code sequences the receiver should select the sequence that exhibits the lowest Hamming distance to the received sequence. If from 2^n possible binary sequences of length n such 2^k ($k < n$) code sequences have been selected that the Hamming distance between any pair of them is not lower than $d_{\min}$, then the decoder operating according to the maximum likelihood decision rule will make a correct decision unless the Hamming distance between the received sequence and the transmitted one exceeds the value

$$t = \left\lfloor \frac{d_{\min} - 1}{2} \right\rfloor \tag{2.2}$$

Let us quote the following example (Clark and Cain 1981).

Example 2.1.1 *Let the length of the binary block be $n = 5$. Then there are $2^5 = 32$ possible binary sequences of length 5. From these sequences we select $2^k = 2^2 = 4$ code sequences. Thus, each code sequence is related to one of four messages or, equivalently, to a particular combination of $k = 2$ binary messages. Let the selected code sequences have the form*

$$00000, \quad 00111, \quad 11100, \quad 11011$$

Comparing the number of positions in which any pair of sequences differs, we conclude that the minimum Hamming distance between them is $d_{\min} = 3$. Therefore, if the decision upon the transmitted sequence made by the receiver is to be correct, the received sequence can differ from the transmitted one in at most $t = \lfloor (3 - 1)/2 \rfloor = 1$ position. Thus, error correction relies on the association of the received 5-bit sequence (out of 32 sequences, 28 are incorrect) with the code sequence that is the closest in the Hamming distance sense. Table 2.1 presents all possible 5-bit sequences in an ordered manner. Let us note that the leader of each column is a code sequence. Binary sequences that differ from a given code sequence in one position are placed below this sequence in the same column, therefore their Hamming distance from the column leader does not exceed t. Under those sequences there are blocks for which the Hamming distance from the column leader is $d = 2$. Unfortunately, some of those sequences are equidistant from two different code sequences, e.g. the sequence 10101 differs in two positions from the code sequence 00111 located in the same column and the code sequence 11011 in the next column. The sequences located

Table 2.1 Assignment of binary sequences to the codewords

Codewords	00000	11100	00111	11011
Correctable sequences	10000	01100	10111	01011
($d = 1$)	01000	10100	01111	10011
	00100	11000	00011	11111
	00010	11110	00101	11001
	00001	11101	00110	11010
Uncorrectable sequences	10001	01101	10110	01010
($d = 2$)	10010	01110	10101	01001

below the line in the table cannot be uniquely associated with any code sequence. We call them uncorrectable sequences, although the errors contained in them are detectable.

Table 2.1 can be directly applied in the maximum likelihood decoding process. Maximum likelihood decoding consists of finding the received sequence in a decoding table such as Table 2.1 and associating it with the code sequence that is located on top of the same column. This can be done only for correctable sequences. When locating the received sequence among uncorrectable sequences the decoder can warn the other part of the receiver about errors in the received sequence.

Such a table decoding is optimal in the sense of maximum likelihood; however, it becomes cumbersome when the length of codewords increases and the number of codewords rises as well. So it is applicable for short codes only.

When we create a decoding table, 2^n possible sequences are distributed in 2^k columns. The number of columns is equal to the number of code sequences. The code is able to correct t errors if the number of N_e sequences placed in each column fulfills the following inequality

$$N_e \geq 1 + \binom{n}{1} + \binom{n}{2} + \cdots + \binom{n}{t} \tag{2.3}$$

The right-hand side of the inequality represents the number of sequences for which the Hamming distance from a given code sequence does not exceed t. So there is the code sequence itself, $\binom{n}{1} = n$ sequences differing from the code sequence in one position, $\binom{n}{2}$ sequences differing in two positions, etc., and finally there are $\binom{n}{t}$ sequences differing from the code sequence in t positions. We conclude that the right-hand side of the above inequality is the number of sequences located in a given column that are correctable.

One can state the following question: How many binary code sequences of length n can be found if the code has to have the ability to correct t errors? The answer to this question is the following. As we know, there are 2^n different binary sequences of length n. Each column of the decoding table in which there are sequences assigned to a single codeword contains N_e sequences, therefore the number N_c of code sequences is given by

$$N_c = \frac{2^n}{N_e} \leq \frac{2^n}{1 + \binom{n}{1} + \binom{n}{2} + \cdots + \binom{n}{t}} \tag{2.4}$$

In most cases an inequality holds, which means that there exist sequences for which the Hamming distance to any code sequence is higher than t, so they cannot be uniquely assigned to any code sequence. The equality is fulfilled if each column of the decoding table consists only of correctable sequences. A code with this feature is called a *perfect code*. As mentioned before, the number of codewords is usually given by the expression $N_c = 2^k$, and then inequality (2.4) achieves the form

$$2^{n-k} \geq \sum_{i=0}^{t} \binom{n}{i} \tag{2.5}$$

For perfect codes the sum on the right-hand side is maximum and equal to 2^{n-k}. We say that in this case the *Hamming upper bound* is reached.

The number of n-bit code sequences is equal to 2^k. This implies that in the coding process a message block of the length of k bits is mapped onto the code sequence of length n bits. Therefore we can conclude that $n - k$ redundant bits have been appended to the original message block. These bits facilitate the differentiation of the code sequences among each other. Instead of transmitting k unprotected bits, the transmitter sends n bits of the code sequence. The ratio $R = k/n$ is called a *code rate*. In Example 2.1.1 there are four 5-bit code sequences, so the code rate is $R = 2/5$.

2.2   Classification of Codes

Codes can be classified in several ways. Below we present a classification based on different criteria.

Let us note that a k-bit message block can be mapped onto an n-bit code sequence in many ways. If the first k bits in the code sequence are the message bits and in the coding process the next $n - k$ redundant bits are added, then such a code is called a *systematic code*. In some cases, in a code sequence there are no message bits in the direct form but there are bits that are combinations of the message bits only. Such a code is described as *nonsystematic*.

So far we have considered codes in which the code symbols are binary. Such codes are called *binary codes*. In some applications the codes that apply code symbols belonging to larger than binary sets are used. Such codes are known as *nonbinary*. Among nonbinary codes the most popular ones are Reed-Solomon codes, which find applications in digital TV systems, CD recording systems and CD players and in many radio systems, including deep-space communications.

Another method of code classification results from operation of the encoder. From this point of view, codes can be divided into *block* and *convolutional* codes. In the first case the n-bit code sequence is the outcome of the mapping performed according to the mathematic rule for creating codewords, based on a single k-bit message block currently given to the encoder input. Such a mapping can be implemented using a combinatorial logic circuit only. In such a circuit memory cells do not need to be applied, although they appear in some specific implementations.

The basic block code parameters are the length of message block, k, and the length of the codeword, n, which are often given in the form of a pair (n, k), the minimum distance between codewords $d_{\min}$ and the code rate $R = k/n$. An important code parameter is also the number of correctable errors t. Typically the length of the message block ranges between three and a few hundred symbols whereas the code rate is contained between $1/4$ and $7/8$. However, due to rapid progress in coding theory and implementation capabilities, the codeword length can even reach a few tens of thousands of bits (see description of turbo codes and LDPC codes at the end of this chapter).

Convolutional codes require a sequential logic circuit and, consequently, memory cells to implement the encoder. Thus, memory cells to implement an encoder are necessary. The generated n-bit codewords are the result not only of the k-bit message sequence currently fed to the encoder input but also of the current state of the encoder, which is determined by the contents of its memory cells. This state depends on the message blocks previously given to the encoder input. Code sequences can often be interpreted as a convolution of the encoder impulse response with the input sequence. This explains the origin of the name of these codes. An encoder impulse response is the response of the encoder to a single "1" followed by a sequence of zeros. As before, the code rate is described by the ratio $R = k/n$. A convolutional code is characterized by the pair (n, k) and the so-called *free distance* d_{free}, which is the lowest Hamming distance between any two sequences of the same length received from the encoder. Another characteristic parameter of a convolutional code is the *constraint length* L, which is equal to the number of input symbols used in the generation of the encoder output sequence. Therefore L is the sum of k (the number of symbols given to the encoder input) and m (the number of memory cells in the encoder). Typically the values of k and n range from 1 to 8, and the code rate is, as before, between $1/4$ and $7/8$. In turn, the number of memory cells in the encoder is contained between 2 and 60.

Codes are also classified according to their algebraic structure. From this point of view, the codes are divided into *linear* and *nonlinear*. Most codes applied in practice belong to the first type. In the algebraic sense, linear codes create a vector space. Linear block codes create an algebraic group with respect to addition. Basic properties of such a group are as follows: The sum of any two codewords is also a codeword of the same code and a zero sequence also belongs to the group. The additive operation is defined as summing two sequences "symbol by symbol". In each linear vector space there exists a small set of vectors, so-called *basis*, i.e. the vectors whose linear combination can synthesize any space element. In our case such a space element is a codeword belonging to the given code. The properties of any linear code can be easily analyzed by considering the transmission of a zero codeword only and observing the properties of the received sequence and the distances of other codewords to the zero codeword. The distance of codewords to the zero codeword, which is equal to the number of "1"s contained in them, is called a *Hamming weight*.

We can also classify codes with respect to the type of errors that the codes are aiming to correct. From this point of view the most popular are *random error correcting codes*. They are designed by assuming that code symbols are transmitted over a memoryless transmission channel. Recall that errors occurring in such channels are mutually statistically independent. Another group of codes resulting from the above classification criterion are *burst error correcting codes*. The channel model used in their design includes a memory of the error source (see Chapter 1). We should mention that the codes correcting burst errors are applied if the error bursts are relatively well defined. In other cases it is much more convenient to apply random error correcting codes supported by an interleaver in the transmitter and a deinterleaver in the receiver (see Chapter 1 and Section 2.10).

Last but not least, we distinguish *error correction* and *error detection* codes. This differentiation results from two basic aims of coding that have already been mentioned. Error correcting codes are characterized by a relatively high number of redundant symbols contained in a codeword. As we have already mentioned, their typical code rate is between 1/4 and 7/8. They are able to detect and subsequently correct errors contained in corrupted codewords. On the contrary, error detection codes only check if errors have occurred in the received codeword. Since the erroneous sequences are not corrected, the blocks must be transmitted again. This, in turn, implies the existence of a feedback channel for sending acknowledgements or repetition requests. In this method, the number of redundant bits is relatively small compared with the codeword length, so the code rate is very high.

2.3 Hard- and Soft-Decision Decoding

The problem of the so-called hard- and soft-decision decoding is strictly connected with the kind of signals available on the decoder input, and with the assumed transmission channel model. In traditional digital transmission systems, in particular those in which modulation and error protection are treated separately, binary sequences appear on the demodulator output. The decoder attempting to recover the transmitted codeword and the message block associated with it has only a binary sequence at its disposal. The decoding process can only rely on the knowledge of algebraic dependencies applied during the code construction in the creation of redundant symbols. The decoding process in which only binary sequences appear on the decoder input is called *hard-decision decoding*. Thus, we can assume that the applied channel model has a binary output. In the previous chapter we have descriptively shown that in the case of binary sequence processing the maximum likelihood decision rule reduces to finding the codeword that is closest to the received sequence in the Hamming distance sense. We will show this in a more formal way now.

Assume that we analyze n-element binary sequences that are transmitted over a binary memoryless channel. As we remember from Chapter 1, the probability of reception of sequence $\mathbf{r}$ under the condition that codeword $\mathbf{c}$ has been transmitted is given by the formula

$$P(\mathbf{r}|\mathbf{c}) = p^D (1 - p)^{n-D} \tag{2.6}$$

where D is the Hamming distance between codeword $\mathbf{c}$ and sequence $\mathbf{r}$. Let us find the codeword that maximizes this probability. Taking the logarithm of both sides in (2.6) we obtain

$$\ln P(\mathbf{r}|\mathbf{c}) = D \ln p + (n - D) \ln(1 - p)$$

$$= D \ln p + n \ln(1 - p) - D \ln(1 - p)$$

$$= D \ln \frac{p}{1 - p} + n \ln(1 - p) \tag{2.7}$$

The expression whose value we intend to maximize consists of two parts: the first component, which is proportional to the Hamming distance D, and the second component, which is proportional to the codeword length n and for a given n is constant. With the assumption that $p < 0.5$, the logarithm of the division $p/(1 - p)$ is negative, therefore $\ln P(\mathbf{r}|\mathbf{c})$ will be maximal if D is minimal. This conclusion confirms what we already know – we should select the codeword $\mathbf{c}$ that is the closest in the Hamming sense to the received sequence $\mathbf{r}$.

In practice, the demodulator can produce not only binary decisions but also some additional signals that enable the decoder to act with enhanced knowledge on the decoded sequences. The simplest case is the application of a multilevel quantizer instead of the two-level decision device producing binary signals on the demodulator output. Owing to this approach we "measure" the signal level using a ruler with a more precise scale. As a result, the signals and whole signal sequences are more distinguishable from each other. It turns out that this approach results in a significant improvement in decoding quality. In the case of an 8-level quantizer, the memoryless channel model that characterizes the whole chain of the communication system blocks, starting from the channel encoder output and finishing at the decoder input, looks like that shown in Figure 1.19c. There also exist other solutions of *soft-decision decoding*, in which binary signals on the decoder output are accompanied by additional signals that are a measure of their reliability. Generally, in modern digital communication systems soft-decision decoding is applied more and more often.

In hard-decision decoding it was the Hamming distance that was used in selection of the codeword by the decoder. Now let us show how the soft-decision decoder should decide upon the transmitted codeword on the basis of signal samples of the received sequence. Assume that vector $\mathbf{r}$ appearing at the decoder input consists of n samples r_i $(i = 1, \ldots, n)$ of the received signal. Each sample is the sum of the data symbol $d_i = \pm 1$ representing the ith bit of codeword $\mathbf{c}$ and a sample v_i of a zero mean Gaussian noise with variance σ^2. We also assume that noise samples are mutually statistically independent. According to the maximum likelihood rule, the decoder searches for such a codeword $\mathbf{c}$ that maximizes the conditional probability density function $p(\mathbf{r}|\mathbf{c}_j)$ $(j = 1, \ldots, 2^k)$. Due to the Gaussian character of the noise samples the conditional probability density function of vector $\mathbf{r}$ is described by the following formula

$$p(\mathbf{r}|\mathbf{c}_j) = \prod_{i=1}^{n} \frac{1}{\sqrt{2\pi}\sigma} \exp\left[-\frac{\left(r_i - d_{j,i}\right)^2}{2\sigma^2}\right]$$

$$= \left(\frac{1}{\sqrt{2\pi}\sigma}\right)^n \exp\left[-\frac{1}{2\sigma^2}\sum_{i=1}^{n}\left(r_i - d_{j,i}\right)^2\right] \tag{2.8}$$

where $d_{j,i}$ denotes the data symbol assigned to the ith element of the jth codeword $\mathbf{c}_j$. Calculation of the logarithm of (2.8) results in an expression that is equivalent to (2.8) with respect to selection of the most likelihood codeword $\mathbf{c}$

$$\ln p(\mathbf{r}|\mathbf{c}_j) = n \ln \frac{1}{\sqrt{2\pi}\sigma} - \frac{1}{2\sigma^2}\sum_{i=1}^{n}\left(r_i - d_{j,i}\right)^2 \tag{2.9}$$

Maximization of $\ln p(\mathbf{r}|\mathbf{c}_j)$ results in finding such a codeword $\mathbf{c}$ for which the sum $S_j = \sum_{i=1}^{n} (r_i - d_{j,i})^2$ is minimized. This means selecting the codeword $\mathbf{c}$ for which the representation using symbols $d_{j,i}$ $(i = 1, \ldots, n)$ is the closest to the received sample vector $\mathbf{r}$ in the Euclidean distance sense. A suboptimal solution is the search for the code-word using basically the same method if the received sample is quantized in the quantizer featuring a small number of quantization levels, e.g. eight. Another simplification that could have an important practical meaning is a replacement of the squared differences $r_i - d_{j,i}$ by their modules. Then the criterion for selection of the codeword $\mathbf{c}$ reduces to searching for such a data sequence associated with the codeword that minimizes the sum

$$S'_j = \sum_{i=1}^{n} |Q(r_i) - d_{j,i}| \tag{2.10}$$

where $Q(r_i)$ is a quantized form the received sample r_i.

2.4 Coding Gain

Coding gain is the concept that has been introduced in order to compare systems with and without error correction coding. Figure 2.2 presents typical error probability curves as a function of the ratio E_b/N_0 for both kinds of systems. The scale of the vertical axis is logarithmic, and the scale of the horizontal axis is expressed in decibels. E_b is the signal energy per transmitted bit, whereas $N_0/2$ is the power spectral density of the additive white Gaussian noise. As we will show in the next chapter, the ratio E_b/N_0 expresses the signal power to noise power ratio per single transmitted bit. As compared with the system without coding, the system with error correction coding that has to transmit the user data at a given data rate has to send a higher number of bits in a time unit. This is a result of appending the message bits with the redundant bits needed for error correction. Although coding ensures higher robustness against errors, the energy per transmitted bit

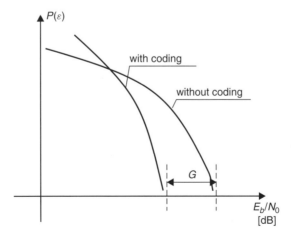

Figure 2.2 Explanation of the concept of coding gain

is lower than in case of transmission without coding. Performance gain that is achieved as a result of redundant coding is a compromise between these two counteracting factors. The higher the ratio E_b/N_0 in the channel, the higher the performance gain; however, for small values of E_b/N_0 we observe a loss in performance instead of gain. Typically both curves intersect at a certain value of E_b/N_0. Application of coding below this value is unreasonable.

Coding gain for a given error probability is the difference between the required value of E_b/N_0 in the system without coding and the value of E_b/N_0 for the system with error correction coding. Depending on the considered level of error rate, the coding gain can have different values. We can also determine the asymptotic value of the coding gain when E_b/N_0 tends to infinity. Such a value is called an *asymptotic coding gain*. In the case of transmission of codewords using bipolar signals (see Chapters 3 and 4) the asymptotic coding gain for the hard-decision decoding is (Clark and Cain 1981)

$$G_t = 10 \log [R(t + 1)] \tag{2.11}$$

whereas for ideal soft-decision decoding it is

$$G_d = 10 \log(R d_{\min}) \tag{2.12}$$

We know from formula (2.2) that $d_{\min} \geq 2t + 1$, so the difference $G_d - G_t$ between asymptotic coding gains for hard- and soft-decision decoding is around 3 dB. In practice, replacement of the hard-decision decoding by its soft version results in an improvement of about 2 dB. Let us stress that it is a significant difference. Lowering of the required value of E_b/N_0 by 2 dB is equivalent to a decrease in the transmitted signal power to 63% of its previous value. This is a significant result not only from the power point of view. For many systems, in particular for radio systems, lowering of the transmitted power results in a decrease in the level of distortions induced to other users of the same system and has a positive influence on the overall system capacity, i.e. the number of users who can simultaneously operate in the given area.

2.5   Block Codes

Recall from Section 2.2 that for block codes the vector of k message symbols uniquely determines the n-symbol codeword generated by the encoder. Among many possible block codes *linear block codes* have a practical meaning. Let us limit ourselves temporarily to binary codes. Codewords of a linear block code constitute an algebraic group with respect to the additive operation. As we have already mentioned, the sum of two codewords is also a codeword. The zero codeword, being a zero element of the algebraic group, also belongs to the codeword set. Denote the codewords $\mathbf{a}$ and $\mathbf{b}$ as vectors

$$\mathbf{a}^T = [a_1, a_2, \ldots, a_n] \quad \text{and} \quad \mathbf{b}^T = [b_1, b_2, \ldots, b_n] \tag{2.13}$$

(T denotes vector transposition). Addition of two vectors is defined as

$$\mathbf{a} + \mathbf{b} = \mathbf{c} = [c_1, c_2, \ldots, c_n]^T, \quad \text{where } c_i = a_i \oplus b_i, \ i = 1, 2, \ldots, n \tag{2.14}$$

The symbol $\oplus$ denotes modulo-2 addition. Let us note that for the additive operation defined in such a way the zero vector is also a codeword, as it is the sum of two identical codewords.

Consider a simple example of the block code that belongs to the class of the so-called generalized parity control codes. Let the code (n, k) be systematic, with $k = 3$ and $n = 6$. Let the first three bits be message bits given to the encoder input, and let the three remaining bits, the parity bits, be the linear combinations of these message bits. Let the codewords be determined by the expression

$$\mathbf{a}^T = [a_1, a_2, a_3, (a_1 \oplus a_2), (a_2 \oplus a_3), (a_1 \oplus a_2 \oplus a_3)] \tag{2.15}$$

which means that

$$a_4 = a_1 \oplus a_2, \quad a_5 = a_2 \oplus a_3, \quad a_6 = a_1 \oplus a_2 \oplus a_3 \tag{2.16}$$

Consider the addition of two codewords, $\mathbf{a}$ and $\mathbf{b}$, both described by expression (2.15). Their sum is a sequence $\mathbf{c}$ of the form

$$c^T = [c_1, c_2, c_3, c_4, c_5, c_6]$$

$$= [(a_1 \oplus b_1), (a_2 \oplus b_2), \ldots, (a_6 \oplus b_6)]$$

where

$$
\begin{aligned}
c_4 &= a_4 \oplus b_4 = a_1 \oplus a_2 \oplus b_1 \oplus b_2 = (a_1 \oplus b_1) \oplus (a_2 \oplus b_2) = c_1 \oplus c_2 \\
c_5 &= a_5 \oplus b_5 = a_2 \oplus a_3 \oplus b_2 \oplus b_3 = (a_2 \oplus b_2) \oplus (a_3 \oplus b_3) = c_2 \oplus c_3 \\
c_6 &= a_6 \oplus b_6 = a_1 \oplus a_2 \oplus a_3 \oplus b_1 \oplus b_2 \oplus b_3 \\
&= (a_1 \oplus b_1) \oplus (a_2 \oplus b_2) \oplus (a_3 \oplus b_3) = c_1 \oplus c_2 \oplus c_3
\end{aligned}
\tag{2.17}
$$

We see from formula (2.17) that all redundant bits of the sequence $\mathbf{c}$ are obtained as a result of the same operations as in formula (2.15), which describes a codeword of the code (6, 3). The sequence $\mathbf{c}$ is therefore also a codeword.

One of the most important features of a block code is the minimum Hamming distance $d_{\min}$ between any pair of its codewords. From this point of view, linear codes have a useful property. Let us note that if the Hamming distance between the codewords $\mathbf{a}$ and $\mathbf{b}$ is $d(\mathbf{a}, \mathbf{b}) = d$, then the Hamming distance between the codewords $\mathbf{a} + \mathbf{a}$ and $\mathbf{a} + \mathbf{b}$ is also d. Let $\mathbf{a} + \mathbf{b} = \mathbf{c}$, so the distance between the codewords $\mathbf{a}$ and $\mathbf{b}$ is the same as the distance between the codeword $\mathbf{c} = \mathbf{a} + \mathbf{b}$ and the zero codeword. In consequence, investigation of the distances between codewords can be limited to checking the distance of nonzero codewords from the zero codeword. The Hamming distance of the codeword $\mathbf{c}$ and the zero codeword is equal to the number of "1"s contained in that codeword. That number is called the *Hamming weight* of the given codeword $\mathbf{c}$ and is denoted as $w(\mathbf{c})$. We will take advantage of this property many times in the analysis of decoders for several linear codes. The lowest Hamming weight, i.e. the lowest number of "1"s contained in any nonzero codeword of a given linear binary code, is equivalent to the minimal distance $d_{\min}$ of this code.

2.5.1 Parity Check Matrix

So far we have described the code by defining the equations determining its redundant bits. One of the basic decoder operations is checking if the received sequence is a codeword. Knowing the equations that govern the production of redundant bits, the decoder can locally generate the redundant bits on the basis of the received message bits. If these bits are the same as the bits received in the redundant bit positions, it is highly probable that the received sequence is a codeword. As an example, let us take the same code as previously. Equations defining the redundant bits are determined by formula (2.16). We can write them in the following equivalent form

$$a_1 \oplus a_2 \oplus a_4 = 0$$

$$a_2 \oplus a_3 \oplus a_5 = 0 \tag{2.18}$$

$$a_1 \oplus a_2 \oplus a_3 \oplus a_6 = 0$$

Checking if the received sequence $\mathbf{r}^T = [r_1, r_2, \ldots, r_6]$ is a codeword is reduced to checking if equation set (2.18) is fulfilled when the symbols r_i of the received sequence replace codeword symbols a_i $(i = 1, 2, \ldots, 6)$ in (2.18). The same operation can also be written in matrix form

$$H\mathbf{a} = \mathbf{0} \tag{2.19}$$

where the form of matrix H results from the equation set (2.18). In our example this is equal to

$$H = \begin{bmatrix} 1 & 1 & 0 & 1 & 0 & 0 \\ 0 & 1 & 1 & 0 & 1 & 0 \\ 1 & 1 & 1 & 0 & 0 & 1 \end{bmatrix} \tag{2.20}$$

Each row of the matrix reflects a single equation of equation set (2.18). Note that "1"s in a given row appear in those positions in which they occur in a given equation. Each equation checks the parity of a certain set of codeword bits, so matrix H is called a *parity check matrix*. In turn, each column of the parity check matrix is related to a single codeword symbol (bit). For a systematic code the first k columns are related to message bits of a codeword, and the $n - k$ remaining columns are associated with the redundant bits. The value of the latter bits ensures parity, i.e. the result of modulo-2 addition of selected message and redundant bits is equal to zero. This is the reason why the redundant bits are also called *parity check bits*. Let us note that if the code is systematic, its matrix H consists of two parts: matrix P_k^T of size $k \times (n - k)$, which is associated with the message bits, and the unity (identity) matrix I_{n-k} of size $(n - k) \times (n - k)$ describing the positions of the parity check bits. Therefore H can take the form

$$H = [P_k^T \mid I_{n-k}] \tag{2.21}$$

Now let us represent the parity check matrix in the form of the column vectors $\mathbf{h}_i$ $(i = 1, 2, \ldots, n)$ where vector $\mathbf{h}_i$ is the ith column of matrix H. Thus, matrix H has the following form

$$H = [\mathbf{h}_1, \mathbf{h}_2, \ldots, \mathbf{h}_6] \qquad (2.22)$$

Then equation (2.19) can take the form

$$[\mathbf{h}_1, \mathbf{h}_2, \ldots, \mathbf{h}_6] \begin{bmatrix} a_1 \\ \vdots \\ a_6 \end{bmatrix} = \sum_{i=1}^{6} \mathbf{h}_i a_i = \mathbf{0} \qquad (2.23)$$

and it can be interpreted in the following way. The parity check matrix is constructed in such a way that its appropriate columns, which are selected by "1"s of a given codeword, result in a zero vector after adding them together. Recall that our considerations on the Hamming distance between codewords have led us to the conclusion that the minimum distance d_{min} can be determined as the lowest number of "1"s in a nonzero codeword. However, this distance can also be determined on the basis of a parity check matrix. Equation (2.23) is also fulfilled for a codeword featuring the lowest Hamming weight. For that codeword, the lowest number of summed columns of matrix H results in a zero vector. That number is simply the minimum distance d_{min} of the given code.

Determine the minimum distance for our example code. All columns of matrix H are different from each other, therefore the sum of any pair of columns is not equal to the zero vector. Hence, we conclude that the minimum distance is higher than 2. However, we can find three columns whose sum is a zero vector. For example, the sum of columns 1, 4 and 6 is a zero vector

$$\begin{bmatrix} 1 \\ 0 \\ 1 \end{bmatrix} + \begin{bmatrix} 1 \\ 0 \\ 0 \end{bmatrix} + \begin{bmatrix} 0 \\ 0 \\ 1 \end{bmatrix} = \begin{bmatrix} 0 \\ 0 \\ 0 \end{bmatrix}$$

In general, one can formulate the following theorem.

Theorem 2.5.1 *If the minimum Hamming distance for a linear block code is equal to d_{min}, then there exists at least one subset of d_{min} columns of parity check matrix H whose sum is the zero vector. Moreover, there does not exist any subset of $d_{min} - 1$ or fewer columns of matrix H whose sum is equal to the zero vector.*

If the columns of matrix H are treated as vectors, one can state that for the code of minimum distance d_{min} all the subsets of $d_{min} - 1$ columns have to be linearly independent.[1]

Denote now the parity check matrix H using the vectors $\mathbf{p}_i^T$ $(i = 1, \ldots, (n-k))$ describing its respective rows, i.e.

$$H = \begin{bmatrix} \mathbf{p}_1^T \\ \vdots \\ \mathbf{p}_{n-k}^T \end{bmatrix} \qquad (2.24)$$

---

[1] Recall that the vectors belonging to a certain set are linearly independent when their weighted sum is equal to zero if and only if all the weighting coefficients are equal to zero.

Each product $\mathbf{p}_i^T \mathbf{a}$ represents a single parity check equation. Equivalently, parity is also checked if the parity check equations are added side by side, e.g. $(\mathbf{p}_i^T + \mathbf{p}_j^T)\mathbf{a} = 0$ $(i, j = 1, \ldots, (n-k))$ is also a parity check equation. Therefore we conclude that based on matrix H one can build another equivalent form of a parity check matrix by creating linear combinations of the rows of the original matrix H. The necessary condition is that the newly created parity check matrix has to apply a certain form of all parity check equations $\mathbf{p}_i^T \mathbf{a} = \mathbf{0}$ $(i = 1, \ldots, (n-k))$. Linear combinations of the parity check matrix rows are applied in some block code decoding algorithms.

2.5.2 Generator Matrix

Previously we have determined a code by giving the equations for parity check bits or, equivalently, the parity check matrix. Another way to describe a code is proposing the construction of its generator matrix. The following subsection is devoted to this issue.

As we remember, the sum of any two codewords of a linear code is also a codeword of this code. Applying this rule repeatedly, we conclude that any combination of the codewords of a linear code is also a codeword of this code. Because the message symbols are selected independently, we can determine the codewords that result from the message block, which consist of a "1" in one of its k positions and zeros in the remaining positions. In this way we obtain the set of k codewords that, through the linear combination thereof, can be applied to synthesize a codeword determined by any k-bit message sequence of the systematic code. The k codewords found in this way constitute a basis of the code vector space. One can imagine these codewords as k unit vectors in k-dimensional space, which can be used in the creation of any vector in this space. Let us mention that the selected basis is not the only one possible, as with a vector description in a given space that can be done in several coordination systems.

As an example, let us consider the systematic code $(7, 4)$ for which the parity bit equations are

$$a_5 = a_1 \oplus a_3 \oplus a_4$$
$$a_6 = a_1 \oplus a_2 \oplus a_4 \qquad (2.25)$$
$$a_7 = a_1 \oplus a_2 \oplus a_3$$

One can easily determine four codewords of this code that have a single "1" in each subsequent message position. The codewords are

$$\mathbf{c}_1^T = [1000111]$$
$$\mathbf{c}_2^T = [0100011]$$
$$\mathbf{c}_3^T = [0010101] \qquad (2.26)$$
$$\mathbf{c}_4^T = [0001110]$$

Let us write them in the compact matrix form as

$$G = \begin{bmatrix} \mathbf{c}_1^T \\ \mathbf{c}_2^T \\ \mathbf{c}_3^T \\ \mathbf{c}_4^T \end{bmatrix} = \begin{bmatrix} 1 & 0 & 0 & 0 & 1 & 1 & 1 \\ 0 & 1 & 0 & 0 & 0 & 1 & 1 \\ 0 & 0 & 1 & 0 & 1 & 0 & 1 \\ 0 & 0 & 0 & 1 & 1 & 1 & 0 \end{bmatrix} \tag{2.27}$$

The above matrix is called the *generator matrix* for the code $(7, 4)$. As the matrix rows are the codewords creating the basis of the code vector space, a linear combination of any rows of this matrix can create any codeword of the code characterized by this matrix. Denoting the k-bit message block by the vector $\mathbf{m} = [m_1, m_2, \dots, m_k]^T$, we receive the following formula for the codeword assigned to the given message block

$$\mathbf{c}^T = \sum_{i=1}^{k} m_i \mathbf{c}_i^T = \mathbf{m}^T G \tag{2.28}$$

Note yet another interesting property. Each row of the generator matrix is a codeword, so according to formula (2.19) multiplication of such a codeword by the parity check matrix results in the zero vector. Performing this operation jointly for all rows of matrix G, we receive the following dependence

$$HG^T = [0]_{k \times (n-k)} \tag{2.29}$$

where $[0]_{k \times (n-k)}$ is a zero matrix of k columns and $n - k$ rows. One can also discover the next dependence between those two matrices. For a systematic code the parity check matrix can have the form (2.21), and the generator matrix of that code has the form $G = [I_k \mid Q]$, so knowing that expression (2.29) holds, we obtain

$$[P_k^T \mid I_{n-k}] \begin{bmatrix} I_k \\ Q^T \end{bmatrix} = P_k^T + Q^T = [0]_k \tag{2.30}$$

which is true only if $Q = P_k$. Therefore the generator matrix of the systematic code (n, k) has the form

$$G = [I_k \mid P_k] \tag{2.31}$$

As we have mentioned, for a given code more than one generator matrix can be constructed by the appropriate choice of a linear combination of the rows of matrix G derived for a systematic code. Other matrices of the same code (which determine the same set of codewords) do not have the form (2.31), therefore we say that they do not have a *canonical form*. However, a characteristic feature of such matrices is that message bits, if appearing directly, are located in those positions for which the columns of the generation matrix contain a single "1" supplemented by zero symbols.

Example 2.5.1 *Consider the code $(7, 4)$ created on the basis of the original systematic code determined by parity check equations (2.25), through their modifications aiming at* *placing the message bits in positions 1, 2, 6 and 7. Therefore independent bits are a_1, a_2, a_6 and a_7, whereas the parity bits are a_3, a_4 and a_5. Parity equations for a_3 and a_4 are received by adding $a_1 \oplus a_2$ to both sides of the second and third equation. The parity equation for a_5 is received on the basis of equations (2.25) by adding the second and third equations side by side and substituting the result in the first equation. We finally have*

$$a_3 = a_1 \oplus a_2 \oplus a_7$$

$$a_4 = a_1 \oplus a_2 \oplus a_6$$

$$a_5 = a_1 \oplus a_6 \oplus a_7$$

The appropriate forms of parity check matrix H' and matrix G' result from the above equations. They are equal to, respectively

$$H' = \begin{bmatrix} 1 & 1 & 1 & 0 & 0 & 0 & 1 \\ 1 & 1 & 0 & 1 & 0 & 1 & 0 \\ 1 & 0 & 0 & 0 & 1 & 1 & 1 \end{bmatrix}, \quad G' = \begin{bmatrix} 1 & 0 & 1 & 1 & 1 & 0 & 0 \\ 0 & 1 & 1 & 1 & 0 & 0 & 0 \\ 0 & 0 & 0 & 1 & 1 & 1 & 0 \\ 0 & 0 & 1 & 0 & 1 & 0 & 1 \end{bmatrix} \quad (2.32)$$

Indeed, the columns of matrix G' of numbers 1, 2, 6 and 7 have a single "1", therefore they reflect the message bit positions in a codeword. Another important fact is that if we generate all codewords on the basis of matrix G described by formula (2.27) and matrix G' given by (2.32), then we discover that both codeword sets are identical; however, in both cases a given codeword is assigned to two different message blocks.

2.5.3 Syndrome

During transmission of codewords through a communication channel errors occur in them. Applying a binary channel model (not necessarily symmetric and memoryless), we can assume that as a result of arising errors codeword **c** is turned into sequence **r**, where **r** = **c** + **e**. As previously, summation of the two vectors is a modulo-2 addition of their appropriate components. Vector **e** is the so-called *error vector*, containing zeros in those positions in which errors did not occur and "1"s in those in which errors appeared. Analyzing the decoding table for a simple code (5, 2) shown in Table 2.1 we see that its first column associated with the zero codeword contains in fact the error vectors. A part of this column contains correctable errors. As we have mentioned, application of the decoding table in the decoding process is feasible only for short and simple codes. For that reason, for longer codes it is necessary to apply other decoding methods. In many cases the first step in the decoding process is checking if the received sequence is a codeword. If the check result is positive, then the decoding process is finished and the message bits are extracted from the received sequence. If the check result is negative, then steps have to be undertaken to find the erroneous positions in the received sequence, followed by their correction (when an error correction code is applied) or only followed by signaling the error event (when the error detection code is used). In the simplest case, in order to verify if the received sequence is a codeword, the parity check equations are applied. In matrix form this operation can be described as

$$\mathbf{s} = H\mathbf{r} \tag{2.33}$$

The result of this operation, realized by multiplication of the received sequence by the parity check matrix or performed in another equivalent way, is called a *syndrome*. The syndrome $\mathbf{s}$ is a vector of length $n - k$. Let us note that because the equality $H\mathbf{c} = \mathbf{0}$ holds, the following property is also true

$$\mathbf{s} = H(\mathbf{c} + \mathbf{e}) = \mathbf{0} + H\mathbf{e} = H\mathbf{e} \tag{2.34}$$

As we see, the syndrome is exclusively determined by the error vector and does not depend on the transmitted codeword. Because the syndrome vector has $n - k$ elements, there are 2^{n-k} different binary syndrome forms. The syndrome is a zero vector if the error vector $\mathbf{e}$ is a zero vector or any nonzero codeword. In the first case it means that the correct codeword has been received. In the second case the syndrome indicates reception of the codeword; however, the decoder does not know that it is not the same codeword that has been transmitted. The error correction code should be constructed in such a way that the latter case has a very low probability of occurrence. As we have mentioned, there are 2^{n-k} different syndromes; however, at the same time there are 2^n different n-bit error sequences. Therefore it follows from equation (2.34) that the same syndrome is related to many different error sequences. A correctly operating decoder is constructed in such a way that, among all possible error vectors $\mathbf{e}$ resulting in the same syndrome, it selects the most probable one. Because in a typical situation the error probability of a single bit is lower than 1/2, for application of a binary memoryless symmetric channel the most probable error sequence among those resulting in the same syndrome is the one that contains the lowest number of "1"s, i.e. the number of error events within the n-bit block is the smallest one. Finding that error sequence on the basis of the previously calculated syndrome is the main task of the decoder. After the error sequence has been determined, correction of the received sequence $\mathbf{r}$ follows. This consists of addition of the estimated error sequence $\widehat{\mathbf{e}}$ to the received sequence $\mathbf{r}$. Such an operation is described by the expression

$$\widehat{\mathbf{c}} = \mathbf{r} + \widehat{\mathbf{e}} = \mathbf{c} + (\mathbf{e} + \widehat{\mathbf{e}}) \tag{2.35}$$

If the correct error sequence has been determined on the basis of the syndrome, the final decision of the decoder is correct.

2.5.4 Hamming Codes

Let us illustrate the application of the parity check and generator matrices by introducing *Hamming codes*. One can state the following question: What should the length n of the codewords be for the given number of parity bits $p = n - k$ so that the minimum distance of the code is $d_{\min} = 3$?

In order to guarantee $d_{\min} = 3$, all column pairs of the parity check matrix of such a code should be linearly independent, so they have to be different from each other and from the zero vector. For p parity bits there exist $2^p - 1$ such columns, so $n = 2^p - 1$.

In general, we receive a code of parameters $(2^p - 1, 2^p - 1 - p)$. Examples of pairs (n, k) fulfilling this condition are: $(7, 4)$, $(15, 11)$, $(31, 26)$, etc. If $d_{min} = 3$, then the error correction capability of such a code is equal to a single error $(t = 1)$. Analyzing condition (2.5) on the Hamming upper bound for the given parameters, we see that these numbers fulfill the equality

$$2^{n-k} = \sum_{i=0}^{1} \binom{n}{i} = 1 + n \tag{2.36}$$

The codes with parameters $(n, k) = (2^p - 1, 2^p - 1 - p)$ that feature a single error correction capability are called *Hamming codes*. They conform to condition (2.36) therefore they are perfect codes. Parity equations result from the above-mentioned parity check matrix, which consists of all nonzero columns of length $p = n - k$. By approriate ordering of the columns a systematic code is received. For $n = 7$ and $k = 4$, matrix H has the form

$$H = \begin{bmatrix} 1 & 1 & 1 & 0 & 1 & 0 & 0 \\ 1 & 1 & 0 & 1 & 0 & 1 & 0 \\ 1 & 0 & 1 & 1 & 0 & 0 & 1 \end{bmatrix} \tag{2.37}$$

and generation matrix G is given by formula (2.27). Therefore, the exemplary code applied before in the description of the generator matrix construction was a Hamming code.

The Hamming code $(7, 4)$ is often applied as an example in handbooks on digital communications and coding theory. Hamming codes of higher length n are applied in some wireless communication systems.

2.5.5 The Iterated Code

Below we present another simple code that is applicable when data can be ordered in the form of a table. A good example is the transmission of a block of ASCII characters. Let each character be represented by a 7-bit block. Each block is appended by a parity bit that checks the parity of all seven bits. In this way a row of the data matrix is completed. Parity is also checked in each column of this matrix, so the block of data rows is supplemented with an additional row, often called the *Frame Control Check* (FCC). Note that the last bit of the FCC block checks the parity of the parity bit column. The configuration of such a frame, with the application of parity bits in each row and the FCC block, can have the following form

$$\begin{bmatrix} a_{1,1} & \cdots & a_{1,7} & p_1 \\ a_{2,1} & \cdots & a_{2,7} & p_2 \\ \cdots & \cdots & \cdots & \cdots \\ a_{n,1} & \cdots & a_{n,7} & p_n \\ q_1 & \cdots & q_7 & q_8 \end{bmatrix}$$

A message block is a bit sequence $a_{i,j}$ $(i = 1, \ldots, n, \ j = 1, \ldots, 7)$. Bit p_i is a modulo-2 sum of all bits of the ith row, whereas bit q_j is a modulo-2 sum of the bits located in the jth position in all rows. The code constructed in this way is able to identify the position of a single error in the table. The decoder calculates the parity bit in each row and modulo-2 accumulates the values of each row. In this way the FCC block is calculated. The error location is determined by the row and column coordinates, for which inconsistencies of the calculated and received parity bits occur. As the code is able to correct a single error, its minimum distance is at least equal to 3. Unfortunately, this code is not able to correct two errors, which is illustrated in Figure 2.3. In the figure a shadowed square denotes the position in which the received and calculated parity bits differ. A single error is identified on the basis of row and column coordinates (Figure 2.3a). Two errors in the same column appear as the inconsistency of two row bits, however the column parity bit in FCC remains unchanged, so it is not possible to identify in which column the errors have occurred (Figure 2.3b). The situation is similar when both errors appear in the same row (Figure 2.3c). Finally, if two errors are located in different columns and rows, their identification is not possible either because there are two possible error locations, resulting in the same parity check bit inconsistency. It can be shown that the minimum distance of the considered code is $d_{\min} = 4$.

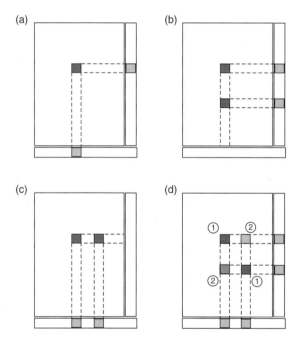

Figure 2.3 Ilustration of possible error cases in the iterative codeword: (a) a single error, (b) a double error in the same column, (c) a double error in the same row, (d) a double error in different rows and columns (two different error pairs giving the same parity check results are denoted by numbers 1 and 2) (Goldsmith and Varaiya (1997)) © 1997 IEEE

2.5.6 Polynomial Codes

The mathematical apparatus applied so far in code description has been limited to vectors and matrices and modulo-2 operations. From the implementation point of view, polynomial code representation is a very convenient way of code description. Block codes defined in the polynomial domain are called *polynomial codes*. Application of polynomial calculus allows many codes to be synthesized and the decoding of these codes can be described in a very clear way.

Let the codeword be given in vector form as

$$\mathbf{c}^T = [c_0, c_1, \ldots, c_{n-1}] \tag{2.38}$$

Equivalently, it can be represented by a polynomial

$$c(x) = c_0 + c_1 x + \cdots + c_{n-1} x^{n-1} \tag{2.39}$$

Polynomial $c(x)$ is called a *code polynomial*.

Thus, the code (n, k) can be determined not only by a list of codewords but also by a full set of code polynomials of degree not higher than $n - 1$. In the case of polynomial codes, the code polynomials have a special property. There exists a certain polynomial $g(x)$ called a *generator polynomial*, which is a common factor of all code polynomials creating the given code. Applying polynomial operations, we can efficiently describe operations of both encoder and decoder. The characteristic feature of the polynomial description is that for binary codes the polynomial coefficients belong to the two-element set $\{0, 1\}$ whereas for nonbinary codes the polynomial coefficients belong to the appropriately selected and finite set of non-negative integers. In general, they belong to a certain finite field. Let us recall the definition of a finite field.

2.5.6.1 Finite Field

Definition 2.5.1 *A finite field* (Galois Field) *$GF(q)$ is a finite set of q elements for which a set of arithmetic rules described by the following properties are defined:*

1. *Two operations (additive and multiplicative) are defined in the field.*
2. *The result of adding or multiplying two elements belonging to the finite field is an element of the same field.*
3. *The field contains an additive identity zero element and a multiplicative identity unit element, for which the following expressions hold*

$$\bigwedge_a a + 0 = a \qquad \bigwedge_a 1 \cdot a = a$$

4. *For each field element there exists an additive inverse element, i.e.*

$$\bigwedge_a \bigvee_{(-a)} a + (-a) = 0$$

and for all nonzero field elements there exists a multiplicative inverse element, i.e.

$$\bigwedge_{a \neq 0} \bigvee_{a^{-1}} a \cdot a^{-1} = 1$$

5. In the set of field elements the associative, commutative and distributive properties apply. This means that if a, b and c are the field elements then

$$a + (b + c) = (a + b) + c \quad a(bc) = (ab)c$$

$$a + b = b + a \qquad\qquad ab = ba$$

$$a(b + c) = ab + bc$$

Finite fields exist only if the number of their elements is a prime number or is a power of a prime number. In the first case we talk about a *prime field* and in the second one about an *extension field*. For each prime number q there exists exactly one finite field. The elements of this prime field denoted as $GF(q)$ are integers in the range 0 to $q - 1$. Addition and multiplication are defined as modulo-q addition and multiplication. In turn, the elements of the extension field whose number $q = p^m$ is a power of a prime number p are all possible polynomials of degree not higher than $m - 1$. The polynomial coefficients belong to the prime field $GF(p)$. Additive and multiplicative operations in the extension field are defined as addition and multiplication of polynomials in the usual sense, recalling that the polynomial coefficients belong to $GF(p)$ and addition and multiplication of these coefficients are performed according to the rules defined for that field. The result of each operation is the subject of reduction modulo the specially selected polynomial $p(x)$ of degree m with the coefficients belonging to the field $GF(p)$. The polynomial $p(x)$ is an *irreducible polynomial*, i.e. it cannot be presented in the form of a product of lower degree polynomials with coefficients belonging to the field $GF(p)$. In the set of polynomials the irreducible polynomials are analogous to the prime numbers in the set of integers. These polynomials are collected in mathematical tables and play a role not only in the description of the polynomial codes but also in the generation of pseudorandom numbers and other operations such as *scrambling* (randomization of a binary sequence).

2.5.6.2 Examples of Operations over Finite Fields

Now we will illustrate basic additive and multiplicative operations for finite prime and extension fields. Assume $GF(7)$ is a prime field and consists of the digits $0, 1, \ldots, 6$. The additive and multiplicative operations over this field are summarized in the tables below.

+	0	1	2	3	4	5	6
0	0	1	2	3	4	5	6
1	1	2	3	4	5	6	0
2	2	3	4	5	6	0	1
3	3	4	5	6	0	1	2
4	4	5	6	0	1	2	3
5	5	6	0	1	2	3	4
6	6	0	1	2	3	4	5

⋆	0	1	2	3	4	5	6
0	0	0	0	0	0	0	0
1	0	1	2	3	4	5	6
2	0	2	4	6	1	3	5
3	0	3	6	2	5	1	4
4	0	4	1	5	2	6	3
5	0	5	3	1	6	4	2
6	0	6	5	4	3	2	1

The table elements are created according to the rules $(a + b) \bmod 7$ and $(ab) \bmod 7$, respectively. Let us present a few typical operations over this field.

$$3 + 5 = 8 \bmod 7 = 1$$

$$4 - 6 = 4 + (-6) = 4 + 1 = 5$$

$$4 \times 4 = 16 \bmod 7 = 2$$

$$\frac{4}{5} = 4 \times 5^{-1} = (4 \times 3) \bmod 7 = 5$$

Regular addition operations do not require any particular comment. Subtraction is performed by finding the opposite element to the subtrahent and adding it modulo-q to the minuend. We proceed similarly in the case of division. First, the inverse element to the divisor is found and then the result is multiplied modulo-q by the dividend.

Let us illustrate operations over the extension field on the example of Galois field $GF(8) = GF(2^3)$ (Clark and Cain 1981). The irreducible polynomial used in this field is $p(x) = 1 + x + x^3$. Note again that operations on the polynomial coefficients are performed over $GF(2)$. The additive operation is then a modulo-2 addition (exclusive-or), and the multiplicative operation is a modulo-2 multiplication, which is equivalent to the logical product. One can easily check that subtraction in $GF(2)$ is identical to addition because the opposite number to the given one is the same number. Let us perform exemplary additions and multiplications over $GF(8)$ for the polynomials $c_1(x) = 1 + x + x^2$ and $c_2(x) = = 1 + x^2$:

$$c_1(x) + c_2(x) = (1 + x + x^2) + (1 + x^2)$$

$$= (1 \oplus 1) + (1 \oplus 0)x + (1 \oplus 1)x^2 = x$$

When performing multiplication of polynomials over $GF(8)$, the distributive law with respect to the modulo-$p(x)$ operation is often applied. According to this law, the remainder from the division by the polynomial $p(x)$ of the sum of polynomials $a(x)$ and $b(x)$ is equal to the sum of remainders from the division of each polynomial separately by the polynomial $p(x)$. Therefore

$$[c_1(x)c_2(x)] \bmod p(x) = \left[(1 + x + x^2)(1 + x^2)\right] \bmod p(x)$$

$$= \left[1 + x + x^2 + x^2 + x^3 + x^4\right] \bmod p(x)$$

$$= \left[1 + x + x^3 + x^4\right] \bmod p(x)$$

Because $x^3 \bmod (x^3 + x + 1) = x + 1$ and $x^4 \bmod (x^3 + x + 1) = x^2 + x$, we obtain

$$[c_1(x)c_2(x)] \bmod p(x) = 1 + x + (1 + x) + (x + x^2)$$

$$= x + x^2$$

There exists a so-called *primitive element* in each finite field. It is a nonzero element that, when raised to successive powers, exhausts all nonzero field elements. Let us consider it with the example of the prime field $GF(7)$. Inspection of the multiplication table over

$GF(7)$ indicates that the primitive element in this field is the number 3, because its subsequent powers in this field are: $3^1 = 3$, $3^2 = 2$, $3^3 = 6$, $3^4 = 4$, $3^5 = 5$, $3^6 = 1 = 3^0$. Then multiplication of two elements of this field can be performed by calculation of the logarithm (i.e. finding the powers of the primitive element in representation of both multiplied factors), adding these logarithms and performing the anti-logarithmic operation. The sense of that kind of procedure is more visible in the case of an extension field.

Consider again the Galois field $GF(2^3)$ generated by the polynomial $p(x) = 1 + x + x^3$. It turns out that the primitive element in this field is the polynomial x. Denote this element as α. As a result, we are able to construct the table of successive powers of the primitive element α supplemented by the zero element. In this way all the elements of the finite field are represented. This assignment is shown in Table 2.2.

Table 2.2 Table of powers of the primitive element of the Galois field $GF(8)$ and their polynomial representation

$GF(8)$	1	x	x^2
0	0	0	0
α^0	1	0	0
α^1	0	1	0
α^2	0	0	1
α^3	1	1	0
α^4	0	1	1
α^5	1	1	1
α^6	1	0	1

Zeroes and "1"s in appropriate columns denote the polynomial coefficients of the appropriate power of the polynomial x. Multiplication of the polynomials can be particularly easily performed applying the above table. The operation can rely on representation of the multipliers by the appropriate powers of the primitive element, modulo-$(q - 1)$ adding these powers and determining the polynomial that is associated with the resulting power of the primitive element. For example,

$$(1 + x + x^2)(1 + x^2) = \alpha^5 \alpha^6 = \alpha^{(5+6) \bmod 7}$$

$$= \alpha^4 = x + x^2$$

2.5.7 Codeword Generation for the Polynomial Codes

As we have already mentioned, the polynomial code (n, k) is determined by listing the full set of code polynomials of degree not higher than $n - 1$, which have a common factor called a generator polynomial $g(x)$ of degree $n - k$. As a result, each code polynomial can be expressed in the form

$$c(x) = a(x)g(x) \tag{2.40}$$

in which $a(x)$ depends on message bits. Let $g(x) = 1 + x + x^3$ and $a(x) = a_0 + a_1 x + a_2 x^2 + a_3 x^3$. The resulting code polynomial is then described by the expression

$$c(x) = (a_0 + a_1 x + a_2 x^2 + a_3 x^3)(1 + x + x^3)$$

$$= a_0 + (a_0 \oplus a_1)x + (a_1 \oplus a_2)x^2 + (a_0 \oplus a_2 \oplus a_3)x^3$$

$$+ (a_1 \oplus a_3)x^4 + a_2 x^5 + a_3 x^6 \tag{2.41}$$

As we see the derived $(7, 4)$ code is nonsystematic. One can easily transform it into a systematic code if the following substitutions are performed

$$b_3 = a_3, \quad b_2 = a_2, \quad b_1 = a_1 \oplus a_3, \quad b_0 = a_0 \oplus a_2 \oplus a_3 \tag{2.42}$$

Thus, if we use the coefficients b_0, b_1, b_2, b_3, calculation of the remaining polynomial coefficients results in the systematic form of the code polynomial, which is further divisible by $g(x) = 1 + x + x^3$:

$$c(x) = (b_0 \oplus b_2 \oplus b_3) + (b_0 \oplus b_1 \oplus b_2)x + (b_1 \oplus b_2 \oplus b_3)x^2$$

$$+ b_0 x^3 + b_1 x^4 + b_2 x^5 + b_3 x^6 \tag{2.43}$$

Consider now possible realizations of the polynomial code encoder. Let us start from the simple case of creation of the codeword described by equation (2.40). Let us turn our attention to the analogy of the polynomial notation for a given sequence with the Z-transform of this sequence if $x = z^{-1}$ is applied. Polynomials with the given coefficients are analogous to the Z-transform of the coefficient sequence, so multiplication of two polynomials results in a new polynomial with coefficients that are the convolution of the coefficients of the multiplied polynomials.[2] Thus, the polynomial $g(x) = 1 + x + x^3$ can be treated as a transform of the filter impulse response of the form 1, 1, 0, 1. The sequence a_3, a_2, a_1, a_0 is fed to the filter input. This implies the encoder implementation shown in Figure 2.4.

As we have already noticed, this encoder generates a nonsystematic code. Below we show how to generate codewords of a systematic code. This structure is often applied in practice. The fact that a code is systematic is particularly advantageous if the code is used only for error detection. The number of redundant bits is then very low compared with the message bits, and placing them at the end of the whole block substantially simplifies the decoder.

As we remember, in polynomial codes, code polynomials are divisible by the generator polynomial. Thus, we can place message bits in the form of the polynomial coefficients at the highest powers of variable x and subsequently calculate the remaining coefficients playing the role of parity bits in such a way that the whole polynomial is divisible by the generator polynomial $g(x)$.

---

[2] Multiplication of the transforms is equivalent to convolution of the related sequences in the time domain.

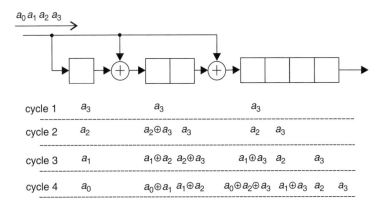

Figure 2.4 cycle content:

cycle 1	a_3	a_3	a_3
cycle 2	a_2	$a_2 \oplus a_3 \quad a_3$	$a_2 \quad a_3$
cycle 3	a_1	$a_1 \oplus a_2 \quad a_2 \oplus a_3$	$a_1 \oplus a_3 \quad a_2 \quad\quad a_3$
cycle 4	a_0	$a_0 \oplus a_1 \quad a_1 \oplus a_2$	$a_0 \oplus a_2 \oplus a_3 \quad a_1 \oplus a_3 \quad a_2 \quad a_3$

Figure 2.4 Encoder scheme for the nonsystematic code with additional memory cells storing the whole codeword

Let us return to the $(7, 4)$ code. Denote message bits as b_0, b_1, b_2, b_3. The associated message polynomial takes the form

$$b(x) = b_3 x^3 + b_2 x^2 + b_1 x + b_0 \qquad (2.44)$$

In general, for a (n, k) code the polynomial $b(x)$ has a degree at most equal to $k - 1$. In order to place the message bits at the highest powers of x in the code polynomial, the polynomial $b(x)$ should be multiplied by x^{n-k}. Let $a(x)$ be the remainder of the division of the polynomial $x^{n-k}b(x)$ by $g(x)$, i.e.

$$a(x) = \left[x^{n-k} b(x) \right] \bmod g(x) \qquad (2.45)$$

This means that the following equation is fulfilled

$$x^{n-k}b(x) = m(x)g(x) + a(x) \qquad (2.46)$$

where $m(x)$ is the result of division of the polynomial $x^{n-k}b(x)$ by $g(x)$. If we add the remainder polynomial $a(x)$ to both sides of (2.46), the result is

$$c(x) = x^{n-k}b(x) + a(x) = m(x)g(x) \qquad (2.47)$$

As we see, the resulting polynomial is a systematic code polynomial. Technical implementation of that kind of encoder is shown in Figure 2.5 and the encoder operates in the following way.

Assume that at the starting moment all memory cells are set to zero. The message bits of the codeword are fed to the encoder input during the first k clock cycles. In this phase of the encoder operation switches P_1 and P_2 remain in position "1". Thus, the message bits are immediately directed to the encoder output and, owing to the linear feedback register determined by the form of the generator polynomial $g(x)$, the remainder bits $a(x)$ are also determined. In the $(k + 1)$st cycle the switches change their positions, which results

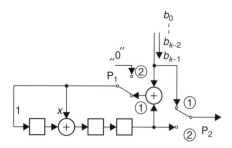

Figure 2.5   Scheme of the encoder for the systematic polynomial code $(7, 4)$

in breaking the feedback connection (switch P_1) in the feedback register and sending the calculated remainder bits through switch P_2 set in position "2" to the encoder output. Let us note that feeding the encoder input with zero bits through switch P_1 starting from the $(k + 1)$st cycle results in gradual filling of the encoder register with zeros. Thus after n cycles the encoder is ready to encode the next message block.

Consider a more general case when the code polynomial coefficients are nonbinary, e.g. q-ary. The generator polynomial has the form

$$g(x) = g_{n-k}x^{n-k} + g_{n-k-1}x^{n-k-1} + \cdots + g_1x + g_0 \tag{2.48}$$

however, this time the polynomial coefficients belong to the Galois field $GF(q)$. All additive and multiplicative operations on the polynomial coefficients are performed in this field. The circuit calculating the remainder resulting from division of the polynomial $x^{n-k}b(x)$ by the generator polynomial $g(x)$ is shown in Figure 2.6. Let us note that the feedback signal in the circuit is introduced in appropriate places of the shift register with the weights that are the opposite numbers to the generator polynomial coefficients in $GF(q)$, while the feedback signal is received as the weighted sum of the contents of the last register cell and the input symbol b_i $(i = k - 1, \ldots, 0)$ in $GF(q)$. The weighting coefficient is the inverse number to the highest power coefficient g_{n-k} of the generator polynomial. We have to stress that each memory cell constituting the shift register stores a single nonbinary symbol. Thus, more than one binary cell is practically required to store each of q-ary symbols.

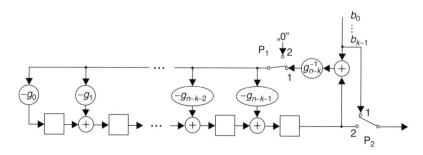

Figure 2.6   Encoder scheme for the nonbinary code with code symbols belonging to the nonbinary Galois field $GF(q)$

2.5.8 Cyclic Codes

For selected values of code length n some polynomial codes display the property of cyclicity. This property means that if the sequence $(c_0, c_1, \ldots, c_{n-1})$ is a codeword of a *cyclic code*, then the sequence $(c_1, c_2, \ldots, c_{n-1}, c_0)$ is a codeword of the same code as well. Applying this property repeatedly, we can observe that the sequence $(c_i, c_{i+1}, \ldots, c_{n-1}, c_0, \ldots, c_{i-1})$ is also a codeword. Thus, we say that cyclic codes are the block polynomial codes whose codewords are cyclic permutations of other codewords.

Let us note that in polynomial calculus the cyclic shift of a codeword by one position to the left can be interpreted as a multiplication of the code polynomial by x followed by reduction of the resulting polynomial modulo-$(x^n - 1)$ $\left[\text{in } GF(2) \text{ also modulo-}(x^n + 1)\right]$. Consider a binary codeword of the form

$$c(x) = c_{n-1}x^{n-1} + c_{n-2}x^{n-2} + \cdots + c_1x + c_0 \tag{2.49}$$

Let us perform the above-described operations. We namely have

$$[xc(x)]\bmod(x^n + 1) = [c_{n-1}x^n + c_{n-2}x^{n-1} + \cdots + c_1x^2 + c_0x]\bmod(x^n + 1)$$

$$= \left[(c_{n-1}x^n + c_{n-2}x^{n-1} + \cdots + c_1x^2 + c_0x) + c_{n-1}(x^n + 1)\right]\bmod(x^n + 1)$$

$$= c_{n-2}x^{n-1} + \cdots + c_1x^2 + c_0x + c_{n-1} \tag{2.50}$$

In deriving (2.50) we took advantage of the already known property, which states that the remainder of division of a polynomial sum by a given polynomial is equal to the sum of remainders resulting from separate division of each sum component by that polynomial. We also use the obvious fact that the remainder of division of the polynomial $c_{n-1}(x^n + 1)$ by $(x^n + 1)$ is equal to zero.

In the context of our current considerations the following question arises: What should be the properties of the generator polynomial for the code to be cyclic? As we know, each code polynomial is divisible by the generator polynomial. Thus, the code polynomial describing the codeword that is a cyclic permutation of the given codeword polynomial also has to be divisible by the generator polynomial. Because the original code polynomial has the property $c(x) = a(x)g(x)$, where the degree of $a(x)$ is at most $k - 1$, the following expression for the cyclic permutation of the codeword has to be true as well

$$[xc(x)]\bmod(x^n + 1) = a_1(x)g(x) \tag{2.51}$$

It turns out that the code is cyclic if the generator polynomial is a divisor of the polynomial $x^n + 1$, which means that there exists such a polynomial $h(x)$ of degree k that $h(x)g(x) = x^n + 1$. Considering from that point of view the calculations shown in (2.50) once more, we have

$$[xc(x)]\bmod(x^n + 1) = \left[xa(x)g(x) + c_{n-1}(x^n + 1)\right]\bmod(x^n + 1)$$

$$= \left[xa(x)g(x) + c_{n-1}h(x)g(x)\right]\bmod(x^n + 1)$$

Two cases are possible. If $c_{n-1} = 0$, the degree of the polynomial $xc(x) = xa(x)c(x)$ is not higher than $n - 1$, thus the remainder of the division of $xc(x)$ by $x^n + 1$ is the divident itself, i.e.

$$[xc(x)] \bmod (x^n + 1) = xc(x) = xa(x)g(x)$$

In consequence, the new polynomial is also a code polynomial, because it is divisible by $g(x)$. A similar case occurs if $c_{n-1} = 1$. We namely have

$$[xc(x)] \bmod (x^n + 1) = \left[xa(x)g(x) + h(x)g(x) \right] \bmod (x^n + 1)$$

$$= \left\{ \left[xa(x) + h(x) \right] g(x) \right\} \bmod (x^n + 1)$$

Since $c_{n-1} = 1$, the polynomial $a(x)$ has to be of the $(k-1)$st degree, so the degree of the polynomial $xa(x)$ is equal to k. As we remember, the degree of the polynomial $h(x)$ is also equal to k, so the sum of $xa(x)$ with $h(x)$ has a degree lower than k. As a result, the degree of the polynomial $\left[xa(x) + h(x) \right] g(x)$ is lower than n. In turn, this imples that

$$[xc(x)] \bmod (x^n + 1) = \left[xa(x) + h(x) \right] g(x) = a_1(x)g(x)$$

so the received polynomial is also a code polynomial of the code generated by $g(x)$.

Summarizing, cyclic (n, k) codes are the polynomial codes, the codewords of which are cyclic permutations of other codewords, and the generator polynomial is a divisor of the polynomial $x^n + 1$.

Although the lengths of cyclic codes, at the assumption of a given generator polynomial $g(x)$, are precisely determined, we can select the required length of a code by applying one of the code modifications, i.e. *code shortening*. This operation is particularly advantageous when error detection codes have a high code rate. A real code length can be selected on the basis of the original length n, setting a certain number of initial message bits permanently to zero so they do not need to be transmitted.

2.5.9 Parity Check Polynomial

As mentioned above, the generator polynomial of the cyclic code (n, k) has the degree $n - k$ and it is a divisor of the polynomial $x^n + 1$. This means that the following property holds true

$$h(x)g(x) = x^n + 1 \tag{2.52}$$

By analogy to the generator matrix G and the parity check matrix H, which when multiplied by each other result in a zero matrix, the polynomial $h(x)$ is called the *parity check polynomial*. Multiplication of the code polynomial $c(x) = a(x)g(x)$ by $h(x)$ results in the expression

$$c(x)h(x) = a(x)g(x)h(x) = a(x)(x^n + 1) = x^n a(x) + a(x) \tag{2.53}$$

Since the polynomial $a(x)$ is of degree at most equal to $k - 1$, on the basis of the right-hand side of (2.53) we can conclude that nonzero coefficients in the polynomial $c(x)h(x)$ can only appear at the components $x^{n+k-1}, x^{n+k-2}, \ldots, x^n$ and $x^{k-1}, x^{k-2}, \ldots, x^0$. The remaining coeffficents at the polynomial terms $x^{n-1}, x^{n-2}, \ldots, x^k$ have to be equal to zero. Let us illustrate this observation with a simple example. Denote the product of the polynomials $c(x)h(x)$ as $q(x)$. Consider the Hamming code $(7, 4)$ generated by the polynomial $g(x) = x^3 + x + 1$. Each code polynomial can be represented by the product

$$c(x) = a(x)g(x) = (a_3 x^3 + a_2 x^2 + a_1 x + a_0)(x^3 + x + 1) \tag{2.54}$$

In turn, the polynomial $q(x)$ becomes

$$q(x) = c(x)h(x) = (x^7 + 1)a(x)$$
$$= a_3 x^{10} + a_2 x^9 + a_1 x^8 + a_0 x^7 + \quad + a_3 x^3 + a_2 x^2 + a_1 x + a_0 \tag{2.55}$$

Knowing from (2.55) simultaneously that

$$q(x) = (c_6 x^6 + c_5 x^5 + \cdots + c_1 x + c_0)(x^4 + h_3 x^3 + h_2 x^2 + h_1 x + 1) \tag{2.56}$$

and calculating those coefficients of $q(x)$ that have to be equal to zero following (2.53), i.e. q_6, q_5 and q_4, we obtain

$$q_6 = c_2 + c_3 h_3 + c_4 h_2 + c_5 h_1 + c_6 = 0$$
$$q_5 = c_1 + c_2 h_3 + c_3 h_2 + c_4 h_1 + c_5 = 0 \tag{2.57}$$
$$q_4 = c_0 + c_1 h_3 + c_2 h_2 + c_3 h_1 + c_4 = 0$$

The parity check polynomial $h(x)$ in equation (2.56) has been presented in the form

$$h(x) = x^4 + h_3 x^3 + h_2 x^2 + h_1 x + 1 \tag{2.58}$$

because in (2.53) the coefficients both at the highest and lowest powers are equal to 1. In the opposite case, $h(x)$ and $g(x)$ would not fulfill (2.52). Let us note that if $k = 4$, then bits at the highest powers of the code polynomial $c(x)$, i.e. c_6, c_5, c_4, c_3, are message bits, so the first equation in equation block (2.57) allows us to find the first parity bit c_2. On the basis of the second equation, knowing already bit c_2, we can recurrently calculate bit c_1; in turn, knowing c_2 and c_1, on the basis of the third equation we are able to determine the last parity bit c_0. Technical implementation of the parity bit calculations on the basis of (2.57), i.e. implementation of the cyclic code encoder using the parity check polynomial $h(x)$, is shown in Figure 2.7.

During the first k clock cycles switch P_1 is in position 1, so the message bits are introduced to the shift register starting from the most meaningful bit. Starting from the $(k + 1)$st cycle, switch P_1 is in position 2. Thus, a feedback loop has been created and

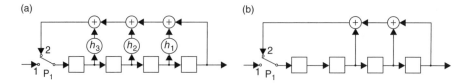

Figure 2.7 Implementation of cyclic code encoder using the parity check polynomial $h(x)$: (a) general case for $(7, 4)$ code, (b) the case for Hamming code $(7, 4)$

the parity bits can be calculated recurrently. After n cycles switch P_1 returns to position 1, the calculated parity bits are gradually shifted out of the encoder register and message bits of the next codeword are simultaneously fed to the encoder input.

Implementation of the cyclic code encoder using the parity check polynomial $h(x)$ is more advantageous than that using the generator polynomial $g(x)$ if the code rate R is lower than $1/2$. For this case the whole structure, in particular the number of memory cells, is lower than if the encoder construction was based on the generator polynomial.

2.5.10 Polynomial Codes Determined by Roots

As we know, a polynomial code is determined by setting the length n of its codewords and the generator polynomial $g(x)$ of degree $n - k$. Each code polynomial is then divisible by $g(x)$. We also know from the polynomial algebra that polynomials can be factored into a product of polynomials of the first degree with coefficients belonging to a certain extension field, e.g.

$$a(x) = (x - \alpha_0)(x - \alpha_1) \ldots (x - \alpha_{k-1}) \qquad (2.59)$$

The elements $\alpha_0, \alpha_1, \ldots, \alpha_{k-1}$ are the roots of the polynomial $a(x)$ and they do not always belong to the same field as the polynomial coefficients. We know that a polynomial cannot always be factored into the product of polynomials of the first degree with the coefficients belonging to the same finite field. In general, if the coefficients of the polynomial $a(x)$ belong to the field $GF(p)$, then the roots of this polynomial belong to the extension field $GF(p^m)$.

The idea of an *irreducible polynomial* is strictly associated with answering the question: To which field do the polynomial roots belong?

Definition 2.5.2 *A polynomial $a(x)$ with the coefficients belonging to a certain finite field is called an **irreducible polynomial** in this field if this polynomial cannot be factored as a product of polynomials of the first degree using the elements of this finite field.*

Property 2.5.1 *If $a(x)$ is an irreducible polynomial with the coefficients selected from the finite field $GF(p)$ and α is the root of this polynomial, then $\alpha^p, \alpha^{p^2}, \alpha^{p^3}, \ldots$ are the roots of this polynomial as well. Moreover, all the roots of this irreducible polynomial can be found in this manner. Polynomial $a(x)$ is called a **minimum function** of the root α.*

Example 2.5.2 *Let the element α belonging to the finite field $GF(8)$ (see Table 2.2) be the root of a certain searched polynomial with coefficients from the finite field $GF(2)$. As we see, in the considered case $p = 2$ and $m = 3$. Therefore if α is the root of this polynomial, then the field elements α^2, α^4 are also the roots. Higher powers of the element α result in the already listed roots, e.g. $\alpha^8 = \alpha^7 \alpha = \alpha$, $\alpha^{16} = \alpha^7 \alpha^7 \alpha^2 = \alpha^2$, because, as it is easy to check, in the finite field $GF(8)$ the element $\alpha^7 = \alpha^0 = 1$. Therefore following Property 2.5.1 the elements α, α^2 and α^4 are all roots of the searched polynomial. In consequence, the polynomial has the following form*

$$a(x) = (x - \alpha)(x - \alpha^2)(x - \alpha^4) \tag{2.60}$$

Since the elements α, α^2 and α^4 belong to the finite field $GF(8)$, they are represented by binary triples according to Table 2.2. Let us stress that the minus signs appearing in formula (2.60) have only a formal meaning, because subtraction in the field $GF(2)$, as well as subtraction of binary triples in the extension field $GF(2^3)$, is equivalent to addition in the respective fields. Determining the polynomial $a(x)$, we obtain

$$a(x) = (x - \alpha)(x - \alpha^2)(x - \alpha^4)$$
$$= x^3 - (\alpha + \alpha^2 + \alpha^4)x^2 + (\alpha^3 + \alpha^5 + \alpha^6)x - \alpha^7$$
$$= x^3 + x + 1 \tag{2.61}$$

because

$$\alpha + \alpha^2 + \alpha^4 = \begin{pmatrix} 0 \\ 1 \\ 0 \end{pmatrix} + \begin{pmatrix} 0 \\ 0 \\ 1 \end{pmatrix} + \begin{pmatrix} 0 \\ 1 \\ 1 \end{pmatrix} = \begin{pmatrix} 0 \\ 0 \\ 0 \end{pmatrix} = 0$$

and

$$\alpha^3 + \alpha^5 + \alpha^6 = \begin{pmatrix} 1 \\ 1 \\ 0 \end{pmatrix} + \begin{pmatrix} 1 \\ 1 \\ 1 \end{pmatrix} + \begin{pmatrix} 1 \\ 0 \\ 1 \end{pmatrix} = \begin{pmatrix} 1 \\ 0 \\ 0 \end{pmatrix} = 1$$

In this way we have calculated the minimum function for the root α and we have shown that the polynomial $a(x) = x^3 + x + 1$ is an irreducible polynomial in the field $GF(2)$ because all its roots belong to the extension field $GF(2^3)$.

Our considerations of polynomial roots and their properties allow us to define polynomial codes in the following way.

Definition 2.5.3 *Polynomials $c_i(x)$ $(i = 1, \ldots, 2^k)$ of degree not higher than $n - 1$ with coefficients belonging to the field $GF(p)$ are the code polynomials of a given polynomial code (n, k) if each code polynomial has the roots $\beta_1, \beta_2, \ldots, \beta_r$ $(r \leq n - k)$ belonging to the field $GF(p^m)$.*

As we remember, each code polynomial of a given code is divisible by the generator polynomial. So if the code polynomials have a common subset of the roots $\beta_1, \beta_2, \ldots, \beta_r$, then these roots are also the roots of the common factor of all code polynomials, namely the generator polynomial. Thus, instead of defining the code by declaring its generator polynomial, we can determine it by listing the common roots of all code polynomials. On the basis of Property 2.5.1 we conclude that besides the roots defining the code directly, there exist code polynomial roots that are the appropriate powers of these roots. Finally, we state the following form of the generator polynomial

$$g(x) = \text{LCM} \left[m_{\beta_1}(x), m_{\beta_2}(x), \ldots, m_{\beta_r}(x) \right] \tag{2.62}$$

where LCM[.] denotes the least common multiple and $m_{\beta_i}(x)$ is a minimum function of the root β_i.

Since $\beta_1, \beta_2, \ldots, \beta_r$ are the roots of code polynomials, the following equalities hold

$$c(\beta_1) = 0, \quad c(\beta_2) = 0, \quad \ldots, c(\beta_r) = 0 \tag{2.63}$$

Thus, if a certain symbol sequence represented by the polynomial $r(x)$ appears at the input of the encoder, then making sure if it is a codeword reduces to inspection if the following equations are fulfilled

$$r(\beta_1) = 0, \quad r(\beta_2) = 0, \quad \ldots, r(\beta_r) = 0 \tag{2.64}$$

Let us note that in matrix notation, checking the validity of all equations is equivalent to multiplying the received sequence vector $\mathbf{r}$ by the parity check matrix, i.e. checking if the following matrix equation holds

$$\begin{bmatrix} \beta_1^0 & \beta_1^1 & \cdots & \beta_1^{n-1} \\ \beta_2^0 & \beta_2^1 & \cdots & \beta_2^{n-1} \\ \cdots & \cdots & \cdots & \cdots \\ \beta_r^0 & \beta_r^1 & \cdots & \beta_r^{n-1} \end{bmatrix} \begin{bmatrix} r_0 \\ r_1 \\ \vdots \\ r_{n-1} \end{bmatrix} = \begin{bmatrix} 0 \\ 0 \\ \vdots \\ 0 \end{bmatrix} \tag{2.65}$$

As an example, consider again the $(7, 4)$ code that is determined by the generator polynomial $g(x) = x^3 + x + 1$. At the same time, this code is determined by the common root α belonging to $GF(2^3)$. As we remember, the minimum function of the root α is just the polynomial $g(x) = x^3 + x + 1$, so besides the root α, the elements α^2 and α^4 are also its roots. The parity check matrix has the form

$$H = \begin{bmatrix} \alpha^0 & \alpha^1 & \alpha^2 & \cdots & \alpha^{n-1} \end{bmatrix} = \begin{bmatrix} 1 & 0 & 0 & 1 & 0 & 1 & 1 \\ 0 & 1 & 0 & 1 & 1 & 1 & 0 \\ 0 & 0 & 1 & 0 & 1 & 1 & 1 \end{bmatrix} \tag{2.66}$$

so it is analogous to the Hamming code parity check matrix considered previously. Taking into account other roots originating from the root α, we obtain a matrix in which some rows are the repeated rows of matrix (2.66). In this way, some parity equations determined by (2.66) are checked a few times. We namely have

$$
\begin{bmatrix}
\alpha^0 & \alpha^1 & \alpha^2 & \cdots & \alpha^{n-1} \\
\left(\alpha^2\right)^0 & \left(\alpha^2\right)^1 & \left(\alpha^2\right)^2 & \cdots & \left(\alpha^2\right)^{n-1} \\
\left(\alpha^4\right)^0 & \left(\alpha^4\right)^1 & \left(\alpha^4\right)^2 & \cdots & \left(\alpha^4\right)^{n-1}
\end{bmatrix}
=
\begin{bmatrix}
1 & 0 & 0 & 1 & 0 & 1 & 1 \\
0 & 1 & 0 & 1 & 1 & 1 & 0 \\
0 & 0 & 1 & 0 & 1 & 1 & 1 \\
1 & 0 & 0 & 1 & 0 & 1 & 1 \\
0 & 0 & 1 & 0 & 1 & 1 & 1 \\
0 & 1 & 1 & 1 & 0 & 0 & 1 \\
1 & 0 & 0 & 1 & 0 & 1 & 1 \\
0 & 1 & 1 & 1 & 0 & 0 & 1 \\
0 & 1 & 0 & 1 & 1 & 1 & 0
\end{bmatrix}
\tag{2.67}
$$

Closer inspection of matrix (2.67) allows us to note that rows 4 and 7 are identical to row 1, row 5 is identical to row 3, the second and the ninth rows are equal to each other, whereas the sixth and the eighth are the sum of the second and the third rows. Thus, it is sufficient to check the parity equations according to matrix (2.66).

As we see, in the above example we have defined the Hamming code in a different way compared with the previous definitions. The current definition is based on selecting the roots of the generator polynomial. In general, the Hamming codes are $(2^m - 1, 2^m - 1 - m)$ polynomial codes, for which all code polynomials have a root that is the primitive element of the finite field $GF(2^m)$. For example, the next Hamming code following the $(7, 4)$ code has the parameters $(n, k) = (15, 11)$, and the root of all its code polynomials is the primitive element of the field $GF(2^4)$ for which the minimum function $m_\alpha(x) = x^4 + x + 1$ is simultaneously the generator polynomial of this code.

2.5.11 Syndrome Polynomial

Recall that a syndrome is the result of multiplication of the parity check matrix by the received sequence vector. If the syndrome is zero, then the received sequence is a codeword. In the case of polynomial codes whose attribute is divisibility of a code polynomial by the generator polynomial, we need to check if this property holds for the received sequence. One can perform this checking in polynomial calculations by derivation of the remainder of the division of the received sequence polynomial by the generator polynomial. If this remainder is equal to zero, then the polynomial representing the received sequence is a code polynomial. In the opposite case the remainder can be associated with the most likelihood sequence, or equivalentlywith the most likelihood error sequence. The polynomial that is the remainder of the division of the received sequence polynomial $r(x)$ by the generator polynomial $g(x)$ is called a *syndrome polynomial* and can be described in the form

$$
s(x) = r(x) \bmod g(x)
\tag{2.68}
$$

Note that because the degree of the generator polynomial $g(x)$ is equal to $n - k$, the degree of the syndrome polynomial can be at most $n - k - 1$, so it has $n - k$ coefficients. Then there are 2^{n-k} possible forms of the syndrome polynomial.

The polynomial describing the received sequence $r(x)$ can be expressed, as in formula (2.34), as the sum of the code polynomial $c(x)$ generated at the transmitter and the polynomial $e(x)$, with the coefficients equal to 1 by those powers of x in which errors occurred. Because $c(x) \bmod g(x) = 0$, we have

$$s(x) = [c(x) + e(x)] \bmod g(x) = c(x) \bmod g(x) + e(x) \bmod g(x)$$

$$= e(x) \bmod g(x) \qquad\qquad (2.69)$$

We conclude from formula (2.69) that the syndrome polynomial is practically the remainder of the division of the error polynomial by the generator polynomial and it does not depend on the code polynomial.

The technical realization of the syndrome calculation according to formula (2.68) is relatively easy and is shown in Figure 2.8. The received sequence is given to the input of the syndrome calculator, starting from the most meaningful position located at the highest power of the received sequence polynomial $r(x)$. The linear feedback shift register (LFSR) has a structure directly resulting from the generator polynomial $g(x)$. After n clock cycles the register memory cells contain the binary coefficients of the syndrome polynomial.

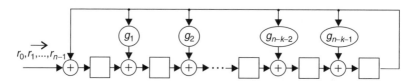

Figure 2.8   General scheme of the syndrome calculator

Example 2.5.3 *Consider the syndrome calculator for the $(7, 4)$ code when the generator polynomial is $g(x) = x^3 + x + 1$. Figure 2.9a presents the decoder scheme, whereas Figure 2.9b shows the mapping of the syndromes onto the correctable error sequences. This mapping has been created by supplying a single "1" to the syndrome calculator input, followed by a sequence of six zeros. The single "1" given as the first input symbol results in the syndrome in the last row of the table, i.e. the one received after seven cycles. Placement of the single "1" in the ith position results in the syndrom located in the $(8 - i)$th table row.*

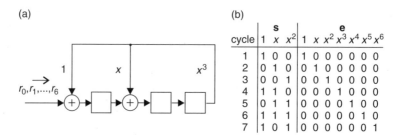

(a)

(b)

cycle	\multicolumn{3}{c}{s}	\multicolumn{7}{c}{e}
	1	x	x^2	1	x	x^2	x^3	x^4	x^5	x^6
1	1	0	0	1	0	0	0	0	0	0
2	0	1	0	0	1	0	0	0	0	0
3	0	0	1	0	0	1	0	0	0	0
4	1	1	0	0	0	0	1	0	0	0
5	0	1	1	0	0	0	0	1	0	0
6	1	1	1	0	0	0	0	0	1	0
7	1	0	1	0	0	0	0	0	0	1

Figure 2.9   (a) Scheme of the syndrome calculator for the $(7, 4)$ code generated by the polynomial $g(x) = x^3 + x + 1$. (b) Mapping of syndromes onto correctable error sequences

The syndrome calculator is a basic element of most block code decoders. Basic difficulty in the decoder design lies in effective implementation of finding the error sequence on the basis of the calculated syndrome. Very often the general decoder scheme looks like the one shown in Figure 2.10. It consists of the syndrome calculator, the circuit recognizing the error sequence on the basis of the calculated syndrome and the block that corrects the received sequence. Decoding methods differ from each other mostly in the block of mapping the syndrome onto the error sequence. In a further part of this chapter we will analyze some examples of the decoders that operate according to the scheme shown in Figure 2.10.

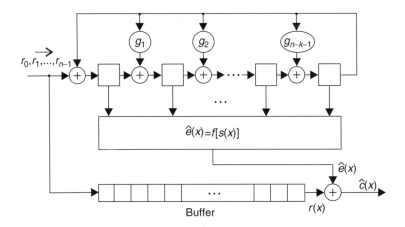

Figure 2.10   General scheme of polynomial code decoding

2.5.12   BCH Codes

BCH codes are an important subclass of the cyclic codes. They are named after their inventors: Bose, Ray-Chaudhuri and Hocquenghem. BCH codes can be considered as a generalization of Hamming codes and have an error correction capability higher than 1. The definition of BCH codes is based on selection of their roots.

Definition 2.5.4 *A primitive BCH code with a correction capability of t errors, built of code symbols belonging to the field $GF(p)$, is a code of codeword length $n = p^m - 1$ that has the following roots of the generator polynomial $g(x)$: $\alpha^{i_0}, \alpha^{i_0+1}, \ldots, \alpha^{i_0+2t-1}$, where α is a primitive element of the field $GF(p^m)$ and i_0 is a certain initial natural number.*

Codes for which the initial number i_0 is 1 are called BCH codes in a narrow sense. We conclude from our previous considerations that the generator polynomial of the BCH code is given by the formula

$$g(x) = \text{LCM}\left[m_{\alpha^{i_0}}(x),\ m_{\alpha^{i_0+1}}(x),\ \ldots,\ m_{\alpha^{i_0+2t-1}}(x)\right] \qquad (2.70)$$

where $m_{\alpha^i}(x)$ is a minimum function of the root α^i.

Example 2.5.4 *Determine the generator polynomial of the BCH code in the narrow sense with a correction capability of t = 2 errors and a codeword length of n = 15. As we see, the codeword length fulfills the condition n = 2⁴ − 1, so p = 2 and m = 4. According to the BCH code definition, the roots of the generator polynomial are equal to* $\alpha, \alpha^2, \alpha^3$ *and* α^4, *where* α *is a primitve element of* $GF(2^4)$. *We would like to determine minimum functions of all the roots, but for that purpose we need the* $GF(16)$ *logarithmic table. This Galois field is generated by the polynomial* $p(x) = x^4 + x + 1$. *Motivated readers will surely be able to determine the polynomial representation of subsequent powers of the primitive element* $\alpha = x$ *as a remainder of division of the polynomial* x^i *by the polynomial* $p(x)$ *for* $i = 0, 1, \ldots, 14$. *Results of these calculations have been summarized in Table 2.3. On this basis we can calculate the minimum functions of the subsequent roots, which are, respectively*

$$m_\alpha(x) = (x - \alpha)(x - \alpha^2)(x - \alpha^4)(x - \alpha^8)$$
$$= (x^2 - \alpha^2 x - \alpha x + \alpha^3)(x^2 - \alpha^8 x - \alpha^4 x + \alpha^{12})$$
$$= (x^2 - \alpha^5 x + \alpha^3)(x^2 - \alpha^5 x + \alpha^{12})$$
$$= x^4 - (\alpha^5 + \alpha^5)x^3 + (\alpha^{12} + \alpha^{10} + \alpha^3)x^2 + (\alpha^2 + \alpha^8)x + \alpha^{15}$$
$$= x^4 + x + 1$$

$$m_{\alpha^2}(x) = (x - \alpha^2)(x - (\alpha^2)^2)(x - (\alpha^2)^4)(x - (\alpha^2)^8)$$
$$= (x - \alpha^2)(x - \alpha^4)(x - \alpha^8)(x - \alpha) = m_\alpha(x)$$
$$m_{\alpha^3}(x) = (x - \alpha^3)(x - (\alpha^3)^2)(x - (\alpha^3)^4)(x - (\alpha^3)^8)$$
$$= (x - \alpha^3)(x - \alpha^6)(x - \alpha^{12})(x - \alpha^9) = x^4 + x^3 + x^2 + x + 1$$

$$m_{\alpha^4}(x) = (x - \alpha^4)(x - (\alpha^4)^2)(x - (\alpha^4)^4)(x - (\alpha^4)^8)$$
$$= (x - \alpha^4)(x - \alpha^8)(x - \alpha)(x - \alpha^2) = m_\alpha(x)$$

Table 2.3 Representation of Galois field $GF(16)$ with the application of the powers of the primitive element α

$GF(16)$	1	x	x^2	x^3		1	x	x^2	x^3
0	0	0	0	0	α^7	1	1	0	1
α^0	1	0	0	0	α^8	1	0	1	0
α^1	0	1	0	0	α^9	0	1	0	1
α^2	0	0	1	0	α^{10}	1	1	1	0
α^3	0	0	0	1	α^{11}	0	1	1	1
α^4	1	1	0	0	α^{12}	1	1	1	1
α^5	0	1	1	0	α^{13}	1	0	1	1
α^6	0	0	1	1	α^{14}	1	0	0	1

so the generator polynomial described by formula (2.70) in our case is

$$g(x) = \text{LCM}\big[m_\alpha(x), m_{\alpha^2}(x), m_{\alpha^3}(x), m_{\alpha^4}(x)\big]$$
$$= (x^4 + x + 1)(x^4 + x^3 + x^2 + x + 1)$$
$$= x^8 + x^7 + x^6 + x^4 + 1 \tag{2.71}$$

Therefore, $n - k = 8$ and we conclude that $k = 7$. As a result, we obtain the $(15, 7)$ BCH code. The assumed number of correctable errors in a codeword has been $t = 2$. In practice it turns out that the code is able to correct not only single and double errors, but also some combinations of triple errors. However, the decoder has to be adjusted to use this potential error correction capability.

BCH codes are a subclass of block codes that are currently often used. For higher lengths BCH codes are one of the best codes with a given code rate.

2.5.13 Reed-Solomon Codes

Reed-Solomon (RS) *codes* are a subclass of BCH codes. They belong to nonbinary codes. The properties of RS codes result from the specific choice of BCH code parameters. For RS codes the values of i_0 and m have been selected as $i_0 = 1$ and $m = 1$ and these parameters determine other properties of RS codes. Thus, the codeword length is $n = p - 1$ and, assuming the error correction capability of t symbols in a codeword, the generator polynomial is given by the formula

$$g(x) = (x - \alpha)(x - \alpha^2) \dots (x - \alpha^{2t}) \tag{2.72}$$

As we see, $2t$ parity symbols are needed to correct up to t errors. It turns out that the codeword length is usually selected in such a way that $n = 2^l - 1$, which means that the RS codes apply nonbinary 2^l-ary symbols that can be represented by l-bit binary blocks. Such blocks are treated as single symbols and all mathematical operations are performed in the Galois field $GF(2^l)$. In consequence, the RS code encoder can be realized according to the scheme shown in Figure 2.6. As l-bit binary blocks are treated as single code symbols, RS codes are applied in systems corrupted by burst errors. They also function as a so-called outer code in *concatenated coding* systems, which will be considered in one of the next sections.

Example 2.5.5 *Let us determine the generator polynomial for the RS code of codeword length $n = 15$ that is able to correct three symbol errors ($t = 3$). From $n = 15$ we conclude that $p = 2^4$ and all calculations are performed in the Galois field $GF(2^4)$ for which the list of elements is given in Table 2.3. Each code symbol is represented by a 4-bit block and all operations are performed on them. Memory cells denoted in the encoder scheme as squares are in reality 4-bit registers. In accordance with formula (2.72) the generator polynomial is given by the formula*

$$g(x) = (x - \alpha)(x - \alpha^2)(x - \alpha^3)(x - \alpha^4)(x - \alpha^5)(x - \alpha^6) \tag{2.73}$$

Using Table 2.3 we achieve the following generator polynomial

$$g(x) = (x - \alpha)(x - \alpha^2)(x - \alpha^3)(x - \alpha^4)(x - \alpha^5)(x - \alpha^6)$$

$$= (x^2 - \alpha^5 x + \alpha^3)(x^2 - \alpha^7 x + \alpha^7)(x^2 - \alpha^9 x + \alpha^{11})$$

$$= x^6 + \alpha^{10}x^5 + \alpha^{14}x^4 + \alpha^4 x^3 + \alpha^6 x^2 + \alpha^9 x + \alpha^6 \qquad (2.74)$$

Recall that if we operate in $GF(2^l)$, then all mathematical operations are performed on l-element binary blocks. Adding two l-bit symbols is realized by bit-wise modulo-2 addition of their l-bit components. We conclude from this observation that the opposite element to any element of the Galois field $GF(2^l)$ is the element itself and subtraction is equivalent to addition. We see from formula (2.74) that the degree of the generator polynomial $g(x)$ is equal to $n - k = 6$. The resulting code is a $(15, 9)$ nonbinary code with 16-level symbols. The code is able to correct three symbol errors, which means that it is able to detect and correct all 8-bit long error bursts and many longer bursts that fit inside three neighboring 4-bit symbols. The encoder scheme is shown in Figure 2.11. The figure also presents schemes of the adder block and the block multiplying by α^j. Let us note that if the multiplicand is equal to zero then also the result of multiplication is zero. If, however, the multiplicand is a nonzero element of $GF(16)$, then the result of its multiplication by α^j is the element α^q whose power fulfills the condition

$$q = (i + j) \bmod n \qquad (2.75)$$

The multiplying circuit can be implemented as a ROM with l-bit address lines and l-bit output, or as an appropriately synthesized combinatorial circuit with l inputs and l outputs.

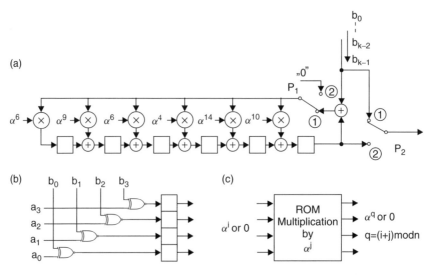

Figure 2.11 (a) Scheme of the encoder for the RS (15,9) code, (b) scheme of the adder block and (c) example of the multiplying block

Currently a few important applications of RS codes are known. The most popular applications are code protection of music files stored on compact disks and outer code in Digital Video Broadcasting (DVB) transmission.

2.5.14 Golay Codes

When we analyze the Hamming upper bound, we can notice that the upper bound is achievable not only for Hamming code parameters but also when $n = 23$, $k = 12$ and $t = 3$. Thus, there exists a perfect code (23, 12) able to correct $t = 3$ errors. Golay showed how to construct it. The code named after him has been synthesized by taking the following elements as the roots of the generator polynomial of the BCH code: $\beta, \beta^2, \beta^3, \beta^4$, where $\beta = \alpha^{89}$. In turn, α is a primitive element of the Galois field $GF(2^{11})$. Deriving the minimum functions of the selected roots, we find the generator polynomial from (2.70) as

$$g(x) = x^{11} + x^9 + x^7 + x^6 + x^5 + x + 1 \tag{2.76}$$

or

$$g(x) = x^{11} + x^{10} + x^6 + x^5 + x^4 + x^2 + 1 \tag{2.77}$$

depending on which of the two possible irreducible polynomials is the basis for creation of the extended Galois field $GF(2^{11})$. Despite the fact that $t = 2$ was assumed in the code construction (note that only four roots have been selected), the code minimum distance is $d_{min} = 7$ so the code is able to correct three errors. Golay also found the ternary code (11, 6), which is an ideal code as well.

2.5.15 Maximum Length Codes

Maximum length codes are cyclic codes that are dual to the Hamming codes. As we remember, the parameters of (n, k) Hamming codes are $(2^m - 1, 2^m - 1 - m)$. Hamming codes are created by taking the irreducible polynomial $p(x)$ for the Galois field $GF(2^m)$ as the generator polynomial. Recall that in the case of cyclic codes a dual code is obtained by exchanging the roles of the generator polynomial $g(x)$ and the parity check polynomial $h(x)$. So in the case of maximum length codes, $h(x) = p(x)$ and the code parameters are $(n, k) = (2^m - 1, m)$. With a higher value of m the code rate significantly decreases.

Consider the maximum length (7, 3) code as an example. It is dual to the Hamming code (7, 4) generated by the polynomial $g(x) = x^3 + x + 1$, therefore for the considered code $h(x) = x^3 + x + 1$. On this basis we can build the encoder shown in Figure 2.12. Because the number of codewords is small, we can simply list all of them and analyze their properties. Table 2.4 presents message blocks and the codewords related to them.

The maximum length code is cyclic, so each nonzero codeword is a cyclic permutation of another codeword. The way the codewords are ordered in Table 2.4 allows us to notice that each codeword is a cyclic shift of the previous codeword. A constant number of "1"s is a characteristic feature of all nonzero codewords. Generally, the number of "1"s is equal to 2^{m-1}, so the number of zeros is smaller by one, i.e. it is equal to $2^{m-1} - 1$. Therefore the code minimum distance is 2^{m-1}. If for the encoder shown in Figure 2.12 more than n clock cycles were applied (remember that n is the codeword length) and

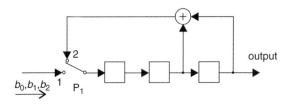

Figure 2.12   Encoder scheme of the maximum length code $(7, 3)$

Table 2.4   Message blocks and codewords for the maximum length code $(7, 3)$

Message blocks			Codewords						
b_0	b_1	b_2	c_0	c_1	c_2	c_3	b_0	b_1	b_2
0	0	0	0	0	0	0	0	0	0
0	1	1	1	0	1	0	0	1	1
0	0	1	1	1	0	1	0	0	1
1	0	0	1	1	1	0	1	0	0
0	1	0	0	1	1	1	0	1	0
1	0	1	0	0	1	1	1	0	1
1	1	0	1	0	0	1	1	1	0
1	1	1	0	1	0	0	1	1	1

the switch remained in position "2", then it would turn out that the generated sequence becomes periodic and the period length is equal to the codeword length $n = 2^m - 1$. The sequence period equal to $2^m - 1$ is the highest possible period that can be achieved in the generation of a sequence using the shift register with $m = 3$ memory cells. This is then the origin of the name of this class of codes.

Maximum length codes have a low code rate and an even value of the minimum distance, so they are not too attractive from the point of view of transmision effectiveness. However, they are used to generate pseudorandom sequences. It turns out that these sequences have multiple desirable statistical properties. For representing binary symbols by bipolar symbols ± 1, the deterministic autocorrelation function of the periodic codewords reaches the maximum of n for zero shift of the correlated codeword with its replica and is equal to -1 if this shift is nonzero. Thus, for long codewords the shape of the autocorrelation function approximates well the ideal autocorrelation function of the sequence of uncorrelated bipolar pulses of zero mean. The maximum length sequences are applied in spread spectrum systems, which will be dealt with in Chapter 7.

2.5.16   Code Modifications

So far we have considered codes that were characterized by specific code lengths n and message block lengths k. Values of n and k for which codes have been designed are not always advantageous from the point of view of a communication system in which the code is to be applied. Very often an adjustment of the code parameters to the communication system requirements is desired even at a certain loss in coding efficiency. Below we show how a code can be modified.

A code is the subject of *extension* if additional parity bits are added in order to improve the codewords' weight structure. The most common way of realizing this task is adding a bit that checks the parity of all remaining bits. In the polynomial notation this bit is the result of division of a codeword polynomial by the polynomial $(x - 1)$. As a result of this operation the code minimum distance is increased by 1. Consider the Hamming code $(7, 4)$ as an example. Recall its parity check matrix:

$$
H = \begin{bmatrix}
1 & 1 & 1 & 0 & 1 & 0 & 0 \\
1 & 1 & 0 & 1 & 0 & 1 & 0 \\
1 & 0 & 1 & 1 & 0 & 0 & 1
\end{bmatrix}
\tag{2.78}
$$

When we supplement a codeword with the additional bit that checks the parity of all other bits, the parity check matrix is extended by one row that is filled with all "1"s and by an additional column because the codeword length increases by 1. The parity check matrix of the extended code has the form H_1, which is equivalent to the canonical form H_2. The latter has been received by replacing the last row in H_1 by the row that is the sum of all rows of matrix H_1.

$$
H_1 = \begin{bmatrix}
1 & 1 & 1 & 0 & 1 & 0 & 0 & 0 \\
1 & 1 & 0 & 1 & 0 & 1 & 0 & 0 \\
1 & 0 & 1 & 1 & 0 & 0 & 1 & 0 \\
1 & 1 & 1 & 1 & 1 & 1 & 1 & 1
\end{bmatrix}
\quad
H_2 = \begin{bmatrix}
1 & 1 & 1 & 0 & 1 & 0 & 0 & 0 \\
1 & 1 & 0 & 1 & 0 & 1 & 0 & 0 \\
1 & 0 & 1 & 1 & 0 & 0 & 1 & 0 \\
0 & 1 & 1 & 1 & 0 & 0 & 0 & 1
\end{bmatrix}
\tag{2.79}
$$

The codeword length can decrease if some parity bits of original codewords are omitted. This process is called *puncturing* and is particularly often applied in the case of convolutional codes, considered later in this chapter.

We say that a code is *expurgated* if some codewords are excised. In the case of cyclic codes for which the generator polynomial is a divisor of $x^n + 1$, the generator polynomial is mostly multiplied by an additional divisor of the polynomial $x^n + 1$, usually $x - 1$. For example, expurgation of the Hamming code $(7, 4)$ leads to the code $(7, 3)$ with the generator polynomial in the form $g(x) = (x + 1)(x^3 + x + 1)$.

If new codewords are added to the original set a code is *augmented*. Consequently, the codeword length does not change, although the number of message bits increases at the cost of parity bits. In case of cyclic codes the code augmentation is achieved by construction of a new generator polynomial by dividing the original generator polynomial by one of its own factors.

A code is *lengthened* if codewords are supplemented with additional message bits. In turn, a code is *shortened* if some message bits are not transmitted. For a systematic code, shortening can be easily implemented by setting a certain number of zeros at the beginning of the message block. As mentioned before, the process of code shortening is particularly often used in cyclic codes applied for error detection.

2.6 Nonalgebraic Decoding for Block Codes

The encoding process is a simpler part of the protection of binary messages against errors occurring in the transmission process. Recovering of the most likelihood transmitted sequence at the receiver is a much more difficult task and requires technical means of much higher complexity than the encoder itself. Making a decision upon the transmitted codeword based on the received sequence is usually much more complicated than encoding. In this section we will present three classical decoding methods for block codes described in the book by Clark and Cain (1981). These methods are conceptually simple but not universal, so they can be applied in the decoding of some codes only. However, due to their simplicity they can be treated as a good introduction before considering more elaborate methods of algebraic decoding.

2.6.1 Meggitt Decoder

The Meggitt decoder (Meggitt 1961) is a circuit that can be applied to decode any cyclic or shortened cyclic code. It was proposed in the early 1960s when microelectronic circuits were still in their introductory phase of development. The decoder is applicable for codes correcting up to three errors and can be used only in a hard-decision mode. The Meggitt decoder takes advantage of the following two properties of cyclic codes:

1. There is a unique relation between each element of the correctable error pattern set and the respective element of the syndrome set.
2. If $s(x)$ is a syndrome polynomial calculated on the basis of the received sequence polynomial $r(x)$ of a cyclic code, then $[xs(x)] \bmod g(x)$ is a syndrome polynomial related to the polynomial of the received sequence $[xr(x)] \bmod (x^n - 1)$ that is cyclically shifted by one position with respect to the received sequence.

To prove property 2, let us present the polynomial of the received sequence in a regular form, i.e.

$$r(x) = r_{n-1}x^{n-1} + r_{n-2}x^{n-2} + \cdots + r_1 x + r_0 \qquad (2.80)$$

Then the polynomial describing a sequence that is a cyclic shift of the sequence $(r_{n-1}, r_{n-2}, \ldots, r_0)$ can be presented as

$$r_1(x) = r_{n-2}x^{n-1} + r_{n-3}x^{n-2} + \cdots + r_0 x + r_{n-1}$$

$$= xr(x) + r_{n-1}(x^n + 1) \qquad (2.81)$$

Both polynomials, $r(x)$ and $r_1(x)$, can be represented as the multiple of the generator polynomial $g(x)$ supplemented by the remainder of the division of these polynomials by $g(x)$, i.e. by the syndrome of the received sequence polynomial $r(x)$ or $r_1(x)$, respectively. So

$$r(x) = a(x)g(x) + s(x) \qquad (2.82)$$

and

$$r_1(x) = x[a(x)g(x) + s(x)] + r_{n-1}g(x)h(x) \qquad (2.83)$$

because we take advantage of (2.81) of the polynomial $r_1(x)$ and the fact that the generator polynomial of cyclic codes is a divisor of $x^n + 1$. Let the remainder of division of the polynomial $r_1(x)$ by $g(x)$, i.e. its syndrome, be $s_1(x)$, i.e.

$$r_1(x) = b(x)g(x) + s_1(x) \qquad (2.84)$$

Let us derive the expression $xs(x)$ from equation (2.83) and calculate the remainder of its division by $g(x)$. We have

$$xs(x) = r_1(x) + xa(x)g(x) + r_{n-1}g(x)h(x) \qquad (2.85)$$

The polynomial $r_1(x)$ can be represented by (2.84), so we obtain

$$xs(x) = b(x)g(x) + s_1(x) + xa(x)g(x) + r_{n-1}g(x)h(x)$$

$$= [b(x) + xa(x) + r_{n-1}h(x)]g(x) + s_1(x) \qquad (2.86)$$

We can conclude from the last equation that $[xs(x)] \bmod g(x) = s_1(x) = r_1(x) \bmod g(x)$. Knowing that a syndrome exclusively depends on the error sequence, we have

$$[xs(x)] \bmod g(x) = e_1(x) \bmod g(x) \qquad (2.87)$$

where $e_1(x) = [xe(x)] \bmod(x^n + 1)$. This ends the proof.

Property 2 indicates that if a given syndrome $s(x)$ corresponds to a certain error sequence, then the error sequence that is cyclically shifted by one position corresponds to the new syndrome that is received in the circuit, calculating the remainder of division by $g(x)$ by advancing the clock by one clock cycle when the shift register of this circuit contains the coefficients of the syndrome $s(x)$. A further implication of Property 2 is the possibility to divide the set of correctable error sequences into classes, each of which contains cyclic shifts of a given error sequence. Owing to Property 2, the error pattern recognition block needs to recognize only one representative in each class on the basis of the calculated syndrome. As a result, the error pattern recognition block structure can be substantially simplified.

The Meggitt decoder, shown in its basic version in Figure 2.13, conforms to the general structure presented in Figure 2.10. Let us consider it in detail in the example of the (15, 7) cyclic code with the generator polynomial $g(x)$ given by formula (2.71). This is a BCH code with the minimum distance $d_{\min} = 5$. It is able to correct all single and double errors and about 30% of triple errors occurring in one block.

The Meggitt decoder operates in the following way. In the first $n = 15$ clock cycles the received bits are fed both to the syndrome calculator and the serial buffer of length n. In the next 15 clock cycles an error correction process is performed along with the

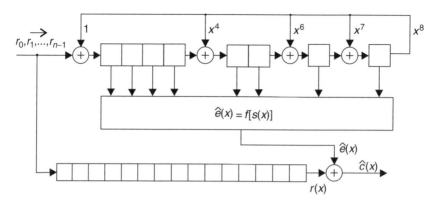

Figure 2.13   Basic scheme of the Meggitt decoder for the (15,7) code with generator polynomial (2.71)

gradual generation of the output sequence. At the same time the syndrome calculator and the buffer are gradually filled with zeros. For each clock cycle the syndrome calculator performs the operation $[xs(x)] \bmod g(x)$. The circuit that recognizes the syndromes related to the specific error patterns generates a binary "1" on its output if the input syndrome is related to the error sequence $[x^i e(x)] \bmod (x^n + 1)$, $i = 0, 1, \ldots, (n-1)$, in which the error is located in the highest, $(n-1)$st, position. Because all double errors are also correctable, another error can simultaneously be located in another position. So the circuit that recognizes the syndrome should generate a binary "1" if the syndrome on its input is related to the error sequence in which there is a single "1" in the 14th position or two binary "1"s where the first is again in the 14th position and the other is in any other position. As we can see, there are 15 different syndrome combinations that should trigger a logical "1" on the output of the syndrome recognizing circuit.

Let us note that in the decoder, except for the syndrome calculator, there are no paths transferring the output signal back to the input. In practical solutions in which the decoding rate is a critical parameter, *pipelining* can be applied. The functional blocks of a particular decoder can be separated by the buffers, which store the operation of a given block until the next clock cycle occurs. Thus, the buffers can be applied between the syndrome calculator and the circuit that recognizes the syndromes and the output of the latter. As a result, the slowest block determines the speed of operation of the whole decoder. Without pipelining, the speed would result from the sum of delays introduced by all blocks operating in cascade.

We propose that the motivated reader traces the decoder operation for the case in which the zero codeword is corrupted by two errors that occurred in the ith and jth positions. To perform this task it is helpful to determine all the syndromes resulting from the single "1" given to the decoder input followed by a sequence of $n-1$ zeros.

2.6.2 Majority Decoder

A *majority decoder* differs from the Meggitt decoder in the way a syndrome is applied to generate correction symbols added modulo-2 to the received sequence. This decoding method can be applied only for some cyclic codes with some specific features; however, if these features occur, the decoder is technically very simple.

Consider the operation of the majority decoder with the example of the $(15, 7)$ code as in the previous section. When the codeword symbols are written in increasing order, for example as $(c_0, c_1, \ldots, c_{14})$, the parity check matrix of the considered code $(15, 7)$ has the following form

$$H = \begin{bmatrix} 1 & 0 & 0 & 0 & 0 & 0 & 0 & 0 & 1 & 1 & 0 & 1 & 0 & 0 & 0 \\ 0 & 1 & 0 & 0 & 0 & 0 & 0 & 0 & 0 & 1 & 1 & 0 & 1 & 0 & 0 \\ 0 & 0 & 1 & 0 & 0 & 0 & 0 & 0 & 0 & 0 & 1 & 1 & 0 & 1 & 0 \\ 0 & 0 & 0 & 1 & 0 & 0 & 0 & 0 & 0 & 0 & 0 & 1 & 1 & 0 & 1 \\ 0 & 0 & 0 & 0 & 1 & 0 & 0 & 0 & 1 & 1 & 0 & 1 & 1 & 1 & 0 \\ 0 & 0 & 0 & 0 & 0 & 1 & 0 & 0 & 0 & 1 & 1 & 0 & 1 & 1 & 1 \\ 0 & 0 & 0 & 0 & 0 & 0 & 1 & 0 & 1 & 1 & 1 & 0 & 0 & 1 & 1 \\ 0 & 0 & 0 & 0 & 0 & 0 & 0 & 1 & 1 & 0 & 1 & 0 & 0 & 0 & 1 \end{bmatrix} \tag{2.88}$$

Each row is related to a single syndrome bit. All these bits appear simultaneously after n clock cycles in the appropriate memory cells of the syndrome calculator. As we remember, the parity check control can also be performed by the modified parity check equations resulting from adding selected rows of the parity check matrix. Massey (1963), the inventor of this decoding method, took advantage of the fact that even if only a small number of parity equations are applied they can be modified in such a way that the most significant codeword position appears in each of them whereas the remaining received bits show their position in the parity check equation set only once at most. Such a construction of the parity check equations is not always possible, therefore the method is used in only some codes.

Let us return to our example and take into account the parity check matrix rows 4 and 8, the sum of rows 2 and 6 and the sum of rows 1, 3 and 7. Denoting the received sequence as $(r_0, r_1, \ldots, r_{14})$, we obtain the following parity check equations

$$P_1 = r_3 + r_{11} + r_{12} + r_{14}$$
$$P_2 = r_7 + r_8 + r_{10} + r_{14}$$
$$P_3 = r_1 + r_5 + r_{13} + r_{14}$$
$$P_4 = r_0 + r_2 + r_6 + r_{14} \tag{2.89}$$

Since these equations are satisfied for codeword bits, their results depend on the error sequence only, so the equations evolve to the form

$$P_1 = e_3 + e_{11} + e_{12} + e_{14}$$
$$P_2 = e_7 + e_8 + e_{10} + e_{14}$$
$$P_3 = e_1 + e_5 + e_{13} + e_{14}$$
$$P_4 = e_0 + e_2 + e_6 + e_{14} \tag{2.90}$$

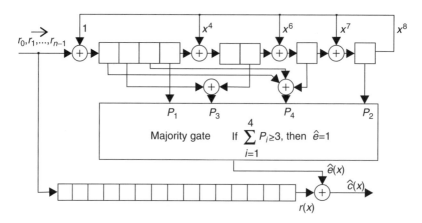

Figure 2.14 Scheme of the majority decoder for the (15,7) code determined by the generator polynomial (2.71)

The majority decoder shown in Figure 2.14 takes advantage of the same properties that are the basis of the Meggitt decoder operation. In the first n clock cycles the syndrome of the received sequence is calculated and the serial buffer is gradually filled with the received sequence bits. During the next n clock cycles the serial buffer is gradually emptied. Simultaneously, correction of the contents at the highest buffer position can be performed. Error detection at the currently highest buffer position is performed by checking the results of the parity check equations. If a single error that has corrupted a codeword is located at the given moment in the highest, i.e. 14th, position then the results $P_i (i = 1, \dots, 4)$ of all parity check equations are equal to logical "1". If a single error is temporarily found in other than the highest position, then at most one equation results in a logical "1", whereas the other equations are satisfied resulting in a logical "zero". A situation in which two errors have occurred in the received sequence is similar to the previous case. If one of the errors shows up in the 14th position, at least three out of four equations result in the symbol 1, because if the second error is in the position participating in one of the parity check equations, then one of these equations is again satisfied. However, if the second error is located in the position that is not taken into account in the parity check equations at all, then all equations result in logical "1". In turn, if none of two errors appeared in the 14th position, then at most one parity check equation results in logical "1". Therefore, through counting the number of "1"s on the outputs P_1, P_2, P_3 and P_4 we can easily detect the presence of an error in the 14th position. Namely, if the arithmetic sum $\sum\limits_{i=1}^{4} P_i \geq 3$, then the error has occurred in the 14th position, so it can be corrected by a logical "1" generated by the arithmetic circuit calculating the sum of outputs P_i. This logical "1" is modulo-2 added to the buffer output.

The decoder described above is able to correct all single and double errors and a certain number of combinations of triple errors. The correction of some triple errors is possible because not all error positions participate in the calculations of the sum of the parity check equation results. If the error occurs in the 14th position and at least in one position that is not taken into account in the calculations of P_1, P_2, P_3 and P_4, then the result given by the majority gate will be correct. In order to support correction of the selected combinations

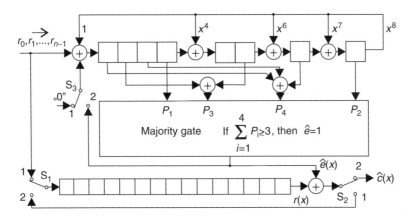

Figure 2.15   Modified majority decoder for the $(15, 7)$ cyclic code

of triple errors, the decoder can be modified to extend its operation on another period of n clock cycles. The scheme of the modified decoder is shown in Figure 2.15.

The modified decoder has three switches, S_1, S_2 and S_3, and the majority gate output is connected with the input of the syndrome calculator. In the first period of n clock cycles switches S_1 and S_3 are in position 1. The state of switch S_2 is irrelevant. In this phase of decoding the syndrome of the received sequence is calculated and the received sequence is simultaneously introduced into the serial buffer. In the second period of n clock cycles switches S_1 and S_3 are set in position 2 and switch S_2 is set in position 1. Now, gradual error correction is performed, complemented by removing the influence of corrected errors from the following form of the syndrome. It is done by feeding the correction symbol through the feedback path from the majority gate output to the input of the syndrome calculator. In the last set of n clock cycles switch S_2 is shifted to position 2 and errors still remaining in the buffer are subsequently corrected when being transported out of the buffer to the decoder output.

It is worth noting that the same modification can also be introduced in the Meggitt decoder so that it will be able to correct selected triple error combinations.

2.6.3   Information Set Decoding

The next decoding method, unlike the majority decoding, can be used for a wide class of block codes. Let us recall our considerations on possible locations of message and parity bits in the codeword (see Example 2.5.1). In Example 2.5.1 we transformed the original code placing the message bits at desired positions without changing the algebraic dependencies among particular bits of a codeword. This property is applied in information set decoding. An information set is a set of bits that are treated as message bits, so the remaining bits in a codeword are considered as parity bits. For each code one can define a certain number of information sets. For each of them, based on bits belonging to this set, we calculate the remaining bits of the codeword related to parity checks.

Assume that a sequence $\mathbf{r} = (r_{n-1}, r_{n-2}, \ldots, r_0)$ appears on the decoder input. Some bits have been received erroneously; however, we assume that the number of erroneous positions does not exceed the error correction capability of the applied code. The *infor-mation set decoding* algorithm can be formulated as follows (Clark and Cain 1981):

1. Select the appropriate number of information sets according to a certain rule.
2. On the basis of the received block, construct hypothetical codewords for each informa-tion set by assuming that message bits in the considered information set are error-free.
3. Compare each of the created candidate codewords with the received block **r**. The decoder's decision is the hypothetical codeword that is closest to the received sequence in the sense of the selected distance measure (e.g. Hamming distance).

As we can see, the algorithm is based on elementary properties of linear block codes, so it is useful in decoding many different codes. The only problem is the right choice of information sets. It is crucial that with a probability tending to 1, at least for one infor-mation set the selected message block contains errorless symbols, so based on them the decoder is able to synthesize the codeword that has actually been sent by the transmitter. This means that among hypothetical codewords there is a correct one. The operation of the decoding algorithm is explained on the examples quoted in Clark and Cain (1981).

Example 2.6.1 *Consider the Hamming code* $(7, 4)$ *again. For this code we select three information sets* $I_1 = \{1, 2, 3, 4\}$, $I_2 = \{4, 5, 6, 7\}$ *and* $I_3 = \{1, 2, 6, 7\}$. *The digits in parentheses denote positions of the message bits in a codeword starting from the left side. Recall that message bits are related to the columns of the generator matrix that have a single 1 in them. Information set* I_1 *is thus related to generator matrix* G_1 *in its canonical form, whereas the other matrices* G_2 *and* G_3 *are achieved by summing the rows of matrix* G_1 *in such a way that the appropriate columns contain a single 1 in appropriate rows. Finally, matrices* G_1, G_2 *and* G_3 *achieve the form*

$$G_1 = \begin{bmatrix} 1 & 0 & 0 & 0 & 1 & 1 & 0 \\ 0 & 1 & 0 & 0 & 0 & 1 & 1 \\ 0 & 0 & 1 & 0 & 1 & 1 & 1 \\ 0 & 0 & 0 & 1 & 1 & 0 & 1 \end{bmatrix} \quad G_2 = \begin{bmatrix} 1 & 1 & 0 & 1 & 0 & 0 & 0 \\ 0 & 1 & 1 & 0 & 1 & 0 & 0 \\ 1 & 1 & 1 & 0 & 0 & 1 & 0 \\ 1 & 0 & 1 & 0 & 0 & 0 & 1 \end{bmatrix} \quad (2.91)$$

$$G_3 = \begin{bmatrix} 1 & 0 & 1 & 1 & 1 & 0 & 0 \\ 0 & 1 & 1 & 0 & 1 & 0 & 0 \\ 0 & 0 & 1 & 1 & 0 & 1 & 0 \\ 0 & 0 & 0 & 1 & 1 & 0 & 1 \end{bmatrix} \quad (2.92)$$

Let the transmitted codeword be $\mathbf{c} = (1011100)^T$, *whereas* $\mathbf{r} = (1010100)^T$ *is the received sequence. By inspection we see that the error has occurred in the fourth position. Its location is obviously unknown to the decoder. Based on the received sequence the decoder determines hypothetical message bit sets for each selected information set:* (1010) *for set* I_1, (0100) *for set* I_2 *and* (1000) *for set* I_3. *In turn, for each of these hypothetical message blocks the decoder calculates hypothetical codewords. These are*

$$\mathbf{c}_1^T = (1010)G_1 = (1010001)$$

$$\mathbf{c}_2^T = (0100)G_2 = (0110100)$$

$$\mathbf{c}_3^T = (1000)G_3 = (1011100)$$

As we can see, among hypothetical codewords that are candidates for the final decision there is one that has actually been generated in the transmitter. From the codewords $\mathbf{c}_1$, $\mathbf{c}_2$ and $\mathbf{c}_3$ the maximum likelihood decoder selects the codeword for which the Hamming distance to the received sequence is the lowest. These distances are, respectively, $d(\mathbf{c}_1, \mathbf{r}) = 2$, $d(\mathbf{c}_2, \mathbf{r}) = 2$ and $d(\mathbf{c}_3, \mathbf{r}) = 1$. Consequently, the decoder makes the right decision that the codeword $\mathbf{c}_3$ has been transmitted.

Example 2.6.2 *Let us illustrate the operation of the same algorithm in the case of soft-decision decoding. Let the received samples related to the particular bits of the transmitted codeword be represented in the form of digits from the range 0–7. Define the distance metric between two sequences – the received sequence and the hypothetical code sequence – as a sum of modules of differences between the received samples and the hypothetical samples representing bits of the hypothetical codeword, i.e.*

$$d(\mathbf{r}, \mathbf{c}_i) = \sum_{k=0}^{n-1} \left| r_k - c_{i,k} \right| \qquad (2.93)$$

where $r_k \in \{0, 1, 2, \ldots, 7\}$ and $c_{k,i} \in \{0, 7\}$. Consider the codeword assumed in the former example; however, this time the received sequence has the form $\mathbf{r} = (7033701)^T$. Let us note that binary decisions made with respect to particular samples would result in a binary sequence $\mathbf{r}' = (1000100)^T$. In this case the Hamming distance of this binary sequence to the transmitted codeword is equal to 2, so the error correction capability of the Hamming code if hard-decision decoding were applied would be exceeded and there is no chance for correct decoding. Let us check how the soft-decision decoder copes with the errors even if a simplified metric (2.93) has been applied. As previously, the decoder synthesizes message bits for each information set by taking advantage of temporary binary decisions shown in the sequence $\mathbf{r}'$. So for the information set I_1 the hypothetical message bits are (1000), for I_2 they are (0100) and for I_3 they are (1000). The codewords related to them are, respectively

$$\mathbf{c}_1^T = (1000)G_1 = (1000110) \quad \text{in representation } \{0, 7\}: (7000770)$$

$$\mathbf{c}_2^T = (0100)G_2 = (0110100) \quad \text{in representation } \{0, 7\}: (0770700)$$

$$\mathbf{c}_3^T = (1000)G_3 = (1011100) \quad \text{in representation } \{0, 7\}: (7077700)$$

If the distances among the hypothetical sequences with symbols from the set $\{0, 7\}$ and the received sequence $\mathbf{r}$ are calculated according to formula (2.93), we obtain $d(\mathbf{r}, \mathbf{c}_1) = 14$, $d(\mathbf{r}, \mathbf{c}_2) = 22$ and $d(\mathbf{r}, \mathbf{c}_3) = 9$. As we can see, this time the decoder decision will also be correct because the sequence $\mathbf{c}_3$ is the closest to the received sequence $\mathbf{r}$ in the sense of the metric defined in (2.93).

Let us note that owing to more accurate quantization the distances among codewords and the received sequence are more distinguishable. For the Hamming code for which the minimum Hamming distance is $d_{\min} = 3$, in the case of soft-decision decoding and 8-level quantization, the codewords presented in 8-level representation differ among themselves by at least 21. This means that all received sequences that differ by no more than 10 will be decoded correctly.

A basic difficulty in the construction of the decoder operating according to this method is the appropriate choice of information sets. They should cover the set of possible codewords and the probability that among the selected hypothetical codewords there is no transmitted codeword should be negligibly small. The choice of the information sets is often made by a computer search.

So far in the information set decoding we have used generator matrices related to the appropriate positions of the message bits associated with a given information set. However, a parity check matrix can also be used in a similar algorithm. Let us note that if message bits for a certain information set are errorless, then the errors have possibly been committed in the parity positions. For parity bits the columns of the parity check matrix contain single "1"s and a syndrome is in fact a linear combination of them. So on the basis of a syndrome one can conclude whether the message bit positions determined by a given information set are error-free. For example, for the Hamming code $(7, 4)$ and the information set $I_1 = \{1, 2, 3, 4\}$ the parity check matrix has the familar form

$$H_1 = \begin{bmatrix} 1 & 0 & 1 & 1 & 1 & 0 & 0 \\ 1 & 1 & 1 & 0 & 0 & 1 & 0 \\ 0 & 1 & 1 & 1 & 0 & 0 & 1 \end{bmatrix}$$

A single correctable error in one of the parity positions results in a tri-bit syndrome containing a single "1" strictly associated with the position in which the error has occurred. Then the decoding algorithm relies on the syndrome calculations for all parity check matrices associated with each information set and testing if the received syndrome has the weight at most equal to 1.

In the general case we search for such parity bit positions that, when the associated parity check matrix is applied, result in a syndrome of the weight not exceeding the correctable number of t errors. Thus, the form of a syndrome determines those parity positions that should be corrected in the received sequence $\mathbf{r}$. The motivated reader could determine other parity check matrices and information sets I_2 and I_3 associated with them for the considered Hamming code. Performing the decoding process similar to that shown in Example 2.6.1 would be an interesting excercise.

The considered information set decoding method using parity check matrices can be substantially simplified for cyclic codes and can be conveniently described using a polynomial notation. The decoding method for these codes, which is a variant of the information set decoding, is called *error trapping decoding*; see (Lin and Costello 2004) for details.

2.7 Algebraic Decoding Methods for Cyclic Codes

The decoding methods considered so far have taken advantage of specific properties of the applied codes, so they have been called nonalgebraic decoding methods. However, an important class of decoding algorithms is the class of *algebraic decoding* methods, relying on efficient solution of a certain equation set. These methods are applicable not only to binary codes, but also for nonbinary ones such as Reed-Solomon codes. Thus, they are important from the application point of view.

Consider the decoding of binary BCH codes. The method that will be presented below can be easily extended to nonbinary codes. As we remember, the roots of the BCH code generator polynomial $g(x)$ are equal to $\alpha^{i_0}, \alpha^{i_0+1}, \ldots, \alpha^{i_0+2t-1}$, where α is a primitive element of the Galois field $GF(p^m)$ and i_0 is a certain initial natural number. Assume without loss of generality that $i_0 = 1$. Recall that the roots of the generator polynomial determine the form of the parity check matrix, which according to formula (2.65) is the following

$$H = \begin{bmatrix} \alpha^0 & \alpha & \ldots & \alpha^{n-1} \\ (\alpha^2)^0 & (\alpha^2)^1 & \ldots & (\alpha^2)^{n-1} \\ \ldots & \ldots & \ldots & \ldots \\ (\alpha^{2t})^0 & (\alpha^{2t})^1 & \ldots & (\alpha^{2t})^{n-1} \end{bmatrix} \tag{2.94}$$

We also remember that $H\mathbf{r} = H\mathbf{e} = \mathbf{s}$, so the syndrome calculated for the received sequence depends exclusively on the error sequence $\mathbf{e}$ or, equivalently, on the error polynomial $e(x)$. Denote the result of the scalar product of the ith row of matrix H given by (2.94) and the error sequence as s_i. One can easily find that it is the ith component of the syndrome vector $\mathbf{s}$. We have

$$\begin{bmatrix} s_1 \\ s_2 \\ \vdots \\ s_{2t} \end{bmatrix} = \begin{bmatrix} \alpha^0 & \alpha & \ldots & \alpha^{n-1} \\ (\alpha^2)^0 & (\alpha^2)^1 & \ldots & (\alpha^2)^{n-1} \\ \ldots & \ldots & \ldots & \ldots \\ (\alpha^{2t})^0 & (\alpha^{2t})^1 & \ldots & (\alpha^{2t})^{n-1} \end{bmatrix} \begin{bmatrix} e_0 \\ e_1 \\ \vdots \\ e_{n-1} \end{bmatrix} \tag{2.95}$$

so in the polynomial notation for $i = 1, \ldots 2t$ the ith syndrome component s_i can be shown in the form

$$s_i = e(\alpha^i) = e_{n-1}(\alpha^i)^{n-1} + e_{n-2}(\alpha^i)^{n-2} + \cdots + e_1\alpha^i + e_0 \tag{2.96}$$

Each syndrome component s_i is a linear combination of the powers of the root α^i and therefore belongs to the Galois field $GF(p^m)$. Assume that $w \leq t$ errors have occurred in the received sequence. Their positions are unknown to the decoder. Denote them as $j_1, j_2, \ldots, j_w$. Therefore the error polynomial is expressed as

$$e(x) = e_{j_w}x^{j_w} + e_{j_{w-1}}x^{j_{w-1}} + \cdots + e_{j_2}x^{j_2} + e_{j_1}x^{j_1} \tag{2.97}$$

In the case of binary codes the coefficients $e_{j_1}, e_{j_2}, \ldots, e_{j_w}$ are equal to binary "1"s. Taking advantage of (2.97), we obtain equation set (2.96) in the following form

$$\begin{aligned} s_1 &= e_{j_w}\alpha^{j_w} + e_{j_{w-1}}\alpha^{j_{w-1}} + \cdots + e_{j_2}\alpha^{j_2} + e_{j_1}\alpha^{j_1} \\ s_2 &= e_{j_w}(\alpha^{j_w})^2 + e_{j_{w-1}}(\alpha^{j_{w-1}})^2 + \cdots + e_{j_2}(\alpha^{j_2})^2 + e_{j_1}(\alpha^{j_1})^2 \\ &\vdots \\ s_{2t} &= e_{j_w}(\alpha^{j_w})^{2t} + e_{j_{w-1}}(\alpha^{j_{w-1}})^{2t} + \cdots + e_{j_2}(\alpha^{j_2})^{2t} + e_{j_1}(\alpha^{j_1})^{2t} \end{aligned} \tag{2.98}$$

so each syndrome component is described by the expression

$$s_i = \sum_{l=1}^{w} e_{j_l}(\alpha^{j_l})^i \tag{2.99}$$

The main goal of the decoder is to identify the error positions $j_1, j_2, \ldots, j_w$. In general, determination of the error positions relies on calculation of the syndrome components $s_1, s_2, \ldots, s_{2t}$, followed by solution of the nonlinear equation set with respect to unknowns $\alpha^{j_1}, \alpha^{j_2}, \ldots, \alpha^{j_w}$. The appropriate powers of the primitive element α are found on the basis of the achieved solutions. The powers indicate error positions in the received sequence. Solution of the nonlinear equation set (2.98) is generally cumbersome. The conceptually simplest approach would be to find the solution by successive substitution of unknowns $\alpha^{j_1}, \alpha^{j_2}, \ldots, \alpha^{j_w}$ by all possible powers of the primitive element α in equation set (2.98). Unfortunately, it is reasonable only if the number of correctable errors t is small. Particular decoding methods basically differ in the method of finding the solution of equation set (2.98). In this section we will present the approach proposed by Berlekamp (1965), modified by Massey (1972) and summarized in a clear way by Lee (2000).

Instead of solving the nonlinear equation set, the Berlekamp-Massey algorithm defines an *error location polynomial* $\Lambda(x)$ and performs some operations on it. The polynomial has the form

$$\Lambda(x) = \Lambda_w x^w + \Lambda_{w-1} x^{w-1} + \cdots + \Lambda_1 x + 1 \tag{2.100}$$

$$= (1 - \alpha^{j_w} x)(1 - \alpha^{j_{w-1}} x) \ldots (1 - \alpha^{j_1} x) = \prod_{l=1}^{w}(1 - \alpha^{j_l} x) \tag{2.101}$$

The roots $(\alpha^{j_l})^{-1}$ $(l = 1, 2, \ldots, w)$ of the error location polynomial are inverses of the searched solutions of the nonliner equation set (2.98). Thus, the solution of (2.98) has been replaced by construction of the polynomial $\Lambda(x)$, followed by finding its roots.

Let us multiply both sides of the polynomial expression $\Lambda(x)$ given by formula (2.100) by $e_{j_l}(\alpha^{j_l})^{k+w}$, where k is a certain natural number,[3] and calculate its value for $x = (\alpha^{j_l})^{-1}$. Because $(\alpha^{j_l})^{-1}$ is a root of the polynomial $\Lambda(x)$, from (2.100) we get the following dependence

$$e_{j_l}(\alpha^{j_l})^{k+w}\left[\Lambda_w(\alpha^{j_l})^{-w} + \Lambda_{w-1}(\alpha^{j_l})^{-w+1} + \cdots + \Lambda_1(\alpha^{j_l})^{-1} + 1\right] = 0$$

or equivalently

$$e_{j_l}\left[\Lambda_w(\alpha^{j_l})^k + \Lambda_{w-1}(\alpha^{j_l})^{k+1} + \cdots + \Lambda_1(\alpha^{j_l})^{k+w-1} + (\alpha^{j_l})^{k+w}\right] = 0 \tag{2.102}$$

---

[3] One should not confuse the natural number k with the message block length of a codeword.

Let us sum both sides of equation (2.102) for all values of the index $l = 1, 2, \ldots, w$. Grouping all components containing Λ_i, we receive the following equation

$$\Lambda_w \sum_{l=1}^{w} e_{j_l} (\alpha^{j_l})^k + \Lambda_{w-1} \sum_{l=1}^{w} e_{j_l} (\alpha^{j_l})^{k+1} + \cdots$$

$$+ \Lambda_1 \sum_{l=1}^{w} e_{j_l} (\alpha^{j_l})^{k+w-1} + \sum_{l=1}^{w} e_{j_l} (\alpha^{j_l})^{k+w} = 0 \qquad (2.103)$$

If we recall formula (2.99) for syndrome components, we see that equation (2.103) can be written in the form

$$\Lambda_w s_k + \Lambda_{w-1} s_{k+1} + \cdots + \Lambda_1 s_{k+w-1} + s_{k+w} = 0 \qquad (2.104)$$

Equation (2.104) contains the existing components of the syndrome calculated from (2.95) if k is located within the interval $[1, w]$, because if, as it is assumed, $w \leq t$, then $k + w \leq 2t$. Substituting $i = k + w$ for $w + 1 \leq i \leq 2w$, we get the following equation

$$\Lambda_w s_{i-w} + \Lambda_{w-1} s_{i-w+1} + \cdots + \Lambda_1 s_{i-1} + s_i = 0 \qquad (2.105)$$

so

$$s_i = - \sum_{l=1}^{w} \Lambda_l s_{i-l} \quad \text{for} \quad i = w + 1, w + 2, \ldots, 2w \qquad (2.106)$$

Based on (2.106) we get the following matrix equation

$$\begin{bmatrix} s_{w+1} \\ s_{w+2} \\ \vdots \\ s_{2w} \end{bmatrix} = - \begin{bmatrix} s_1 & s_2 & \cdots & s_w \\ s_2 & s_3 & \cdots & s_{w+1} \\ \cdots & \cdots & \cdots & \cdots \\ s_w & s_{w+1} & \cdots & s_{2w-1} \end{bmatrix} \begin{bmatrix} \Lambda_w \\ \Lambda_{w-1} \\ \vdots \\ \Lambda_1 \end{bmatrix} \qquad (2.107)$$

This is a linear equation set that needs to be solved in the finite field with respect to the coefficient set $\{\Lambda_1, \Lambda_2, \ldots, \Lambda_w\}$ of the error location polynomial. Equation (2.106) implies that the subsequent syndrome components can be found by applying the feedback shift register shown in Figure 2.16 if the coefficients $\Lambda_1, \Lambda_2, \ldots, \Lambda_w$ are known. The solution of equation set (2.107) is then equivalent to the design of such a feedback shift register that is able to generate the sequence of the syndrome components. The iterative method of deriving the coefficients $\Lambda_1, \Lambda_2, \ldots, \Lambda_w$ was given by Massey (1972). His method is related to Berlekamp's method (Berlekamp 1965), so the name of the decoding algorithm contains the names of both scientists. Let us note that the decoder does not know the number of errors w that have occurred in the received sequence. Assuming that the probability of a single error in the received symbol is smaller than $1/2$, the case in which fewer errors have occurred is more probable than the case in which the number of errors is higher. Thus, we should search for the register with the lowest degree that correctly generates the sequence of the syndrome components calculated for the received sequence.

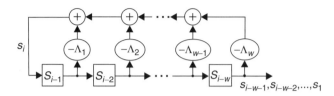

Figure 2.16   Linear feedback register synthesized in the Massey algorithm

The Massey algorithm is thus a method of feedback register synthesis that determines the register of minimum length. This register generates the required sequence of syndrome components. During operation of the algorithm the sequence of syndrome components generated by the current form of the feedback register is subsequently compared with the desired syndrome components calculated on the basis of the received sequence. This is done step by step until all the syndrome components are correctly produced by the register or the divergence between a component calculated by the register and that calculated on the basis of the received sequence is observed. In the case of divergence, the feedback register is modified so as to remove it. The syndrome components are generated again until their number is exhausted or the next divergence between the calculated and generated sequences appears again.

Denote the polynomial describing the correction of the feedback taps as $D(x)$. L is the current degree of the synthesized connection polynomial $\Lambda(x)$. Let i be the number of the subsequent syndrome component. The algorithm performing the synthesis of the feedback register leading to determination of the error location polynomial coefficients can be formulated in the following steps (Michelson and Levesque 2003).

1. Derive the syndrome components s_i, $i = 1, 2, \ldots, 2t$.
2. Initialize the variables applied in the algorithm: $i = 1$, $\Lambda(x) = 1$, $D(x) = x$, $L = 0$.
3. For a new syndrome component s_i calculate the discrepancy

$$\delta = s_i + \sum_{l=1}^{L} \Lambda_l s_{i-l} \tag{2.108}$$

4. Check the calculated value of discrepancy δ. If $\delta = 0$, go to step 9, otherwise go to step 5.
5. Modify connection polynomial $\Lambda(x)$. Let $\Lambda^*(x) = \Lambda(x) - \delta D(x)$.
6. Test the length of the feedback register. If $2L \geq i$, go to step 8, i.e. do not extend the register length, otherwise go to step 7.
7. Increase the register length and update the correction polynomial $L := i - L$, $D(x) = \Lambda(x)\delta^{-1}$
8. Update the connection polynomial: $\Lambda(x) := \Lambda^*(x)$.
9. Update the correction polynomial: $D(x) := x D(x)$.
10. Update the counter of the syndrome components: $i := i + 1$.
11. Check if the counter of the syndrome component has reached the final value, i.e. if $i > 2t$. If not, go to step 3; otherwise stop the procedure.

Let us note that discrepancy δ is defined in such a way that in the case of the first realization of step 3 it is equal to the first syndrome component s_1. Synthesis of the correction polynomial $D(x)$ is performed not only to set discrepancy δ to zero but also to modify the polynomial $\Lambda(x)$ in such a manner that the feedback register configured according to this polynomial would generate all preceding syndrome components. Thus it is not then necessary to check the correctness of generation of the previous syndrome components by the modified feedback register. This property has a crucial influence on the algorithm complexity, which, as a result, depends linearly on the number of correctable errors.

After finding the coefficients of the polynomial $\Lambda(x)$ we have to find its roots $\left(\alpha^{j_l}\right)^{-1}$, which, as we remember, are inverses of the Galois field elements indicating the error locations. Searching for the roots is often performed by substitution of each nonzero element of the extension field $GF(p^m)$, in which the primitive element α is defined, to the polynomial $\Lambda(x)$ determined by formula (2.100), and testing if $\Lambda(x) = 0$. If this is true, the tested element is a root of the polynomial $\Lambda(x)$, so the power of the primitive element related to the root inverse determines the error location in the received sequence.

Concluding, the Berlekamp-Massey algorithm of decoding of BCH codes can be summarized in the following steps.

1. Derive the syndrome components $s_1, s_2, \ldots, s_{2t}$ related to the received sequence described by the polynomial $r(x)$.
2. Apply the received syndrome components $s_1, s_2, \ldots, s_{2t}$ in the Berlekamp-Massey algorithm, which calculates the coefficients of the error location polynomial $\Lambda(x)$.
3. Find the roots of the polynomial $\Lambda(x)$.
4. On the basis of the inverses of the roots of $\Lambda(x)$, determine the error polynomial $e(x)$.
5. Correct the received sequence, i.e. add the error polynomial to the received sequence polynomial $r(x)$.

Let us illustrate the operation of the Berlekamp-Massey algorithm by the following example taken from Lee (2000).

Example 2.7.1 *Consider decoding of the BCH* $(15, 5)$ *code of correction capability of* $t = 3$ *errors. The generator polynomial is* $g(x) = x^{10} + x^8 + x^5 + x^4 + x^2 + x + 1$. *This polynomial has the roots* $\alpha, \alpha^2, \alpha^3, \alpha^4, \alpha^5, \alpha^6$, *where* α *is the primitive element of the field* $GF(2^4)$ *generated by the polynomial* $p(x) = x^4 + x + 1$. *The list of field elements represented as the powers of the primitive element has been shown in Table 2.3. Assume that the zero codeword has been transmitted, i.e.* $c(x) = 0$; *however, the received sequence polynomial has the form* $r(x) = x^{12} + x^5 + x^3$. *In reality it is an error polynomial* $e(x)$, *but this fact is not known to the decoder. In the first phase of the decoding algorithm the syndrome components have to be derived by calculating* $s_i = r(\alpha^i)$, $i = 1, 2, \ldots, 6$. *On the basis of Table 2.3 we get*

$$s_1 = r(\alpha) = \alpha^{12} + \alpha^5 + \alpha^3$$

$$= \begin{pmatrix} 1 \\ 1 \\ 1 \\ 1 \end{pmatrix} + \begin{pmatrix} 0 \\ 1 \\ 1 \\ 0 \end{pmatrix} + \begin{pmatrix} 0 \\ 0 \\ 0 \\ 1 \end{pmatrix} = \begin{pmatrix} 1 \\ 0 \\ 0 \\ 0 \end{pmatrix} = 1$$

The other syndrome elements calculated in a similar way are

$$s_2 = r(\alpha^2) = 1 \qquad s_3 = r(\alpha^3) = \alpha^{10} \qquad s_4 = r(\alpha^4) = 1$$
$$s_5 = r(\alpha^5) = \alpha^{10} \qquad s_6 = r(\alpha^6) = \alpha^5$$

Knowing the syndrome components we have to determine the coefficients of the error location polynomial $\Lambda(x)$, taking advantage of the iterative Berlekamp-Massey algorithm. Table 2.5 presents subsequent steps of this algorithm.

Table 2.5   Results of subsequent steps of the Berlekamp-Massey iterative procedure (Lee 2000)

i	δ	$D(x)$	$\Lambda^*(x)$	$\Lambda(x)$	L
0	−	x	−	1	0
1	$s_1 = 1$	x	$1 + x$	$1 + x$	1
2	0	x^2	$1 + x$	$1 + x$	1
3	α^5	$\alpha^{10}x^2 + \alpha^{10}x$	$\alpha^5x^2 + x + 1$	$\alpha^5x^2 + x + 1$	2
4	0	$\alpha^{10}x^3 + \alpha^{10}x^2$	$\alpha^5x^2 + x + 1$	$\alpha^5x^2 + x + 1$	2
5	α^{10}	$\alpha^{10}x^3 + \alpha^5x^2 + \alpha^5x$	$\alpha^5x^3 + x + 1$	$\alpha^5x^3 + x + 1$	3
6	0	$\alpha^{10}x^4 + \alpha^5x^3 + \alpha^5x^2$	$\alpha^5x^3 + x + 1$	$\alpha^5x^3 + x + 1$	3

Analyze the operation of the algorithm. At the initial moment the variables used by the algorithm are initialized, i.e. $D(x) = x$, $\Lambda(x) = 1$, $L = 0$. In the first iteration the discrepancy δ is nonzero and equal to the first syndrome component $s_1 = 1$. Thus, the connection polynomial is modified by setting $\Lambda^*(x) = 1 - \delta x = 1 + x$. Because the mathematical operations are performed in $GF(2^4)$, subtraction of 4-bit blocks representing the elements of this field is equivalent to modulo-2 addition of their components. Therefore, we will consequently apply signs of mathematical addition. Because $L = 0$, $i = 1$, and so far $2L < i$, the feedback register length is increased ($L := i - L = 1$) and the correction polynomial is updated accordingly, i.e. $D(x) = \Lambda(x)\delta^{-1} = 1$. Next the connection polynomial is also updated, so $\Lambda(x) := \Lambda^*(x) = 1 + x$, and the correction polynomial is changed again, $D(x) := xD(x) = x$. At the end of this iteration the syndrome element counter is increased ($i := i + 1 = 2$). In the next iteration the discrepancy is calculated again: $\delta = s_2 - \sum_{l=1}^{1} \Lambda_l s_{2-l} = 1 - 1 = 0$. Consequently, in this iteration we go directly to step 9 and the feedback register length is not changed because s_2 has been generated correctly by the feedback register in the current form. In turn, updating of the correction polynomial is performed, i.e. $D(x) := xD(x) = x^2$, and the iteration counter is increased by 1. The reader is encouraged to trace the next iterations of this algorithm, Finally, in the sixth iteration the following connection polynomial is received

$$\Lambda(x) = \alpha^5x^3 + x + 1 \tag{2.109}$$

Now the polynomial roots have to be found by substituting subsequent nonzero elements of Galois field $GF(2^4)$ into equation (2.109) and checking if the result is equal to zero. It turns out that the polynomial roots are α^3, α^{10} and α^{12}. However, we are interested in their inverses because the latter, expressed as the powers of the primitive elements, indicate the *error locations in the received sequence. The calculated root inverses are α^{12}, α^5 and α^3, respectively. The error polynomial $e(x)$ achieves the form $e(x) = x^{12} + x^5 + x^3$. The final step of the decoder is correcting the errors by performing the following polynomial calculation*

$$\widehat{c}(x) = r(x) + e(x) = (x^{12} + x^5 + x^3) + (x^{12} + x^5 + x^3) = 0$$

As we can see, the final decoder decision is correct.

The example presented above illustrates the process of decoding of a binary code. However, the Berlekamp-Massey algorithm is often applied for decoding nonbinary BCH codes, such as Reed-Solomon codes. In the tutorial by Michelson and Levesque (2003) the example for RS code decoding is presented.

One could wonder why such a complicated decoding method is used. The answer is simple. If the error correction capability is relatively large (e.g. if six errors can be corrected) and the codewords are long, then nonalgebraic decoding methods become too complex. Direct search of the solution of the nonlinear equation set by checking all nonzero elements of the field $GF(p^m)$ exhausting the set of all possible error combinations is also too complex. Thus, the Berlekamp-Massey algorithm becomes one of possible solutions for decoding such codes.

The Berlekamp-Massey algorithm is a classical solution of the BCH code decoding. In recent years many new soft-decision decoding algorithms have been developed; however, their presentation is beyond the scope of this introductory chapter.

2.8 Convolutional Codes and Their Description

Convolutional codes have become an important class of error correction codes owing to their simplicity, high coding gains and an effective decoding method invented by Viterbi (1967). Nowadays the convolutional codes are often applied as channel codes in digital communication systems. These codes can be found in digital TV broadcasting, cellular GSM radio systems, cellular spread spectrum systems (such as cdmaOne, cdma2000 and UMTS), wireless local area networks (WLANs) and others.

Considering code classification, we mentioned that a convolutional code encoder is a finite state machine featuring a certain number of memory cells and generating the output sequence depending on the input sequence and the current contents of the memory cells. Below we present a few approaches for describing convolutional codes that clearly result from the above observation.

2.8.1 Convolutional Code Description

There exist a few ways to describe convolutional codes. The simplest one is to present the encoder scheme. Figure 2.17a shows an exemplary encoder, featuring two memory cells, of the code with the code rate $R = 1/3$. For each information bit supplied to the encoder input, three output bits are generated. Each of them appears in each one-third of the input bit duration. The first bit is the input bit, the second bit is a modulo-2 sum of the current input bit and the one delayed by two time instants, and the third bit is a modulo-2 sum

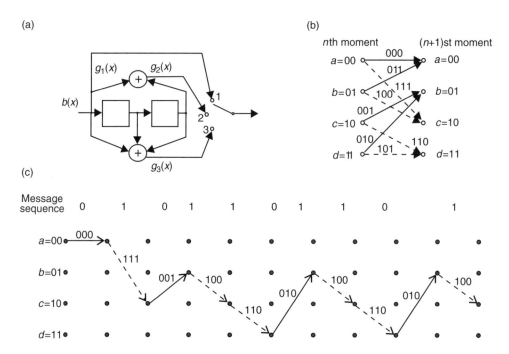

Figure 2.17 Example of the convolutional encoder (a), its trellis diagram (b) and an example of the path on the trellis diagram implied by a specific message sequence (c)

of the current input bit, the preceding bit and the bit delayed by two time instants. The number of input bits that participate in the generation of the output bits determines the size of encoder memory and is called the *constraint length* of a code. It is often denoted by L. An encoder, as in the case of block codes, is frequently described by defining its generator polynomials. For the encoder shown in Figure 2.17a these polynomials have the form

$$g_1(x) = 1 \quad g_2(x) = 1 + x^2 \quad g_3(x) = 1 + x + x^2 \qquad (2.110)$$

Let us write the input sequence in the polynomial form

$$b(x) = b_0 + b_1 x + b_2 x^2 + b_3 x^3 + \dots \qquad (2.111)$$

As we can see, for convolutional codes the input sequence can be infinitely long, although in practical systems its length results from the higher layer system structures, e.g. the information frame length. The polynomials describing the outputs of subsequent encoder branches are

$$w_1(x) = b(x)g_1(x) \quad w_2(x) = b(x)g_2(x) \quad w_3(x) = b(x)g_3(x) \qquad (2.112)$$

Then it is clear that the branch output signals are convolutions of the input sequence and the binary sequences determined by the taps of each encoder branch, which at the same time can be interpreted as binary impulse responses of each encoder branch. Next, the branch output signals are multiplexed, so the joint binary impulse response can be determined for the encoder. This response can be easily generated, feeding the encoder input with the sequence $(1000000\ldots)$ and observing the encoder output. It can be easily verified that for the considered encoder the impulse response is

$$\mathbf{h} = (111\ 001\ 011\ 000\ 000\ldots)$$

Let us note that equations (2.112), which determine the output signals, indicate that the encoder is a linear system, i.e. the superposition principle holds for it. For that reason, based on the encoder impulse response the generator matrix can be constructed for the considered code. This matrix consists of rows, each of which is a replica of the one directly above it, shifted by three positions to the right. The first row exactly describes the encoder impulse response. For the considered code the generator matrix has the form

$$G = \begin{bmatrix} 111 & 001 & 011 & 000 & 000 & \ldots & \ldots & \ldots \\ 000 & 111 & 001 & 011 & 000 & 000 & \ldots & \ldots \\ 000 & 000 & 111 & 001 & 011 & 000 & 000 & \ldots \\ \ldots & \ldots & \ldots & \ldots & \ldots & \ldots & \ldots & \ldots \\ \ldots & \ldots & \ldots & \ldots & \ldots & \ldots & \ldots & \ldots \end{bmatrix} \qquad (2.113)$$

As we can see, the generator matrix is infinite, which corresponds to infinitely long input sequences. It would be a regular matrix if finite input sequences were considered. A codeword generated by the considered encoder fulfills the expression

$$\mathbf{c}^T = \mathbf{b}^T G \qquad (2.114)$$

where $\mathbf{b}$ is the vector of a message block, possibly infinitely long, and $\mathbf{c}$ is a codeword.

It is worth mentioning that polynomial description is only one possible description of the encoder structure. An equivalent description is based on binary vectors assigned to each output branch for which the presence of the tap in a given branch is denoted by a binary "1". In such notation the considered code is determined by the vectors

$$\mathbf{g}_1 = [100] \quad \mathbf{g}_2 = [101] \quad \mathbf{g}_3 = [111] \qquad (2.115)$$

In turn, in convolutional code tables often found in books on coding theory one can find encoder description in which the tap vectors are presented in an octal form. Thus, our code would be denoted by the triple $(4, 5, 7)$.

Figure 2.18 presents an example of another convolutional code. This code is applied in the American cellular system conforming to the ANSI standard IS-95. It has the code rate $R = 1/2$ and the constraint length $L = 9$, and the generator polynomials are given by

$$g_1(x) = 1 + x + x^2 + x^3 + x^5 + x^7 + x^8$$
$$g_2(x) = 1 + x^2 + x^3 + x^4 + x^8 \qquad (2.116)$$

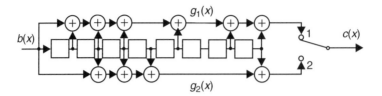

Figure 2.18 Convolutional code encoder applied in the IS-95 cellular telephony (cdmaOne)

In binary notation they are

$$\mathbf{g}_1 = [111101011] \quad \mathbf{g}_2 = [101110001] \tag{2.117}$$

whereas in octal notation the code is denoted as $(753, 561)$.

Let us analyze Figure 2.17b, which shows a trellis diagram for the encoder shown in Figure 2.17a. A trellis diagram is a kind of state diagram, which is a convenient tool for description of finite state machines. A trellis diagram differs from the state diagram in the presentation of the encoder. The trellis diagram presents the encoder (automaton) operation between the nth and $(n + 1)$st moment. The encoder states are determined by the contents of its memory cells and are denoted by characters a, b, c and d. So, state a is related to zeros in both memory cells, state b is related to zero in the first cell and 1 in the second cell, state c to 1 in the first cell and zero in the second cell, and finally state d to "1"s in both memory cells. For each trellis state at the nth moment there are paths to the selected states at the $(n + 1)$st moment. If these paths result from the zero symbol given to the encoder input they are denoted by a vector with a solid line, otherwise a dashed line is used. The path vectors are accompanied by the encoder output symbols above them. Figure 2.17c presents a whole route on the trellis diagram passed by the encoder as a result of a certain input symbol sequence. This route is uniquely associated with the message sequence. Let us note that the decoding task can be interpreted as finding a route on the trellis diagram passed by the encoder that has been caused by a given input sequence. Finding this route is equivalent to the determination of the message sequence the decoder is looking for.

There exists another equivalent means of graphical presentation of the encoder operation, the so-called *code tree diagram*. The code tree diagram for the considered code is presented in Figure 2.19 and has been created in the following way. The tree grows from a root related to the initial moment and the zero encoder state. A branch growing in the upper direction corresponds to the zero symbol given to the encoder input, whereas a branch growing in the lower direction reflects the symbol 1 at the encoder input. At subsequent moments the tree diagram illustrating all possible input symbol sequences grows in upper and lower directions. The current encoder state is denoted under each branch, whereas the output sequence generated at this moment is denoted above that branch. A characteristic feature of a tree diagram is the recurrence of its blocks. The size of these repeatable blocks depends on the encoder memory, i.e. on the number of possible encoder states. Supplying the encoder with a certain input symbol sequence results in a certain route along the branches of the tree diagram. Similarly to the case when a trellis diagram is used, decoding of the received sequence can be interpreted as finding the most likely

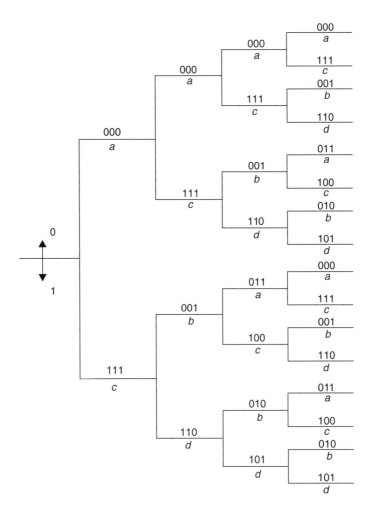

Figure 2.19 Tree diagram for the code shown in Figure 2.17

route on the tree diagram. There exist algorithms of convolutional code decoding in which the above rule is applied.

2.8.2   *Code Transfer Function*

A code transfer function is a very useful tool in investigation of the properties of a given code and evaluation of its decoding error rate. If a code is linear then, similarly as for block codes, its properties, in particular the minimum distance, can be found by considering transmission of the zero codeword and testing how much the received sequence differs from it. In order to derive the code transfer function we apply a traditional state diagram. Figure 2.20a presents a state diagram of the code shown in Figure 2.17. Its form is equivalent to the trellis diagram. Figure 2.20b shows, in turn, the modified state diagram in which the state *a* related to zeros in the encoder memory cells is split into two states, *a* and *e*. Assuming the transmission of the zero codeword, the encoder does not leave state *a*, performing a loop on the state diagram that begins and ends in state *a*. On the basis of the received sequence the decoder should track the state sequence and decide that the encoder remained all the time in state *a*. However, due to channel distortions the received sequence can differ from the all-zero sequence and the decoder can make wrong decisions. The splitting of state *a* into two states, *a* and *e*, allows tracking of a possible divergence from the correct state *a* and merging with it again after a few time instants to reach state *e*. The modified diagram shown in Figure 2.20 can be treated as a graph with the nodes determined by possible encoder states that are connected by the branches with appropriately selected transfer functions, as in electrical circuit analysis. A branch transfer function may consist of three symbols: *J*, *N* and *D* raised to appropriate powers. Symbol *J* appears in each branch. It is introduced to enable calculation of the number of steps in which diverging from and subsequently merging the all-zero route (or approaching state *a*) is possible. Symbol *N* is placed in the transfer function of those branches that result from feeding an information symbol 1 to the encoder input. The power of symbol *D* is, in turn, determined by the Hamming weight of the codeword associated with the path between two states. For example, transfer between state *a* and state *c* results from giving a single 1 to the encoder input (symbol *N* appears in the branch transfer function) and generation of the output sequence 111, whose Hamming weight is equal to 3. Consequently, this branch transfer function is JND^3.

A transfer function is meant as a transmittance between nodes *a* and *e*. Denoting it as $T(D, N, J)$, we have

$$T(D, N, J) = \frac{X_e}{X_a} \qquad (2.118)$$

where X_e is a signal seen in node *e* and X_a is a signal given to node *a*. We can determine the equation describing signals in each state. Thus, for the considered code we receive

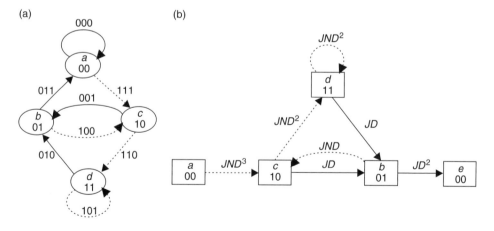

Figure 2.20 State diagram (a) and diagram used for derivation of the transfer function for the code shown in Figure 2.17 (b)

the following equation set

$$X_c = JND^3 X_a + JNDX_b$$

$$X_b = JDX_c + JDX_d$$

$$X_d = JND^2 X_c + JND^2 X_d$$

$$X_e = JD^2 X_b \qquad\qquad (2.119)$$

The transfer function $T(D, N, J)$ can be determined from these equations in a traditional way or using the Mason rule (known from the circuit theory). The solution, using any of these methods, leads to the following formula

$$T(D, N, J) = \frac{J^3 N D^6}{1 - JND^2(1 + J)} \qquad\qquad (2.120)$$

Let us note that the transfer function has a form similar to the infinite sum of a geometrical series $a_0, a_0 q, a_0 q^2, a_0 q^3, \ldots$ that is described by the formula

$$S = \frac{a_0}{1 - q} \qquad\qquad (2.121)$$

In our case $a_0 = J^3 N D^6$, and $q = JND^2(1 + J)$. Therefore the transfer function may be expressed as a sum of infinite geometrical series

$$T(D, N, J) = J^3 N D^6 + J^4 N^2 D^8 + J^5 N^2 D^8 + J^5 N^3 D^{10}$$

$$+ 2J^6 N^3 D^{10} + J^7 N^3 D^{10} + \cdots \qquad\qquad (2.122)$$

Let us interpret the derived formula. The subsequent components of the series expansion are related to the given paths diverging from the all-zero route and merging with it again later. The first expansion component is associated with the route performed in three cycles (J^3), which, if selected by the decoder, would be associated with a single "1" in the decided information sequence (N), whereas the Hamming distance of the zero path and the code sequence associated with this route, i.e. between states a and e, would be $d_H = 6$ (D^6). We say that the code has a *free distance* $d_{\text{free}} = 6$. Analysis of the trellis diagram in Figure 2.17c indicates that this component is related to the path $(a, c, b, a = e)$. Consider now the second expansion term equal to $J^4 N^2 D^8$. The path diverging from the all-zero path and merging with it again is performed in four cycles. It would be associated with two "1"s in the decided sequence if this path were selected by the decoder. The Hamming weight of the decided code sequence would be equal to 8. We can easily show that this route is determined by the sequence of states $(a, c, d, b, a = e)$.

Let us note that the transfer function in the series expansion form allows us to inspect the code weighting structure and indirectly helps us to determine one of the most important code parameters, i.e. the free distance d_{free}. In the case of the considered code the free distance is equal to 6 and there is only a single path diverging from and merging with the all-zero path with such weight. The higher the number of minimum weight paths, the higher the mean error probability of the decoder. The maximum likelihood decoder will select a wrong nonzero path of the lowest weight if at least three errors occur in appropriate positions within the fragment of the code sequence of length 9 bits corresponding to three cycles of the input signals.

2.8.3 Convolutional Codes with Rate k/n

The code that we used in our considerations was characterized by the code rate $R = 1/n$, where $n = 3$. Transmission systems often require other code rates. In general, we are interested in application of a convolutional code of the code rate $R = k/n$. There are two method of achieving codes with such a code rate. The first one is a construction of such an encoder that accepts k new information bits in each clock cycle with simultaneous generation of n bits of a code sequence. An example of such an approach is shown in Figure 2.21. The encoder with the code rate $R = 2/3$ has two parallel delay lines storing a pair of information bits. In each cycle $n = 3$ code sequence bits are generated. Let us note that at the ith moment the encoder can evolve from each state of the trellis diagram to one of four states at the $(i + 1)$st moment.

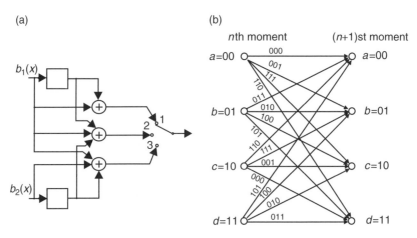

Figure 2.21 Scheme of the convolutional code encoder of the code rate $R = 2/3$ (a) and the associated trellis diagram (b)

The second approach is known as code *puncturing*. It relies on skipping some selected code bits according to the prescribed pattern. This method is often applied if the system requirements strongly depend on the channel conditions or if there is a long list of required data rates and several levels of decoding quality. Figure 2.22a presents the scheme of the so-called RCPC (*Rate Compatible Punctured Convolutional Code*) encoder with the code rate $R = 4/5$. Its core is a regular convolutional code encoder with the code rate $R = 1/2$. Some bits produced by the RCPC encoder are periodically omitted, which is reflected in the *puncturing table* of the following form

$$\mathbf{a} = \begin{bmatrix} 1 & 1 & 1 & 0 \\ 1 & 0 & 0 & 1 \end{bmatrix} \qquad (2.123)$$

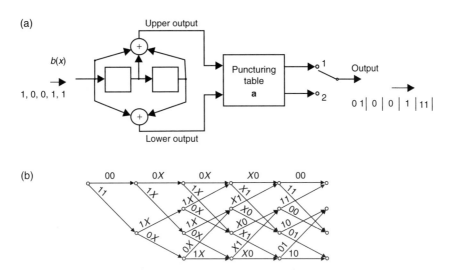

Figure 2.22 RCPC encoder of the code rate $R = 4/5$ with exemplary input and output sequences for the puncturing table given by (2.123) (a) and the corresponding trellis diagram (b)

The symbol 1 in table **a** denotes a transfer of the $R = 1/2$ encoder output bit to the RCPC encoder output, whereas the symbol 0 indicates that the bit generated by the $R = 1/2$ encoder is omitted. So for every four encoder input bits there are five RCPC encoder output bits and we conclude that the code rate is $R = 4/5$. Figure 2.22a also presents the response of the RCPC encoder to the specific information input sequence. The advantage of bit puncturing is the application of a single encoder with the code rate $R = 1/n$ supplemented with an easily modifiable puncturing table. Figure 2.22b, in turn, shows the trellis diagram of the code modified by the appropriate puncturing. The punctured bits have been denoted by X. Let us note that the number of trellis chain elements shown in the trellis diagram must match the number of columns in the puncturing table. As each column of the puncturing table reflects one encoder cycle, the operation of the encoder repeats after the number of cycles, which is equal to the number of columns in **a**.

2.9   Convolutional Code Decoding

There are several known methods of convolutional code decoding. The oldest ones are similar to some block code decoding methods applying a syndrome. However, convolutional codes gained popularity partially due to an efficient decoding algorithm invented by Viterbi. This performs maximum likelihood decoding both in hard- and soft-decision versions.

2.9.1   Viterbi Algorithm

Decoding of convolutional codes is historically the first application of the Viterbi algorithm. Other applications of this algorithm will be discussed in Chapters 3 and 6. The

Viterbi algorithm was first published in 1967 (Viterbi 1967). As we have mentioned, it implements maximum likelihood decoding. We have shown in the previous chapter that for hard-decision decoding this criterion reduces to selection of the codeword that is the closest in the Hamming distance sense to the received sequence. If the distance between two blocks of the same length is defined in another way, soft-decision decoding is realized. Comparison of the received sequence with all possible codewords of the same length and finding the closest one becomes infeasible already at moderate sequence lengths. Thus, there is a need for a method that will do it in an effective way, e.g. by using in the current comparisons the results of comparisons of those parts of the codewords with the appropriate parts of the received sequence obtained previously. Let us note that selection of the optimum codeword performed by the maximum likelihood decoder is equivalent to finding the sequence of states in which the convolutional code encoder subsequently resided while generating this codeword. The algorithm is based on the following core observation: The shortest route at the nth moment (in the sense of the selected metric measuring the distance between the received sequence and the hypothetical codeword) to the appropriate state (e.g. the kth one) consists of the shortest path to one of the states (e.g. the jth) at the $(n-1)$st moment (called a *survivor*) and the path from this state to the considered state at the nth moment. Consequently, each route to the considered kth state from the jth state that does not contain the shortest route to the jth state at the $(n-1)$st moment will feature a higher metric, so it will not be optimal. The following fundamental conclusion results from the above observation.

Conclusion 2.9.1 *If at the nth moment the shortest routes (called survivors) to each of the trellis states are known, then the shortest route to each trellis state at the $(n+1)$st moment can be determined by searching for the path to the currently considered state from one of the states at the previous moment, for which the sum of the state survivor metric at the previous moment and the path metric from that state at the previous moment to the currently considered state is minimum.*

Thus, searching for the sequence of states on the trellis diagram that is associated with the codeword closest to the received sequence is a recurrent process in which the results obtained at the previous time instant are used in the next time instants. Consequently the algorithm is computationally efficient and the number of operations depends linearly on the codeword length. However, it strongly depends on the number of trellis states.

The Viterbi algorithm can be divided into two phases. The first one is its initialization and lasts until the time instant in which the metrics of the paths to all trellis states are determined. Assuming the convolutional code of the constraint length equal to L, the initialization phase lasts for $L-1$ algorithm cycles. At the Lth moment the second phase starts in which the regular algorithm begins. In this phase, selection of the shortest route to each trellis state is performed. As we have mentioned, two cases are possible. In the first one the codeword is finite. Sometimes the generated codeword is appended by the bit sequence of such length that uniquely determines the final encoder state. Thus, when decoding the received sequence, the decoder starts and ends its operation in a known (usually zero) trellis state. In the second case the decoded sequence is so long that it can be considered as almost infinite. Waiting for decoding of the whole sequence is not feasible any more and a partial decision upon a part of the decoded sequence is necessary, along with extension of the shortest routes to each trellis state in each timing instant.

The Viterbi algorithm for a code of constraint length equal to L and the number of states equal to 2^{L-1} can be formulated in the following steps:

1. *Algorithm initialization* at the moments $i = 1, 2, \ldots, L-1$. Calculate path metrics to each trellis state determining the distances between the received sequence and the codewords associated with the given path on the trellis diagram. As a result of initialization, unique paths from the initial state at the zero moment to each trellis state $k = 1, 2, \ldots, 2^{L-1}$ denoted by $D_k^{L-1} = (\sigma_{k_1}, \sigma_{k_2}, \ldots, \sigma_{k_{L-1}})$ are found. Each path is associated with the accompanied metric M_k^{L-1}. The subscript is related to the state number, whereas the superscipt denotes the current timing instant. In turn, σ_{k_i} symbolizes the state in which the encoder is found at the ith moment when travelling through the path ending in the kth state at the $(L-1)$st moment.

2. *Recurrent phase* (at the moments $i = L, L+1, \ldots$). Knowing the shortest routes D_k^i to each trellis state $k = 1, 2, \ldots, 2^{L-1}$ at the ith moment and the metrics M_k^i associated with them, determine the shortest route and the metric associated with it for each trellis state at the $(i+1)$st moment according to the following rule. Let the kth state at the ith moment be reachable from the states indexed by j_1 and j_2 at the $(i-1)$st moment. The shortest route to the kth ($k = 1, 2, \ldots, 2^{L-1}$) state at the ith moment has the metric

$$M_k^i = \min_{\{j_1, j_2\}} \left\{ \left[M_{j_1}^{i-1} + d(\mathbf{r}_i, \mathbf{c}_{j_1,k}) \right], \left[M_{j_2}^{i-1} + d(\mathbf{r}_i, \mathbf{c}_{j_2,k}) \right] \right\} \tag{2.124}$$

where $d(\mathbf{r}_i, \mathbf{c}_{j_m,k})$ ($m = 1, 2$) denotes the Hamming (or other) distance between the received sequence $\mathbf{r}_i$ at the ith moment and the codeword associated with the path on the trellis diagram between the j_mth state at the $(i-1)$st moment and the kth state at the ith moment. The choice of state with the index $\widehat{j} = j_1$ or j_2 from the previous moment determines the shortest route to the kth state at the ith moment

$$D_k^i = (\sigma_{\widehat{j_1}}, \sigma_{\widehat{j_2}}, \ldots, \sigma_{\widehat{j_{i-1}}}, \sigma_k) \quad \text{for } k = 1, 2, \ldots, 2^{L-1} \tag{2.125}$$

The path, described by formula (2.125) containing the sequence of states from the initial moment to the current one, is uniquely associated with the appropriate codeword and the information symbol sequence.

It has been noticed that the shortest routes to all trellis states starting in the zero timing instant are common, with the probability close to unity up to the moment delayed by $3L$ to $5L$ timing instants, as compared with the current timing instant. Therefore, producing finite decisions upon the partially decoded sequence is possible with such a delay. It is a particularly important observation in the case of long codewords. For such codewords waiting for their end would require an unacceptably long delay. The delay in obtaining the final decision from the decoder is called a *decoding depth*.

Consider the operation of the Viterbi decoder on the example of the code shown in Figure 2.17a. Figure 2.23 presents the trellis diagram achieved as a result of the algorithm operation after three different numbers of decoder cycles. Figure 2.23a shows the end of the initialization phase. Each state diagram can be reached by only one route from the initial state. The path metric is the Hamming distance between the received sequence and the codeword associated with the given route finishing at the appropriate state. The

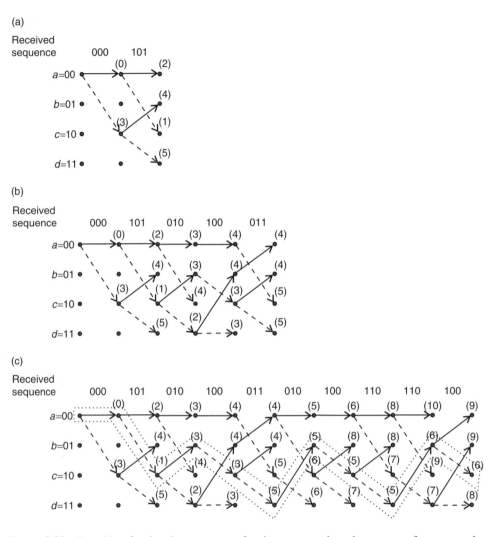

Figure 2.23 Searching for the shortest routes for the states at the nth moment after two cycles (a), five cycles (b) and ten cycles (c)

recurrent phase of the algorithm starts in the third timing instant. From this moment the selection of the shortest route to each trellis state is performed. As an example, let us consider determining the shortest routes to each state in the third timing instant. After finishing the initialization phase we have the following metrics

$$M_1^2 = 2, \quad M_2^2 = 4, \quad M_3^2 = 1, \quad M_4^2 = 5$$

and corresponding sequences of states indicating the routes to each state are

$$D_1^2 = (a, a, a), \quad D_2^2 = (a, c, b), \quad D_3^2 = (a, a, c), \quad D_4^2 = (a, c, d)$$

At the third moment a sequence $\mathbf{r}_3 = (010)$ has been received. Let us start from selection of the shortest route to state a. It is possible to achieve it from states a or b. The path from state a to state a at the next moment is associated with the code sequence 000, whereas the path from state b to a results in the code sequence 011. As a result, the Hamming distances between the candidate code sequences and the received sequence are, respectively,

$$d(\mathbf{r}_3, \mathbf{c}_{a,a}) = d(010, 000) = 1 \quad d(\mathbf{r}_3, \mathbf{c}_{b,a}) = d(010, 011) = 1$$

Thus, the minimum cost of approaching the state a at the third cycle is

$$M_1^3 = \min_{a,b} \left\{ \left[M_1^2 + d(\mathbf{r}_3, \mathbf{c}_{a,a}) \right], \left[M_2^2 + d(\mathbf{r}_3, \mathbf{c}_{b,a}) \right] \right\}$$
$$= \min_{a,b} [(2+1), (4+1)] = 3$$

The shortest route to state a in the third timing instant leads through state a from the second timing instant, so

$$D_1^3 = (a, a, a, a)$$

Performing similar steps for the remaining states, we receive

$$M_1^3 = 3, \quad M_2^3 = 3, \quad M_3^3 = 4, \quad M_4^3 = 2$$

and

$$D_1^3 = (a, a, a, a), \quad D_2^3 = (a, a, c, b), \quad D_3^3 = (a, a, a, c), \quad D_4^3 = (a, a, c, d)$$

In each timing instant similar calculations to those shown above are performed. Figure 2.23c presents the result of the Viterbi algorithm operation after ten cycles. We can find the route featuring the lowest metric at the final moment. It has been denoted by an envelope drawn with dashed lines. If only 30 bits were transmitted, then the code sequence associated with this route would be the decoder decision upon the transmitted codeword and information sequence.

The Viterbi algorithm shown above performs hard-decision decoding. Let us recall, however, that in current digital communication systems mostly soft-decision decoding is applied. In the simplest version, binary sequences at the decoder input are replaced by the samples (e.g. 8-level) of the received symbols. The operation of the Viterbi algorithm differs from that shown above only in definition of the applied distance. It can be a Euclidean distance or another distance, e.g. that expressed by formula (2.10).

A convolutional code is often used as an *inner code* in a coding system in which two codes are applied in a cascade. To enable applying soft-decision decoding in an *outer code* the convolutional code decoder should not only generate binary decisions but also provide the measure of their reliability. The algorithm that supplies both values is called the SOVA (*Soft-Output Viterbi Algorithm*) (Hagenauer and Hoeher 1989) and will be the subject of our considerations in the next section.

The Viterbi algorithm has a very regular structure, which makes it very attractive for hardware implementation. Many operations can be performed in parallel, which results in a high decoding speed. For example, each state can be served by a separate circuit that calculates the metrics of the routes leading to it and by selecting the shortest one among them.

The Viterbi algorithm is not the only algorithm for convolutional code decoding, although it is certainly the most popular. The Fano algorithm is another famous algorithm. Similarly to the Viterbi algorithm, it finds the optimal code sequence analyzing the sequence of the hypothetical encoder states, so it belongs to the sequential algorithms. However, unlike the Viterbi algorithm, the basis for the search for the state sequence is a code tree diagram. In general, the algorithm measures the distance between the received sequence and the code sequence associated with the given route on the code tree along which the algorithm "moves". In each cycle the route is extended along a single tree branch only. It is done as long as the distance between the associated codeword and the received sequence does not exceed a predetermined threshold value. If the threshold value is reached, the algorithm goes a few steps back and tries to select another route. Thus, the delay introduced by the Fano algorithm is random. If the symbol error probability is low, the event of going back and selecting a new route is rare. However, the randomness of the decoding delay is a disadvantage of this algorithm.

If the number of encoder states grows, the Viterbi algorithm becomes so complicated that suboptimal solutions must be sought. The number of required calculations stays under control at the expense of a certain loss in the decoding performance. Suboptimal algorithms are commonly used when the Viterbi algorithm is used to detect the signals corrupted by intersymbol interference. This problem will be discussed further in Chapter 6.

2.9.2 Soft-Output Viterbi Algorithm (SOVA)

As we remember, the regular Viterbi algorithm decides about the transmitted codeword upon the received sequence by using the maximum likelihood criterion. Our considerations on the SOVA will be presented in a wider perspective using the *Maximum a Posteriori* (MAP) criterion applied in the decision process for the whole received sequences. As in the regular Viterbi algorithm, the algorithm will find the optimal codeword but, unlike the latter, possibly unequal probabilities of the codewords are taken into account as well. Equivalently, unequal probabilities of particular message sequences influence the decoder operation.

Assume the channel model shown in Figure 1.19a. Thus, the codeword $\mathbf{c}_1^i$ transmitted from the initial moment up to the ith time unit is represented by a bipolar sequence $\mathbf{d}_1^i$ that, in the case of the convolutional code of code rate $R = 1/n$, has the form

$$\mathbf{d}_1^i = (\mathbf{d}_1, \mathbf{d}_2, \ldots, \mathbf{d}_i) \tag{2.126}$$

The jth vector element in (2.126) is a vector of bipolar symbols characterizing the codeword generated in the jth time unit

$$\mathbf{d}_j = (d_{j,1}, d_{j,2}, \ldots, d_{j,n}), \ d_{j,k} = \pm\sqrt{E_c}, \quad j = 1, \ldots, i, \quad k = 1, \ldots, n \tag{2.127}$$

where E_c is the signal energy per single code symbol. Let $d_{j,k} = -\sqrt{E_c}$ if $c_{j,k} = 0$ and let $d_{j,k} = \sqrt{E_c}$ if $c_{j,k} = 1$. Let us also note that assuming a particular initial state of the encoder, both the codeword $\mathbf{c}_1^i$ and its bipolar version $\mathbf{d}_1^i$ are uniquely associated with the message sequence

$$\mathbf{m}_1^i = (m_1, m_2, \ldots, m_i) \qquad (2.128)$$

The transmitted vector $\mathbf{d}_1^i$ is subject to disturbance by additive white Gaussian noise, so at the decoder input it has the form

$$\mathbf{r}_1^i = (\mathbf{r}_1, \mathbf{r}_2, \ldots, \mathbf{r}_i) \qquad (2.129)$$

where

$$\mathbf{r}_j = (r_{j,1}, r_{j,2}, \ldots, r_{j,n}), \quad r_{j,k} = d_{j,k} + v_{j,k}, \quad j = 1, \ldots, i, \quad k = 1, \ldots, n \qquad (2.130)$$

and $v_{j,k}$ is a white Gaussian noise sample added to the kth element of the bipolar codeword symbol in the jth time unit. As the Gaussian noise source is white, any different noise samples are statistically independent. Let us now formulate the MAP criterion for finding the codeword $\mathbf{c}_1^i$ on the basis of the received sequence $\mathbf{r}_1^i$ or, equivalently, finding the sequence $\mathbf{d}_1^i$, both uniquely associated with the transmitted message sequence $\mathbf{m}_1^i$. The codeword $\mathbf{c}_{1,\text{opt}}^i$ or its bipolar version $\mathbf{d}_{1,\text{opt}}^i$ is searched according to the MAP criterion, which results from the maximized *a posteriori* probability

$$\mathbf{d}_{1,\text{opt}}^i = \arg\max_{\mathbf{d}_1^i} P(\mathbf{d}_1^i | \mathbf{r}_1^i) \qquad (2.131)$$

Recalling Bayes' formula, we have

$$\mathbf{d}_{1,\text{opt}}^i = \arg\max_{\mathbf{d}_1^i} P(\mathbf{d}_1^i | \mathbf{r}_1^i) = \arg\max_{\mathbf{d}_1^i} \frac{p(\mathbf{r}_1^i | \mathbf{d}_1^i) P(\mathbf{d}_1^i)}{p(\mathbf{r}_1^i)}$$

$$= \arg\max_{\mathbf{d}_1^i} p(\mathbf{r}_1^i | \mathbf{d}_1^i) P(\mathbf{d}_1^i) = \arg\max_{\mathbf{d}_1^i} p(\mathbf{r}_1^i | \mathbf{d}_1^i) P(\mathbf{m}_1^i) \qquad (2.132)$$

We have used the observation that the denominator in Bayes' formula is common for all possible bipolar codewords $\mathbf{d}_1^i$, so it does not influence the choice of the best codeword. We also applied the fact that the probability of the codeword $\mathbf{d}_1^i$ is equal to the probability of the message sequence $\mathbf{m}_1^i$. Instead of comparing the probabilities we can compare their logarithms, so the MAP criterion evolves to the form

$$\mathbf{d}_{1,\text{opt}}^i = \arg\max_{\mathbf{d}_1^i} \ln \; p(\mathbf{r}_1^i | \mathbf{d}_1^i) P(\mathbf{m}_1^i) \qquad (2.133)$$

Let us consider the term that is the subject of maximization in detail. Because noise samples are statistically independent, we can write this term in the form

$$\ln \ p(\mathbf{r}_1^i|\mathbf{d}_1^i)P(\mathbf{m}_1^i) = \ln \prod_{l=1}^{i} p(\mathbf{r}_l|\mathbf{d}_l)P(m_l)$$

$$= \ln \prod_{l=1}^{i} \left[\prod_{k=1}^{n} p(r_{l,k}|d_{l,k}) \right] P(m_l) \qquad (2.134)$$

The inner product reflects the conditional probabilities of particular n samples received within the lth timing instant. We have assumed in (2.134) that subsequent message symbols m_l are statistically independent, although their probabilities can have different values. For our convenience we recall the formula describing the conditional probability $p(r_{l,k}|d_{l,k})$ for the white Gaussian noise channel, which is described by the expression

$$p(r_{l,k}|d_{l,k}) = \frac{1}{\sqrt{2\pi}\sigma} \exp\left[-\frac{1}{2\sigma^2}(r_{l,k} - d_{l,k})^2 \right], \quad l = 1, \ldots, i, \quad k = 1, \ldots, n \quad (2.135)$$

where σ^2 is the noise variance. We will prove in Chapter 3 that for the white Gaussian noise channel and the optimum receiver $\sigma^2 = N_0/2$ (where $N_0/2$ is the power spectral density of additive white Gaussian noise on the receiver input). After substituting (2.135) in (2.134) we receive

$$\ln \ p(\mathbf{r}_1^i|\mathbf{d}_1^i)P(\mathbf{m}_1^i) = \ln \left(\frac{1}{\sqrt{2\pi}\sigma} \right)^{ni}$$

$$+ \left[\sum_{l=1}^{i} \left\{ \left[-\frac{1}{2\sigma^2}\sum_{k=1}^{n}(r_{l,k} - d_{l,k})^2 \right] + \ln P(m_l) \right\} \right] \qquad (2.136)$$

The first term of the right-hand side in (2.136) does not depend on the searched codeword. It linearly grows with the length of the codeword, so it does not influence the maximized logarithm of the probability $p(\mathbf{r}_1^i|\mathbf{d}_1^i)P(\mathbf{m}_1^i)$ and can be omitted. Thus, we can write (2.133) in a new form

$$\mathbf{d}_{1,\text{opt}}^i = \arg\max_{\mathbf{d}_1^i} \ln \ p(\mathbf{r}_1^i|\mathbf{d}_1^i)P(\mathbf{m}_1^i)$$

$$= \arg\max_{\mathbf{d}_1^i} \left\langle \sum_{l=1}^{i} \left\{ \left[-\frac{1}{2\sigma^2}\sum_{k=1}^{n}(r_{l,k} - d_{l,k})^2 \right] + \ln P(m_l) \right\} \right\rangle \qquad (2.137)$$

Maximization of the term in curly brackets is equivalent to minimization of the sum of terms that consist of the squared errors between the received sample $r_{l,k}$ and the bipolar symbol $d_{l,k}$ of a hypothetical codeword and the logarithms of the message probabilities m_l ($l = 1, \ldots, i$). Let us note that the noise variance is used in the minimization process and it influences its result. If probabilities of all message symbols are equal, then their logarithms do not influence the choice of the decoded codeword and can be omitted. Consequently, the criterion reduces to the well-known result considered in the previous section

$$\mathbf{d}^i_{1,\text{opt}} = \arg\min_{\mathbf{d}^i_1} \sum_{l=1}^{i}\sum_{k=1}^{n} (r_{l,k} - d_{l,k})^2 \tag{2.138}$$

However, let us come back to the more general case shown in (2.137). Expanding the squared errors in square brackets, we have

$$\sum_{l=1}^{i}\left\{\left[-\frac{1}{2\sigma^2}\sum_{k=1}^{n}(r_{l,k} - d_{l,k})^2\right] + \ln P(m_l)\right\}$$

$$= \sum_{l=1}^{i}\left\{\left[-\frac{1}{2\sigma^2}\sum_{k=1}^{n}(r_{l,k}^2 - 2r_{l,k}d_{l,k} + d_{l,k}^2)\right] + \ln P(m_l)\right\}$$

$$= \sum_{l=1}^{i}\left\{\left[-\frac{1}{2\sigma^2}\sum_{k=1}^{n}(r_{l,k}^2 + d_{l,k}^2) + \frac{1}{\sigma^2}\sum_{k=1}^{n}r_{l,k}d_{l,k}\right] + \ln P(m_l)\right\}$$

$$= \sum_{l=1}^{i}\left[C_l + L_v\sum_{k=1}^{n}r_{l,k}d_{l,k} + \ln P(m_l)\right] \tag{2.139}$$

where

$$C_l = -\frac{1}{2\sigma^2}\sum_{k=1}^{n}(r_{l,k}^2 + d_{l,k}^2) \tag{2.140}$$

is a common term in all possible codewords and does not influence the choice of the decoded codeword. Denote $L_v = 1/\sigma^2$. In consequence, the criterion achieves the simplified form

$$\mathbf{d}^i_{1,\text{opt}} = \arg\max_{\mathbf{d}^i_1}\left\{\sum_{l=1}^{i}\left[L_v\sum_{k=1}^{n}r_{l,k}d_{l,k} + \ln P(m_l)\right]\right\} \tag{2.141}$$

Let us now assume that the message symbols are bipolar as well. Without changing the decoder decision we can add a certain value dependent on the current time index l to each term summed in subsequent time units, i.e. instead of $L_v\sum_{k=1}^{n}r_{l,k}d_{l,k} + \ln P(m_l)$ we write

$$2L_v\sum_{k=1}^{n}r_{l,k}d_{l,k} + 2\ln P(m_l) - \ln\Pr\{m_l = 1\} - \ln\Pr\{m_l = -1\}$$

$$= L'_v\sum_{k=1}^{n}r_{l,k}d_{l,k} + m_l\Lambda(m_l) \tag{2.142}$$

where $L'_v = 2/\sigma^2$ and $\Lambda(m_l)$ is the log-likelihood ratio (LLR) of the symbol m_l, i.e.

$$\Lambda(m_l) = \ln \frac{\Pr\{m_l = 1\}}{\Pr\{m_l = -1\}} \tag{2.143}$$

In deriving (2.142) we used the observation that

$$2 \ln P(m_l) - \ln \Pr\{m_l = 1\} - \ln \Pr\{m_l = -1\}$$

$$= \begin{cases} \ln \Pr\{m_l = 1\} - \ln \Pr\{m_l = -1\} & \text{if } m_l = 1 \\ \ln \Pr\{m_l = -1\} - \ln \Pr\{m_l = 1\} & \text{if } m_l = -1 \end{cases} = m_l \Lambda(m_l) \tag{2.144}$$

Finally, the criterion achieves the useful form

$$\mathbf{d}^i_{1,\text{opt}} = \arg \max_{\mathbf{d}^i_1} \left[M(\mathbf{r}^i_1 | \mathbf{d}^i_1) \right] \tag{2.145}$$

where

$$M(\mathbf{r}^i_1 | \mathbf{d}^i_1) = \sum_{l=1}^{i} \left[L'_v \sum_{k=1}^{n} r_{l,k} d_{l,k} + m_l \Lambda(m_l) \right] \tag{2.146}$$

is a maximized metric. Searching for the best codeword reduces to finding such a code-word (or message sequence) for which the accumulated sum of the cross-correlation between the received samples and the hypothetical codewords in bipolar form weighted by L'_v and the LLRs of the hypothetical message symbols weighted by their bipolar values is maximized. Let us note that metric (2.146) can be calculated recurrently using the formula

$$M(\mathbf{r}^i_1 | \mathbf{d}^i_1) = M(\mathbf{r}^{i-1}_1 | \mathbf{d}^{i-1}_1) + L'_v \sum_{k=1}^{n} r_{i,k} d_{i,k} + m_i \Lambda(m_i) \tag{2.147}$$

The Viterbi algorithm calculates the metric $M(\mathbf{r}^i_1 | \mathbf{d}^i_1)$ for each trellis state in each time unit, trying to determine the survival path to each trellis state s_j ($j = 1, \ldots, 2^{L-1}$). Consider such a calculation for the jth state at the ith moment. Let this state be accessible from states l_1 and l_2 from the previous moment. Denote the survival path metrics for states l_1 as $M_{l_1}(\mathbf{r}^{i-1}_1 | \mathbf{d}^{i-1}_1)$ and $M_{l_2}(\mathbf{r}^{i-1}_1 | \mathbf{d}^{i-1}_1)$, respectively, and the metrics associated with the paths between the pairs of states (s_{l_1}, s_j) and (s_{l_2}, s_j) as

$$d(\mathbf{r}_i, s_{l_1}, s_j) = L'_v \sum_{k=1}^{n} r_{i,k} d^{(l_1,j)}_{i,k} + m^{(l_1,j)}_i \Lambda \left[m^{(l_1,j)}_i \right] \tag{2.148}$$

and

$$d(\mathbf{r}_i, s_{l_2}, s_j) = L'_v \sum_{k=1}^{n} r_{i,k} d^{(l_2,j)}_{i,k} + m^{(l_2,j)}_i \Lambda \left[m^{(l_2,j)}_i \right] \tag{2.149}$$

where $d_{i,k}^{(l_1,j)}$ and $d_{i,k}^{(l_2,j)}$ are the bipolar codeword sequences associated with the path between pairs of states (s_{l_1}, s_j) and (s_{l_2}, s_j), respectively, whereas $m_i^{(l_1,j)}$ and $m_i^{(l_2,j)}$ are the message symbols associated with these paths. Thus, for each trellis state s_j at the ith moment the decoder selects the path for which the following expression holds

$$\max_{(l_1,l_2)} \left\{ M_{l_1,j} \left[\mathbf{r}_1^i \,|\, (\mathbf{d}_1^i)^{(l_1,j)} \right], M_{l_2,j} \left[\mathbf{r}_1^i \,|\, (\mathbf{d}_1^i)^{(l_2,j)} \right] \right\} \tag{2.150}$$

where

$$M_{l_1,j} \left[\mathbf{r}_1^i \,|\, (\mathbf{d}_1^i)^{(l_1,j)} \right] = M_{l_1}(\mathbf{r}_1^{i-1} | \mathbf{d}_1^{i-1}) + d(\mathbf{r}_i, s_{l_1}, s_j)$$

$$M_{l_2,j} \left[\mathbf{r}_1^i \,|\, (\mathbf{d}_1^i)^{(l_2,j)} \right] = M_{l_2}(\mathbf{r}_1^{i-1} | \mathbf{d}_1^{i-1}) + d(\mathbf{r}_i, s_{l_2}, s_j)$$

The vectors $(\mathbf{d}_1^i)^{(l_1,j)}$ and $(\mathbf{d}_1^i)^{(l_2,j)}$ denote the codewords associated with the paths reaching state s_j through states s_{l_1} and s_{l_2}, respectively, and the new survival path metric for the state s_j is

$$M_j(\mathbf{r}_1^i | \mathbf{d}_i^i) = \max_{(l_1,l_2)} \left\{ M_{l_1,j} \left[\mathbf{r}_1^i \,|\, (\mathbf{d}_1^i)^{(l_1,j)} \right], M_{l_2,j} \left[\mathbf{r}_1^i \,|\, (\mathbf{d}_1^i)^{(l_2,j)} \right] \right\} \tag{2.151}$$

The above procedure is illustrated in Figure 2.24. We still need to assign a certain measure of reliability to the decision upon the path selection. This is necessary for generation of a soft decoder output for each message element. It is intuitively clear that if the candidate metrics $M_{l_1,j}(\mathbf{r}_1^i | \mathbf{d}_1^i)$ and $M_{l_2}, j(\mathbf{r}_1^i | \mathbf{d}_1^i)$ do not differ much, then selection of the correct path is unreliable, whereas when there is a large difference between them, the probability of selecting a wrong path is low. In this context let us choose the measure of reliability of reaching the state s_j as

$$\Delta_{i-1}(s_j) = \frac{1}{2} \left\{ M_{l_1,j} \left[\mathbf{r}_1^i \,|\, (\mathbf{d}_1^i)^{(l_1,j)} \right] - M_{l_2,j} \left[\mathbf{r}_1^i \,|\, (\mathbf{d}_1^i)^{(l_2,j)} \right] \right\} \tag{2.152}$$

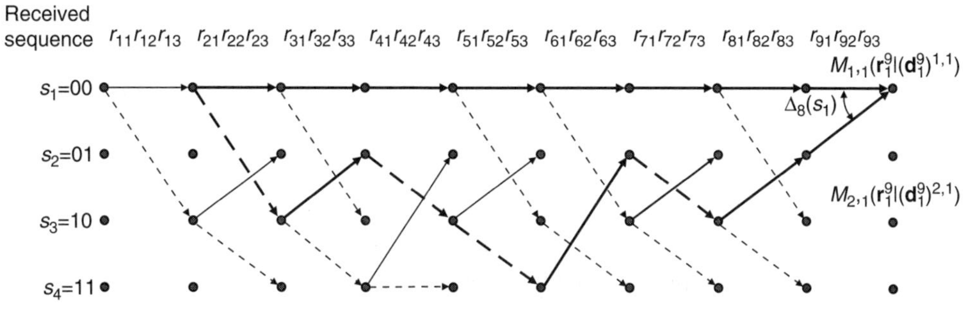

Figure 2.24 Selection of the survival path for state s_1 at the ninth moment, accompanied by calculation of the metric difference $\Delta_8(s_1)$

Let us arbitrarily assume that the correct path is the one that reaches state s_j from state s_{l_1} together with its survival path. Then the probability of the correct path selection can be expressed in the form of the MAP probabilities associated with the candidate paths reaching state s_j; i.e. in the form

$$P_c(s_j) = \frac{P\left[(\mathbf{d}_1^i)^{(l_1,j)} | \mathbf{r}_1^i\right]}{P\left[(\mathbf{d}_1^i)^{(l_1,j)} | \mathbf{r}_1^i\right] + P\left[(\mathbf{d}_1^i)^{(l_2,j)} | \mathbf{r}_1^i\right]} \tag{2.153}$$

Recalling Bayes' theorem, we have

$$P_c(s_j) = \frac{p\left[\mathbf{r}_1^i | (\mathbf{d}_1^i)^{(l_1,j)}\right] P\left[(\mathbf{m}_1^i)^{(l_1,j)}\right]}{p\left[\mathbf{r}_1^i | (\mathbf{d}_1^i)^{(l_1,j)}\right] P\left[(\mathbf{m}_1^i)^{(l_1,j)}\right] + P\left[\mathbf{r}_1^i | (\mathbf{d}_1^i)^{(l_2,j)}\right] P\left[(\mathbf{m}_1^i)^{(l_2,j)}\right]} \tag{2.154}$$

However, our previous analysis allows us to express the probabilities in such a form that the probability of selecting the correct path takes the following shape

$$P_c(s_j) = \frac{C \exp\left\{\frac{1}{2} M_{l_1,j}\left[\mathbf{r}_1^i | (\mathbf{d}_1^i)^{(l_1,j)}\right]\right\}}{C \exp\left\{\frac{1}{2} M_{l_1,j}\left[\mathbf{r}_1^i | (\mathbf{d}_1^i)^{(l_1,j)}\right]\right\} + C \exp\left\{\frac{1}{2} M_{l_2,j}\left[\mathbf{r}_1^i | (\mathbf{d}_1^i)^{(l_2,j)}\right]\right\}} \tag{2.155}$$

where the constant C accumulates all the components in the logarithm domain that do not influence the choice of path [see the first component of (2.136), C_l in (2.139) and (2.140)]. On the other hand, the scaling factor $\frac{1}{2}$ reverts the influence of multiplication of the original metric by 2 performed in (2.143) and (2.144). After multiplying the nominator and denominator of (2.155) by $\exp\left\{-\frac{1}{2} M_{l_2,j}\left[\mathbf{r}_1^i | (\mathbf{d}_1^i)^{(l_2,j)}\right]\right\}$ we obtain

$$P_c(s_j) = \frac{\exp[\Delta_{i-1}(s_j)]}{\exp[\Delta_{i-1}(s_j)] + 1} \tag{2.156}$$

Finally, the log-likelihood ratio or *reliability* of the path decision concerning reaching state s_j at the ith moment is

$$\ln \frac{P_c}{1 - P_c} = \Delta_{i-1}(s_j) \tag{2.157}$$

It still remains to describe how the reliability of a path decision given by (2.157) is associated with the hard-decision output of the Viterbi decoder. As we know, the decoder produces hard decisions $\widehat{m}_i$ and the reliabilities associated with them. Recall that we have assumed that $(\mathbf{d}_1^i)^{(l_1,j)}$ is associated with the correct survival path for state s_j. The codeword $(\mathbf{d}_1^i)^{(l_1,j)}$ is in turn uniquely associated with the message sequence $(\mathbf{m}_1^i)^{(l_1,j)}$, whereas the second competing path is associated with the message sequence $(\mathbf{m}_1^i)^{(l_2,j)}$. The choice of the survival path and the reliability associated with it affects only those positions in the message sequence in which the candidate sequences are different. The calculated reliability becomes important after the initialization phase of the algorithm. Consider the first moment of the regular phase of the algorithm, which occurs at the time unit $i = L$ (L is a code constraint length). At this moment the path selection is performed for the first time for each trellis state and reliabilities $\Delta_{L-1}(s_j)$ are calculated for $j = 1, \ldots, 2^{L-1}$. As we can see, each state s_j is characterized not only by the path metric $M_j(\mathbf{r}_1^i | \mathbf{d}_j^i)$ and the state sequence D_j^i, as in a regular Viterbi algorithm, but also by a reliability vector described by the expression

$$\mathbf{R}_L(s_j) = [R_1(s_j), R_2(s_j), \ldots, R_{L-1}(s_j)] \tag{2.158}$$

where

$$R_l(s_j) = \begin{cases} \Delta_{L-1}(s_j) & \text{if } m_l^{(l_1,j)} \neq m_l^{(l_2,j)} \\ \infty & \text{if } m_l^{(l_1,j)} = m_l^{(l_2,j)} \end{cases}, \quad \text{for } l = 1, \ldots, L-1 \tag{2.159}$$

In this way, the reliability vectors are initialized for each trellis state. In subsequent time instants of the SOVA recurrent phase, the metrics for each survival paths are updated using formula (2.151) and the state sequence vectors are updated accordingly. Additionally, the reliability vectors are modified using the following rule

$$\mathbf{R}_{i+1}(s_j) = [R_1(s_j), R_2(s_j), \ldots, R_i(s_j)] \tag{2.160}$$

where

$$R_l(s_j) = \begin{cases} \min[\Delta_{L-1}(s_j), R_l(s_j)] & \text{if } m_l^{(l_1,j)} \neq m_l^{(l_2,j)} \\ R_l(s_j) & \text{if } m_l^{(l_1,j)} = m_l^{(l_2,j)} \end{cases}, \quad \text{for } l = 1, \ldots, i \tag{2.161}$$

As we can see from (2.161), elements of the reliability vector at the $(i+1)$st moment preserve their previous value if two candidate paths have the same message symbol on the appropriate position. If the message symbols differ, then the minimum of the currently calculated reliability and of the previous vector entry is selected. The operation of the algorithm is completed at the end of the received vector. The algorithm produces the decided message sequence with the attached reliabilities that are the final values of the reliability vector elements for the state featuring the maximum path metric. In the case of very long codewords, for which waiting for the processing of the whole sequence of symbols is infeasible, the appropriately long decoding depth is applied and all processed vectors are truncated to the selected length.

At the end of our considerations let us note the potential role of the *a priori* term $m_l \Lambda(m_l)$ in the metric calculated according to (2.146). If some extra knowledge on the *a priori* probabilities of the message symbols is available, then it can be applied for improving decoding quality as compared with the case in which it is more or less arbitrarily assumed that $\Pr\{m_l = 1\} = \Pr\{m_l = -1\}$ regardless of the real values of these probabilities. This potential improvement ability is utilized in iterative decoding, in which in each decoding iteration the *a priori* LLR term $\Lambda(m_l)$ gets more and more precise. Iterative decoding is a subject of our considerations in one of the next sections of this chapter.

2.9.3 Error Probability Analysis for Convolutional Codes

Consider the problem of decoding of convolutional codes using a hard-decision algorithm, e.g. the Viterbi algorithm. Let us estimate the probability of error of the information symbols on the output of the Viterbi algorithm. The code transfer function appears to be useful for that purpose. Calculate the error probability for the code shown in Figure 2.17a. As we remember, its transfer function is expressed in the form of an infinite series presented by formula (2.122). Substitution of $J = 1$ in it results in the simplified form

$$T(D, N) = ND^6 + 2N^2D^8 + 3N^3D^{10} + \cdots \qquad (2.162)$$

Knowing that the considered code is linear, assume without loss of generality that an all-zero codeword has been transmitted. We say that at a given jth moment an *error event* has occurred if the all-zero path on the trellis diagram has been eliminated in favor of another path merging with the all-zero path at that moment. If the decoder has decided to select the path featuring the Hamming weight $w_H = 6$, then the error event has occurred if, among six positions in which both paths differ, the received sequence agrees with the path of weight $w_H = 6$ in four or more positions. Note that errors occurring in the positions in which both paths do not differ have no influence on the decoder decision, as they equally increase the distance of the received sequence from codewords associated with both candidate paths. Let us additionally assume that if errors have occurred exactly in three positions out of six meaningful positions determined by the incorrect codeword of weight $w_H = 6$, then the error event occurs with probability $1/2$. If a memoryless binary symmetric channel model is assumed, then binary errors are statistically independent and their probability is equal to p. As a result, an incorrect codeword will be chosen with the probability given by the formula

$$P_6 = \frac{1}{2}\binom{6}{3}p^3(1-p)^3 + \sum_{i=4}^{6}\binom{6}{i}p^i(1-p)^{6-i} \qquad (2.163)$$

In the general case in which a codeword of weight $w_H = k$ is selected instead of the all-zero codeword, we have

$$P_k = \begin{cases} \sum_{i=(k+1)/2}^{k}\binom{k}{i}p^i(1-p)^{k-i} & \text{for } k \text{ odd} \\ \frac{1}{2}\binom{k}{k/2}p^{k/2}(1-p)^{k/2} + \sum_{i=k/2+1}^{k}\binom{k}{i}p^i(1-p)^{k-i} & \text{for } k \text{ even} \end{cases} \qquad (2.164)$$

The probability of the first error event can be upper-bounded by the sum of probabilities of selection of particular incorrect codewords (paths on the trellis diagram)

$$P_E(j) \le \sum_{k=d_{\text{free}}}^{\infty} L_k P_k \qquad (2.165)$$

where L_k is the number of codewords of weight $w_H = k$. Analysis of (2.162) for our code indicates that $L_6 = 1$, $L_7 = 0$, $L_8 = 2$, $L_9 = 0$, $L_{10} = 3$, etc. The bound shown in (2.165) does not depend on any particular moment j, therefore formula (2.165) can be presented in the form

$$P_E \leq \sum_{k=d_{\text{free}}}^{\infty} L_k P_k \tag{2.166}$$

The formulae describing the probabilities P_k can be upper-bounded as follows.
For k odd we have

$$P_k = \sum_{i=(k+1)/2}^{k} \binom{k}{i} p^i (1-p)^{k-i} < \sum_{i=(k+1)/2}^{k} \binom{k}{i} p^{\frac{k}{2}} (1-p)^{\frac{k}{2}}$$

$$= p^{\frac{k}{2}} (1-p)^{\frac{k}{2}} \sum_{i=(k+1)/2}^{k} \binom{k}{i} < p^{\frac{k}{2}} (1-p)^{\frac{k}{2}} \sum_{i=0}^{k} \binom{k}{i} = 2^k p^{\frac{k}{2}} (1-p)^{\frac{k}{2}} \tag{2.167}$$

The last equality sign in (2.167) results from the fact that

$$\sum_{i=0}^{k} \binom{k}{i} = 2^k$$

In turn, for an even value of k we have

$$P_k = \frac{1}{2} \binom{k}{k/2} p^{\frac{k}{2}} (1-p)^{\frac{k}{2}} + \sum_{i=k/2+1}^{k} \binom{k}{i} p^i (1-p)^{k-i}$$

$$< \sum_{i=k/2}^{k} \binom{k}{i} p^i (1-p)^{k-i} < \sum_{i=k/2}^{k} \binom{k}{i} p^{\frac{k}{2}} (1-p)^{\frac{k}{2}}$$

$$< p^{\frac{k}{2}} (1-p)^{\frac{k}{2}} \sum_{i=0}^{k} \binom{k}{i} = 2^k p^{\frac{k}{2}} (1-p)^{\frac{k}{2}} \tag{2.168}$$

Therefore

$$P_E < \sum_{k=d_{free}}^{\infty} L_k \left[2\sqrt{p(1-p)} \right]^k = T(D) \Big|_{D=2\sqrt{p(1-p)}} \tag{2.169}$$

For small values of probability p the sum (2.169) is dominated by its first component and then we have

$$P_E \simeq L_{d_{\text{free}}} \left[2\sqrt{p(1-p)} \right]^{d_{\text{free}}} \simeq L_{d_{\text{free}}} 2^{d_{\text{free}}} p^{d_{\text{free}}/2} \tag{2.170}$$

Each error event that is interpreted as diverging from and merging with the all-zero codeword implies at least one error in the decoded message sequence. On the basis of the estimated probability of an error event we are able to evaluate the error probability for message bits on the decoder output. As we remember, the number of "1"s in the message sequence resulting from selection of the path different from the all-zero path can be deduced from the code transfer function $T(D, N, J)$. This number is in fact a power of variable N in each component of the series expansion of this function. Each such component characterizes a certain path different from the all-zero path. For a given error event a number of incorrectly decoded message symbols can be estimated as the weighted sum of probabilities P_k of selection of the route on the trellis diagram with the weight k, where the weights are numbers B_k of message symbol errors resulting from the selection of a given route, i.e.

$$P_b < \sum_{k=d_{free}}^{\infty} B_k P_k \tag{2.171}$$

We have already shown that probability P_k can be upper-bounded using the formula

$$P_k < \left[2\sqrt{p(1-p)} \right]^k$$

Notice also that when we calculate the derivative of the code transfer function $T(D, N)$ in the series expansion form with respect to N and we substitute $N = 1$, we receive the sum from formula (2.171). Namely, we have from (2.162)

$$\frac{\partial T(D, N)}{\partial N} = D^6 + 4N D^8 + 9N^2 D^{10} + \cdots \tag{2.172}$$

Substituting $N = 1$ and $D = 2\sqrt{p(1-p)}$ in (2.172), we obtain

$$P_b < \left. \frac{\partial T(D, N)}{\partial N} \right|_{N=1, D=2\sqrt{p(1-p)}} \tag{2.173}$$

For small values of a single codeword bit error probability p, the sum (2.173) is dominated by its first component. Then the probability of a single message symbol can be approximated by the formula

$$P_b \simeq B_{d_{free}} \left[2\sqrt{p(1-p)} \right]^{d_{free}} \simeq B_{d_{free}} 2^{d_{free}} p^{d_{free}/2} \tag{2.174}$$

For the considered code we have $B_{d_{free}} = B_6 = 1$, which for $p = 0.01$ results in the message error bit probability equal to about 6.4×10^{-5}. For small values of binary error probabilities the shortest error events dominate. These events, in turn, cause single errors in decoded message sequences.

One can show that due to hard-decision Viterbi decoding and application of bipolar modulation (see Chapter 3) the asymptotic coding gain expressed in decibels is about $10 \log_{10} \frac{R d_{free}}{2}$.

Let us extend our considerations on decoding performance to the soft-decision decoding implemented by the soft-input Viterbi algorithm. We will apply our results to the SOVA presented in the previous section. Let us again assume that all message symbols are statistically independent and equiprobable. Thus, the LLR function of the message symbols $\Lambda(m_l)(l = 1, 2, \ldots, i)$ is equal to zero and does not influence the metric values. The Viterbi algorithm reduces to the application of the maximum likelihood decision rule. Consequently, the coefficient L'_v does not have an impact on the choice of the decided sequence and can be omitted. Using (2.146) we can describe the maximized metric by the formula

$$M(\mathbf{r}_1^i | \mathbf{d}_1^i) = \sum_{l=1}^{i}\sum_{k=1}^{n} r_{l,k} d_{l,k} \tag{2.175}$$

Recall that index i denotes the current time instant, $1/n$ is the coding rate, $d_{l,k}$ is a bipolar code symbol in the ith time unit appearing in the kth position of the codeword and $r_{l,k}$ is the additive Gaussian noise channel output when $d_{l,k}$ is given to the channel input, i.e., $r_{l,k} = d_{l,k} + v_{l,k}$. As previously, consider the probability of the error event for diverging from and then merging with the all-zero path at the ith moment. Counting time units starting from the moment of divergence from the all-zero path, the metric of the all-zero route, denoted as $M^{(0)}(\mathbf{r}_1^i | \mathbf{d}_1^i)$, takes the form (recall that $d_{l,k} = -1$ for a zero codeword symbol)

$$M^{(0)}(\mathbf{r}_1^i | \mathbf{d}_1^i) = -\sum_{l=1}^{i}\sum_{k=1}^{n} r_{l,k} = -\sum_{j=1}^{in} r_j, \quad r_j = r_{l,k}, \quad j = k + n(l-1),$$

$$k = 1, \ldots, n \tag{2.176}$$

The decoder will commit an error if an incorrect path different from the all-zero route is decided. Denote the metric of this path as $M^{(1)}(\mathbf{r}_1^i | \mathbf{d}_1^i)$. Thus, the probability of an error event P_E is

$$P_E = \Pr\{M^{(1)}(\mathbf{r}_1^i | \mathbf{d}_1^i) > M^{(0)}(\mathbf{r}_1^i | \mathbf{d}_1^i)\}$$

$$= \Pr\{M^{(1)}(\mathbf{r}_1^i | \mathbf{d}_1^i) - M^{(0)}(\mathbf{r}_1^i | \mathbf{d}_1^i) > 0\} \tag{2.177}$$

Let us note that, as in the hard-decision analysis, the result of metric comparison is influenced only by those positions and signal samples in which codewords associated with the all-zero and the other candidate path differ. Let them differ in at least $d = w_H$ positions. Thus, the probability of the error event when two paths differ in d positions can be expressed by the formula

$$P_d = \Pr\{\sum_{j=k}^{d} r_{j_k} > 0\} \tag{2.178}$$

where the set $\{j_1, j_2, \ldots, j_d\}$ lists all the sample indices in which two candidate codewords differ. Recall that due to the fact that the all-zero path is the correct one $r_{j_k} = -\sqrt{E_c} + v_{j_k}$, where the noise samples are statistically independent Gaussian zero-mean variables. As a result, the sum $\sum_{j=k}^{d} r_{jk}$ is a Gaussian random variable whose deterministic component is equal to $-d\sqrt{\mathcal{E}_c}$ whereas its variance is the sum of variances of each component r_{jk}, i.e. it is equal to $d\sigma^2$. The probability distribution function of the random variable $U = \sum_{j=k}^{d} r_{jk}$ is then given by the formula

$$p_U(u) = \frac{1}{\sqrt{2\pi}\sqrt{d}\sigma} \exp\left[-\frac{(u + d\sqrt{\mathcal{E}_c})^2}{2d\sigma^2}\right] \tag{2.179}$$

Thus, the desired probability of error event P_d is

$$P_d = \Pr\{\sum_{j=k}^{d} r_{jk} > 0\} = \int_0^\infty p_U(u)\,du \tag{2.180}$$

Let us apply the function $Q(x)$ that describes the area under the tail of the normalized Gaussian distribution and is often found in the tables. This function is given by the formula

$$Q(x) = \frac{1}{\sqrt{2\pi}} \int_x^\infty \exp\left(-\frac{t^2}{2}\right) dt \tag{2.181}$$

We can easily show, using appropriate substitutions, that

$$P_d = Q\left(\sqrt{\frac{\mathcal{E}_c d}{\sigma^2}}\right) = Q\left(\sqrt{d\frac{2E_c}{N_0}}\right) \tag{2.182}$$

The meaning of the Q-function is shown in Figure 2.25. As we know, the path differing in d positions from the all-zero path is not the only one that can appear. The possible values of d can be found from the code transfer function, which is expressed in the form of an expansion series. In general, formula (2.166) can be applied as an upper bound of the probability of an error event resulting in

$$P_E \leq \sum_{d=d_{\text{free}}}^\infty L_d P_d = \sum_{d=d_{\text{free}}}^\infty L_d Q\left(\sqrt{\frac{\mathcal{E}_c d}{\sigma^2}}\right) \tag{2.183}$$

where, as before, L_d denotes the number of paths differing from the all-zero path in d positions. When the argument x of the Q-function is growing then the function can be tightly upper-bounded by an exponential function of the form

$$Q(x) \leq \frac{1}{2} \exp\left(-\frac{x^2}{2}\right)$$

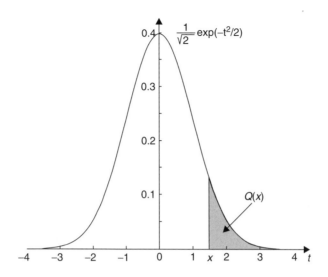

Figure 2.25 Illustration of the Q-function

Then, as in (2.169), formula (2.183) reduces to

$$P_E \leq \sum_{d=d_{\text{free}}}^{\infty} L_d P_d = \frac{1}{2} D^d \bigg|_{D=\exp[E_c/2\sigma^2]} = \frac{1}{2} \sum_{d=d_{\text{free}}}^{\infty} T(D) \bigg|_{D=\exp[E_c/2\sigma^2]} \qquad (2.184)$$

By analogy to (2.173), the message bit error probability can be upper-bounded in the following way

$$P_b < \frac{1}{2} \frac{\partial T(D, N)}{\partial N} \bigg|_{N=1, D=\exp[E_c/2\sigma^2]} \qquad (2.185)$$

Finally, for small noise variance the shortest route featuring the Hamming distance from the all-zero route equal to d_{free} dominates and then, as in (2.174), the message bit probability can be approximated in the following way

$$P_b \simeq \frac{1}{2} B_{d_{\text{free}}} \exp\left(-\frac{d_{\text{free}} E_c}{2\sigma^2}\right) = \frac{1}{2} B_{d_{\text{free}}} \exp\left(-\frac{d_{\text{free}} E_c}{N_0}\right) \qquad (2.186)$$

To end our considerations, let us illustrate the gain of soft-decision decoding over hard-decision decoding by giving some quantitative examples based on the derived approximations for high signal-to-noise ratios.

Recall the example of bit error probability for hard-decision decoding when the probability of a single code symbol is $p = 0.01$. Using (2.174) and substituting for our code $d_{\text{free}} = 6$, $B_{d_{\text{free}}} = 1$, we again have $P_{b,\text{hard}} \simeq 6.4 \times 10^{-5}$. As we will learn from Chapter 3, the probability of an error in bipolar transmission for high signal-to-noise ratios is

$$p = Q\left(\sqrt{\frac{E_c}{\sigma^2}}\right) \simeq \frac{1}{2} \exp\left(-\frac{1}{2} \frac{E_c}{\sigma^2}\right) = \frac{1}{2} \exp\left(-\frac{E_c}{N_0}\right) \qquad (2.187)$$

Substituting (2.187) into (2.186), we have

$$P_{b,\text{soft}} \simeq \frac{1}{2} B_{d_{\text{free}}} \left[\frac{1}{2} \exp\left(-\frac{E_c}{2\sigma^2} \right) \right]^{d_{\text{free}}} \left(\frac{1}{2} \right)^{-d_{\text{free}}} = \frac{1}{2} B_{d_{\text{free}}} 2^{d_{\text{free}}} p^{d_{\text{free}}}$$

Thus, if $p = 0.01$, then $P_{b,\text{soft}} \simeq 3.2 \times 10^{-11}$. As we can see, the difference in performance is significant. Let us also inspect the difference in the required E_c/σ^2 for a given probability of bit error P_b for hard- and soft-decision decoding. Let us stay at $P_{b,\text{hard}} = P_{b,\text{soft}} \simeq 6.4 \times 10^{-5}$, so for hard-decision decoding $p = 0.01$. Using the approximation applied in (2.187) for hard-decision decoding we have

$$\left. \frac{E_c}{N_0} \right|_{\text{hard}} = -\ln 2p = 3.91 = 5.92 \text{ dB}$$

In turn, using (2.186) we obtain

$$\left. \frac{E_c}{N_0} \right|_{\text{soft}} = -\frac{1}{d_{\text{free}}} \ln 2 P_{b,\text{soft}} = 1.49 = 1.74 \text{ dB}$$

so the gain achieved by application of soft-decision decoding instead of hard-decision decoding is of the order of 4 dB! Let us note that this quantitative result is not very precise, as only approximations of the bit error probabilities have been applied for both types of decoding. Typically, we can expect about 2 dB gain of soft-decision decoding over its hard-decision version. Anyway, one can also easily notice that the code-free distance d_{free} plays a crucial role in the overall decoding performance.

2.10 Concatenated Coding

Some communication systems require very high transmission quality in terms of low error probability; however, error detection codes and automatic repetition of erroneous blocks cannot be applied. Therefore, very strong FEC coding is a must. A good example is the system applied for communication with very distant space aircraft (*deep space communications*), and for transmission of telemetric and control signals for space satellites travelling towards other planets. Another example in which very high transmission quality is necessary is broadcasting of DVB (*Digital Video Broadcast*) signals. Here, efficient source coding of video and audio signals has been achieved by strong compression of the digital stream representing both types of signals. As a result, the decompression process performed by the receiver is very sensitive to binary errors in the compressed data stream. Thus, the applied FEC code should ensure a very low error probability. As we know from information theory and the Shannon theorem on channel coding, if the code rate does not exceed the channel capacity, it is possible to construct an error correcting code of appropriately large length for which the probability of erroneous decoding of a codeword can be arbitrarily low. However, very high codeword length would result in high

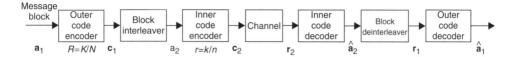

Figure 2.26   Scheme of concatenated coding with interleaving

complexity of encoders and a very difficult hardware and algorithmic implementation of decoders. Moreover, finding good codes of high length is itself a difficult task. Theerfore, instead of searching for and subsequently applying such codes, Forney (1966) proposed to construct a channel coding system by concatenation of two codes, an *outer code* and an *inner code*. Both encoders are separated by an interleaver, whereas the related decoders are separated by a dual block to the interleaver, i.e. by a deinterleaver. The basic scheme of concatenated coding is shown in Figure 2.26.

Input message blocks denoted as vectors $\mathbf{a}_1$ are coded in the outer code encoder of the code rate $R = K/N$. Codewords $\mathbf{c}_1$ are obtained. Subsequently, bits of codewords $\mathbf{c}_1$ are interleaved. As a result, the received output block consists of the input bits whose time sequence is changed with respect to the input. The resulting binary block constitutes the input sequence of the inner code encoder with the code rate equal to $r = k/n$. As we can see, the code rate of the code concatenation is in fact $\frac{Kk}{Nn}$. The receiver performs the operations that are dual to those made in the transmitter. Thus, the decoder of the inner code decides upon the input message sequence on the basis of the sequence corrupted by the channel and noise. The resulting sequence is a subject of deinterleaving in which the original time sequence of message bits is recovered. Finally, the outer code decoder decodes the deinterleaved sequence.

Motivation for application of interleaving is the following. In many practical situations errors arising in a channel have a bursty character; however the codes are usually designed to correct independent errors. By reordering bits in the original sequence, the deinterleaver disrupts error bursts created in the channel. The errors become quasi-statistically independent. This in turn makes it possible to utilize the full correction capabilities of the applied code.

So far we have considered the so-called serial concatenation, in which a codeword of the outer code is a message sequence of the inner code. However, there also exists parallel concatenated coding which has been applied in turbo-codes. Turbo codes will be considered in detail later on. The interleaver applied in the encoder plays a crucial role in achieving a high quality of turbo code decoding.

Let us focus on basic structures of the interleaver and deinterleaver. In general, we differentiate between a *block* interleaver and a *convolutional* interleaver. A basic scheme of a block interleaver and deinterleaver is shown in Figure 2.27. The interleaver consists of a memory block, presented symbolically as a matrix, featuring a certain number of rows and columns. Bits, or generally symbols, are written horizontally in subsequent rows, and they are read vertically from subsequent columns. In the receiver, the incoming data are written into the deinterleaver matrix in the vertical direction and are read in the horizontal

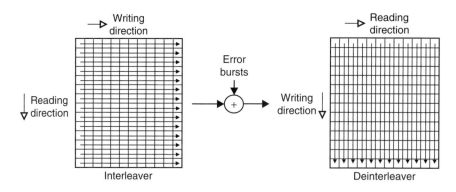

Figure 2.27 Scheme of block interleaver and deinterleaver

direction. Burst errors recorded one after the other in the columns of the deinterleaver are
spread when data are read in the horizontal direction. If the error burst does not exceed
the length of a column, then particular errors are separated by correctly received bits,
whose number is equal to the deinterleaver matrix row length.

The ordered recording in row-column directions is not the only one possible. Data may
be written and read from the interleaver/deinterleaver matrix in a pseudorandom way.
Matrices of the interleaver and deinterleaver shown in Figure 2.27 in reality must be
doubled to enable writing to one of them while the other is being read. The matrices
change their roles after each cycle of reading/writing.

Figure 2.28 presents, in turn, a convolutional interleaver and an appropriate deinterleaver. The interleaver consists of $B - 1$ serial registers of length $M, 2M, \ldots, (B - 1)M$,
respectively. Input sequence bits are supplied to the subsequent register inputs via a commutator or demultiplexer. The first branch transfers the input signals directly to the output
without delay. Other registers introduce delays that are multiples of the number M. Such
a construction of parallel registers with input signals supplied sequentially by a commutator causes reordering of the signals appearing at the output of the output commutator.
Commutators in the deinterleaver have to be synchronized with those in the interleaver.
The input commutator feeds the received symbols to the register inputs; however, this
time they are placed in the opposite order, i.e. starting from the register introducing the
largest delay and ending with the branch without delay. Owing to such a configuration
and synchronization of the commutators, delays between the input of the registers in the
interleaver and the output of the appropriate registers in the deinterleaver remain constant.

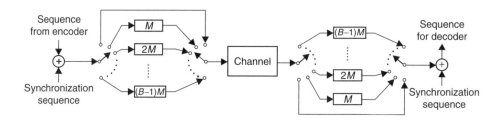

Figure 2.28 Scheme of the convolutional interleaver and deinterleaver

Thus, on the output of the deinterleaver the transmitted data have their original ordering, whereas burst errors that are subject to deinterleaving only are spread. It is worth mentioning that if a nonbinary outer code is applied, whole data symbols, e.g. binary blocks represented by symbols selected from $GF(2^m)$, are the subject to interleaving and deinterleaving.

2.11 Case Studies: Two Examples of Concatenated Coding

2.11.1 Concatenated Coding in Deep Space Communications

A concatenated coding system for deep space communication has been standardized by the CCSDS (*Consultative Committee for Space Data Systems*). Using the coding system conforming to this standard, telemetric data can be transmitted between earth stations and remote space aircrafts (Heegard and Wicker 1999). Figure 2.29 presents the architecture of this system.

The function of an outer code is performed by the (255, 223) Reed-Solomon code. According to the theory already known to us, its code length is $n = q - 1$, where q is a number of Galois field elements in which all operations on code symbols are made. We conclude that $q = 256 = 2^8$. Therefore, code symbols are binary blocks of 8 bits and all operations are performed in $GF(2^8)$. The Galois field has been defined on the basis of the primitive polynomial $p(x) = x^8 + x^7 + x^2 + x + 1$, and the generator polynomial is given by the formula

$$g(x) = \prod_{j=112}^{143} (x - \alpha^{11j}) \tag{2.188}$$

where α is the root of $p(x)$. The (255, 223) RS code is able to correct up to 16 8-bit erroneous symbols in a codeword that has the length of $255 \times 8 = 2040$ bits. Between outer code and inner code encoders a block interleaver is applied. This spans from two to eight codewords of the RS code. As an inner code, the convolutional code has been selected. Its

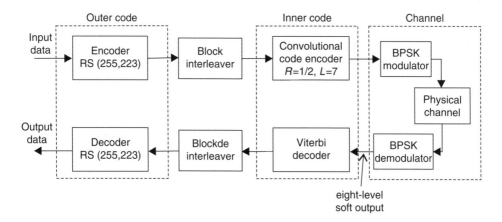

Figure 2.29 Channel coding in telemetric standard CCSDS

constraint length is $L = 7$ and the code rate is equal to $R = 1/2$. Generator polynomials are expressed in binary form as the vectors $\mathbf{g}_1 = [1111001]$ and $\mathbf{g}_2 = [1011011]$. On the receiver side, the Viterbi decoder is typically applied. The decoder accepts eight-level signal samples. Note that the code rate of the overall concatenated coding system is equal to $R = (223/255) \times (1/2) = 0.4373$.

2.11.2   Channel Coding in the DVB Satellite Segment

Below we present channel coding in the satellite segment of DVB (*Digital Video Broadcasting*) standardized by the European Telecommunications Standards Institute (ETSI 1997). This system applies many theoretical ideas considered so far in our chapter. Figure 2.30 shows a functional block diagram of the blocks on the satellite DVB transmit side. TV video, audio and data signals are first encoded in the appropriate source encoders. The resulting data streams and the streams produced by some other services are multiplexed, creating a transport stream. All these blocks jointly constitute a superblock called the MPEG-2 source coding and multiplexing block[4] (see Figure 2.30). This block produces the so-called multiplex packets of length of 188 bytes. The first byte is a synchronization pattern containing the sync word 01000111 (i.e. 47_{HEX}). Eight packets constitute a frame. In the first packet of the frame the sync word has a negated form, i.e. it is $B8_{\text{HEX}}$. The framing structure is shown in Figure 2.31.

The first block of the channel coding part performs randomization of the data stream contained in the packets. It is desirable that transmitted data stream looks like a random binary stream, i.e. binary transitions occur in adequate numbers. As we will learn later, the randomization ensures appropriate synchronization at the receiver and prohibits concentration of energy in small subranges of the signal spectrum in the RF band. Randomization is performed by a block called a *scrambler*. The scheme of the scrambler applied in a DVB satellite system transmitter is shown in Figure 2.32. Its main part is a Pseudo-Random Binary Sequence (PRBS) generator, which is a linear feedback shift

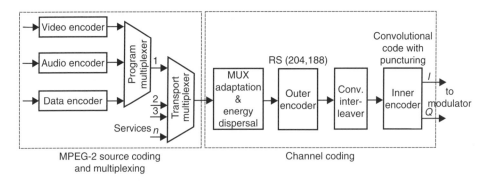

Figure 2.30   Source and channel coding structure in a DVB satellite transmitter © European Telecommunications Standards Institute 1997. Further use, modification, redistribution is strictly prohibited. ETSI standards are available from http://pda.etsi.org/pda.

---

[4] Moving Picture Expert Group (MPEG) is an international group working within the International Standardization Organization (ISO), focused on development and standardization of video and audio encoding methods. MPEG-2 is a widely used standard in DVB and other systems.

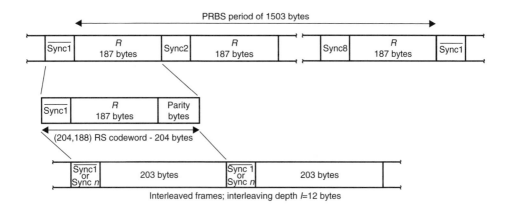

Figure 2.31 Framing structure of DVB satellite data stream © European Telecommunications Standards Institute 1997. Further use, modification, redistribution is strictly prohibited. ETSI standards are available from http://pda.etsi.org/pda.

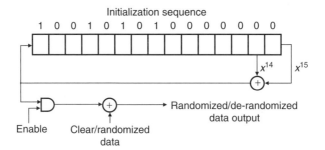

Figure 2.32 Diagram of scrambler and descrambler applied in a transmitter/receiver of a DVB satellite segment

register (LFSR) of structure similar to that featured by maximum length codes. The LFSR polynomial is $h(x) = 1 + x^{14} + x^{15}$. The pseudorandom binary sequence is uniquely set at the beginning of each frame by the initialization sequence shown in Figure 2.32. Then the binary sequence generated by the initialized LFSR is modulo-2 added to each transport frame, starting from the first information byte of the first packet. Although timing signals are fed to the LFSR during the whole frame, the negative enable signal prohibits modulo-2 addition of the LFSR output signal when sync patterns appear at the scrambler input. Thus, synchronization words remain unchanged.

Obviously, a dual operation to scrambling has to be done in the receiver. Fortunately, both scrambler and descrambler have an identical construction. The data sequence scrambled in the transmitter and recovered in the receiver is modulo-2 added to the binary sequence generated by the LFSR of the descrambler. Owing to synchronization bytes introduced at the beginning of each transport frame, frame and bit synchronizations are possible.

The scrambled data packets are the subject of outer code encoding. The outer code is the (204, 188) shortened Reed-Solomon code. It has been created by substitution of 51 zero symbols at the beginning of each codeword of the full (255, 239) RS code. Consequently,

51 first codeword symbols (bytes) do not need to be transmitted and in this way the code structure has been matched to the transport packet structure. Message blocks consist of 184-byte long packets, including sync patterns. The codeword symbols are selected from the extended finite Galois field $GF(2^8)$ generated by polynomial $p(x) = x^8 + x^4 + x^3 + x^2 + 1$. The generator polynomial of the applied RS code is given by the formula

$$g(x) = (x - \beta^0)(x - \beta^1)(x - \beta^2) \cdot \ldots \cdot (x - \beta^{15}) \qquad (2.189)$$

where $\beta = 02_{HEX}$.

As a result of outer coding, 204-byte long packets are created. In order to increase immunity of the coded packets against burst errors, a convolutional interleaver is applied. Its structure is shown in Figure 2.33 and it is very similar to that shown in Figure 2.28; however, this time all operations are performed on bytes. The scrambler consists of input and output switches (commutators) that direct codeword symbols sequentially to 12 parallel branches. The first branch is always used by a syncronization byte. All others contain FIFO (First-In First-Out) registers of length being a multiple of 17 bytes. As in a typical interleaver/deinterleaver pair, the structure of the deinterleaver is dual to that of the interleaver, so the first branch used to transport the synchronization byte is the longest one and introduces a delay of 11×17 bytes. Delays of the other branches are also a multiple of 17 bytes, but in descending order.

The interleaved bytes are coded by the inner code. The latter is a convolutional code of constraint lenght $L = 7$ and code rate $R = 1/2$. Its generator polynomials are

$$g_1(x) = 1 + x^2 + x^3 + x^5 + x^6$$

$$g_2(x) = 1 + x + x^2 + x^3 + x^6 \qquad (2.190)$$

Two parallel encoder outputs are given to the input of the puncturing block. Its structure must ensure the required final code rate equal to 1/2, 2/3, 3/4, 5/6 or 7/8. It also determines the free distance of a resulting punctured code. Table 2.6 presents possible code rates, associated puncturing tables, free distances of the resulting punctured codes and values of E_b/N_0 required to achieve Quasi-Error Free (QEF, $10^{-10} - 10^{-11}$) performance on the output of the outer RS decoder when the interleaver presented in Figure 2.33 is applied.

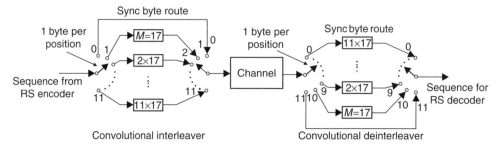

Figure 2.33 Diagram of interleaver and deinterleaver applied in the transmitter/receiver of a DVB satellite segment © European Telecommunications Standards Institute 1997. Further use, modification, redistribution is strictly prohibited. ETSI standards are available from http://pda.etsi.org/pda.

Table 2.6 Convolutional code parameters applied in DVB satellite transmission

R	d_{free}	Puncturing table	Required E_b/N_0 [dB]
1/2	10	$\begin{bmatrix} 1 \\ 1 \end{bmatrix}$	4.5
2/3	6	$\begin{bmatrix} 1 & 0 \\ 1 & 1 \end{bmatrix}$	5.0
3/4	5	$\begin{bmatrix} 1 & 0 & 1 \\ 1 & 1 & 0 \end{bmatrix}$	5.5
5/6	4	$\begin{bmatrix} 1 & 0 & 1 & 0 & 1 \\ 1 & 1 & 0 & 1 & 0 \end{bmatrix}$	6.0
7/8	3	$\begin{bmatrix} 1 & 0 & 0 & 0 & 1 & 0 & 1 \\ 1 & 1 & 1 & 1 & 0 & 1 & 0 \end{bmatrix}$	6.4

As the energy per codeword symbol is equal to energy per bit, in Table 2.6 we have replaced E_c by E_b.

The number of possible code rates allows us to select the most appropriate level of error protection for a given DVB service and data rate. As shown in Figure 2.30, two codeword symbols appear in parallel on the channel encoder output. They are denoted as I and Q and are related to the upper and lower row of the selected puncturing table.

At the receiver side dual blocks to those contained in the transmitter are applied. Figure 2.34 shows the block diagram of the channel decoding part of the receiver. Two parallel samples from the demodulator and filtering block are supplied to the depuncturing block and the inner decoder. In order for the other blocks to operate correctly, the bit error rate of the hard-decision samples should be at the level of 10^{-1}–10^{-2}, depending on the applied code rate. The bit error rate on the inner decoder output should be about 2×10^{-4} or lower. Although the ETSI Standard (ETSI 1997) does not precisely define the receiver side, Annex B of this standard states that the inner code decoder implemented with the application of the Viterbi algorithm makes use of soft-decision information, i.e. it is a soft-input decoder. An additional task of this block is automatic recognition of the applied code rate and of the puncturing configuration. The decoder is able to try all possible puncturing combinations in order to find the correct one.

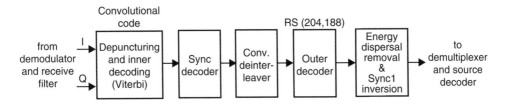

Figure 2.34 Channel decoder on the DVB satellite receiver side

The next block decodes MPEG-2 synchronization bytes, thus providing synchronization information needed for correct operation of the deinterleaver. The synchronized deinterleaver reorders received bytes so as to obtain their primary sequence, and the outer decoder finds message sequences representing MPEG-2 transport packets. As we have mentioned before, assuming a bit error rate on the output of the inner code Viterbi decoder at the level of 2×10^{-4} or lower, the RS decoder output should be practically quasi error free.

The last block performs descrambling and $\overline{\text{Sync1}}$ inversion, producing original transport packets. The packets are then used by the source decoder.

As usual, the ETSI Standard (ETSI 1997) precisely defines the operation of a transmitter, leaving the structure and algorithms of a receiver to the system designer. Since topics associated with modulation and signal reception have not been considered in the course of this book so far, we have not explained them in this case study. However, hopefully the main ideas of channel coding applied in the DVB satellite segment have been clarified and practical applications of some theoretical issues presented in this chapter have been shown.

2.12   Turbo Codes

In the section on concatenated coding we mentioned that both serial and parallel code concatenation is possible. Let us now analyze parallel code concatenation. As previously, on the transmit side two component code encoders separated by an interleaver participate in the encoding process. This configuration of parallel concatenated coding was first presented in 1993 by Berrou, Glavieux and Thitimajshima (Berrou *et al.* 1993). They also proposed an original decoding method well fitted to the applied codes. The core of this method is the use of reliability information about temporary decisions worked out by one component code decoder to improve decision likelihood in the second component decoder. Reliability information derived in the second decoder is in turn fed back to the first one, implying the improvement of decision reliability. This process is continued in a closed loop, as is done in a car engine with turbo loading. The similarity justifies the name of codes proposed by their inventors, i.e. *turbo codes*. Since the first article by Berrou and co-authors (Berrou *et al.* 1993), hundreds of papers have been published on turbo codes. There are some specialized books as well, including those by Heegard and Wicker (1999) and Vucetic and Yuan (2000). The reason for the huge interest in turbo codes is their excellent quality, a fraction of dB worse than the theoretical Shannon limit. However, before we present the basic structure of a turbo code, let us focus on *Recursive Systematic Convolutional Codes* (RSCC), which constitute a basic functional block of turbo codes.

2.12.1   RSCC Code

So far we have considered nonsystematic convolutional codes. Recall polynomial description of a convolutional code of code rate $R = 1/2$. An encoder of such a code generates two output signals related to a single input signal. They can be described in polynomial notation as

$$w_1(x) = b(x)g_1(x) \quad w_2(x) = b(x)g_2(x) \tag{2.191}$$

where the polynomial $b(x)$ represents the input signal, and the polynomials describing outputs from both encoder branches are given by the formulae

$$g_1(x) = g_{1,1} + g_{1,2}x + \cdots + g_{1,L-1}x^{L-1}$$
$$g_2(x) = g_{2,1} + g_{2,2}x + \cdots + g_{2,L-1}x^{L-1}$$

A systematic code featuring the same set of codewords and thus preserving the properties of the original code (e.g. the same free distance) can be obtained by division of the right-hand side of both polynomial products in (2.191) by either $g_1(x)$ or $g_2(x)$. If $g_1(x)$ is used in the division, the upper and lower branch output polynomials receive the following form characteristic for an equivalent systematic code

$$w_1(x) = b(x) \quad w_2(x) = \frac{b(x)g_2(x)}{g_1(x)} \tag{2.192}$$

If the following polynomial is denoted as $d(x)$

$$d(x) = \frac{b(x)}{g_1(x)} \tag{2.193}$$

then the branch output polynomials are described by the expressions

$$w_1(x) = b(x) \quad w_2(x) = d(x)g_2(x) \tag{2.194}$$

In the time domain marked by index k, assuming $g_{1,1} = 1$, these equations are equivalent to the following expressions

$$w_{1,k} = b_k$$

$$w_{2,k} = \sum_{i=0}^{L-1} g_{2,i}d_{k-i} \tag{2.195}$$

$$d_k = b_k + \sum_{i=1}^{L-1} g_{1,i}d_{k-i}$$

The scheme of an exemplary nonsystematic convolutional code encoder is shown in Figure 2.35a and its equivalent recursive form can be found in Figure 2.35b. Note the feedback that results from the last equation in formulae (2.195).

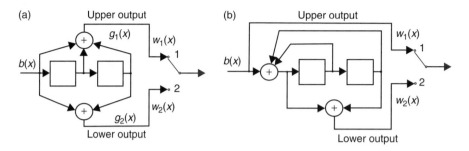

Figure 2.35 Convolutional code encoder of a systematic code (a) and of an equivalent recursive systematic convolutional code (b)

2.12.2 Basic Turbo Code Encoder Scheme

As mentioned before, an RSCC encoder is the basic element of a turbo code encoder. A basic scheme of the turbo code encoder is shown in Figure 2.36. The encoder consists of two identical RSCC encoders and an interleaver. The first encoder generates two parallel streams. The first one is a message stream, directly transferred from the encoder input. The second one consists of parity bits generated in the circuit with feedback. Message bits are a subject of deep interleaving in a block interleaver. In practice a turbo code encoder works in a block manner. As a result, interleaved message bits are fed to the second RSCC encoder. Only parity bits of the second encoder are further transferred to the turbo code encoder output. Therefore, the basic scheme of a turbo code encoder realizes coding of the code rate $R = 1/3$. The generated binary stream consists of a stream of message bits

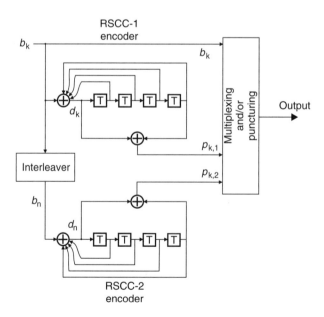

Figure 2.36 Scheme of the turbo code encoder (Berrou *et al.* 1993)

and two streams of parity bits. These streams are multiplexed and punctured, if required. Puncturing allows higher code rates to be achieved.

A block interleaver applied in the turbo code encoder plays a special role. The interleaving depth is typically very high, i.e. it amounts to hundreds or even thousands of bits. Therefore a message bit and a parity bit of the first component encoder are in the time sequence next to a parity bit of the second component encoder, which is related to a very distant message bit. In this way very long codewords are produced in a simple way. As we will see later, such codewords can be efficiently decoded using the turbo decoding principle. As already mentioned, encoding is performed in frames of a given length and the interleaving depth is matched to it.

2.12.3 RSCC Code MAP Decoding

Excellent decoding quality of turbo codes is achieved owing to the iterative procedure applied in the receiver. In order to present this procedure, we have to describe a decoding algorithm that can be applied in a component RSCC code decoder.

One such algorithm is the BCJR algorithm named after its inventors Bahl, Cocke, Jelinek and Raviv (Bahl *et al.* 1974). Berrou, Glavieux and Thitimajshima modified this algorithm and matched it to a turbo code. The modified BCJR algorithm determines each message symbol of a codeword using the *Maximum a Posteriori* (MAP) criterion. In such a sense it is optimal. In practice, however, suboptimal algorithms are mostly used, such as SOVA (Hagenauer and Hoeher 1989) or the so-called Max-Log-MAP algorithm.

Consider the MAP algorithm of recursive decoding of an RSCC code. In the presentation of this algorithm the author has used the book by Vucetic and Yuan (2000) as guidance. It also be found elsewhere but the derivation of the algorithm presented in by Vucetic and Yuan (2000) is the most clear for the reader. Despite this, a less mathematically experienced reader can find the algorithm derivation quite complicated, although only basic probability knowledge is required in the course of its presentation.

Assume a discrete memoryless channel model with a continuous signal amplitude on its output. Let codeword bits be represented by bipolar symbols ± 1. Bipolar symbols are distorted by additive Gaussian noise of zero mean and variance σ^2. Let the length of a message bit block be n. For each message bit b_k ($k = 1, \ldots, n$) the RSCC encoder generates a parity bit p_k related to it. In turn, let a_k be a bipolar symbol reflecting the bit b_k, and let r_k be a bipolar symbol related to the parity bit p_k. A codeword can be described with the following vector form

$$\mathbf{x}_1^n = [\mathbf{x}_1, \mathbf{x}_2, \ldots, \mathbf{x}_n] \tag{2.196}$$

where each vector entry is itself a two-element vector

$$\mathbf{x}_k = \begin{bmatrix} a_k \\ r_k \end{bmatrix} \quad k = 1, 2, \ldots, n \tag{2.197}$$

As in the description of the SOVA decoding algorithm, in (2.196) we directly denoted the first and last time index of the analyzed sequence. It will be very useful in the description of the BCJR decoding algorithm.

Let vector $\mathbf{x}_1^n$ be transmitted through an additive Gaussian noise channel. Then we receive the following vector on the channel output

$$\mathbf{y}_1^n = [\mathbf{y}_1, \mathbf{y}_2, \ldots, \mathbf{y}_n] \qquad (2.198)$$

Each vector element is a sum of input symbol $\mathbf{x}_k$ and noise vector $\mathbf{v}_k$, i.e.

$$\mathbf{y}_k = \begin{bmatrix} y_{a,k} \\ y_{r,k} \end{bmatrix} = \begin{bmatrix} a_k + v_{a,k} \\ r_k + v_{r,k} \end{bmatrix} = \mathbf{x}_k + \mathbf{v}_k \qquad (2.199)$$

The MAP algorithm makes a decision upon the kth message symbol by selecting from two possible values $+1$ or -1 the value (denoted by symbol i) for which *a posteriori* probability is maximum, i.e. it determines the value of decision $\widehat{a}_k$ according to the following rule

$$\widehat{a}_k = \arg\max_i \Pr\{a_k = i | \mathbf{y}_1^n\} \qquad (2.200)$$

Note that despite the fact that decisions are made symbol by symbol, the coding interrelation is embedded in the received sequence $\mathbf{y}_1^n$. Equivalently, a decision upon each symbol can be determined by calculation of the quotient of *a posteriori* probabilities for two possible values of data symbols, followed by calculation of the logarithm of this quotient. The sign of the logarithm will be equivalent to decision (2.200), i.e.

$$\widehat{a}_k = \operatorname{sgn}\left[\Lambda(a_k)\right] \qquad (2.201)$$

where

$$\Lambda(a_k) = \ln \frac{\Pr\{a_k = +1 | \mathbf{y}_1^n\}}{\Pr\{a_k = -1 | \mathbf{y}_1^n\}} \qquad (2.202)$$

When *a posteriori* probabilities in the nominator and denominator are identical, $\Lambda(a_k)$ is equal to zero. The higher the difference between them, the more $\Lambda(a_k)$ differs from zero. The value of $\Lambda(a_k)$ is therefore a likelihood measure of the decision upon the symbol a_k. It follows from our considerations on soft-decision decoding that it is a *soft decision* upon this symbol.

Let the considered RSCC encoder have M states. Assume that the encoder has generated the sequence $\mathbf{x}_1^n$ and the initial and final states are known and equal to zero state ($s_0 = s_n = 0$). The decoder makes decisions about subsequent message symbols of the sequence $\mathbf{x}_1^n$ on the basis of the whole received sequence $\mathbf{y}_1^n$. As we know, the decoder operation can be interpreted as a search for the best route on the code trellis diagram. For example, consider the path on the trellis diagram, which starts from state u at the $(k-1)$st moment and leads to state v at the kth moment. Each pair of states (u, v) is uniquely associated with a symbol $a_k = \pm 1$ generated by the encoder. Let B_{+1} denote the set of state pairs (u, v) for which during the transition from state u to v the symbol $a_k = 1$ is generated.

Let B_{-1} be the set of state pairs (u, v) that are associated with the symbol $a_k = -1$. Thus, the conditional probabilities in formula (2.202) can be presented as

$$\Pr\{a_k = +1 | \mathbf{y}_1^n\} = \sum_{(u,v) \in B_{+1}} \Pr\{s_{k-1} = u, s_k = v | \mathbf{y}_1^n\} \tag{2.203}$$

and

$$\Pr\{a_k = -1 | \mathbf{y}_1^n\} = \sum_{(u,v) \in B_{-1}} \Pr\{s_{k-1} = u, s_k = v | \mathbf{y}_1^n\} \tag{2.204}$$

Applying Bayes' formula on the right-hand side of both equations, we receive

$$\Pr\{a_k = +1 | \mathbf{y}_1^n\} = \frac{\displaystyle\sum_{(u,v) \in B_{+1}} \Pr\{s_{k-1} = u, s_k = v, \mathbf{y}_1^n\}}{p(\mathbf{y}_1^n)} \tag{2.205}$$

and

$$\Pr\{a_k = -1 | \mathbf{y}_1^n\} = \frac{\displaystyle\sum_{(u,v) \in B_{-1}} \Pr\{s_{k-1} = u, s_k = v, \mathbf{y}_1^n\}}{p(\mathbf{y}_1^n)} \tag{2.206}$$

Consequently, formula (2.202) can be presented in the form

$$\Lambda(a_k) = \ln \frac{\displaystyle\sum_{(u,v) \in B_{+1}} \Pr\{s_{k-1} = u, s_k = v, \mathbf{y}_1^n\}}{\displaystyle\sum_{(u,v) \in B_{-1}} \Pr\{s_{k-1} = u, s_k = v, \mathbf{y}_1^n\}} \tag{2.207}$$

Consider a single component of the sums appearing both in the nominator and denominator of (2.207). We are interested in the operation of the algorithm at the kth moment. The reception of sequence $\mathbf{y}_1^n$ can be treated as reception of the sequence $\mathbf{y}_1^{k-1}$, symbol $\mathbf{y}_k$ (at the kth moment) and sequence $\mathbf{y}_{k+1}^n$ [from the $(k + 1)$st moment to the nth moment]. Then

$$\Pr\{s_{k-1} = u, s_k = v, \mathbf{y}_1^n\} = \Pr\{\mathbf{y}_1^{k-1}, \mathbf{y}_k, \mathbf{y}_{k+1}^n, s_{k-1} = u, s_k = v\}$$

$$= \Pr\{\mathbf{y}_{k+1}^n | \mathbf{y}_k, \mathbf{y}_1^{k-1}, s_{k-1} = u, s_k = v\} \Pr\{\mathbf{y}_k, \mathbf{y}_1^{k-1}, s_{k-1} = u, s_k = v\} \tag{2.208}$$

Since the channel through which the sequence $\mathbf{x}_k$ is transmitted is memoryless, received sequence $\mathbf{y}_{k+1}^n$ starting from the $(k + 1)$st moment depends only on the state on the trellis diagram in which the encoder was at the kth moment. Therefore

$$\Pr\{s_{k-1} = u, s_k = v, \mathbf{y}_1^n\} = \Pr\{\mathbf{y}_{k+1}^n | s_k = v\} \Pr\{\mathbf{y}_k, \mathbf{y}_1^{k-1}, s_{k-1} = u, s_k = v\}$$

$$= \Pr\{\mathbf{y}_{k+1}^n | s_k = v\} \Pr\{s_k = v, \mathbf{y}_k | s_{k-1} = u, \mathbf{y}_1^{k-1}\} \Pr\{s_{k-1} = u, \mathbf{y}_1^{k-1}\} \tag{2.209}$$

Apply the following notation

$$\beta_k(v) = \Pr\{\mathbf{y}_{k+1}^n | s_k = v\} \tag{2.210}$$

$$\alpha_{k-1}(u) = \Pr\{s_{k-1} = u, \mathbf{y}_1^{k-1}\} \tag{2.211}$$

Thus, variable $\beta_k(v)$ represents the probability of reception of sequence $\mathbf{y}_{k+1}^n$ from the $(k+1)$st moment to the nth moment under the condition that the encoder was in state v at the kth moment. Variable $\alpha_{k-1}(u)$ denotes the probability of reaching state u in the $(k-1)$st moment when the sequence $\mathbf{y}_1^{k-1}$ has been received. We know that appearance of the symbol $\mathbf{y}_k$ at the kth moment on the decoder input and finding the decoder itself in state v at the same moment depends exclusively on the data symbol $a_k = i$ $(i = \pm 1)$ and on the state in which the encoder was at the $(k-1)$st moment. Therefore, we can write the following equality

$$\Pr\{s_k = v, \mathbf{y}_k | s_{k-1} = u, \mathbf{y}_1^{k-1}\} = \Pr\{s_k = v, \mathbf{y}_k | s_{k-1} = u\} \tag{2.212}$$

Applying the notation

$$\gamma_k^i(u, v) = \Pr\{a_k = i, s_k = v, \mathbf{y}_k | s_{k-1} = u\} \tag{2.213}$$

we can present the probability given in formula (2.209) in the following form

$$\Pr\{s_{k-1} = u, s_k = v, \mathbf{y}_1^n\} = \alpha_{k-1}(u) \sum_{i \in \{-1, +1\}} \gamma_k^i(u, v)\beta_k(v) \tag{2.214}$$

Finally, on the basis of (2.207) and (2.214) we have

$$\Lambda(a_k) = \ln \frac{\sum\limits_{(u,v) \in B_{+1}} \alpha_{k-1}(u)\gamma_k^{+1}(u, v)\beta_k(v)}{\sum\limits_{(u,v) \in B_{-1}} \alpha_{k-1}(u)\gamma_k^{-1}(u, v)\beta_k(v)} \tag{2.215}$$

It seems that formula (2.215) is more complicated than its primary form (2.202); however, as we shall see, this new form of expression enables us to calculate soft decisions for subsequent data symbols recurrently. Namely, probabilities $\alpha_{k-1}(u)$ and $\beta_k(v)$ can be derived recurrently, the first one in ascending order and the second one in descending order of index k.

First, derive a recurrence for $\alpha_k(v)$. According to (2.211) we have

$$\alpha_k(v) = \Pr\{s_k = v, \mathbf{y}_1^k\} = \sum_{u=0}^{M-1} \Pr\{s_{k-1} = u, s_k = v, \mathbf{y}_1^k\}$$

$$= \sum_{u=0}^{M-1} \Pr\{s_{k-1} = u, s_k = v, \mathbf{y}_1^{k-1}, \mathbf{y}_k\}$$

$$= \sum_{u=0}^{M-1} \Pr\{s_k = v, \mathbf{y}_k | s_{k-1} = u, \mathbf{y}_1^{k-1}\} \Pr\{s_{k-1} = u, \mathbf{y}_1^{k-1}\} \tag{2.216}$$

The last factor of the sum (2.216) is $\alpha_{k-1}(u)$. Therefore

$$\alpha_k(v) = \sum_{u=0}^{M-1} \alpha_{k-1}(u) \Pr\{s_k = v, \mathbf{y}_k | s_{k-1} = u, \mathbf{y}_1^{k-1}\}$$

Appearance of the symbol $\mathbf{y}_k$ at the kth moment and transition to state v does not depend on the sequence of preceding signals, but exclusively on the state u in which the encoder found itself previously, therefore

$$\alpha_k(v) = \sum_{u=0}^{M-1} \alpha_{k-1}(u) \Pr\{s_k = v, \mathbf{y}_k | s_{k-1} = u\} \qquad (2.217)$$

We can, in turn, take advantage of the dependence

$$\Pr\{s_k = v, \mathbf{y}_k | s_{k-1} = u\} = \sum_{i \in \{-1,+1\}} \Pr\{a_k = i, s_k = v, \mathbf{y}_k | s_{k-1} = u\}$$

$$= \sum_{i \in \{-1,+1\}} \gamma_k^i(u, v) \qquad (2.218)$$

Finally,

$$\alpha_k(v) = \sum_{u=0}^{M-1} \alpha_{k-1}(u) \sum_{i \in \{-1,+1\}} \gamma_k^i(u, v) \qquad (2.219)$$

As we see, probabilities $\alpha_k(v)$ can be calculated recurrently with appropriate initialization at the zero moment, e.g. in agreement with the following rule

$$\alpha_0(0) = 1, \quad \alpha_0(u) = 0, \quad u = 1, 2, \ldots, M - 1 \qquad (2.220)$$

since we assumed that the initial state is known and it is a zero state. After reception of the full symbol sequence $\mathbf{y}_1^n$, we can determine subsequent values of probabilities $\alpha_k(v)$ in ascending order of time index k. However, it is necessary to determine $\gamma_k^i(u, v)$ for each state pair (u, v) and each value of the data signal $i = \pm 1$. We will now derive the formula for $\gamma_k^i(u, v)$.

Let us first derive the probabilities $\beta_k(v)$. From the definition of $\beta_k(v)$ and Bayes' formula we obtain

$$\beta_k(v) = \Pr\{\mathbf{y}_{k+1}^n | s_k = v\} = \sum_{w=0}^{M-1} \Pr\{s_{k+1} = w, \mathbf{y}_{k+1}^n | s_k = v\}$$

$$= \sum_{w=0}^{M-1} \frac{\Pr\{s_{k+1} = w, \mathbf{y}_{k+1}^n, s_k = v\}}{\Pr\{s_k = v\}} \qquad (2.221)$$

Extracting $\mathbf{y}_{k+1}$ from sequence $\mathbf{y}_{k+1}^n$, we can write

$$\beta_k(v) = \sum_{w=0}^{M-1} \frac{\Pr\{\mathbf{y}_{k+1}, \mathbf{y}_{k+2}^n, s_k = v, s_{k+1} = w\}}{\Pr\{s_k = v\}}$$

$$= \sum_{w=0}^{M-1} \frac{\Pr\{\mathbf{y}_{k+2}^n | \mathbf{y}_{k+1}, s_k = v, s_{k+1} = w\} \Pr\{\mathbf{y}_{k+1}, s_k = v, s_{k+1} = w\}}{\Pr\{s_k = v\}} \qquad (2.222)$$

For the memoryless channel, appearance of sequence $\mathbf{y}_{k+2}^n$ on the channel output depends only on the state in which the encoder was at the $(k+1)$st moment, so

$$\beta_k(v) = \sum_{w=0}^{M-1} \frac{\Pr\{\mathbf{y}_{k+2}^n | s_{k+1} = w\} \Pr\{\mathbf{y}_{k+1}, s_k = v, s_{k+1} = w\}}{\Pr\{s_k = v\}}$$

$$= \sum_{w=0}^{M-1} \frac{\beta_{k+1}(w) \Pr\{\mathbf{y}_{k+1}, s_{k+1} = w | s_k = v\} \Pr\{s_k = v\}}{\Pr\{s_k = v\}}$$

$$= \sum_{w=0}^{M-1} \beta_{k+1}(w) \Pr\{\mathbf{y}_{k+1}, s_{k+1} = w | s_k = v\}$$

$$= \sum_{w=0}^{M-1} \beta_{k+1}(w) \sum_{i \in \{-1,+1\}} \Pr\{a_{k+1} = i, \mathbf{y}_{k+1}, s_{k+1} = w | s_k = v\}$$

$$= \sum_{w=0}^{M-1} \beta_{k+1}(w) \sum_{i \in \{-1,+1\}} \gamma_{k+1}^i(v, w) \qquad (2.223)$$

Owing to (2.223) it is possible to derive $\beta_k(v)$ in a backward direction, starting from the last symbol and finishing at the first symbol. However, knowledge of probabilities $\gamma_{k+1}^i(v, w)$ is also needed. Assuming that the final state of the trellis diagram is a zero state, we can initialize the values of $\beta_k(v)$ in the following way

$$\beta_n(0) = 1, \quad \beta_n(v) = 0, \quad v = 1, 2, \ldots, M - 1 \qquad (2.224)$$

Bringing the RSCC encoder with feedback into the zero state can be troublesome, therefore we can disconnect the feedback in the last cycles, subsequently introducing zero symbols to the encoder memory. This idea is presented in Figure 2.37. In normal operation the switch is in position 1. For the last L steps, where L is the number of encoder memory cells, the switch is placed in position 2, so the input of the delay line is supplied with the modulo-2 sum of two identical symbols resulting in zero symbols. These symbols are simultaneously the message symbols on the encoder output.

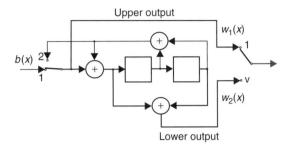

Figure 2.37 Scheme of the encoder from Figure 2.35 supplemented by the switch forcing the zero state at the end of the sequence

We still have to derive the probabilities $\gamma_k^i(u, v)$. Recall that they have been given by formula (2.213). Applying the conditional probability formula, we obtain

$$
\gamma_k^i(u, v) = \frac{\Pr\{a_k = i, \mathbf{y}_k, s_k = v, s_{k-1} = u\}}{\Pr\{s_{k-1} = u\}}
$$

$$
= \frac{p(\mathbf{y}_k | a_k = i, s_k = v, s_{k-1} = u)\Pr\{a_k = i, s_k = v, s_{k-1} = u\}}{\Pr\{s_{k-1} = u\}} \tag{2.225}
$$

The conditional probability density function that appears in formula (2.225) can be represented as a product of two probability density functions. The first function is related to the message component $y_{a,k}$ of the received pair $\mathbf{y}_k$ at the kth moment. The message component $y_{a,k}$ depends exclusively on symbol a_k. The second function is related to parity component $y_{r,k}$, which results from the transition from state u to v at the kth moment. Thus, for a memoryless channel with additive Gaussian noise with variance σ^2 we have

$$
p(\mathbf{y}_k | a_k = i, s_k = v, s_{k-1} = u) = p(y_{a,k} | a_k = i) p(y_{r,k} | x_{r,k}^{(u,v)})
$$

$$
= \frac{1}{2\pi\sigma^2} \exp\left[-\frac{(y_{a,k} - i)^2}{2\sigma^2}\right] \exp\left[-\frac{(y_{r,k} - x_{r,k}^{(u,v)})^2}{2\sigma^2}\right] \tag{2.226}
$$

where $x_{r,k}^{(u,v)}$ is a bipolar parity signal generated at the encoder transition from state u to v. Consequently

$$
\gamma_k^i(u, v) = p(y_{a,k} | a_k = i) p(y_{r,k} | x_{r,k}^{(u,v)})
$$

$$
\times \frac{\Pr\{a_k = i | s_k = v, s_{k-1} = u\}\Pr\{s_k = v, s_{k-1} = u\}}{\Pr\{s_{k-1} = u\}}
$$

$$
= p(y_{a,k} | a_k = i) p(y_{r,k} | x_{r,k}^{(u,v)}) \times
$$

$$
\times \Pr\{a_k = i | s_k = v, s_{k-1} = u\}\Pr\{s_k = v | s_{k-1} = u\} \tag{2.227}
$$

Transition from state u to state v uniquely results from the data symbol $a_k = i$ fed to the encoder input. Therefore

$$\Pr\{s_k = v | s_{k-1} = u\} = \begin{cases} \Pr\{a_k = i\} & \text{for } (u, v) \in B_i \\ 0 & \text{otherwise} \end{cases} \tag{2.228}$$

where $\Pr\{a_k = i\}$ is the *a priori* probability of generation of data symbol $a_k = i$ by the transmitter. In turn, the probability of appearance of data symbol $a_k = i$ under the condition that transition from state u to state v has occurred is

$$\Pr\{a_k = i | s_k = v, s_{k-1} = u\} = \begin{cases} 1 & \text{if } (u, v) \in B_i \\ 0 & \text{otherwise} \end{cases} \tag{2.229}$$

Taking into account expressions (2.227)–(2.229), probability $\gamma_k^i(u, v)$ can be expressed in the following form

$$\gamma_k^i(u, v) = \begin{cases} \Pr\{a_k = i\} p(y_{a,k} | a_k = i) p(y_{r,k} | x_{r,k}^{(u,v)}) & \text{for } (u, v) \in B_i \\ 0 & \text{otherwise} \end{cases} \tag{2.230}$$

Substituting $\gamma_k^i(u, v)$ determined by formulae (2.230) and (2.226) to formula (2.215) describing a soft decision related to the symbol a_k, we get

$$\Lambda(a_k) = \ln \frac{\Pr\{a_k = +1\} p(y_{a,k} | a_k = +1) \displaystyle\sum_{(u,v) \in B_{+1}} \alpha_{k-1}(u) p\left(y_{r,k} | x_{r,k}^{(u,v)}\right) \beta_k(v)}{\Pr\{a_k = -1\} p(y_{a,k} | a_k = -1) \displaystyle\sum_{(u,v) \in B_{-1}} \alpha_{k-1}(u) p\left(y_{r,k} | x_{r,k}^{(u,v)}\right) \beta_k(v)} \tag{2.231}$$

This can be divided into three parts

$$\Lambda(a_k) = \ln \frac{\Pr\{a_k = +1\}}{\Pr\{a_k = -1\}} + \ln \frac{p(y_{a,k} | a_k = +1)}{p(y_{a,k} | a_k = -1)}$$

$$+ \ln \frac{\displaystyle\sum_{(u,v) \in B_{+1}} \alpha_{k-1}(u) p\left(y_{r,k} | x_{r,k}^{(u,v)}\right) \beta_k(v)}{\displaystyle\sum_{(u,v) \in B_{-1}} \alpha_{k-1}(u) p\left(y_{r,k} | x_{r,k}^{(u,v)}\right) \beta_k(v)}$$

$$= \Lambda^{ap}(a_k) + \Lambda^{int}(a_k) + \Lambda^{ext}(a_k) \tag{2.232}$$

The first component is the so-called *a priori information* and its value depends on inequality of probabilities of the input symbol $a_k = i$. If the symbols ± 1 are equiprobable, then this component is zero. The second component is called *intrinsic information* and its value results exclusively from the generated data symbol a_k and the component $y_{a,k}$ related to it, received at the kth moment on the channel output. For the assumed channel model, intrinsic information according to formula (2.226) is

$$\Lambda^{int}(a_k) = \ln \frac{p(y_{a,k}|a_k = +1)}{p(y_{a,k}|a_k = -1)} = \ln \frac{\exp\left[-\dfrac{(y_{a,k} - 1)^2}{2\sigma^2}\right]}{\exp\left[-\dfrac{(y_{a,k} + 1)^2}{2\sigma^2}\right]}$$

$$= \ln \exp\left[-\frac{(y_{a,k} - 1)^2}{2\sigma^2} + \frac{(y_{a,k} + 1)^2}{2\sigma^2}\right] = \frac{2}{\sigma^2} y_{a,k} \qquad (2.233)$$

The third component is the so-called *extrinsic information*, which allows us to improve a decision upon the data symbol a_k on the basis of all remaining signals contained in the received block, i.e.

$$\Lambda^{ext}(a_k) = \ln \frac{\displaystyle\sum_{(u,v)\in B_{+1}} \alpha_{k-1}(u) p\left(y_{r,k}|x_{r,k}^{(u,v)}\right)\beta_k(v)}{\displaystyle\sum_{(u,v)\in B_{-1}} \alpha_{k-1}(u) p\left(y_{r,k}|x_{r,k}^{(u,v)}\right)\beta_k(v)}$$

$$= \ln \frac{\displaystyle\sum_{(u,v)\in B_{+1}} \alpha_{k-1}(u)\gamma_k^{+1}(y_{r,k}, u, v)\beta_k(v)}{\displaystyle\sum_{(u,v)\in B_{-1}} \alpha_{k-1}(u)\gamma_k^{-1}(y_{r,k}, u, v)\beta_k(v)} \qquad (2.234)$$

where

$$\gamma_k^i(y_{r,k}, u, v) = \exp\left[-\frac{(y_{r,k} - x_{r,k}^{(u,v)})^2}{2\sigma^2}\right] \quad \text{for } (u, v) \in B_i, \ i = \pm 1 \qquad (2.235)$$

The division of the soft decision $\Lambda(a_k)$ into three separate components is the most important aspect of the presented algorithm. The importance of each component will be clarified when we describe the turbo decoding algorithm.

Let us stress again that the algorithm presented above is a MAP algorithm. It is optimal in this sense; however, its computational complexity is very high. In practical systems suboptimum algorithms such as Max-Log-MAP (see Vucetic and Yuan 2000) or SOVA (see this chapter) are often applied at the cost of a fraction of dB loss in the performance.

2.12.4 Turbo Decoding Algorithm

As we said earlier, codewords generated by a turbo code encoder are usually very long because of the large size of the applied interleaver, therefore the optimal decoder would be extremely complex. For that reason a suboptimum decoder structure has to be applied. Owing to the applied algorithm, the difference in decoding quality is insignificant compared with the optimal decoder. This result is achieved by using two component decoders

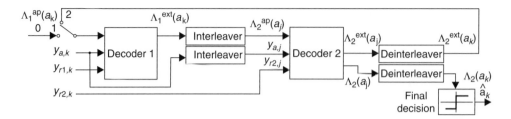

Figure 2.38 Block diagram of a turbo decoder

that mutually improve each other's decision quality. A basic scheme of the turbo code decoder is shown in Figure 2.38.

The first decoder makes soft decisions on subsequent message bits contained in the received block. For this purpose it uses the received message symbols $y_{a,k}$ and parity symbols $y_{r1,k}$ related to the signals generated by the first component encoder. Signals $y_{a,j}$ that are related to interleaved message symbols (the time index has been changed due to interleaving) and symbols $y_{r2,j}$ corresponding to parity symbols generated by the second encoder appear at the input of the second component decoder. Additionally, the second decoder is fed by *a priori* information achieved after interleaving of extrinsic information from the first decoder. After the symbol sequence has been reordered to the original time sequence by the deinterleaver, extrinsic information from the second decoder becomes *a priori* information in the subsequent iteration of the first decoder.

We can formally present the operation of a turbo code decoder in the following steps:

1. *First iteration*
   (a) *Decoder 1*. If the statistics of the message symbols a_k are unknown, then it is assumed that message symbols a_k are equiprobable. Then

$$\Lambda^{ap}(a_k) = \ln \frac{\Pr\{a_k = 1\}}{\Pr\{a_k = -1\}} = 0$$

The switch shown in Figure 2.38 is in position 1. On the basis of the vector of the received samples

$$\mathbf{y}_1^n = [\mathbf{y}_1, \mathbf{y}_2, \ldots, \mathbf{y}_n], \quad \mathbf{y_k} = \begin{bmatrix} y_{a,k} \\ y_{r1,k} \\ y_{r2,k} \end{bmatrix}, \quad k = 1, \ldots, n \qquad (2.236)$$

metrics $\gamma_k^{+1}(y_{r1,k}, u, v)$ and $\gamma_k^{-1}(y_{r,k}, u, v)$ of the transition from state u to state v at the $(k+1)$st moment ($u, v = 0, 1, \ldots, M-1$) are calculated for each symbol a_k ($k = 1, 2, \ldots, n$). It is also assumed that the initial state of the encoder is known, e.g. $u = 0$. Then for $k = 1$ transitions from the initial state to two different states are possible. They are determined by $a_1 = 1$ or $a_1 = -1$, respectively. Using the probabilities $\gamma_k^{+1}(y_{r1,k}, u, v)$ and $\gamma_k^{-1}(y_{r,k}, u, v)$, we determine the values of $\alpha_{k-1}(u)$ and $\beta_k(v)$ for subsequent time instants k, applying iterative formulae (2.219) and (2.223). Knowing the full set of the above probabilities, we calculate extrinsic information $\Lambda_1^{ext}(a_k)$ for each message symbol a_k. This constitutes the main output signal of the first decoder.

(b) *Interleaver*. After message bits have been interleaved, they are used to generate parity bits by the second component encoder. Consequently, the process of interleaving obviously implies the sequence of subsequent parity symbols used in the second turbo code decoder. If the results produced by the first decoder are to be used in the second decoder, this sequence has to be reordered in conformance with the parity symbols of the second encoder. The same applies to message symbols. Thus, the second decoder input is a triple: *a priori* information achieved owing to interleaving of the extrinsic information, interleaved sequence of the message symbols and the sequence of parity symbols matched in time to the interleaved sequence of message symbols.

(c) *Decoder 2*. Decoder 2 uses (besides message and parity symbols) the symbols that play the role of *a priori* information. This function is performed by extrinsic information received from Decoder 1. In this way correlation between *a priori* information and input symbols applied in Decoder 2 is avoided. Decoder 2, which functions identically as Decoder 1, determines extrinsic information $\Lambda_2^{ext}(a_j)$.

(d) *Deinterleaver*. This block matches the order of extrinsic information sequence obtained from Decoder 2 to the order of the sequence processed in Decoder 1. Extrinsic information $\Lambda_2^{ext}(a_j)$ is applied again in Decoder 1 as *a priori* information (the switch is in position 2).

2. *Subsequent iterations*. Starting from the second iteration, each component decoder receives *a priori* information from the other decoder. Owing to this information the decoding quality improves. The value of *a priori* information related to each message symbol gradually stabilizes its sign and increases its module.

3. *The last iteration*. After a number of iterations the decoding process achieves a state close to optimum and a final decision is made for each message bit. This decision is based on the soft decision $\Lambda(a_k)$ given by formula (2.232), i.e.

$$\widehat{a}_k = \text{sgn}\left[\Lambda(a_k)\right] \qquad (2.237)$$

As we said earlier, knowledge of the final state of encoders, which is often assumed, improves the final decoding quality.

Figure 2.39 shows the simulation results presented by Berrou, Glavieux and Thitimajshima in their original paper (Berrou *et al.* 1993). They are in the form of error probability curves versus signal-to-noise ratio. It is clear that decoding quality significantly improves with the increasing number of iterations. We can also notice that the most meaningful improvement is achieved after performing the first few iterations. Other investigations have shown that decoding quality strongly depends on the length of the decoded sequence or, equivalently, on the size and structure of the applied block interleaver.

Turbo codes are believed to be one of the most important achievements of communication theory in the 1990s. Their inventors have shown how to get significantly closer to the Shannon limit. It has turned out that only about 0.7 dB is left to achieve the theoretical limit. Invention of turbo codes triggered an enormous interest by code theorists and communication engineers in iterative processing. They developed other applications

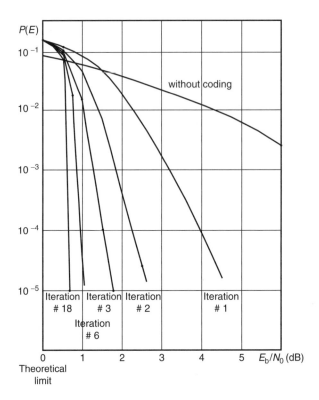

Figure 2.39   Error rate versus E_b/N_0 as a function of the number of iterations in the turbo decoder for the turbo code with the encoder presented in Figure 2.36. Reproduced by permission of IEEE (Berrou *et al.* (1993)) © 1993 IEEE

of the concept of turbo decoding. In practice, turbo codes are applied in systems in which decoding delay is not a critical parameter; however, a low error probability is of primary importance. Therefore turbo codes are useful in data transmission systems. They are applied e.g. in the UMTS (*Universal Mobile Telecommunication System*) cellular standard in which data transmission is one of the offered services.

2.13   LDPC Codes

LDPC (*Low-Density Parity Check*) codes are another class of codes that feature high decoding quality. They were invented by Gallager in the 1960s (Gallager 1968) but they did not gain attention at that time. They focused the interest of code theorists and practitioners again in the 1990s owing to the works of MacKay and Neal (1996) and MacKay (1999).

LDPC codes are (n, k) block codes whose parity check matrix is *sparse*, i.e. contains a small number of "1"s compared with the matrix size. A (J, K)-*regular* LDPC code is characterized by the parity check matrix in which there are J "1"s in each column and K "1"s in each row. For given code parameters (n, k) the assumed column weight J is larger than 3 and a parity check matrix H with a given number of "1"s in each column and row is generated in a pseudorandom manner. If the parity check matrix H is a low density matrix but the number of "1"s per column is not constant, the code is called an *irregular* LDPC code. The codeword length is typically very high and the code rate can be close to 1, although lower code rates are often applied too. Examples of code parameters (n, k) of high code rate LDPC codes presented in Benvenuto and Cherubini (2002) are $(495, 433)$, $(1998, 1777)$, $(4376, 4095)$. The related code rates are 0,8747, 0,8894, 0,9358.

Let us present three examples of parity check matrices for LDPC codes.

Consider a $(10, 5)$ linear block code with four "1"s in a row and two "1"s in each column. Its parity check matrix is the following

$$H = \begin{bmatrix} 1 & 1 & 1 & 1 & 0 & 0 & 0 & 0 & 0 & 0 \\ 1 & 0 & 0 & 0 & 1 & 1 & 1 & 0 & 0 & 0 \\ 0 & 1 & 0 & 0 & 1 & 0 & 0 & 1 & 1 & 0 \\ 0 & 0 & 1 & 0 & 0 & 1 & 0 & 1 & 0 & 1 \\ 0 & 0 & 0 & 1 & 0 & 0 & 1 & 0 & 1 & 1 \end{bmatrix} \quad (2.238)$$

The $(10, 5)$ linear code is not a typical LDPC code but it can serve as a simple example for educational reasons. Let us note that the parity check matrix is not in a canonical form. The density of "1"s in H is equal to 0.4, so the parity check matrix is not sparse in a strict sense.

Let us show another matrix created according to the rules valid for (J, K)-regular LDPC codes. Let the parameters for that matrix be $J = K = 4$ and $n = 15$. The parity check matrix is (Lin and Costello 2004):

$$H = \begin{bmatrix} 0 & 0 & 0 & 0 & 0 & 0 & 0 & 1 & 1 & 0 & 1 & 0 & 0 & 0 & 1 \\ 1 & 0 & 0 & 0 & 0 & 0 & 0 & 0 & 1 & 1 & 0 & 1 & 0 & 0 & 0 \\ 0 & 1 & 0 & 0 & 0 & 0 & 0 & 0 & 0 & 1 & 1 & 0 & 1 & 0 & 0 \\ 0 & 0 & 1 & 0 & 0 & 0 & 0 & 0 & 0 & 0 & 1 & 1 & 0 & 1 & 0 \\ 0 & 0 & 0 & 1 & 0 & 0 & 0 & 0 & 0 & 0 & 0 & 1 & 1 & 0 & 1 \\ 1 & 0 & 0 & 0 & 1 & 0 & 0 & 0 & 0 & 0 & 0 & 0 & 1 & 1 & 0 \\ 0 & 1 & 0 & 0 & 0 & 1 & 0 & 0 & 0 & 0 & 0 & 0 & 0 & 1 & 1 \\ 1 & 0 & 1 & 0 & 0 & 0 & 1 & 0 & 0 & 0 & 0 & 0 & 0 & 0 & 1 \\ 1 & 1 & 0 & 1 & 0 & 0 & 0 & 1 & 0 & 0 & 0 & 0 & 0 & 0 & 0 \\ 0 & 1 & 1 & 0 & 1 & 0 & 0 & 0 & 1 & 0 & 0 & 0 & 0 & 0 & 0 \\ 0 & 0 & 1 & 1 & 0 & 1 & 0 & 0 & 0 & 1 & 0 & 0 & 0 & 0 & 0 \\ 0 & 0 & 0 & 1 & 1 & 0 & 1 & 0 & 0 & 0 & 1 & 0 & 0 & 0 & 0 \\ 0 & 0 & 0 & 0 & 1 & 1 & 0 & 1 & 0 & 0 & 0 & 1 & 0 & 0 & 0 \\ 0 & 0 & 0 & 0 & 0 & 1 & 1 & 0 & 1 & 0 & 0 & 0 & 1 & 0 & 0 \\ 0 & 0 & 0 & 0 & 0 & 0 & 1 & 1 & 0 & 1 & 0 & 0 & 0 & 1 & 0 \end{bmatrix} \quad (2.239)$$

The size of the above parity check matrix is (15×15) so obviously some parity check equations represented by the rows of this matrix are redundant. It can be proved that after performing reduction of the parity check equations we end up with a $(15,7)$ cyclic BCH code considered earlier in this chapter. The density of matrix (2.239) is equal to 0.267, so the matrix can be considered sparse.

Finally, let us present matrix H designed according to the procedure originally proposed by Gallager. A code determined by such a parity check matrix is called a *Gallager code*.

The parity check matrix consists of J submatrices of dimensions $m \times mK$, denoted as $H_1, H_2, \ldots, H_J$. The full parity check matrix has the form

$$H = \begin{bmatrix} H_1 \\ H_2 \\ \vdots \\ H_J \end{bmatrix} \tag{2.240}$$

Each row of a constituent submatrix has K and each column has a single "1". In particular, in the ith row ($1 \le i \le m$) matrix H_1 contains a block of "1"s in the columns with numbers between $(i-1)K+1$ and iK. The remaining matrices are created by permutation of matrix H_1. However, not all permutations result in a good code that features a high value of minimum distance. An intensive computer search is needed to find such a code. Formula (2.241) presents an example of the parity check matrix of the Gallager code, characterized by $m = 5$, $K = 4$ and $J = 3$.

$$H = \left[\begin{array}{cccc|cccc|cccc|cccc|cccc}
1 & 1 & 1 & 1 & 0 & 0 & 0 & 0 & 0 & 0 & 0 & 0 & 0 & 0 & 0 & 0 & 0 & 0 & 0 & 0 \\
0 & 0 & 0 & 0 & 1 & 1 & 1 & 1 & 0 & 0 & 0 & 0 & 0 & 0 & 0 & 0 & 0 & 0 & 0 & 0 \\
0 & 0 & 0 & 0 & 0 & 0 & 0 & 0 & 1 & 1 & 1 & 1 & 0 & 0 & 0 & 0 & 0 & 0 & 0 & 0 \\
0 & 0 & 0 & 0 & 0 & 0 & 0 & 0 & 0 & 0 & 0 & 0 & 1 & 1 & 1 & 1 & 0 & 0 & 0 & 0 \\
0 & 0 & 0 & 0 & 0 & 0 & 0 & 0 & 0 & 0 & 0 & 0 & 0 & 0 & 0 & 0 & 1 & 1 & 1 & 1 \\
\hline
1 & 0 & 0 & 0 & 1 & 0 & 0 & 0 & 1 & 0 & 0 & 0 & 1 & 0 & 0 & 0 & 0 & 0 & 0 & 0 \\
0 & 1 & 0 & 0 & 0 & 1 & 0 & 0 & 0 & 1 & 0 & 0 & 0 & 0 & 0 & 0 & 1 & 0 & 0 & 0 \\
0 & 0 & 1 & 0 & 0 & 0 & 1 & 0 & 0 & 0 & 0 & 0 & 0 & 1 & 0 & 0 & 0 & 1 & 0 & 0 \\
0 & 0 & 0 & 1 & 0 & 0 & 0 & 0 & 0 & 0 & 1 & 0 & 0 & 0 & 1 & 0 & 0 & 0 & 1 & 0 \\
0 & 0 & 0 & 0 & 0 & 0 & 0 & 1 & 0 & 0 & 0 & 1 & 0 & 0 & 0 & 1 & 0 & 0 & 0 & 1 \\
\hline
1 & 0 & 0 & 0 & 0 & 1 & 0 & 0 & 0 & 0 & 0 & 1 & 0 & 0 & 0 & 0 & 0 & 1 & 0 & 0 \\
0 & 1 & 0 & 0 & 0 & 0 & 1 & 0 & 0 & 0 & 1 & 0 & 0 & 0 & 0 & 0 & 1 & 0 & 0 & 0 \\
0 & 0 & 1 & 0 & 0 & 0 & 0 & 1 & 0 & 0 & 0 & 0 & 1 & 0 & 0 & 0 & 0 & 0 & 1 & 0 \\
0 & 0 & 0 & 1 & 0 & 0 & 0 & 0 & 1 & 0 & 0 & 0 & 0 & 1 & 0 & 0 & 1 & 0 & 0 & 0 \\
0 & 0 & 0 & 0 & 1 & 0 & 0 & 0 & 0 & 1 & 0 & 0 & 0 & 0 & 1 & 0 & 0 & 0 & 0 & 1
\end{array}\right]$$
$$\tag{2.241}$$

Since reinvention of the LDPC codes in the 1990s, a large number of methods for their construction have been proposed. They are beyond the scope of this chapter but the interested reader may explore the rich literature on this subject, starting from Lin and Costello (2004), Moon (2005) or Ryan (2004).

2.13.1 Tanner Graph

A *Tanner graph* is a crucial graphical tool used in analysis of LDPC codes and their decoding algorithms. In a sense a Tanner graph representing an LDPC code is analogous to a trellis diagram representing convolutional codes. Its structure fully reflects the code properties. The Tanner graph is a *bipartite graph*, which means that its nodes are divided into two separate sets and each edge of the graph connects a pair of nodes, each of which belongs to a different set. One set consists of *variable nodes* (called *v-nodes*) whereas the other one contains *check nodes* (denoted as *c-nodes*). Each v-node is associated with a specific codeword element, so for an (n, k) code there are n v-nodes. Each c-node reflects a single row of the parity check matrix, and the edges connecting a c-node with the v-nodes result from a particular parity check equation described by the appropriate row of matrix H. Strictly speaking, check node j is connected to variable node i if element h_{ji} of matrix H is equal to 1. Figure 2.40 presents Tanner graphs for the codes determined by the parity check matrices given by (2.238) and (2.239), respectively. We will not present the Tanner graph for the code determined by (2.240) due to its complexity.

The edges of the Tanner graph that join a given check node s_j indicate which received symbols participate in the generation of the jth element s_j of the syndrome vector $\mathbf{s}$. On the other hand, the edges departing from a given variable node v_i inform which syndrome symbols are influenced by the ith symbol v_i of the received block $\mathbf{v}$. The Tanner graph helps to manage iterative decoding algorithms relying on the process of passing messages back and forth between v-nodes and c-nodes, which leads to gradual improvement of the selected decoding criterion, eventually resulting in the decoded codeword.

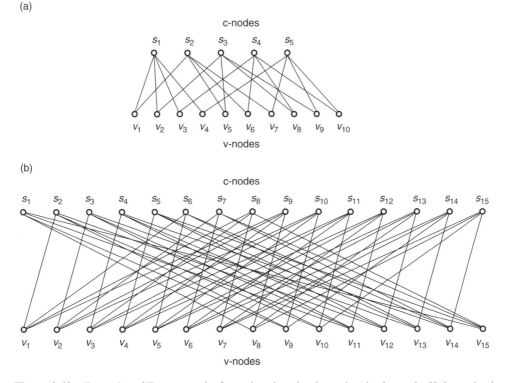

Figure 2.40 Examples of Tanner graphs for codes given by the parity check matrix H determined by (2.238) (a) and (2.239) (b)

2.13.2 Decoding of LDPC Codes

As in the case of other codes, LDPC codes can be decoded using hard- and soft-decision algorithms. There is a large variety of decoding algorithms. Let us look at some representative examples, i.e.:

- Bit-Flipping Algorithm (BPA): hard-decision decoding;
- Weighted Bit-Flipping Algorithm (WBPA): partially soft-decision decoding;
- Sum-Product Algorithm (SPA): fully soft-decision decoding.

2.13.2.1 Bit-Flipping Algorithm

The bit-flipping algorithm was proposed by Gallager in his early works on LDPC codes. It entirely relies on calculations of syndrome elements s_j, $(j = 1, \ldots, m)$, where m is the number of rows in the parity check matrix H. As before, denote the received block as $\mathbf{r} = (r_1, \ldots, r_n)^T$. First, the syndrome is calculated according to the well-known formula

$$\mathbf{s} = H\mathbf{r} \tag{2.242}$$

If $\mathbf{s} = 0$ then the received block is assumed to be correct and the decoder's decision is $\widehat{\mathbf{c}} = \mathbf{r}$. Otherwise the algorithm takes into account the nonzero syndrome elements and changes the bit in the received block $\mathbf{r}$ that participates in at least L failed parity check equations, i.e. for which $s_j = 1$. After that change, the modified received vector $\mathbf{r}'$ is the base for subsequent calculation of syndrome (2.242). Such a procedure is performed iteratively in the loop until the calculated syndrome is zero. Formally, the hard-decision bit-flipping algorithm can be defined in the following steps (Lin and Costello 2004):

1. Compute the syndrome bits s_j, $(j = 1, \ldots, m)$ from equation (2.242) for the received block $\mathbf{r}$. If all of them are equal to zero, the received block $\mathbf{r}$ is the decided codeword $\widehat{\mathbf{c}}$ and end the decoding procedure. Otherwise go to step 2.
2. Find the number f_i of failed syndrome bits for every received symbol r_i $(i = 1, \ldots n)$.
3. Determine the set S of bits for which f_i is the largest.
4. Negate the bits in set S.
5. Repeat steps 1–4 until all syndrome bits are zero or the maximum allowable number of iterations is reached.

Let us illustrate the operation of the bit-flipping algorithm with an example.

Example 2.13.1 *Consider the* (15, 7) *BCH code, which when treated as an LDPC code is determined by matrix H described by formula (2.239). We have considered this code when describing the Meggitt and majority decoders. The code is able to correct t = 2 errors and a certain number of combinations of triple errors. For simplicity assume that an all-zero codeword has been transmitted and the received block has the form* $\mathbf{r} = (110010000000000)^T$. *Thus, "1"s in the received block clearly indicate errors that have to be corrected by the decoder. The algorithm proceeds in the following steps:*

- *First, syndrome* $\mathbf{s} = H\mathbf{r}$ *is calculated, resulting in the sequence*

$$\mathbf{s} = (011000110001100)^T$$

- *For each received bit the number* f_i *of failed syndrome bits is calculated, giving the following vector* $\mathbf{f}$:

$$\mathbf{f} = (221122211222112)$$

- *The largest number* f_i *occurs already for the first bit, so the change of its value in the received block* $\mathbf{r}$ *results in a new block* $\mathbf{r}' = (010010000000000)^T$
- *Syndrome* $\mathbf{s} = H\mathbf{r}'$ *is calculated again, giving the vector*

$$\mathbf{s} = (001001101001100)^T$$

- *For each element of vector* $\mathbf{r}'$ *the number of failed syndrome bits is calculated again, resulting in the vector*

$$\mathbf{f} = (230232120121221)$$

- *This time the largest number of failed parity check equations is due to the second bit* ($f_2 = 3$), *so the second bit of vector* $\mathbf{r}'$ *is negated and the new vector* $\mathbf{r}''$ *is*

$$\mathbf{r}'' = (000010000000000)^T$$

- *Syndrome* $\mathbf{s} = H\mathbf{r}''$ *is calculated once more, resulting in the vector*

$$\mathbf{s} = (000001000101100)^T$$

Let us note that the syndrome is the fifth column of matrix H, which means that only one error, i.e. the error in the fifth position of the candidate codeword, remains to be corrected.

- *Inspection of the bits influencing "1"s in syndrome* $\mathbf{s}$ *results in the following vector* $\mathbf{f}$

$$\mathbf{f} = (111141111011110)$$

Clearly $f_5 = 4$ *indicates the bit that has to be flipped. If it is done, the resulting vector is*

$$\mathbf{r}''' = (000000000000000)^T$$

which agrees with the codeword produced by the encoder.

- *The resulting syndrome* $\mathbf{s} = H\mathbf{r}'''$ *is finally equal to zero, which terminates the decoding algorithm.*

For the considered code $d_{min} = 5$ so all combinations of two errors are correctable. Let us note, however, that the selected received block contained three errors, which means that the considered combination of three errors belongs to the correctable error patterns.

The above decoding algorithm is fully hard-decision. As we said earlier, its performance can be improved if the algorithm takes advantage of more detailed knowledge about the received symbols obtained from the channel.

2.13.2.2 Weighted Bit-Flipping Algorithm

Consider bipolar transmission of the codewords. Let $\mathbf{c} = (c_1, \ldots, c_n)^T$ be the codeword generated in the LDPC code encoder. The codeword $\mathbf{c}$ is subsequently transmitted in the bipolar form $\mathbf{x} = (x_1, \ldots, x_n)^T$, where $x_i = a(2c_i - 1)$ and a is the magnitude of bipolar pulses. The received block is denoted as $\mathbf{y} = (y_1, \ldots, y_n)$, where $y_i = x_i + v_i$ and v_i is a Gaussian noise sample, assumed to be statistically independent of other noise samples. We have already noticed that a simple measure of reliability of the received symbol y_i is its magnitude $|y_i|$. The larger the magnitude $|y_i|$, the more reliable is the hard-decision symbol r_i associated with the sample y_i. Denote the set of hard-decision bits r_i that participate in the calculation of syndrome bit s_j as $\mathcal{N}(j)$, so $\mathcal{N}(j) = \{i : h_{ji} = 1\}$. Similarly, let $\mathcal{M}(i) = \{j : h_{ji} = 1\}$ denote the set of parity checks in which bit r_i participates. First, reliability measures of syndrome components are calculated as

$$|y|_{min}^{(j)} = \min_{i:i\in\mathcal{N}(j)} |y_i| \quad j = 1, \ldots, m \tag{2.243}$$

In other words we assume that the reliability measure of syndrome bit s_j is the lowest magnitude of that received sample y_i for which the corresponding hard-decision decoded bit r_i is used to calculate syndrome bit s_j. After calculation of the reliability measures for each syndrome bit the main part of the weighted bit flipping algorithm is performed in the following steps:

1. On the basis of (2.242) calculate the syndrome, using hard-decision vector $\mathbf{r}$.
2. For each received bit r_i $(i = 1, \ldots, n)$ calculate

$$E_i = \sum_{j\in\mathcal{M}(i)} (2s_j - 1)|y|_{min}^{(j)}, \quad i = 1, \ldots, n \tag{2.244}$$

3. Flip bit r_k located in position k for which $k = \arg \max_{1 \leq i \leq n} E_i$.
4. Repeat steps 1–3 until all parity check equations are satisfied or the maximum number of iterations is reached.

Let us note that in step 3 position k is identified by finding the maximum value of E_i $(i = 1, \ldots, n)$. This value indicates that the selected kth bit influences the highest number of syndrome bits equal to 1. Note that the correct syndrome bit $(s_j = 0)$, if its index belongs to set $\mathcal{M}(i)$, implies that appropriate E_i decreases.

Example 2.13.2 *Let us consider a similar example to that used to illustrate the regular bit-flipping algorithm for the (15,7) BCH code determined by the parity check matrix H*

*given by (2.239). Again let the all-zero codeword be generated by the encoder, which
results in the bipolar vector*

$$\mathbf{x} = (-1, -1, -1, -1, -1, -1, -1, -1, -1, -1, -1, -1, -1, -1, -1)^T$$

At the output of the AWGN channel let the received block be

$$\mathbf{y} = (0.05, 0.12, -0.8, -1.2, 0.07, -2.1, -0.9, -0.1, -1.0, -0.5,$$
$$-0.09, -0.7, -0.99, -0.32, -1.2)^T$$

Hard-decision decoding would produce the received vector $\mathbf{r} = (110010000000000)^T$
*identical to that in the example illustrating the hard-decision bit-flipping algorithm.
Inspection of parity check matrix (2.239) leads to the sets* $\mathcal{N}(j)$ $(j = 1, \ldots, m)$ *and*
$\mathcal{M}(i)$ $(i = 1, \ldots, n)$, *which are shown in matrix form as*

$$
\mathcal{N} =
\begin{bmatrix}
8 & 9 & 11 & 15 \\
1 & 9 & 10 & 12 \\
2 & 10 & 11 & 13 \\
3 & 11 & 12 & 14 \\
4 & 12 & 13 & 15 \\
1 & 5 & 13 & 14 \\
2 & 6 & 14 & 15 \\
1 & 3 & 7 & 15 \\
1 & 2 & 4 & 8 \\
2 & 3 & 5 & 9 \\
3 & 4 & 6 & 10 \\
4 & 5 & 7 & 11 \\
5 & 6 & 8 & 12 \\
6 & 7 & 9 & 13 \\
7 & 8 & 10 & 14
\end{bmatrix}
\qquad
\mathcal{M} =
\begin{bmatrix}
2 & 6 & 8 & 9 \\
3 & 7 & 9 & 10 \\
4 & 8 & 10 & 11 \\
5 & 9 & 11 & 12 \\
6 & 10 & 12 & 13 \\
7 & 11 & 13 & 14 \\
8 & 12 & 14 & 15 \\
1 & 9 & 13 & 15 \\
1 & 2 & 10 & 14 \\
2 & 3 & 11 & 15 \\
1 & 3 & 4 & 12 \\
2 & 4 & 5 & 13 \\
3 & 5 & 6 & 14 \\
4 & 6 & 7 & 15 \\
1 & 5 & 7 & 8
\end{bmatrix}
$$

Each row of matrices $\mathcal{N}$ *and* $\mathcal{M}$ *corresponds to* $\mathcal{N}(j)$ $(j = 1, \ldots, m)$ *and* $\mathcal{M}(i)$ $(i = 1, \ldots, n)$, *respectively. They are useful in calculations of (2.243) and (2.244). The search
according to (2.243) for the smallest magnitude of those received samples that take part
in parity check giving syndrome element* s_j $(j = 1, \ldots, m)$ *leads to the vector*

$$\mathbf{y}_{\min} = \{|y|_{\min}^{(j)} : j = 1, \ldots m\}$$
$$= (0.09, \ 0.05, \ 0.09, \ 0.09, \ 0.70, \ 0.05, \ 0.12, \ 0.05, \ 0.05, \ 0.07,$$
$$0.50, \ 0.07, \ 0.07, \ 0.90, \ 0.10)^T$$

*After computing reliability measures of each syndrome bit we can start the iterative part
of the weighted bit-flipping algorithm.*

- *As usual, the syndrome is calculated first:*

$$\mathbf{s} = H\mathbf{r} = (011000110001100)^T$$

- *Then, for each bit r_i $(i = 1, \ldots, n)$ its reliability measure E_i is computed according to (2.244), giving the vector of values*

$$\mathbf{E} = \{E_i : i = 1, \ldots, n\}$$
$$= (0, \ 0.09, \ -0.61, \ -1.18, \ 0.02, \ -1.21, \ -0.88, \ -0.17, \ -1.01, \ -0.46,$$
$$-0.02, \ -0.67, \ -1.56, \ -0.12, \ -0.62)^T$$

- *The largest value E_i is located at position $k = 2$ (step 3), so r_2 in vector $\mathbf{r}$ is flipped, resulting in a new vector*

$$\mathbf{r}' = (100010000000000)^T$$

- *For this vector the syndrome is computed again, resulting in the block*

$$\mathbf{s} = H\mathbf{r}' = (010000011101100)^T$$

- *Consequently, vector $\mathbf{E}$ for the new syndrome is (step 2):*

$$\mathbf{E} = (0.10, \ -0.09, \ -0.47, \ -1.08, \ 0.16, \ -1.45, \ -0.88, \ -0.07, \ -0.87, \ -0.64,$$
$$-0.20, \ -0.67, \ -1.74, \ -0.36, \ -0.86)^T$$

- *This time the largest E_i can be found at position $k = 5$, so r_5 in vector $\mathbf{r}$ is flipped, giving a new candidate for the decided codeword*

$$\mathbf{r}'' = (100000000000000)^T$$

- *We come back to step 1 again and calculate the new syndrome*

$$\mathbf{s} = H\mathbf{r}'' = (010001011000000)^T$$

- *The calculated syndrome is still different from zero, so the algorithm is continued. The newly computed vector $\mathbf{E}$ is*

$$\mathbf{E} = (0.20, \ -0.23, \ -0.61, \ -1.22, \ -0.16, \ -1.59, \ -1.02, \ -0.21, \ -1.01, \ -0.64,$$
$$-0.34, \ -0.81, \ -1.64, \ -0.26, \ -0.86)^T$$

- *Clearly, the highest value of E_i is at position $k = 1$, so we flip bit r_1. The resulting vector $\mathbf{r}$ is an all-zero vector $\mathbf{r}'''$.*
- *The algorithm is stopped after checking that this time the syndrome is equal to zero.*

As we can see, the algorithm has succesfully decoded the received block. In the considered example, calculation of all vectors by hand during the algorithm operation can be cumbersome, so a simple Matlab program has been written by the author to implement the weighted bit-flipping algorithm. The reader is advised to write his/her own program as an exercise.

2.13.2.3 Sum-Product Algorithm

The *Sum-Product Algorithm* is also called the *Belief-Propagation Algorithm* and is a representative example of soft-decision decoding of LDPC codes. Let the codeword $\mathbf{c}$ of length n generated by the encoder be represented by bipolar symbols. These symbols are distorted by additive white Gaussian noise, which results in reception of the sample block $\mathbf{y}$. Generally, the optimal decoding should be performed using the *Maximum a Posteriori* criterion. According to the MAP criterion the algorithm searches for the sequence $\widehat{\mathbf{c}} = [\widehat{c}_1, \ldots, \widehat{c}_n]$, which maximizes the *a posteriori* probability $\Pr\{\mathbf{c}|\mathbf{y}\}$ and fulfills the syndrome condition $\mathbf{s} = H\widehat{\mathbf{c}} = \mathbf{0}$. Thus, the criterion used by the MAP decoder can be formulated as

$$\widehat{\mathbf{c}} = \arg\max_{\mathbf{c}} \Pr\{\mathbf{c}|\mathbf{y}, H\mathbf{c} = \mathbf{0}\} \qquad (2.245)$$

However, the task defined by (2.245) is computationally complex to such an extent that it is replaced by the following n suboptimal criteria

$$\widehat{c}_i = \arg\max_{c_i} \Pr\left\{c_i|\mathbf{y}, \{s_j = 0, \ j \in \mathcal{M}(i)\}\right\}, \quad i = 1, \ldots, n \qquad (2.246)$$

where $\mathcal{M}(i) = \{j : h_{ji} = 1\}$ is the set of indices of check bits in which codeword bit c_i is used. Condition $\{s_j = 0, j \in \mathcal{M}(i)\}$ means that all checks involving c_i are satisfied. Denote

$$Q_i(b) = \Pr\left\{c_i = b|\mathbf{y}, \{s_j = 0, \ j \in \mathcal{M}(i)\}\right\}, \ b \in \{0, 1\}, \quad i = 1, \ldots, n \qquad (2.247)$$

The above *a posteriori* probability is the base for hard decisions upon transmitted codeword bits, i.e.

$$c_i = \begin{cases} 0 & \text{if } Q_i(0) > Q_i(1) \\ \\ 1 & \text{otherwise} \end{cases} \qquad (2.248)$$

The sum-product decoder estimates $Q_i(0)$ and $Q_i(1)$ iteratively and checks if the derived decision $\widehat{\mathbf{c}}$ implies the zero syndrome. If it does, the decoding process is considered as successfully finished. As we see, the main problem is to estimate the probabilities (2.247). In order to perform this task let us introduce the auxiliary probabilities

$$Q_{ji}(b) = \Pr\left\{c_i = b|\mathbf{y}, \text{ all checks involving } c_i \text{ except } s_j \text{ are satisfied}\right\}$$
$$= \Pr\left\{c_i = b|\mathbf{y}, \{s_{j'} = 0, \ j' \in \mathcal{M}(i)\backslash j\}\right\}, \ i = 1, \ldots, n, \quad j = 1, \ldots, m \qquad (2.249)$$

where $\mathcal{M}(i)\backslash j$ denotes the set of those syndrome bit indices, except the jth syndrome bit, in which the codeword bit c_i is involved. We also introduce the second auxiliary probability

$$R_{ji}(b) = \Pr\{s_j = 0|c_i = b, \mathbf{y}\}, \quad b \in \{0, 1\} \qquad (2.250)$$

This is the probability that the jth syndrome check is fulfilled under condition that the ith codeword bit $c_i = b$ ($b \in \{0, 1\}$) and the sample block $\mathbf{y}$ has been received.

An iterative process is performed in order to gradually improve the quality of $Q_i(0)$ and $Q_i(1)$ estimates. First, information from the received data is used to calculate probabilities $R_{ji}(b)$ ($b = 0, 1$), referring to the parity check bits. These probabilities are derived for those pairs (j, i) for which the parity check matrix entries are equal to 1, i.e. $h_{ji} = 1$ ($i = 1, \ldots, n$, $j = 1, \ldots, m$). Based on these probabilities, extrinsic information $Q_{ji}(b)$ ($b = 0, 1$; $i = 1, \ldots, n$, $j = 1, \ldots, m$) about transmitted codeword bits is calculated. The extrinsic information is applied in turn to compute the updated version of the probabilities referring to the check bits. As we have mentioned, the iterative procedure is performed in a loop until $Q_i(b)$ ($i = 1, \ldots, n$) achieves such values that hard decisions (2.248) performed on them result in the zero syndrome.

Now let us derive probabilities $Q_i(b)$, $Q_{ji}(b)$ and $R_{ji}(b)$ applied in the decoding algorithm. We start with $Q_i(b)$. Using the definition of $Q_i(b)$ and Bayes' rule a few times, we obtain

$$Q_i(b) = \Pr\left\{c_i = b|\mathbf{y}, \{s_j = 0, \ j \in \mathcal{M}(i)\}\right\} = \frac{\Pr\left\{c_i = b, \mathbf{y}, \{s_j = 0, \ j \in \mathcal{M}(i)\}\right\}}{\Pr\{\mathbf{y}, \{s_j = 0, \ j \in \mathcal{M}(i)\}\}}$$

$$= \frac{\Pr\left\{c_i = b, \{s_j = 0, \ j \in \mathcal{M}(i)\}|\mathbf{y}\right\} p(\mathbf{y})}{\Pr\{\{s_j = 0, \ j \in \mathcal{M}(i)\}|\mathbf{y}\} p(\mathbf{y})}$$

$$= \frac{1}{\Pr\{\{s_j = 0, j \in \mathcal{M}(i)\}|\mathbf{y}\}} \cdot \frac{\Pr\left\{c_i = b, \{s_j = 0, j \in \mathcal{M}(i)\}, \mathbf{y}\right\}}{p(\mathbf{y})}$$

$$= \frac{1}{\Pr\{\{s_j = 0, j \in \mathcal{M}(i)\}|\mathbf{y}\}} \cdot \frac{\Pr\left\{\{s_j = 0, j \in \mathcal{M}(i)\}|c_i = b, \mathbf{y}\right\}\Pr\{c_i = b|\mathbf{y}\}p(\mathbf{y})}{p(\mathbf{y})}$$

$$= \frac{1}{\Pr\{\{s_j = 0, \ j \in \mathcal{M}(i)\}|\mathbf{y}\}} \cdot \Pr\left\{\{s_j = 0, \ j \in \mathcal{M}(i)\}|c_i = b, \mathbf{y}\right\} \Pr\{c_i = b|\mathbf{y}\}$$

$$\tag{2.251}$$

Assuming independence of subsequent check bits, we may write

$$Q_i(b) = \frac{1}{\Pr\{\{s_j = 0, \ j \in \mathcal{M}(i)\}|\mathbf{y}\}} \cdot \Pr\{c_i = b|\mathbf{y}\} \prod_{j \in \mathcal{M}(i)} \Pr\{s_j = 0|c_i = b, \mathbf{y}\} \tag{2.252}$$

Recalling (2.250) we can formulate $Q_i(b)$ as

$$Q_i(b) = \frac{1}{\Pr\{\{s_j = 0, \ j \in \mathcal{M}(i)\}|\mathbf{y}\}} \cdot \Pr\{c_i = b|\mathbf{y}\} \prod_{j \in \mathcal{M}(i)} R_{ji}(b) \tag{2.253}$$

Due to the statistical independence of codeword bits and noise samples we have

$$\Pr\{c_i = b|\mathbf{y}\} = \Pr\{c_i = b|y_i\}$$

Let us note that the denominator in (2.253) can be expressed as

$$\Pr\{\{s_j = 0, \ j \in \mathcal{M}(i)\}|\mathbf{y}\}$$

$$= \Pr\{c_i = 0|y_i\} \prod_{j \in \mathcal{M}(i)} \Pr\{s_j = 0|c_i = 0, \mathbf{y}\} + \Pr\{c_i = 1|y_i\} \prod_{j \in \mathcal{M}(i)} \Pr\{s_j = 0|c_i = 1, \mathbf{y}\}$$

$$(2.254)$$

As we can see from (2.253) and (2.254), the value of the denominator in (2.253) is such that $Q_i(0) + Q_i(1) = 1$, so instead of calculation of this denominator we can apply the formula

$$Q_i(b) = \alpha_i \Pr\{c_i = b|\mathbf{y}\} \prod_{j \in \mathcal{M}(i)} R_{ji}(b) \qquad (2.255)$$

Coefficients α_i are selected to ensure that $Q_i(0) + Q_i(1) = 1$.
Similar calculations lead to the following formula for $Q_{ji}(b)$

$$Q_{ji}(b) = \Pr\left\{c_i = b|\mathbf{y}, \{s_{j'} = 0, \ j' \in \mathcal{M}(i)\backslash j\}\right\}$$

$$= \alpha_{ji} \Pr\{c_i = b|y_i\} \prod_{j' \in \mathcal{M}(i)\backslash j} R_{j'i}(b) \qquad (2.256)$$

Let us now derive $R_{ji}(b) = \Pr\{s_j = 0|c_i = b, \mathbf{y}\}$. Let us consider the case for which $c_i = 0$, i.e. we derive $R_{ji}(0)$. For that purpose we apply the following theorem.

Theorem 2.13.1 *Consider a sequence of K independent binary symbols a_i featuring probability $\Pr\{a_i = 1\} = p_i$. The probability that the set $\{a_i, \ i = 1, \ldots, K\}$ contains an even number of "1"s is equal to*

$$\frac{1}{2} + \frac{1}{2}\prod_{i=1}^{K}(1 - 2p_i) \qquad (2.257)$$

One can prove by induction that the above result holds true. It can be directly applied in the calculation of $R_{ji}(0)$, because we know that syndrome bit s_j is equal to zero if there is an even number of received symbols equal to 1 used to calculate this syndrome check. Therefore, after replacing p_i by $Q_{ji}(1)$ in (2.257) we can write

$$R_{ji}(0) = \Pr\{s_j = 0|c_i = 0, \mathbf{y}\}$$

$$= \frac{1}{2} + \frac{1}{2} \prod_{i' \in \mathcal{N}(j)\backslash i} \left[1 - 2Q_{ji'}(1)\right] \qquad (2.258)$$

Consequently

$$R_{ji}(1) = 1 - R_{ji}(0) \qquad (2.259)$$

In order to be able to perform the sum-product algorithm we still have to derive the probabilities $\Pr\{c_i = b|y_i\}$, where $b = 0, 1$. Recall our assumption that the received block **y** is the result of transmission of bipolar block **x** through the channel distorted by white Gaussian noise, i.e. $y_i = x_i + v_i$, $i = 1, \ldots, n$, where $x_i = a(2c_i - 1)$ and a is, as before, the magnitude of a bipolar pulse representing a single bit. Let us start from $\Pr\{c_i = 1|y_i\}$. We have

$$\Pr\{c_i = 1|y_i\} = \frac{p(y_i|x_i = a)\Pr\{x_i = a\}}{p(y_i)}$$

$$= \frac{p(y_i|x_i = a)\Pr\{x_i = a\}}{p(y_i|x_i = a)\Pr\{x_i = a\} + p(y_i|x_i = -a)\Pr\{x_i = -a\}} \quad (2.260)$$

Assuming *a priori* probabilities $\Pr\{x_i = a\} = \Pr\{x_i = -a\} = 1/2$, we receive

$$\Pr\{c_i = 1|y_i\} = \frac{p(y_i|x_i = a)}{p(y_i|x_i = a) + p(y_i|x_i = -a)} \quad (2.261)$$

Knowing that for Gaussian additive noise the conditional probability densities are described by formula

$$p(y_i|x_i) = \frac{1}{2\sqrt{2\pi}\sigma}\exp\left[-\frac{(y_i - x_i)^2}{2\sigma^2}\right] \quad (2.262)$$

we get

$$\Pr\{c_i = 1|y_i\} = \frac{1}{1 + \exp\left(-2ay_i/\sigma^2\right)} \quad (2.263)$$

The sum-product algorithm can be summarized in the following steps:

1. For $i = 1, \ldots, n$ calculate $\Pr\{c_i = 1|y_i\}$ according to (2.263). Initiate $Q_{ji}(1) = \Pr\{c_i = 1|y_i\}$ and $Q_{ji}(0) = 1 - Q_{ji}(1)$ for all i and j for which $h_{ji} = 1$.
2. Update the values of $R_{ji}(b)$ ($b = 0, 1$) using equations (2.258) and (2.259) for all i and j for which $h_{ji} = 1$.
3. Actualize the values of $Q_{ji}(b)$ ($b = 0, 1$) using equation (2.256) and calculate the normalizing constants α_{ji}.
4. For $i = 1, \ldots, n$ calculate

$$Q_i(1) = \alpha_i \Pr\{c_i = 1|\mathbf{y}\} \prod_{j \in \mathcal{M}(i)} R_{ji}(1) \quad (2.264)$$

and

$$Q_i(0) = \alpha_i \Pr\{c_i = 0|\mathbf{y}\} \prod_{j \in \mathcal{M}(i)} R_{ji}(0) \quad (2.265)$$

and the normalizing coefficient α_i, taking into account the fact that $Q_i(0) + Q_i(1) = 1$.

5. Find tentative hard decisions using (2.248) and check if $\mathbf{s} = H\widehat{\mathbf{c}} = 0$. If the syndrome
   is zero, stop the algorithm and deliver its decision $\widehat{\mathbf{c}}$. Otherwise go to step 2.

The sum-product algorithm has been presented above in its basic form. It has been
performed in the probability domain. There are more computationally efficient versions
of it in which, instead of probabilities, the likelihoods or log-likelihoods are applied;
however, this is beyond the scope of this introductory chapter. The interested reader is
recommended to study Lin and Costello (2004), Moon (2005) or Ryan (2004).

At the end of this section let us illustrate the operation of the sum-product algorithm
for the code used in the previous examples.

Example 2.13.3 *Assume bipolar transmission with symbol amplitude $a = 1$ distorted by
statistically independent additive Gaussian noise samples. Let the signal-to-noise ratio be
on the level of 1.2 dB, so if the signal power is equal to unity the noise variance is $\sigma^2 =
0.759$. As in the previous examples, let the all-zero codeword represented by all-minus-one
symbol block of length $n = 15$ be transmitted. The received block is*

$$\mathbf{y} = [0.168, -0.662, -0.658, -2.487, -0.802, -0.403, -1.555, -1.873, -1.162,$$

$$-1.918, -1.062, -0.757, 0.196, -0.843, -1.472]$$

so hard decisions performed on it result in the binary block

$$\mathbf{r} = [100000000000100]$$

*with two errors, in the first and thirteen position, respectively. Syndrome calculation results
in the vector*

$$\mathbf{s} = H\mathbf{r} = [011010011000010]$$

which is nonzero, so the error correction procedure has to be applied.

*First the algorithm has to be initialized. Based on the received block $\mathbf{y}$, probabilities
$\Pr\{c_i = 1 | y_i\}$ $(i = 1, \ldots, n)$ are calculated according to (2.263), resulting in the following
vector*

$$\Pr\{c_i = 1 | y_i\}|_{i=1}^{n} = [0.391, 0.851, 0.850, 0.999, 0.892, 0.743, 0.984, 0.993, 0.955,$$

$$0.994, 0.943, 0.880, 0.374, 0.902, 0.980]$$

*The above vector entries are used to initialize $Q_{ji}(1)$ and $Q_{ji}(0) = 1 - Q_{ji}(1)$ $(i =
1, \ldots, n, j = 1, \ldots, m)$ on those positions (i, j) for which $h_{ji} = 1$. Other positions (i, j)
in $Q_{ji}(1)$ and $Q_{ji}(0)$ matrices remain equal to zero.*

*After the above introductory steps the main iterative part of the decoding algorithm
can be started. First, according to step 2, values of $R_{ji}(b)$ $(b = 0, 1)$ are updated using*

equations (2.258) and (2.259), resulting in the matrix

$$[R_{ji}(0)]_{m \times n}$$

$$= \begin{bmatrix}
0 & 0 & 0 & 0 & 0 & 0 & 0 & 0.887 & 0.919 & 0 & 0.931 & 0 & 0 & 0 & 0.897 \\
0.842 & 0 & 0 & 0 & 0 & 0 & 0 & 0 & 0.418 & 0.425 & 0 & 0.402 & 0 & 0 & 0 \\
0 & 0.389 & 0 & 0 & 0 & 0 & 0 & 0 & 0 & 0.421 & 0.412 & 0 & 0.807 & 0 & 0 \\
0 & 0 & 0.771 & 0 & 0 & 0 & 0 & 0 & 0 & 0 & 0.714 & 0.714 & 0 & 0.736 & 0 \\
0 & 0 & 0 & 0.408 & 0 & 0 & 0 & 0 & 0 & 0 & 0 & 0.379 & 0.864 & 0 & 0.404 \\
0.420 & 0 & 0 & 0 & 0.522 & 0 & 0 & 0 & 0 & 0 & 0 & 0 & 0.431 & 0.522 & 0 \\
0 & 0.688 & 0 & 0 & 0 & 0.771 & 0 & 0 & 0 & 0 & 0 & 0 & 0 & 0.664 & 0.637 \\
0.825 & 0 & 0.399 & 0 & 0 & 0 & 0.427 & 0 & 0 & 0 & 0 & 0 & 0 & 0 & 0.426 \\
0.845 & 0.393 & 0 & 0.425 & 0 & 0 & 0 & 0.424 & 0 & 0 & 0 & 0 & 0 & 0 & 0 \\
0 & 0.750 & 0.751 & 0 & 0.724 & 0 & 0 & 0 & 0.693 & 0 & 0 & 0 & 0 & 0 & 0 \\
0 & 0 & 0.739 & 0.668 & 0 & 0.845 & 0 & 0 & 0 & 0.670 & 0 & 0 & 0 & 0 & 0 \\
0 & 0 & 0 & 0.836 & 0.927 & 0 & 0.846 & 0 & 0 & 0 & 0.878 & 0 & 0 & 0 & 0 \\
0 & 0 & 0 & 0 & 0.682 & 0.794 & 0 & 0.645 & 0 & 0 & 0 & 0.688 & 0 & 0 & 0 \\
0 & 0 & 0 & 0 & 0 & 0.389 & 0.444 & 0 & 0.441 & 0 & 0 & 0 & 0.714 & 0 & 0 \\
0 & 0 & 0 & 0 & 0 & 0 & 0.892 & 0.884 & 0 & 0.884 & 0 & 0 & 0 & 0.971 & 0
\end{bmatrix}$$

As each nonzero element $R_{ji}(1) = 1 - R_{ji}(0)$, i.e. it is a complement of $R_{ji}(0)$ with respect to unity, we omit the explicit presentation of matrix $[R_{ji}(1)]_{m \times n}$. Knowing already matrices $[R_{ji}(0)]_{m \times n}$ and $[R_{ji}(1)]_{m \times n}$, the algorithm is able to update the values of $Q_{ji}(b)$ ($b = 0, 1$) using equation (2.256), which results in the matrix

$$[Q_{ji}(0)]_{m \times n}$$

$$= \begin{bmatrix}
0 & 0 & 0 & 0 & 0 & 0 & 0 & 0.868 & 0.451 & 0 & 0.890 & 0 & 0 & 0 & 0.363 \\
0.991 & 0 & 0 & 0 & 0 & 0 & 0 & 0 & 0.991 & 0.985 & 0 & 0.958 & 0 & 0 & 0 \\
0 & 0.960 & 0 & 0 & 0 & 0 & 0 & 0 & 0 & 0.985 & 0.999 & 0 & 0.986 & 0 & 0 \\
0 & 0 & 1.000 & 0 & 0 & 0 & 0 & 0 & 0 & 0 & 1.000 & 0.998 & 0 & 1.000 & 0 \\
0 & 0 & 0 & 0.984 & 0 & 0 & 0 & 0 & 0 & 0 & 0 & 0.973 & 0.985 & 0 & 0.990 \\
0.997 & 0 & 0 & 0 & 0.995 & 0 & 0 & 0 & 0 & 0 & 0 & 0 & 0.995 & 0.998 & 0 \\
0 & 0.987 & 0 & 0 & 0 & 0.999 & 0 & 0 & 0 & 0 & 0 & 0 & 0 & 1.000 & 0.996 \\
1.000 & 0 & 1.000 & 0 & 0 & 0 & 1.000 & 0 & 0 & 0 & 0 & 0 & 0 & 0 & 0.999 \\
0.997 & 0.989 & 0 & 0.993 & 0 & 0 & 0 & 1.000 & 0 & 0 & 0 & 0 & 0 & 0 & 0 \\
0 & 0.993 & 0.999 & 0 & 1.000 & 0 & 0 & 0 & 0.999 & 0 & 0 & 0 & 0 & 0 & 0 \\
0 & 0 & 0.991 & 0.977 & 0 & 0.993 & 0 & 0 & 0 & 0.985 & 0 & 0 & 0 & 0 & 0 \\
0 & 0 & 0 & 0.883 & 0.978 & 0 & 0.973 & 0 & 0 & 0 & 0.994 & 0 & 0 & 0 & 0 \\
0 & 0 & 0 & 0 & 0.956 & 0.874 & 0 & 0.963 & 0 & 0 & 0 & 0.423 & 0 & 0 & 0 \\
0 & 0 & 0 & 0 & 0 & 0.998 & 0.997 & 0 & 0.994 & 0 & 0 & 0 & 0.995 & 0 & 0 \\
0 & 0 & 0 & 0 & 0 & 0 & 0.994 & 0.998 & 0 & 0.981 & 0 & 0 & 0 & 0.997 & 0
\end{bmatrix}$$

Similarly, $[Q_{ji}(1)]_{m \times n}$ has been omitted. The next step is the calculation of the estimates of a posteriori *probabilities according to (2.264) and (2.265). The resulting vectors are*

$$[Q_i(1)]|_{i=1}^{n} = [0.015, 0.060, 0.009, 0.000, 0.002, 0.008, 0.001, 0.000, 0.003, 0.001,$$
$$0.000, 0.048, 0.032, 0.001, 0.003]$$

$$[Q_i(0)]|_{i=1}^{n} = [0.985, 0.940, 0.991, 1.000, 0.998, 0.992, 0.999, 1.000, 0.997, 0.999,$$
$$1.000, 0.952, 0.968, 0.999, 0.997]$$

In the next step the algorithm finds tentative hard decisions using (2.248). The resulting vector is

$$\mathbf{r} = [000000000000000]$$

which is identical to the codeword generated in the transmitter. Clearly, checking the parity $H\mathbf{r}$ results in the syndrome $\mathbf{s} = \mathbf{0}$, which ends the decoding algorithm.

Let us note that the algorithm corrected the received sequence in a single step by finding two erroneous positions at once. The considered numerical example is very simple. If the codeword length is of the order of a few thousand, the algorithm needs many iterations to correctly decode the received sequence. Also, since computational complexity becomes excessive, the sum-product algorithm is performed in the log-likelihood domain.

2.14   Error Detection Structures and Algorithms

As we have mentioned, in cases in which irregularity in data block transfer rate can be tolerated and there is a simultaneous requirement for a very low error probability, error detection with some mechanism of block retransmission is a common solution. We say that the data transmission system is *nontransparent*. Data blocks arrive at the recipient irregularly; however, their reliability is very high. A necessary condition for application of an error detection procedure is the possibility of establishing a feedback channel from the receiver to the transmitter. This channel is used to inform the transmitter about the results of checking the correctness of the received data block. The traditional configuration of an error detection system is shown in Figure 2.41.

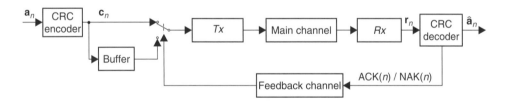

Figure 2.41   System configuration with error detection coding and block repetition

User data are first formed in blocks, denoted in Figure 2.41 by a symbol $\mathbf{a}_n$. These blocks are then coded using a selected shortened cyclic code. Codewords $\mathbf{c}_n$ are subsequently transmitted over the *main channel* to the receiver. For each received block a syndrome is calculated. When the syndrome is nonzero, the receiver signals detection of an erroneous data block by sending a so-called negative acknowledgment NAK (*Not-Acknowledged*) indicating the necessity of repeating the block. If the syndrome calculated by the receiver is zero, it means that the decoder located in the receiver has not detected any errors in the received block and a positive acknowledgment ACK (*Acknowledge*) is sent to the transmitter. Retransmission of erroneous blocks is automatic, therefore this technique is often denoted by the acronym ARQ (*Automatic-Repeat-Request*).

The motivation for using a shortened cyclic code is as follows. The length of the information block is usually determined by a particular application. Applying a polynomial code for error detection results in calculation of the remainder from division of the message polynomial by the generator polynomial $g(x)$. In the case of cyclic codes the generator polynomial is a divisor of $x^n - 1$, where n is the codeword length of a full-length cyclic code. The applied code is often denoted by the abbreviation CRC (*Cyclic Redundancy Check*). The remainder from division of the message polynomial by the generator polynomial calculated by the encoder is attached to the end of the message block and constitutes a parity block. A few standardized generator polynomials $g(x)$ are applied in practice. The most important among them are (Held 1999)

$$
\begin{aligned}
&\text{CRC-16 (ANSI)} && g(x) = x^{16} + x^{15} + x^2 + 1 \\
&\text{CRC (ITU-T)} && g(x) = x^{16} + x^{12} + x^5 + 1 \\
&\text{CRC-12} && g(x) = x^{12} + x^{11} + x^3 + 1 && (2.266) \\
&\text{CRC–32} && g(x) = x^{32} + x^{26} + x^{23} + x^{22} + x^{16} + x^{12} + x^{11} \\
& && \qquad\quad + x^{10} + x^8 + x^7 + x^5 + x^4 + x^2 + x + 1
\end{aligned}
$$

Let us check the error detection oportunities offered by a CRC code. For that purpose we formally define an error burst.

Definition 2.14.1 *An error burst of length b is a sequence of b bits in which the first and the bth bit are erroneous and among the remaining b − 2 bits some other bits can be erroneous too.*

The following theorem states abilities of error burst detection using a linear block code.

Theorem 2.14.1 *The application of a linear code with b parity bits is the necessary and sufficient condition of detection of all error bursts of length b or smaller in a binary block of length n.*

Let us note that error burst detection capability does not depend on the block length n, so the method is very useful when very long message blocks are applied. The blocks can have different lengths, which can also be useful in some applications. In order to make error detection possible, the receiver requires determination of the start and the end of the block.

Now we will show that b parity bits are sufficient to detect all error bursts not longer than b bits. Let us group all message bits into b-bit subblocks. Let the block of b parity bits be attached to the end of each message block. In this way we have created a systematic code with parameters $[n, (n - b)]$. Let the ith parity bit be a modulo-2 sum of the ith bits of each b-bit subblock. Let us note that only one bit in any b-bit or shorter error burst will influence a given parity bit, so the presence of the error burst will be detected. Such a situation occurs in cases when an error burst is contained in a single subblock or when it partially overlaps two neighboring subblocks. This case is illustrated in Figure 2.42.

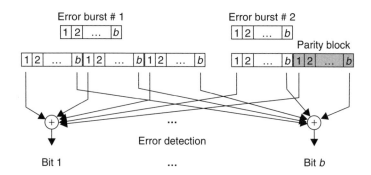

Figure 2.42   Detection of error bursts by a linear code of b parity bits

It turns out that application of a linear code, in particular a shortened cyclic code, allows the detection of most error bursts longer than the number of parity bits. In order to show this, we will prove the following theorem.

Theorem 2.14.2 *A fraction of error bursts of length $b > r$, which remain undetected by a cyclic code (n, k), where $r = n - k$, is 2^{-r} when $b > r + 1$ and is equal to $2^{-(r-1)}$ when $b = r + 1$.*

Proof. Assume without loss of generality that an error burst of length b bits appeared starting from the ith bit in a block. This implies that it ends in the $(i + b - 1)$st bit. In polynomial notation this error burst can be given in the form

$$e(x) = x^i b(x) \tag{2.267}$$

where $b(x) = x^{b-1} + \cdots + 1$ is a polynomial of degree $(b - 1)$ that describes the error burst. Because the error burst starts in the ith position and ends in the $(i + b - 1)$st position, the coefficients of $b(x)$ are equal to 1 for the polynomial components with the power equal to $(b - 1)$ and to zero. The remaining coefficients can take values equal to zero or 1 depending on the current pattern of the error burst. Therefore, there exist 2^{b-2} different error bursts of length b bits. Because error detection is performed by dividing the polynomial representing the received sequence by generator polynomial $g(x)$, an error burst will remain undetected if and only if the polynomial describing it, i.e. $b(x)$, is divisible by $g(x)$. This implies that such an error burst can be written in the form

$$b(x) = g(x)Q(x) \tag{2.268}$$

where $Q(x)$ is a polynomial of degree $\big[(b - 1) - r\big]$. Consider three possible cases:

1. The degree of the polynomial $b(x)$ is lower than the degree of the polynomial $g(x)$, which means that the error burst is shorter than r. Consequently, there is no such polynomial $Q(x)$, which when divided by $g(x)$ results in polynomial $b(x)$. As a result, all error bursts shorter than r will be detected.

2. The degree of the polynomial $b(x)$ is equal to the degree of the polynomial $g(x)$, i.e. $b - 1 = r$. Then there exists exactly one polynomial $b(x)$ that is divisible by the polynomial $g(x)$. This polynomial has the form $b(x) = g(x)$. Then $Q(x) = 1$ and there is a single undetected error burst. Because the number of different polynomials describing error bursts of length b is equal to 2^{b-2}, the fraction of undetected error bursts is $1/2^{b-2} = 2^{-(r-1)}$.

3. The degree of the polynomial $b(x)$ is higher than the degree of the polynomial $g(x)$. The error bursts for which the polynomial $b(x)$ is divisible by $g(x)$ will remain undetected. The result of such a division, i.e. the polynomial $Q(x)$, is of $[(b-1) - r]$ degree. In consequence, the polynomial $Q(x)$ has $b - r$ coefficients and has the form

$$Q(x) = x^{(b-1)-r} + \cdots + 1 \tag{2.269}$$

The coefficient at the highest power of x has to be equal to 1; otherwise, the polynomial degree would be lower. In turn the "1" at zero power results from the fact that both $b(x)$ and $g(x)$ have "1" in this position. Therefore there exist 2^{b-r-2} different polynomials featuring this property. As a result, the fraction of undetected error bursts with respect to all possible error bursts of length b is

$$\frac{2^{b-r-2}}{2^{b-2}} = 2^{-r} \quad \text{for} \quad b - 1 > r \tag{2.270}$$

Let us illustrate the meaning of the above theorem. If we assume that bit errors in a transmitted block occur independently and their probability is denoted as p, then, if the length n of a tranmitted data block is high enough and probability p is small, the probability of an erroneous block can be well approximated by $P_B \approx np$. If a CRC code with r parity bits is applied, the fraction of undetected error bursts will be 2^{-r}, therefore the probability of undetected erroneous block will be

$$P_e = np2^{-r} \tag{2.271}$$

Table 2.7 presents the values of probability P_e for different block lengths n and degrees $r = n - k$ of the polynomial $g(x)$. These values have been achieved under the assumption that the probability of a single bit error is $p = 10^{-5}$. As we can see, error detection using a CRC code ensures very low block error probability; however, in order to be useful, a possibility to retransmit erroneous blocks has to exist.

Table 2.7 Probability of undetected error burst for different block lengths n and parity block lengths r

n	r	P_e
500	8	2×10^{-5}
500	16	10^{-7}
1000	16	2×10^{-7}
1000	32	4×10^{-12}

2.15 Application of Error Detection – ARQ Schemes

We have already learnt about effective ways to perform error detection in received data blocks. Consider now basic strategies of nontransparent transmission, in which after detection of an erroneous block a feedback message is sent to the transmitter, informing it about necessity of block retransmission. There are many data transmission services that take advantage of this kind of strategy – mostly those services that do not require sending data at constant rate or constant delay – but their crucial features are reliability and low probability of undetected block error. An example of such a service is the transmission of banking data. Data transmission in mobile networks is also often organized in this way. The information exchange with block retransmission upon request is applied in many protocols of the data link layer of the *Open System Interconnection* (OSI) description model (Wesołowski 2002).

Consider the system shown in Figure 2.41. We will present three basic ARQ strategies (although others exist as well) that differ in the required size of buffers applied in a transmitter and receiver and, what is more important, in the data transfer efficiency. The three basic ARQ strategies, presented in Figure 2.43, are (Lin and Costello 2004):

- stop-and-wait ARQ (Fig. 2.43a);
- go-back-N ARQ (Fig. 2.43b);
- selective-repeat ARQ (Fig. 2.43c).

The first technique is very easy to implement but its efficiency can be low if it is applied in a transmission system in which a significant delay occurs in the transmission chain. The transmitter sends a data block supplemented with parity bits and waits for acknowledgment, i.e. the message ACK, which has to be transmitted over the feedback channel. If this message does not arrive within a given maximum time period or if the transmitter receives negative acknowledgment (NAK), the transmitter sends the same block again. When the ACK message is received by the transmitter, the next data block is sent. The time between the end of transmission of a given block and the beginning of transmission of the next block is wasted. This is the main reason for the potentially low efficiency of this strategy. The efficiency can be improved by lengthening the data blocks; however, as a result, the probability of their failure increases (cf. the formula $P_B \approx np$), so the frequency of block retransmission also increases. Block lengthening may be impossible in some applications.

Let us estimate the transmission efficiency of the above strategy. An example of a single data block transmission and its acknowledgment is shown in Figure 2.44. The following notation has been applied for respective time intervals: T_p – propagation time between transmitter and receiver of a data block or an acknowledgment block (in reverse direction), T_b – data block duration in time units, T_{bp} – delay due to data block processing in the receiver (e.g. due to applying the error detection procedure), T_a – duration of an acknowledgment block, T_{ap} – delay due to acknowledgment block processing in the transmitter. In practice, when the stop-and-wait ARQ technique is applied, processing times of data blocks and acknowledgment blocks are very short compared with the duration T_b of data transmission block. Also the duration of an acknowledgment block is much shorter than the time needed for transmission of a data block. Consequently, the

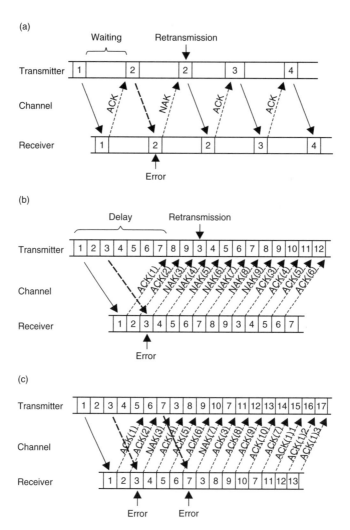

Figure 2.43 Illustration of basic ARQ techniques: (a) stop-and-wait, (b) go-back-N, (c) selective-repeat (Lin and Costello 2004)

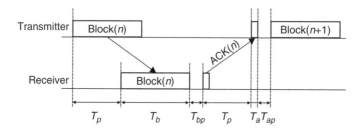

Figure 2.44 Timing scheme of stop-and-wait ARQ strategy

time span between transmission of subsequent blocks can be well approximated by the sum $T_b + 2T_p$. Therefore, transmission efficiency is equal to

$$\eta = \frac{T_b}{T_b + 2T_p} = \frac{1}{1 + 2\alpha}, \quad \text{where } \alpha = \frac{T_p}{T_b} \tag{2.272}$$

Note that we have calculated the transmission efficiency for an ideal situation in which block retransmissions do not occur. In reality, the probability of an erroneous block occurrence is nonzero, which implies a nonzero probability of block retransmission, which in turn leads to a decrease in transmission efficiency. Let N_p attempts of transmission of a single block occur. In a typical situation this number is only slightly higher than 1, which means that a small number of data blocks need to be retransmitted. As we remember, the probability of an erroneous block is $P_B \approx np$. Thus, the probability of a correct block reception is $1 - P_B$. Therefore the mean number of a single block transmission attempts is

$$N_p = \frac{1}{1 - P_B} \tag{2.273}$$

Then from (2.272) we have

$$\eta = \frac{T_b}{N_p T_b + 2N_p T_p} = \frac{1}{N_p(1 + 2\alpha)} = \frac{1 - P_B}{1 + 2\alpha} \tag{2.274}$$

This formula has been achieved under the assumption that the probability of erroneous reception of the acknowledgment (positive or negative) block is vanishingly low compared with the probability of data block error. This assumption is fully justified. As one can conclude from formulae (2.272) and (2.274), transmission efficiency of the stop-and wait ARQ strategy with waiting for acknowledgment strictly depends on the ratio of data block propagation time to block transmission time and the probability of erroneous reception of a data block. Let us determine transmission efficiency for a few representative examples of transmission systems (Halsall 1996).

Example 2.15.1 *Data blocks of length 1000 bits are transmitted using the stop-and-wait ARQ strategy. Determine a transmission efficiency η in the following data links for two data rates: (a) 1 kbit/s, (b) 1 Mbit/s. Assume for simplicity that the signal propagation velocity in the channel is $v = 2 \times 10^8$ m/s and the binary error probability is very small. Consider three data links: (1) a twisted copper pair of length $l = 1$ km, (2) a leased line of length $l = 200$ km, (3) a satellite link of length $l = 50000$ km. Duration of a data block is determined by the formula*

$$T_b = \frac{\text{Number of bits in the data block } N}{\text{Transmission velocity [bit/s]}} \tag{2.275}$$

whereas signal propagation time results from the expression

$$T_p = \frac{\text{Link length } l \text{ [m]}}{\text{Propagation velocity } v \text{ [m/s]}} \tag{2.276}$$

Table 2.8 Effciency of the stop-and-wait ARQ strategy for different link lengths and data transmission rates

Kind of link	Transmission rate	Parameter α	Efficiency η
Twisted copper pair	1 kbit/s	5×10^{-6}	≈ 1
$l = 1000$ m	1 Mbit/s	5×10^{-3}	≈ 1
Leased line	1 kbit/s	1×10^{-3}	≈ 1
$l = 200 \times 10^3$ m	1 Mbit/s	1	0.33
Satellite link	1 kbit/s	0.25	0.67
$l = 50 \times 10^6$ m	1 Mbit/s	250	0.002

Substituting data values for transmission rates and data link lengths, we achieve the efficiency values presented in Table 2.8.

Based on these simple examples one can easily conclude when the application of the stop-and-wait ARQ strategy is reasonable. This technique is very useful if the duration of a data block is significantly longer than the signal propagation time in the channel.

Consider now the second ARQ strategy, i.e. go-back-N. This time, the transmitter does not wait for acknowledgment of the transmitted block, but it subsequently sends the next blocks. Till the moment of acknowledgment of correct reception a given block is stored in a transmitter buffer. The size of the buffer has to be selected by taking into account a maximum time period that can pass until the acknowledgment related to a given transmitted block is received by the transmitter. This time period is determined both by signal propagation through the channel and duration of a data block. If the maximum number of data blocks stored in a buffer is denoted as K, then the buffer memory is sufficient if the following condition is fulfilled

$$KT_b > T_b + 2T_p \quad \text{i.e.} \quad K > 1 + 2\alpha \tag{2.277}$$

A given data block is deleted from the buffer after reception of positive acknowledgment (ACK). If a negative confirmation is received, the transmitter sends subsequent blocks once more, starting from that one for which a positive acknowledgment has not been received. Consequently, it is sufficient to apply the buffer in the transmitter only. The receiver marks the blocks received after the erroneous one as erroneous blocks by sending the NAK block until correct reception of the data block that had been previously received in error occurs. The inefficiency of this technique results from repeated transmission of correctly received blocks if they were preceded by a damaged one.

The third strategy, i.e. the selective-repeat ARQ strategy, does not have the disadvantage of the go-back-N technique considered above. However, the price to pay is the necessity of the data block buffer both in the transmitter and the receiver. It is also necessary to number transmitted blocks modulo-$N_{\max}$ in order for the transmitter to know which data block needs to be retransmitted. The transmitter learns it by reception of the NAK block with the number of the data block that has to be transmitted again. The buffer in the receiver is indispensable because it enables the correct ordering of the data blocks in case some of them are retransmitted.

Consider now the transmission efficiency for the selective-repeat ARQ strategy. This strategy and the previous one do not differ in efficiency when the data blocks are received without errors. Assume for a moment that the probability of data block damage is negligible and the buffer in the transmitter allows K recent data blocks to be stored. If the number of stored blocks is sufficiently high so that it compensates for such a delay i.e. that an acknowledgment block related to the oldest stored data block arrives within this time period, then transmission efficiency is $\eta = 1$. However, if the signal propagation time is so high that the buffer is filled before the acknowledgment block related to the oldest data block stored in the buffer arrives, then transmission has to be halted until the expected acknowledgment block is received. Thus, only part of the time between subsequent blocks can be utilized for transmission and only K blocks can be sent in the round trip delay time. As a result, transmission efficiency is given by the formula

$$\eta = \frac{K T_b}{T_b + 2T_p} = \frac{K}{1 + 2\alpha} \quad \text{if } K < 1 + 2\alpha \qquad (2.278)$$

Now let us extend our considerations to the case when the block error probability cannot be neglected. First consider the selective-repeat ARQ strategy. If the buffer capacity is sufficiently large to allow continuous operation of the transmitter until acknowledgment of a given data block is received, then transmission efficiency is $\eta = 1 - P_B$, where, as before, P_B is the probability of reception of an erroneous data block. However, if the buffer capacity is not sufficient, then we calculate the efficiency as in formulae (2.273) and (2.274) and we obtain

$$\eta = \frac{K T_b}{N_p(T_b + 2T_p)} = \frac{K(1 - P_B)}{1 + 2\alpha} \qquad (2.279)$$

The author suggests that the reader would derive the transmission efficiency for the go-back-N ARQ strategy when the probability of erroneous block reception is not negligible. The results of these calculations would be the following

$$\eta = \begin{cases} \dfrac{K(1 - P_B)}{(1 + 2\alpha)\left(1 + P_B(K - 1)\right)} & \text{for } K < 1 + 2\alpha \\[4mm] \dfrac{(1 - P_B)}{1 + P_B(K - 1)} & \text{for } K \geq 1 + 2\alpha \end{cases} \qquad (2.280)$$

Example 2.15.2 (Halsall 1996) *A sequence of 1000-bit blocks is transmitted over the link of length 100 km at 20 Mbit/s data rate. Let the signal propagation velocity be $v = 2 \times 10^8$ m/s, and let the probability of binary error be $p = 4 \times 10^{-5}$. Derive the transmission efficiency for all three considered ARQ strategies. In the case of transmission with go-back-N or selective-repeat strategies the size of buffer K is assumed to be 10. Substituting the parameters given in this example to formulae (2.275) and (2.276) we receive the transmission time for a data block and the signal propagation delay, which are $T_b = 5 \times 10^{-5}$ s and $T_p = 5 \times 10^{-4}$ s, respectively, so the value of the parameter* *α is 10. We conclude that $1 + 2\alpha = 21$, i.e. $K < 1 + 2\alpha$. Probability of an erroneous data block reception can be approximated by the formula $P_B \approx np$, so we receive $P_B \approx 1000 \times 4 \times 10^{-5} = 4 \times 10^{-2}$. In turn, $1 - P_B = 0.96$. Substituting the calculated parameters to formulae describing transmission efficiency, we obtain*

$$\eta = \begin{cases} \dfrac{1 - P_B}{1 + 2\alpha} = 0{,}046 & \textit{for stop-and-wait ARQ} \\[2em] \dfrac{K(1 - P_B)}{(1 + 2\alpha)[1 + P_B(K - 1)]} = 0{,}336 & \textit{for go-back-N ARQ} \\[2em] \dfrac{K(1 - P_B)}{1 + 2\alpha} = 0{,}46 & \textit{forselective-repeat ARQ} \end{cases}$$

As one could expect, for the given transmission and propagation parameters, the selective-repeat ARQ strategy features the highest transmission efficiency. The stop-and-wait ARQ technique results in very poor link utilization because the signal propagation delay is ten times longer than the time needed for transmission of a single data block and waiting for acknowledgment of a transmitted block takes most of the time.

Despite disadvantages of the stop-and-wait AQR technique, it seems to be very useful if there are N stop-and-wait ARQ processes applied in an interleaved manner. In this way the idle time period in which the transmitter waits for the positive acknowledgment of the transmitted block is used by other ARQ processes performed in parallel.

2.16 Hybrid ARQ

So far we have considered the application of cyclic codes in error detection and we have concentrated on three basic strategies of block exchange associated with CRC codes. However, one could think about more sophisticated schemes in which both error detection and correction are applied. This leads us to hybrid ARQ configurations. The general idea is to apply an error correction code that is able to correct a limited number of typical error patterns; however, the detection capability of the same code or of the outer code is additionally used. Although for small values of bit error probabilities this scheme has lower efficiency than a scheme in which only error detection is used, for higher values of bit error probabilities the block retransmission is expected to be much less frequent compared with a pure ARQ scheme because after correction of typical error patterns retransmissions are not needed so often.

Hybrid ARQ schemes are divided into two types.

2.16.1 Type-I Hybrid ARQ

The type-I hybrid ARQ is the simpler type of the two hybrid ARQ schemes. It can be implemented by applying either two codes or a single code. First let us consider a two-code system. Let the length of the transmitted block be equal to k. Blocks of length k are first encoded using an (n, k) error detecting code. Typically, a CRC code is applied. The resulting packets of length n are subsequently encoded by an FEC (n_1, n) code. At the receiver the blocks are first decoded by the FEC decoder. Due to its operation typical error patterns can be corrected and the resulting blocks of length n are sent to the error detection code decoder. If, despite error correction in the FEC decoder, further errors are detected, the feedback message requesting retransmission is sent to the transmitter and the whole encoding process starts again. In the case of no error detection, the k-bit packet is transferred to the recipient.

The type-I hybrid ARQ can also be applied using a single FEC code. Typically, a block code is applied. Let the block code have the minimum distance $d_{\min}$. Its value can be partitioned into two parts, l and λ, such that $d_{\min} \geq l + \lambda + 1$, where λ is the number of correctable errors and l is the number of detectable errors in a received block ($l > \lambda$). As we can see, the error correction capability is not fully exploited but the coding capability is partially used in error detection. The ability to correct λ errors is achieved by construction of a specific decoding algorithm that is based on the appropriately selected set of parity check equations. After error correction in the FEC decoder (usually requiring syndrome calculation) the syndrome of the corrected block is calculated again. If it is zero, this means that either there were no errors or they have been corrected by the FEC decoding algorithm. If the recalculated syndrome is nonzero, the retransmission request message is sent back to the transmitter.

Another approach in using a single code for the type-I hybrid ARQ relies on application of a block code that is not perfect. Thus, some error patterns can be detected, although they are beyond the correction capabilities of the decoder. In the case of occurrence of such errors, retransmission of the codeword is requested. All correctable error patterns are processed in the decoder and retransmission is not required. Reed-Solomon codes are an excellent example of block codes in such application (Wicker 1995). Another example of a single code that can be used in the type-I hybrid ARQ is given in (Lin and Costello 2004). This code is the $(1023, 923)$ BCH code. Its minimum distance is $d_{\min} = 21$. It can be applied to correct at most five errors, although it is able to detect all error patterns featuring more than five and less than sixteen errors. Thus, retransmission requests will take place if an error pattern containing more than five errors occurs.

2.16.2 Type-II Hybrid ARQ

Let us note that in type-I hybrid ARQ systems retransmission of the whole codeword occurs if uncorrected errors are detected in the receiver. The type-II hybrid ARQ goes one step further in achieving higher system throughput. It is obtained by using the *incremental redundancy* approach. Both convolutional and block codes can be applied to implement this ARQ scheme. Generally, as a result of a retransmission request message the transmitter does not send the whole codeword again, but it sends some additional parity bits that have been calculated but not sent to the receiver. They allow the decoder to decode the previously received block again. The block is now supplemented by newly added parity bits. Thus, a stronger code has been applied that has higher error correction capabilities than its previously used punctured version. This strategy can be realized a few times until the full number of parity bits is used by the decoder. If errors are still detected, the whole procedure starts from the beginning.

As an example consider a classical type-II ARQ scheme using two codes. The first one, denoted as $\mathbf{C}_1$, is a high rate (n, k) error detection code, e.g. CRC code. The second code, denoted as C_2, is a 1/2-rate $(2k, k)$ block code used both for detection and correction. The applied code is called *invertible*. This feature means that knowing only the parity check symbols of a codeword, the message symbols that have implied these parity check symbols can be uniquely determined from them in an *inversion process*.

Let us present the operation of the type-II hybrid ARQ scheme with an invertible code. Let the subject of transmission be a k-bit message block $\mathbf{m}$. First, this block is encoded using the error detection code C_1, which results in a codeword $\mathbf{v} = [p(\mathbf{m}), \mathbf{m}]$, where $p(\mathbf{m})$ is a block of $n - k$ parity check bits. The codeword is transmitted to the receiver. However, the transmitter simultaneously calculates parity bits of the $(2k, k)$ inverse code C_2, so a codeword $[q(\mathbf{m}), \mathbf{m}]$ is generated. The calculated parity bits $q(\mathbf{m})$ are stored in the transmitter in case they are needed in the retransmission procedure. Denote the received block as $\widetilde{\mathbf{v}} = [\widetilde{p}(\mathbf{m}), \widetilde{\mathbf{m}}]$. First, based on the received block $\widetilde{\mathbf{v}}$ the receiver calculates the syndrome of the code C_1. If the syndrome is zero the receiver accepts the message block $\widetilde{\mathbf{m}}$ and sends ACK to the transmitter. On the other hand, a nonzero syndrome indicates errors contained in the received block $\widetilde{\mathbf{v}}$. The received message block $\widetilde{\mathbf{m}}$, possibly containing errors, is stored in the receiver and the negative acknowledgment NAK block is sent back to the transmitter. In this case the transmitter encodes the k-bit parity block $q(\mathbf{m})$ previously generated by the inverse code C_2 using the error detection code C_1. As a result the codeword $\mathbf{v}' = \{p[q(\mathbf{m})], q(\mathbf{m})\}$ is transmitted. Let us note that in fact for a moment a parity block $q(\mathbf{m})$ from the $(2k, k)$ inverse code plays the role of a message block. Let the received block be denoted as $\widetilde{\mathbf{v}}' = \{\widetilde{p}[q(\mathbf{m})], \widetilde{q}(\mathbf{m})\}$. If the syndrome calculated on the basis of $\widetilde{\mathbf{v}}'$ is zero then one can assume that the block $\widetilde{q}(\mathbf{m})$ does not contain errors, so the message block $\mathbf{m}$ can be recovered from it in the inversion process. On the other hand, if the syndrome is nonzero, the message block $\widetilde{\mathbf{m}}$ received in the previous step and the block $\widetilde{q}(\mathbf{m})$ received recently are concatenated into a single block $[\widetilde{q}(\mathbf{m}), \widetilde{\mathbf{m}}]$ of the $(2k, k)$ code. This block is subsequently decoded. If the error correction process is successful, the obtained message block is accepted and the ACK message is sent to the transmitter. If the C_2 decoder detects an uncorrectable error pattern, the received message block $\widetilde{\mathbf{m}}$ is deleted but the parity check block $\widetilde{q}(\mathbf{m})$ is stored for further processing. A NAK is also sent to the transmitter. After receiving the NAK again the transmitter repeats transmission of the codeword $\mathbf{v} = [p(\mathbf{m}), \mathbf{m}]$. The procedure is similar to the previous one. Again, if the syndrome based on the received block $\widetilde{\mathbf{v}} = [\widetilde{p}(\mathbf{m}), \widetilde{\mathbf{m}}]$ is zero, the block $\widetilde{\mathbf{m}}$ is accepted as a message block and the ACK message is issued to the transmitter. If it is not, the recently received block $\widetilde{\mathbf{m}}$ and the previously received block $\widetilde{q}(\mathbf{m})$ form a received block being the subject of error correction decoding. If the error pattern contained in the block $[\widetilde{q}(\mathbf{m}), \widetilde{\mathbf{m}}]$ is correctable, the message block is recovered in the decoding process and the ACK message is sent to the transmitter. If the detectable but uncorrectable errors are contained in the block $[\widetilde{q}(\mathbf{m}), \widetilde{\mathbf{m}}]$, the NAK message is sent again and the parity block $q(\mathbf{m})$ is coded by the C_1 code. Thus, the retransmission process has an alternating character. It is continued until the message $\mathbf{m}$ is correctly received or the number of retransmissions reaches the allowable maximum value indicating that the link quality is unsatisfactory.

At this state of our description of the type-II hybrid ARQ scheme let us show how the inversion process of a $(2k, k)$ code can be implemented. Shortened cyclic codes are applied as inversion codes. In (Lin and Costello 2004) an example of the $(1023, 523)$ BCH code shortened to the $(1000, 500)$ code is considered. The inversion process of a half-rate block code is based on the theorem that states that there are no two codewords in a half-rate shortened cyclic code that have the same parity check blocks (Lin and Costello 2004). Thus, the message part of the codeword is uniquely related to the parity check part. As we remember from our considerations on the generation of a codeword of a systematic polynomial code, the parity check bits are received by division of the message polynomial $x^{n-k}m(x)$ by the generator polynomial $g(x)$, resulting in the remainder $p(x)$, i.e.

$$x^{n-k}m(x) = a(x)g(x) + r(x) \qquad (2.281)$$

We want to recover the message polynomial from the parity polynomial $r(x)$. Let us multiply both sides of (2.281) by x^k, which results in

$$x^n m(x) = a(x)g(x)x^k + r(x)x^k \qquad (2.282)$$

Adding $m(x)$ twice[5] to the left side of (2.282), we have

$$(x^n + 1)m(x) + m(x) = a(x)g(x)x^k + r(x)x^k \qquad (2.283)$$

However, for the (n, k) cyclic code the generator polynomial is a divisor of $x^n + 1$ so $x^n + 1 = g(x)h(x)$. From (2.283) we conclude that

$$r(x)x^k = [m(x)h(x) + a(x)x^k]g(x) + m(x) \qquad (2.284)$$

The last equation indicates that we can recover the message polynomial $m(x)$ by dividing the parity check polynomial $r(x)$ multiplied by x^k by the generator polynomial $g(x)$. The remainder of this division is the desired message polynomial $m(x)$. This operation can be performed in a typical circuit of a polynomial code encoder similar to that shown in Figure 2.5. Let us note that $n = 2k$ in our case, so $x^{n-k} = x^k$.

At the end of our description of the type-II hybrid ARQ let us consider application of convolutional codes in this type of ARQ strategy (Figure 2.45). This approach has been realized in data transmission over GSM/GPRS/EDGE cellular networks and is conceptually very simple. Let the data block of length k be first encoded using an error detecting code such as a CRC code, which results in the block of length n bits. Next, the n-bit block is 1/3-rate convolutionally encoded. This leads to the block of length $3n$. Let us note that transmission over wireless channels very often requires strong FEC coding anyway, in order to achieve satisfactory bit error rate performance. In our case the FEC code is utilized not only in error correction but also in the hybrid ARQ mechanism. Subsequently the $3n$-bit block is a subject of two types of puncturing according to patterns P_1 and P_2, respectively. Each of them leaves n bits out of a $3n$-bit block and they are disjoint. Thus, the n-bit block resulting from the puncturing pattern P_1 is transmitted first. As we can see, the code rate of the punctured convolutional code created in this way is equal to 1. The received soft-decision samples are subject to FEC decoding (mostly using the Viterbi algorithm). The punctured sample positions are filled with zeros. The resulting n-bit block is checked for correctness using the CRC error

---

[5] Recall that $m(x) + m(x) = 0$.

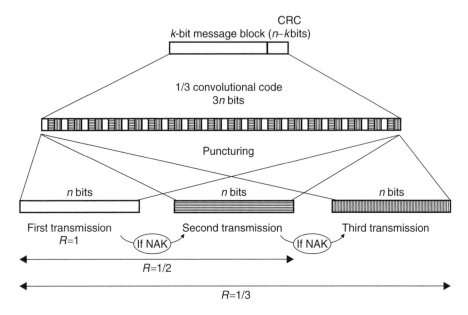

Figure 2.45 Type-II hybrid ARQ with application of CRC and convolutional codes (illustration of incremental redundancy principle)

detecting code decoder. If the syndrome is zero, the ACK message is issued for the transmitter. If the syndrome is nonzero, the NAK message is sent back to the transmitter and the second n-bit block resulting from puncturing using pattern P_2 is transmitted. The received n samples partially fill the zeros in the stored sample vector received in the previous step of the ARQ transmission. Thus, $2n$ out of $3n$ samples are now available for the Viterbi decoder. As we can see, the code rate is now equal to 1/2. After repeated convolutional code decoding and checking the correctness of the decoded block by the CRC decoder, the ACK or NAK message is sent to the transmitter. In the latter case the remaining n-bit block is sent to the receiver, so finally the decoder can take advantage of $3n$ samples and the applied convolutional code has the original code rate of 1/3. If errors are still detected, the whole procedure starts from the beginning. Figure 2.45 illustrates the incremental redundancy principle in the hybrid ARQ procedure with a convolutional code. Let us note that due to memorizing the received samples for later use the application of the soft-decision decoding algorithm is a natural consequence thereof.

Let us also note that in general the hybrid ARQ scheme is accompanied in its acknowledgment message exchange part by one of the above-described regular ARQ schemes such as stop-and-wait, go-back-N or selective-repeat strategies.

Despite the relatively large size of this chapter devoted to the protection of digital messages against errors, we have only sketched basic topics of the channel coding theory. After considering classical topics in coding theory we attempted to clarify the idea of turbo coding and presented basic information on the idea and decoding of LDPC codes. Channel coding is the subject of advanced academic handbooks and monographs, so the author encourages a more interested reader to study them individually.

Problems

Problem 2.1 *A transmitter sends binary blocks of length n over a transmission channel that can be modeled as a binary symmetric memoryless channel featuring the error probability p. What is the probability of the event that the number of errors occurring during transmision of a single block does not exceed k?*

Problem 2.2 *Consider the (5, 2) code described in Example 2.1.1 and the channel model from Problem 2.1. Let the error probability of a single bit be $p = 0.001$.*

1. *Calculate the probability of errorless reception of a codeword.*
2. *Calculate the probability that the decoder will correct a transmission error.*
3. *Calculate the probability that the decoder will detect transmission errors.*
4. *Calculate the probability that the decoder will commit an error during the decoding process, i.e. it will assign a wrong codeword to the received sequence.*
5. *Is the code from Example 2.1.1 linear? What is the minimum distance of the code?*

Problem 2.3 *Plot the probability of reception of a particular sequence $\mathbf{r}$ that differs in D concrete positions from the transmitted sequence $\mathbf{c}$. Make the plots for several lengths of sequences n as a function of D. Let n be equal to 10, 50 and 100. Let the parameter D vary between 0 and 5. Write a computer program that calculates the required probabilities.*

Problem 2.4 *Consider the bipolar input memoryless channel model shown in Figure 1.19a with the Gaussian noise as an additive distortion. Before implementing suboptimal soft-decision decoding, the signals are quantized by a uniform 8-level quantizer with the thresholds shown in Figure 1.19b. Assume signal power $P = A^2$ and the noise variance σ^2. For a given $SNR = P/\sigma^2$ construct a binary input 8-ary output memoryless channel model, shown in Figure 1.19c, and calculate the transition probabilities $\Pr\{i|A\}$ and $\Pr\{i| - A\}$ $(i = 0, \ldots, 7)$ between inputs and outputs. If you like to use Matlab for calculation of these probabilities you can use the function $y = qfunc(x)$, which calculates the value of 1 minus the cumulative distribution function of the normalized Gaussian random variable $(\sigma^2 = 1)$ for the argument x, according to the formula*

$$Q(x) = \frac{1}{\sqrt{2\pi}} \int_{x}^{\infty} \exp[-t^2/2]dt$$

Otherwise use the table of Q-function, available in the Appendix. The resulting discrete memoryless channel model describes the distribution of symbols at the input of the suboptimum soft-decision decoder in the system shown in Figure 2.46.

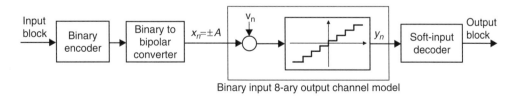

Figure 2.46 System with a binary input 8-ary output channel and soft-input decoder

Problem 2.5 *Let the codewords of the block code be described by the following formula (Clark and Cain 1981)*

$$\mathbf{c} = (c_1, c_2, c_3, c_4, c_5, \ c_1 + c_2 + c_4 + c_5, \ c_1 + c_3 + c_4 + c_5,$$

$$c_1 + c_2 + c_3 + c_5, \ c_1 + c_2 + c_3 + c_4)$$

Determine the code parameters (n, k), construct the parity check matrix H and find the minimum distance d_{min} of this code.

Problem 2.6 *Construct the generator matrix for the code from Problem 2.5.*

Problem 2.7 *Construct the parity check and generator matrices for the repetition codes of lengths $n = 3$, 5 and 7. What is the minimum distance of these codes?*

Problem 2.8 *Consider the $(7, 4)$ Hamming code with the parity check matrix given by formula (2.37).*

1. *Find the generator matrix of this code.*
2. *Find all the codewords of this code.*
3. *Construct the optimum decoding table for the code. Show that this code is perfect.*
4. *Find the syndrome of the received sequence $\mathbf{r} = (0111011)$.*
5. *Draw a logical diagram of the syndrome calculator that uses only basic properties of the parity check matrix.*
6. *As we know, the $(7, 4)$ Hamming code is also generated by the polynomial $g(x) = x^3 + x + 1$. Draw the schemes of the encoder for the nonsystematic and systematic versions of the code.*

Problem 2.9 *Consider the cyclic $(15, 11)$ Hamming code generated by the polynomial $g(x) = x^4 + x + 1$.*

1. *Create a generator matrix G for this code.*
2. *Transform the received generator matrix to the canonical form and find the corresponding parity check matrix H.*
3. *Check if the polynomials $r_1(x) = x^{13} + x^{10} + x^7 + x^6 + x^5 + x^3 + x + 1$ and $r_2(x) = x^{12} + x^{11} + x^7 + x^6 + x^5 + x^4 + x^3 + x + 1$ describe codewords of the $(15, 11)$ Hamming code. Calculate syndromes for them.*

Problem 2.10 *Construct multiplication and addition tables for $GF(5)$ and find the multiplicative and additive inverse elements for each nonzero element of $GF(5)$.*

Problem 2.11 *Solve Problem 2.10 for $GF(7)$.*

Problem 2.12 *$GF(32)$ is generated using the primitive irreducible polynomial $p(x) = x^5 + x^2 + 1$. Generate the table of powers of a primitive element α of this field and find their polynomial representation.*

Problem 2.13 *Consider three block codes of length $n = 31$. Using the table of $GF(32)$ derived in Problem 2.12 find the generator polynomial $g(x)$ for the following codes:*

1. *The Hamming code.*
2. *The maximal-length code.*

3. *The BCH code able to correct two errors.*

Problem 2.14   *Consider the* (15, 7) *BCH code whose generator polynomial* $g(x)$ *was derived in Example 2.5.4.*

1. *Draw the scheme of the encoder based on the generator polynomial* $g(x)$.
2. *Determine the parity check matrix and generator matrix of this code.*
3. *Check if this code is a cyclic code. If it is, determine the parity check polynomial* $h(x)$ *and sketch the encoder based on it.*

Problem 2.15   *Use* $GF(7)$ *from Problem 2.11 with the primitive element* $\alpha = 5$ *to construct a Reed-Solomon code of length* $n = 6$ *that is able to correct a single error* ($t = 1$).

1. *Check if* $\alpha = 5$ *is really a primitive element of* $GF(7)$.
2. *Derive the generator polynomial* $g(x)$ *for this RS code and determine code parameters* (n, k).
3. *Draw the scheme of a codeword generator based on division by* $g(x)$ *and calculation of the remainder. Alternatively, find the scheme of a codeword generator performing multiplication of the information block polynomial by the generator polynomial.*

Problem 2.16   *Consider the Meggitt decoder of the* (15, 7) *BCH code from Problem 2.14 in the version shown in Figure 2.13.*

1. *Determine all syndromes that have to be recognized by the logical circuit denoted in Figure 2.13 as* $\widehat{e}(x) = f[s(x)]$ *resulting in the symbol 1, which corrects the shift register output.*
2. *Let the all-zero sequence of the* (15, 7) *BCH code be transmitted and let the received sequence have the polynomial form* $r(x) = x^{13} + x^9$. *Trace the operation of the Meggitt decoder and show how the errors are gradually corrected.*
3. *Supplement the decoder scheme with the feedback from the output of the logical circuit recognizing the syndromes to the decoder input, such as that shown in Figure 2.15. Determine which syndromes have to be recognized by the logical circuit producing "1"s at its output and trace the operation of the Meggitt decoder for the input sequence* $r(x) = x^{13} + x^8 + x^4$ *if the all-zero sequence was transmitted. Check if the decoder is able to correct this triple-error sequence.*

Problem 2.17   *Consider the majority decoder with the feedback shown in Figure 2.15 for the* (15, 7) *BCH code* [$g(x) = x^8 + x^7 + x^6 + x^4 + 1$]. *Track the operation of the decoder for the input sequence* $r(x) = x^{13} + x^8 + x^4$ *if the all-zero sequence was transmitted.*

Problem 2.18   *Repeat the calculations performed in Example 2.6.2 for soft-decision information set decoding of the Hamming code codewords if the received sequence has the form* $\mathbf{r} = (2760321)$.

Problem 2.19   *Consider the* (15, 7) *cyclic BCH code* [$g(x) = x^8 + x^7 + x^6 + x^4 + 1$] *again. This time we analyze decoding based on information set decoding in the version called error trapping decoding. The decoder is shown in Figure 2.47 and its operation is the following. During the first n clock cycles the received sequence is fed simultaneously to the input of the syndrome calculator and to the shift register. During the next n cycles the subsequent versions of the syndrome for the rotated input sequence are calculated and their weight is checked to find whether it is at most equal to 2. If such a syndrome is*

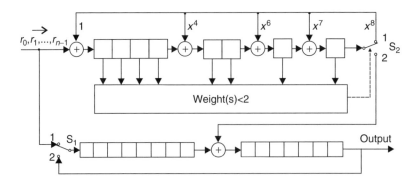

Figure 2.47   Error trapping decoder for the (15,7) BCH code

observed, it indicates errors on current parity check positions. At this moment the switches are shifted to Position 2 and the sequence is shifted out of the register with simultaneous error correction implemented by adding modulo-2 the current syndrome bits with the bits contained in the part of the shift register following the adder. Assume that the sequence $r(x) = x^{13} + x^9$ is received. Trace the operation of this decoder.

Problem 2.20   *Repeat the procedure of decoding the all-zero codeword for (15, 5) BCH code considered in Example 2.7.1 using the iterative Berlekamp-Massey algorithm when the received sequence has the polynomial form $r(x) = x^{11} + x^6 + x$. Perform the subsequent decoding steps and write them in a table similar to Table 2.5.*

Problem 2.21   *Consider the convolutional code of the coding rate $R = 1/3$ that is determined by the generator polynomials $g_1(x) = 1 + x^2$, $g_2(x) = 1 + x + x^2$ and $g_3(x) = 1 + x + x^2$.*

1. Draw the tree, state diagram and trellis for this code.
2. Determine the generator matrix.
3. Draw the diagram used for derivation of the transfer function and determine the transfer function for this code.
4. What is the free distance of this code?
5. Show a few shortest error events on the trellis diagram of the code and compare them with the few first terms of the code transfer function expressed in the form of a series.

Problem 2.22   *Solve Problem 2.21 for a systematic convolutional code of the coding rate $R = 1/2$ determined by the generator polynomials $g_1(x) = 1$ and $g_2(x) = 1 + x + x^3$.*

Problem 2.23   *Show the subsequent steps of the Viterbi algorithm working in the hard-decision mode and decoding the received sequence*

$$111\ 100\ 111\ 101\ 100\ 011\ 111\ 111\ 011\ 111$$

when the encoder is determined by the generator polynomials given in Problem 2.21. If the costs of two paths reaching the same trellis state are identical, use the same rule to choose the survivor, e.g. always choose the upper state. Assume that the zero state is the initial state of the encoder and that this fact is known to the decoder.

Problem 2.24   *Solve Problem 2.23 for the same code and the Viterbi algorithm working in the soft-decision mode when the received sequence is*

$$565\ 421\ 455\ 714\ 424\ 245\ 675\ 657\ 245\ 444$$

Problem 2.25   *Solve Problem 2.23 for the RCPC code of the coding rate $R = 4/5$ shown in Figure 2.22a, whose puncturing table is given by formula (2.123). We assume that the Viterbi decoder works in the hard-decision mode. The received sequence is*

$$11\ 1\ 1\ 0\ 11\ 0\ 1\ 0\ 10\ 0\ 1\ 1$$

What is the free distance of this code?

Problem 2.26   *Consider two equivalent convolutional codes – a nonsystematic code and a recursive systematic code – whose encoders are shown in Figure 2.35. Determine their trellis diagrams, compare them and draw conclusions.*

Problem 2.27   *Consider the (15, 7) BCH code whose parity check matrix in the extended form is given by formula (2.239). Assume that the encoder has generated the all-zero codeword. Write computer programs performing the LDPC code decoding for the following algorithms:*

1. Bit-flipping algorithm,
2. Weighted bit-flipping algorithm,
3. Sum-product algorithm.

Test the programs performing the respective algorithms for the following received sequences:

1. $\mathbf{r} = (010001000000000)$ for bit-flipping algorithm,
2. $\mathbf{r} = (-0.1, 0.2, -0.8, -0.6, -0.3, 0.3, -0.9, -0.1, -1.0, -0.5, -0.09, -0.7, -0.99, -0.32, -1.2)$ for weighted bit-flipping and sum-product algorithms.

3

Digital Baseband Transmission

3.1 Introduction

In this chapter we will consider the ways in which a binary stream generated by a binary message source such as a PCM codec, a sound or video encoder, or a computer terminal can be transformed into a sequence of signals transmitted through the channel. Signals carrying digital messages can be transmitted by a passband or baseband channel. Methods and ways of transmission over passband channels, in particular digital modulations, will be the subject of our considerations in the next chapter. The current chapter is devoted to baseband transmission methods. A channel that can be used for baseband transmission passes spectral components of the signal in the range of frequencies from around DC up to a certain limit frequency W. Examples of such channels are a copper wire pair and a coaxial cable. For technical reasons the DC signal component is often eliminated and very low frequencies contained in the signal are attenuated. This is often done to preserve galvanic separation of the receiver and the channel. A transformer is often applied for this purpose to ensure safety against possible supertensions that can occur in the channel (e.g. due to short circuit to the power line or induction of charge during an atmospheric storm). Figure 3.1 presents an example of the baseband channel characteristics. The frequency range that is cut off by the transformer is also shown symbolically.

3.2 Shaping of Elementary Signals

The form of signals that represent particular bits (or bit blocks) should be well fitted to the channel properties. We understand the best fit as the one that leads to the highest robustness against distortions. In the case of digital transmission, robustness can be understood as the *Bit Error Rate* (BER) measured in the receiver or a maximum range of transmission achievable at the required BER level. Assignment of binary signals to the elementary pulses that are subsequently transmitted over the channel is sometimes called digital baseband modulation.

Let us first consider the simplest case in which subsequent bits of the data stream determine the transmitted elementary pulses. Let elementary pulse $s_0(t)$ be related to bit "0" whereas $s_1(t)$ is selected by bit "1" in the binary stream. Assume for a moment that subsequent bits $a_i (i = -\infty, \ldots, \infty)$ last T seconds each, and that the duration of elementary pulses does not exceed T seconds. In such a case the transmitted signal can

Introduction to Digital Communication Systems Krzysztof Wesołowski
© 2009 John Wiley & Sons, Ltd

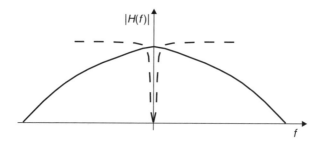

Figure 3.1   Characteristics of an exemplary baseband channel with a DC separating circuit

be described by the formula

$$x(t) = \sum_{i=-\infty}^{\infty} s_k(t - iT), \quad \text{where } k = a_i \qquad (3.1)$$

If the above assumptions are made subsequent pulses do not overlap. The following question can be stated: What are the properties of signal $x(t)$ defined by (3.1)? From the point of view of transmission over the channel, spectral properties of signal $x(t)$ are of crucial importance. The features that enable robust synchronization of the receiver with the received signal are also important. Very often the receiver has to recover the timing clock that determines the moments of start and end of a modulation period, i.e. the moments in which elementary pulses possibly change. Determination of those timing instants is needed to make proper decisions upon the received pulses, and in consequence to find the original binary stream that determined the sequence of pulses generated in the transmitter. Let us note that the binary stream can be considered as random by an external observer, so signal $x(t)$ is random too. Thus, calculation of the spectral density of signal $x(t)$ determined by a specific data sequence, which can be treated as a sample function of the stochastic process, does not characterize spectral properties of this signal. Instead, the power spectral density of random signal $x(t)$ most often represents its spectral features. Power spectral density can be calculated under the appropriate statistical assumptions. In particular, it is assumed that statistical properties of the signal remain constant in time. Moreover, the zero moment on the observer's time axis with respect to the observed signal $x(t)$ can be treated as a random variable that is uniformly distributed in the time period $[0, T]$. Let us give the formula for power spectral density for a specific form of signal $x(t)$ that is often observed in reality. Let the signal be described by the formula

$$x(t) = \sum_{i=-\infty}^{\infty} d_i s(t - iT) \qquad (3.2)$$

where there is a mutually unique mapping between the bit sequence $\{a_i\}$ and data symbol sequence $\{d_i\}$. For example, for bipolar signals $d_i = 2a_i - 1$, so $s_0(t) = -s(t)$, whereas $s_1(t) = s(t)$. In Appendix 3.8 we show that in this case the power spectral density of signal $x(t)$ is expressed by

$$G_x(f) = \frac{1}{T}|S(f)|^2 G_d(f) \tag{3.3}$$

where $S(f) = \mathcal{F}[s(t)]$ is the spectral density of pulse $s(t)$. In turn, $G_d(f)$ is the power spectral density of the data stream d_i and is given by the expression

$$G_d(f) = \sum_{n=-\infty}^{\infty} R_d(n) \exp(-j2\pi f n T) \tag{3.4}$$

$R_d(n)$ is the autocorrelation function of data sequence d_i, i.e.

$$R_d(n) = E[d_i d_{i-n}] \tag{3.5}$$

where $E[.]$ denotes ensemble average. Analyzing formula (3.3) we conclude that the power spectral density of random signal $x(t)$ depends both on the spectral properties of the chosen pulse $s(t)$ and, through the factor $G_d(f)$, on the correlation properties of the data sequence. Consider a simple case when data symbols are uncorrelated, i.e.

$$R_d(n) = \begin{cases} E[d_i^2] = \sigma_d^2 + \mu_d^2 & \text{for } n = 0 \\ E[d_i d_{i-n}] = E[d_i]E[d_{i-n}] = \mu_d^2 & \text{for } n \neq 0 \end{cases} \tag{3.6}$$

where $\mu_d = E[d_i]$ is a mean value (strictly speaking, an expectation or ensemble average) of data symbols, whereas $\sigma_d^2 = E[(d_i - \mu_d)^2]$ is the variance of data symbols. Let us note that the mean value can be interpreted as a constant (DC) component of the data sequence whereas the variance can be interpreted as the mean power of the AC component. Thus, for $n = 0$ the autocorrelation function of the data signal can be interpreted as the mean power of the signal, i.e. the sum of the mean powers of the DC and AC components. For $n \neq 0$ the autocorrelation function is in fact the mean power of the DC component of the data sequence only. So for the case described by (3.6) the power spectral density of the data sequence is given by the formula

$$G_d(f) = \sigma_d^2 + \mu_d^2 \sum_{n=-\infty}^{\infty} \exp(-j2\pi f n T) \tag{3.7}$$

Let us note that the expression

$$\sum_{n=-\infty}^{\infty} \exp(-j2\pi f n T)$$

can be interpreted as a Fourier expansion of such a periodic function of frequency for which the harmonic coefficients are constant and equal to unity. The Dirac pulse sequence along the frequency axis with the period $1/T$ and intensity $1/T$ features such an expansion, i.e. it is a signal of the form

$$\frac{1}{T} \sum_{n=-\infty}^{\infty} \delta\left(f - \frac{n}{T}\right) \tag{3.8}$$

Indeed, the Fourier series expansion coefficients of the periodic signal (3.8) can be determined from the formula

$$c_n = \frac{1}{1/T} \int_{-1/2T}^{1/2T} \frac{1}{T} \delta(f) \exp(-j2\pi f nT) df = 1 \tag{3.9}$$

therefore

$$\sum_{n=-\infty}^{\infty} \exp(-j2\pi f nT) = \frac{1}{T} \sum_{n=-\infty}^{\infty} \delta\left(f - \frac{n}{T}\right) \tag{3.10}$$

As a result, formula (3.7) is transformed to the expression

$$G_d(f) = \sigma_d^2 + \frac{\mu_d^2}{T} \sum_{n=-\infty}^{\infty} \delta\left(f - \frac{n}{T}\right) \tag{3.11}$$

The final form of the power spectral density of signal $x(t)$ with the signal assumed properties is

$$G_x(f) = \frac{\sigma_d^2}{T} |S(f)|^2 + \frac{\mu_d^2}{T^2} \sum_{n=-\infty}^{\infty} \left|S\left(\frac{n}{T}\right)\right|^2 \delta\left(f - \frac{n}{T}\right) \tag{3.12}$$

Let us discuss formula (3.12). The power spectral density of signal $x(t)$ definitely depends on the shape (and also on the spectral density) of the elementary pulse $s(t)$. The pulse shape determines the bandwidth of signal $x(t)$. However, the mean μ_d of data symbols also influences the power spectral density. If the mean is nonzero and the spectral density of the elementary pulse is nonzero for frequency $f = 0$, the spectral line for $f = 0$ appears in the power spectral density of signal $x(t)$, which indicates the nonzero mean of signal $x(t)$. The same can happen at multiple frequencies $f = n/T$. However, as we have stated before, it is very often required that the mean value of data symbols is equal to zero. In this case, the power spectral density of signal $x(t)$ reduces to the form

$$G_x(f) = \frac{\sigma_d^2}{T} |S(f)|^2 \tag{3.13}$$

Consider a bipolar pulse stream in which each pulse lasts for T seconds and has a rectangular shape. Assume that subsequent data bits are mutually uncorrelated. Figure 3.2 presents the elementary pulse shape and the power spectral density of $x(t)$.

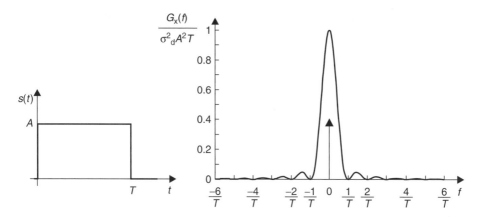

Figure 3.2 Rectangular pulse $s(t)$ and power spectral density $G_x(f)$ of signal $x(t)$ when pulse $s(t)$ is applied

The spectral density of the rectangular pulse shown in Figure 3.2 is given by the formula

$$S(f) = AT\operatorname{sinc}(\pi f T)\exp(-j\pi f T) \tag{3.14}$$

Substituting (3.14) into (3.12), we obtain the power spectral density of signal $x(t)$

$$G_x(f) = \sigma_d^2 A^2 T \left[\operatorname{sinc}(\pi f T)\right]^2 + A^2 \mu_d^2 \delta(f) \tag{3.15}$$

where the second component appears only if the mean value of data symbols d_i is nonzero. Let us note that the remaining possible spectral lines at multiples of $1/T$ do not appear because, independently of the data symbol mean value μ_d, the spectral density of the elementary pulse $s(t)$ takes a zero value for these frequencies. As seen in Figure 3.2, the power spectral density of signal $x(t)$ decreases relatively slowly along the frequency axis (i.e. with its square; cf. the definition of sinc function). As a result, a relatively significant part of the signal power is contained in the sidelobes of the power density spectrum. Cutting off these sidelobes by the channel will result in signal distortions, which in turn will lead to the bit error rate increase. On the contrary, if we wish to receive the signal transmitted over the channel of bandwidth W in an undistorted form, the signaling rate $R = 1/T$ must be significantly lower than W. As we can see, the channel bandwidth cannot be used in an optimal way. If we want to increase the data rate for a given channel bandwidth, we must select such an elementary signal shape $s(t)$ whose spectral density does not exceed the channel bandwidth W or exceeds it in an insignificant manner.

Next we will consider spectral properties of signal $x(t)$ if the shape of elementary signal $s(t)$ is determined by the function called a *raised cosine* curve, which is given by the formula

$$s(t) = \frac{A}{2}\left[1 + \cos\frac{\pi}{T}(t - T)\right] \quad \text{for } 0 \le t \le 2T \tag{3.16}$$

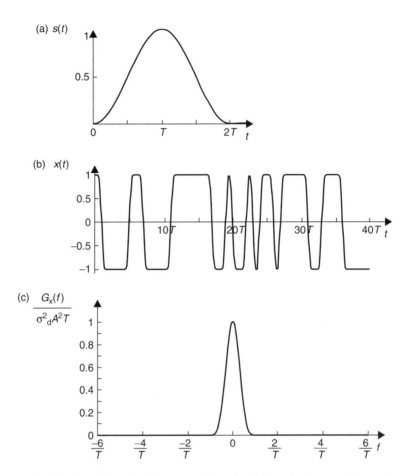

Figure 3.3 Application of the raised cosine pulse: (a) single pulse shape, (b) exemplary signal waveform $x(t)$, (c) power spectral density of signal $x(t)$

shown in Figure 3.3a. Let us note that elementary signal $s(t)$ given by (3.16) lasts twice as long as the modulation period; however, it is sufficiently short to ensure that the samples of the pulse sequence taken at the moments nT depend on a single pulse only. Figure 3.3b presents an example of signal $x(t)$. In this case the power spectral density of $x(t)$ is determined by the following formula

$$G_x(f) = \sigma_d^2 A^2 T \frac{\text{sinc}^2(2\pi f T)}{(1 - 4T^2 f^2)^2} + \mu_d^2 A^2 \delta(f) \tag{3.17}$$

and is shown in Figure 3.3c. Derivation of (3.17) is listed as a problem to solve at the end of this chapter. It can be concluded from (3.17) that the power spectral density decreases with the sixth power of frequency, i.e. much faster than in the case of applying a rectangular shape of $s(t)$. Figure 3.4 compares power spectral densities in logarithmic scale for bipolar random signals $x(t)$ for rectangular and raised cosine pulses. We notice a significant difference in the sidelobe levels of power spectral densities of both signals.

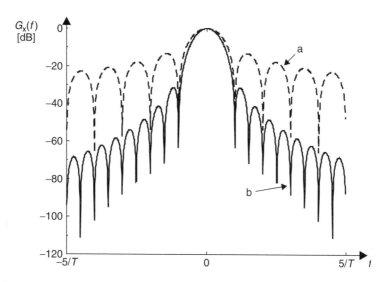

Figure 3.4 Comparison of power spectral densities on the logarithmic scale for a bipolar random signal with the rectangular (a) and raised cosine (b) shape of elementary signal $s(t)$

Application of a smoother shape of the elementary signal and lengthening it up to $2T$ results in a significant improvement in concentration of power in the mainlobe.

Assume that the transmission channel has bandwidth W, its amplitude characteristic is flat and its phase characteristic is linear. This means that if the spectrum of the transmitted signal is fully contained in the channel passband, then the signal is the subject of constant attenuation and delay but it remains undistorted. The channel bandwidth will be fully used if the power spectral density of the transmitted signal has the shape determined by the formula

$$G_x(f) = B \operatorname{rect}\left(\frac{f}{2W}\right) \tag{3.18}$$

In other words, the spectral density of the elementary signal should preserve the rectangular shape of the channel transfer function. Therefore the signal should have the following form

$$s(t) = 2WA \operatorname{sinc}(2W\pi t) \tag{3.19}$$

where $B = A^2/T$ (see Figure 3.5). In theory, the duration of this signal is infinite, and its amplitude decreases inversely proportionally with time. However, in practice only its approximated version, lasting a few modulation periods, is applied. In such cases the truncated signal loses good spectral properties compared with its ideal form. The latter is rather obvious because a shaping filter featuring rectangular characteristic is not physically realizable.

Let us turn our attention to an interesting and important feature of signal $s(t)$ determined by (3.19). Except $t = 0$, the signal waveform has zeros in equal distances every $1/2W$ seconds. Therefore, if every $T = 1/2W$ seconds we send subsequent elementary pulses

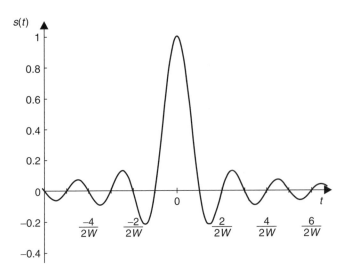

Figure 3.5   Signal waveform (3.19)

(3.19) whose polarization depends on the current data symbol, and if we select sampling instants appropriately, we are able to sample signal $x(t)$ in the receiver in such a way that the received samples depend exclusively on a single data symbol. The waveforms that are the response to other data symbols are equal to zero exactly at these moments. Thus, the channel of W Hz bandwidth could support transmission at the symbol rate of $2W$ symbol/s.

Unfortunately, our considerations have only a theoretical character. As we have already mentioned, the synthesis of signal $s(t)$ featuring spectral properties given by (3.18) is not possible due to the required steepness of the slope in characteristics of the filter that should be used to shape the waveform of signal $s(t)$. In practice the following procedure takes place. The symbol data rate is moderately decreased below $2W$ symbol/s so the smooth transition from passband to stopband is realizable. It turns out that at the fulfillment of certain requirements it is possible to shape the spectral density of $s(t)$ so that the zeros of the waveform occur every T seconds (sometimes between those moments too). The requirements are stated in the form of the following theorem.

Theorem 3.2.1 *Consider the filter called a* Nyquist filter *with the transfer function of the form*

$$H(f) = \begin{cases} \text{rect}\left(\frac{f}{2B}\right) + Y(f) & \text{for } |f| < 2B \\ 0 & \text{otherwise} \end{cases} \quad (3.20)$$

where $Y(f)$ is a real function that is even-symmetric about $f = 0$, i.e.

$$Y(-f) = Y(f), \quad \text{for } |f| < 2B$$

and odd-symmetric about $f = B$, *i.e.*

$$Y(-f + B) = -Y(f + B), \quad \text{for } |f| < B$$

If elementary pulse $s(t)$ *applied in digital transmission is the impulse response of the Nyquist filter conforming to (3.20), then pulse* $s(t)$ *has zeros at the moments that are nonzero multiples of* $1/2B$, *and for the data sequence transmitted in the form*

$$x(t) = \sum_{i=-\infty}^{\infty} d_i s(t - iT)$$

it is possible to find sampling moments at which the samples of $x(t)$ *contain information on a single data symbol. Therefore transmission at the symbol rate* $R = 1/T = 2B$ *is possible.*

   Proof. We will show that the impulse response of the filter described by (3.20) is zero at the moments $t = n/2B$, $n \neq 0$. Let us calculate the inverse Fourier transform of (3.20). We have

$$h(t) = \int_{-2B}^{B} Y(f)e^{j2\pi ft}\mathrm{d}f + \int_{-B}^{B} [1 + Y(f)]e^{j2\pi ft}\mathrm{d}f + \int_{B}^{2B} Y(f)e^{j2\pi ft}\mathrm{d}f$$

or equivalently

$$h(t) = \int_{-B}^{B} e^{j2\pi ft}\mathrm{d}f + \int_{-2B}^{2B} Y(f)e^{j2\pi ft}\mathrm{d}f$$

$$= 2B\frac{\sin 2\pi Bt}{2\pi Bt} + \int_{-2B}^{0} Y(f)e^{j2\pi ft}\mathrm{d}f + \int_{0}^{2B} Y(f)e^{j2\pi ft}\mathrm{d}f \qquad (3.21)$$

Let us substitute $\lambda = f + B$ in the first integral of (3.21) and $\lambda = f - B$ in the second one. Thus, we obtain

$$h(t) = 2B\frac{\sin 2\pi Bt}{2\pi Bt} + e^{-j2\pi Bt}\int_{-B}^{B} Y(\lambda - B)e^{j2\pi \lambda t}\mathrm{d}f + e^{j2\pi Bt}\int_{-B}^{B} Y(\lambda + B)e^{j2\pi \lambda t}\mathrm{d}f$$

$$(3.22)$$

However, we know that $Y(\lambda - B) = -Y(\lambda + B)$, so using this property in (3.22) we have

$$h(t) = 2B\frac{\sin 2\pi Bt}{2\pi Bt} + j2\sin 2\pi Bt \int_{-B}^{B} Y(\lambda + B)e^{j2\pi \lambda t}\mathrm{d}f \qquad (3.23)$$

It is clear from (3.23) that impulse response $h(t)$ indeed has zeros for $t = n/2B$ ($n \neq 0$) due to $\sin 2\pi Bt$ appearing in both components of (3.23), so we can apply $s(t) = h(t)$ for data transmission at the rate $R = 2B = 1/T$. Thus, the thesis of the theorem has been proven.

An additional conclusion from the above theorem and formula (3.23) is that there are many possible shapes of the filter slope, i.e. function $Y(f)$ can be described by different formulae. However, the symmetry conditions have to be preserved.

Pulse $s(t)$, which conforms to the Nyquist criterion, is the most often applied in practice and allows for transmission at the rate $1/T$ not too much lower than $2W$, is the so-called signal with the *raised cosine characteristics*. Its spectral characteristics are determined by the formula

$$S(f) = \begin{cases} T & \text{for } 0 \leq |f| \leq \dfrac{1-\alpha}{2T} \\[2mm] \dfrac{T}{2}\left\{1 + \cos\left[\dfrac{\pi T}{\alpha}\left(|f| - \dfrac{1-\alpha}{2T}\right)\right]\right\} & \text{for } \dfrac{1-\alpha}{2T} \leq |f| \leq \dfrac{1+\alpha}{2T} \\[2mm] 0 & \text{for } |f| \geq \dfrac{1+\alpha}{2T} \end{cases} \quad (3.24)$$

whereas its corresponding time function is described by expression

$$s(t) = \operatorname{sinc}(\pi t/T)\frac{\cos(\pi \alpha t/T)}{1 - 4\alpha^2 t^2/T^2} \quad (3.25)$$

Analyzing formula (3.25) we find that pulse $s(t)$ is similar to that described by (3.19), but the sinc function is additionally multiplied by a fraction with the numerator introducing additional zeros besides the moments $t = n/T$ ($n \neq 0$) and with the denominator pressing faster suppression of the pulse in time (with the third power of time variable t). Figure 3.6 presents the characteristics $S(f)$ and the corresponding pulse $s(t)$ for different values of the so-called *roll-off factor* α. Let us note that in a channel of bandwidth W Hz it is possible to transmit data symbols at such a rate $1/T$ for which $(1 + \alpha)/2T \leq W$, or equivalently $(1 + \alpha)/T \leq 2W$. For values of coefficient α contained in the range $0.1-0.5$ the symbol data rate is in the range of $2/3$ up to 0.9 of the theoretical maximum value equal to $2W$.

Despite its slightly more complex mathematical description, elementary pulse $s(t)$ shown in Figure 3.6 is often applied in the baseband transmission and it also serves as a shaping pulse in digital modulations of sinusoidal carriers. In practice, pulse (3.25) is approximated by a pulse of finite duration, however, due to a fast decrease of function (3.25) in time, a good approximation is relatively easily achievable. Let us stress once more that elementary pulse $s(t)$ features zeros each T seconds (except the zero moment in which the pulse achieves its maximum), which allows such pulses to be sent from the transmitter every T seconds and also that such a sampling phase is found at the receiver (assuming that the channel does not distort these pulses) for which the samples depend on a single data symbol only. This case has been illustrated in Figure 3.7. We see signal $x(t)$, which consists of a sequence of pulses s(t) conforming to (3.25).

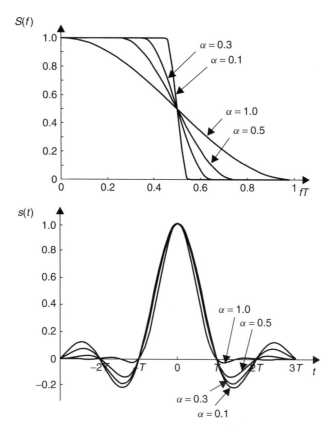

Figure 3.6 Spectral density of signal with raised cosine characteristics and its waveform for a number of values of roll-off factor α

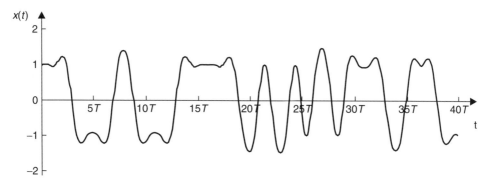

Figure 3.7 An example of signal waveform of $x(t)$ when pulses of raised cosine characteristics (roll-off factor $\alpha = 0.5$) are applied

An *eye pattern* is a certain visual measure of a signal shape or signal distortions that occur if a channel changes its form. It is obtained on the oscilloscope display if signal $x(t)$ is fed to its input. The time base of the oscilloscope is triggered in such a way that we observe signal $x(t)$ during two modulation periods ($2T$), whereas the screen displays the past signal transitions for such a long time that we see overlapping signal transitions that have occurred during many modulation periods. Figure 3.8 shows examples of eye patterns for signal $x(t)$ with the elementary pulse having the raised cosine characteristics with the roll-off factor $\alpha = 0.5$ and $\alpha = 1.0$. The eye has the upper lid and lower lid and the internal distance between the maximum of the upper lid and minimum of the lower lid is called a *maximum eye opening*. Lines drawing the eye pattern can be blurred due to additive noise occurring in the channel. The points on the time axis for which maximum eye opening occurs are often selected as optimal moments for sampling signal $x(t)$. Due to distortions introduced by the channel, the eye pattern can change to such an extent that the eye is partially or even fully closed and decisions based on such samples of signal $x(t)$ are highly unreliable. Quite often the maximum eye opening is determined by finding half of the time distance between averaged signal zero crossings. We see in Figure 3.8 that when the roll-off factor of the applied pulse α is 0.5, then zero crossings of signal $x(t)$ fluctuate. In order to recover the stable timing signal that would allow the optimal sampling phase to be determined a phase-locked loop (PLL) can be applied that operates by averaging the fluctuating zero crossings. Let us note that relaxing the requirements on the slope width (e.g. by selection of $\alpha = 1.0$) results in a decrease of zero crossing fluctuations and, in consequence, simplified operation of the timing recovery system.

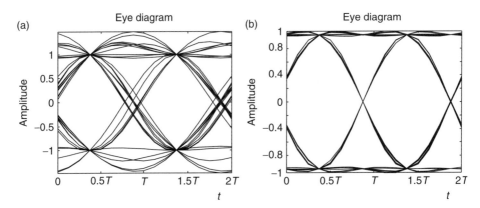

Figure 3.8 Eye diagrams for signal $x(t)$ for elementary pulse $s(t)$ with the raised cosine characteristics; the roll-off factor: (a) $\alpha = 0.5$ and (b) $\alpha = 1.0$

3.3   Selection of the Data Symbol Format

So far we have considered such a baseband modulation in which a data bit simply determines polarization of the applied elementary pulse $s(t)$. In such case we talk about PAM – *Pulse Amplitude Modulation* – or simply about the application of bipolar signaling. We limit the current considerations to signals of a rectangular shape; however, shapes other than rectangular ones are also possible. Selection of a specific data format allows the spectral properties of the transmitted signal to be influenced through appropriate shaping of data symbol power spectral density $G_d(f)$ applied in formulae (3.3) and (3.4). Figure 3.9 presents a few typical formats of data symbols representing binary data. The first of them is called unipolar NRZ (*Non-Return-to-Zero*). Bit "1" determines the application of a rectangular signal with positive polarization, whereas for bit "0" the data signal has a zero level. As a result, the data signal has a nonzero mean, which can often be considered a drawback.

The second format is a bipolar NRZ, considered in the previous section. If probabilities of bits "1" and "0" are the same, the signal featuring this format has a zero mean; however, in short time intervals the mean can be different from zero.

The third format shown in Figure 3.9 is called *pseudoternary* – despite binary assignment of data bits to data symbols, the resulting signal may have three levels $+1$, 0 and -1. This is a consequence of alternating polarity of a unipolar signal in subsequent nonzero pulses. This kind of signal processing is an example of line coding. The signal presented in Figure 3.9c conforms to AMI (*Alternate Mark Inversion*) coding. For this format the signal mean is zero also in short time intervals. However, there is still a danger, as in the case of the signal shown in Figure 3.9a, that long zero bit sequences will result in a long interval of the zero level, which is disadvantageous from the synchronization point of view. In reality much more sophisticated coding schemes than a regular AMI are often applied. We will show them on the example of ISDN line coding options at the end of this chapter.

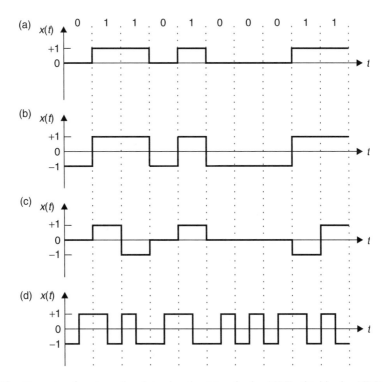

Figure 3.9   Formats of rectangular data signals: (a) unipolar NRZ, (b) bipolar NRZ, (c) pseudoternary (AMI), (d) Manchester (biphase)

Finally, the fourth data format shown in Figure 3.9d is the *Manchester code* (biphase format). Each data symbol is represented by a double pulse with alternating polarity. For example, bit "1" lasting T seconds is associated with a data symbol consisting of a positive pulse of duration $T/2$ seconds, followed by a negative pulse of the same magnitude lasting for the remaining second half of the modulation period T. In turn, bit "0" is represented by the data symbol that has a reversed shapes compared with the symbol representing bit "1". In this format the signal features a zero mean independently of the data stream. In the middle of each signaling period the signal changes its polarization to the reversed one. This fact is very advantageous for the timing recovery process performed in the receiver. However, the drawback of this format is a wider bandwidth of the power density spectrum compared with the formats considered previously, because an elementary signal $s(t)$ is a pair of bipolar pulses each of length equal to $T/2$.

Let us show now how the choice of the data format influences the spectral properties of the transmitted signal $x(t)$. Assume a rectangular shape of the elementary pulses. Assume again that probabilities of bits "0" and "1" are the same and equal to 1/2, and particular bits are mutually statistically independent. Correlation among data symbols can be the result of the applied data format only.

First consider unipolar NRZ signals. Let the data symbols take value $d_i = 1$ or $d_i = 0$. Calculate the autocorrelation function $R_d(n)$ of data signals that is needed for determination of the data signal power density spectrum $G_d(f)$. For that purpose we calculate [cf. formula (3.6)]

$$R_d(0) = E\left[d_i^2\right] = \sigma_d^2 + \mu_d^2 = (0)^2 \Pr\{d_i = 0\} + (1)^2 \Pr\{d_i = 1\} = \frac{1}{2} \qquad (3.26)$$

and

$$R_d(n) = E\left[d_i d_{i-n}\right] = \mu_d^2 = 3 \cdot 0 \cdot \Pr\{d_i d_{i-n} = 0\}$$
$$+1 \cdot (1)^2 \cdot \Pr\{d_i d_{i-n} = (1)^2\} = \frac{1}{4} \text{ for } n \neq 0 \qquad (3.27)$$

Formula (3.27) results from the fact that three out of four possible values of the product $d_i d_{i-n}$ are equal to zero and the probability of each data combination is 1/4 (which is the consequence of the statistical independence of data bits and their respective probabilities equal to 1/2). Finally, because the spectral density of the rectangular pulse of duration T is given by formula (3.14) we obtain the expression for power density spectrum $G_x(f)$ of the signal from Figure 3.9a in the form

$$G_x(f) = \frac{A^2 T}{4} \left[\text{sinc}(\pi f T)\right]^2 + \frac{A^2}{4}\delta(f) \qquad (3.28)$$

As we can see in (3.28), the signal has a nonzero mean value. The mean power of the DC component is equal to $A^2/4$. Similar calculations performed for the bipolar data format from Figure 3.9b lead to the expression

$$G_x(f) = A^2 T \left[\text{sinc}(\pi f T)\right]^2 \qquad (3.29)$$

This results from the fact that for assumed statistical independence of data bits and their probabilities the values of the data symbol autocorrelation function are respectively equal to $R_d(0) = 1$ and $R_d(n) = 0$ for $n \neq 0$.

Consider now the pseudoternary format with alternating polarizations of "1" s. It is obvious that the current value of the data symbol reflecting bit "1" depends on the previous data symbols (in particular on whether the previous "1" has been represented by a symbol with positive or negative polarization). Therefore, in this case we deal with the introduction of correlation among data symbols. Assuming equal probabilities of bits "1" and "0", we see that $\Pr\{d_i = 1\} = \Pr\{d_i = -1\} = 1/4$ and $\Pr\{d_i = 0\} = 1/2$. Therefore

$$R_d(0) = (1)^2 \Pr\{d_i = 1\} + (0)^2 \Pr\{d_i = 0\} + (-1)^2 \Pr\{d_i = -1\} = \frac{1}{2} \qquad (3.30)$$

For $n = \pm 1$ the pair of bits (called a *dibit*) that is represented by a pair of data symbols takes the possible values $(0, 0)$, $(0, 1)$, $(1, 0)$ and $(1, 1)$. Thus, the values of the product $d_i d_{i\pm1}$ are $0, 0, 0$ and -1, respectively. The last value results from the fact that two subsequent ones have inverse polarities. Then

$$R_d(n = \pm 1) = 3 \times (0) \times \frac{1}{4} + (-1) \times 1 \times \frac{1}{4} = -\frac{1}{4} \qquad (3.31)$$

We can easily check that for $|n| > 1$ $R_d(n) = 0$. Finally, after simple derivation and application of formulae (3.3) and (3.4), we obtain the following form of the power density spectrum of the AMI coded signal (Figure 3.9, curve c)

$$G_x(f) = A^2 T \left[\text{sinc}(\pi f T) \right]^2 \sin^2(\pi f T) \qquad (3.32)$$

The Manchester line code is a variant of the previously considered bipolar PAM modulation, in which an elementary pulse $s(t)$ given by

$$s(t) = \text{rect}\left(\frac{t - T/4}{T/2} \right) - \text{rect}\left(\frac{t - 3T/4}{T/2} \right) \qquad (3.33)$$

is applied. After calculation of the spectral density $S(f)$ of elementary signal $s(t)$, inserting the result into formula (3.13) and knowing that $\sigma_d^2 = 1$, we obtain

$$G_x(f) = A^2 T \text{sinc}^2(\pi f T/2) \sin^2(\pi f T/2) \qquad (3.34)$$

Power density spectra of signal $x(t)$ as a function of frequency normalized with respect to modulation period T are shown in Figure 3.10 for different data formats sketched in Figure 3.9. We see that bipolar and unipolar formats have a substantial part of their power in the vicinity of the DC, so application of a separating transformer or other separating circuit cutting off the DC and low frequency components will cause signal distortions. Spectral properties of a pseudoternary signal are relatively advantageous because a significant part of its power is contained in the frequency interval up to $1/T$ Hz. Good synchronization properties of the Manchester code result in its wide power density spectrum. Plots (a) and (c) shown in Figure 3.10 should be scaled if we want to have identical

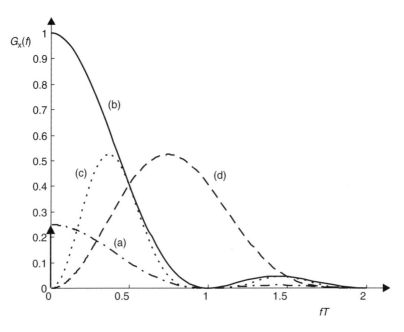

Figure 3.10 Power density spectra for signal $x(t)$ using several data formats: (a) unipolar NRZ,
(b) bipolar NRZ, (c) pseudoternary (AMI), (d) Manchester

mean power for all considered data formats. However, even in the current figure we
observe characteristic spectral features of particular signals.

So far we have considered binary baseband modulation. Having at our disposal a
channel with a limited bandwidth and wishing to send the data stream at the bit rate
higher than $1/T$ bit/s we can proceed in the following way. We divide the data stream into
disjunctive k-bit blocks. Each block can be mapped onto one of 2^k possible amplitudes of
the elementary signal. As a result, we receive $M = 2^k$-ary pulse amplitude modulation,
so k/T bits per second are transmitted over the channel. Unfortunately the data rate
increase is not for free! Comparing two- or higher level PAM signals we should assume
their equal mean power. Typically noise is added to the signal during transmission, which
results in a certain error rate level. As we can observe, the difference between two
neighboring multilevel PAM signals compared with the difference between two binary
PAM signals featuring the same mean power is much smaller. Thus, multilevel PAM
signals are much more sensitive to noise, resulting in a higher error probability. Such a
comparison is often illustrated by the error probability versus signal-to-noise ratio curves.
However, for fair comparison, the curves are displayed as a function of signal energy
per bit E_b divided by power spectral density N_0 of white noise. Figure 3.11 presents
error probability curves as a function of E_b/N_0 for two-level PAM ($k = 1$, $M = 2$),
4-PAM ($k = 2$, $M = 4$), 8-PAM ($k = 3$, $M = 8$) and 16-PAM ($k = 4$, $M = 16$). Symbol
error probability does not directly translate itself onto bit error probability. The latter
depends on the bit block-to-symbol mapping. Table 3.1 presents two possible ways of bit
block-to-symbol mapping for 4-PAM. One of them is called *natural encoding* and the
other is called *Gray encoding*.

Table 3.1 Mapping of binary blocks onto 4-PAM data symbols

Symbol d_i	Natural encoding	Gray encoding
−3	00	00
−1	01	01
1	10	11
3	11	10

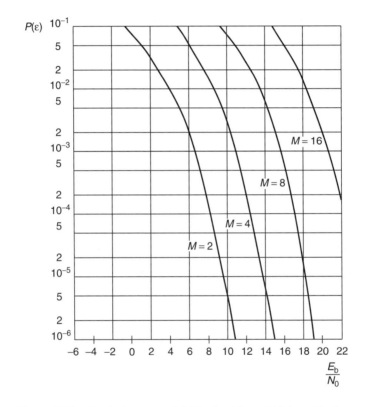

Figure 3.11 Symbol error probability for M-ary PAM modulations

Gray encoding features smaller bit error probability compared with natural encoding. This results from the fact that the most probable errors lead the receiver to select the neighboring symbol to the one that has been actually transmitted. Let us note that in the case of Gray encoding a single bit error will be committed, whereas if natural encoding is applied, in some cases double binary errors will be made.

3.4   Optimal Synchronous Receiver

Error probability is the basic measure of digital transmission system quality. It depends on the choice of elementary signals, the mapping of binary blocks onto them and signal-to-noise ratio in the channel, and also on the applied method of signal reception. Therefore, it is necessary to derive the receiver structure that ensures minimal probability of erroneous signal reception. The error probability curves shown in Figure 3.11 have been plotted for the case when the optimal receiver derived in this section is applied.

Below we determine the optimum receiver structure for digital signals transmitted over a channel introducing an additive noise. First, we will consider the general case of an optimum receiver for binary data symbols distorted by additive Gaussian noise characterized by a given power density spectrum. After that we will consider some particular cases for which the optimal receiver scheme is considerably simplified. In particular, we will assume that the additive Gaussian noise is white, the data symbols are equiprobable or that PAM is applied. We will also consider the optimal receiver for multilevel PAM signals.

3.4.1 Optimal Reception of Binary Signals

Assume that transmission is performed using two elementary signals, $s_0(t)$ and $s_1(t)$. We temporarily assume that the duration of the elementary signals does not exceed the modulation period T, and that binary data are statistically independent. These assumptions allow us to perform analysis of the received signal within a single modulation period. Let the received signal have the following form

$$y(t) = s_i(t) + n(t) \qquad (3.35)$$

where $n(t)$ is the additive Gaussian noise.

One of the basic receiver optimization criteria is minimization of a single data symbol error probability at its output. This is not the only criterion possible. If the receiver operating according to this criterion is difficult to implement, other criteria can be applied, e.g. minimum mean square error at the receiver output or minimum error probability for a whole received data sequence. Let us concentrate on the first criterion. We conclude from Chapter 1 that minimization of the error probability at the receiver output is equivalent to maximization of the *a posteriori probability* that a given signal has been transmitted after observation of the signal at the receiver input. Recall that this criterion is called MAP (*Maximum a Posteriori Probability*).[1] The optimization criterion receives the following form

$$\max_i P[i|y(t)] \quad \text{where } i = 0 \text{ or } 1 \qquad (3.36)$$

Let $P[s_i(t)] = P_i$ be the probability of transmission of signal $s_i(t)$. Following the assumption that the duration of an elementary signal does not exceed T seconds, it is sufficient for the receiver to observe the received signal $y(t)$ in that time interval. Instead of considering the continuous signal $y(t)$, let us consider its samples $y(t_k)$ taken with an arbitrarily small time step Δt within the interval $[0, T]$, where $t_k = t_{k-1} + \Delta t$, $k = 1, \ldots, K$. Denote the vectors of the received signal, noise and the transmitted signal samples as $\mathbf{y}$, $\mathbf{n}$ and $\mathbf{s}_i$,

---

[1] The word *a posteriori* means "after experience (experiment)".

respectively, so

$$\mathbf{y} = [y(t_1), y(t_2), \ldots, y(t_K)]$$
$$\mathbf{n} = [n(t_1), n(t_2), \ldots, n(t_K)] \tag{3.37}$$
$$\mathbf{s}_i = [s_i(t_1), s_i(t_2), \ldots, s_i(t_K)]$$

If we analyze vectors instead of continuous time functions, the MAP probability $P[i|y(t)]$ receives the form $P(i|\mathbf{y})$. Thanks to Bayes' formula this can be expressed in the form

$$P(i|\mathbf{y}) = \frac{p(\mathbf{y}|\mathbf{s}_i)P(\mathbf{s}_i)}{P(\mathbf{y})} \tag{3.38}$$

Recall that we search for the data symbol i, or equivalently the elementary signal $s_i(t)$ represented by vector $\mathbf{s}_i$, that maximizes the probability expressed by formula (3.38). The denominator in (3.38) is common for all elementary signals (it refers to the signal observed at the receiver input), so it does not influence the search result for the best data symbol i in the MAP sense. Thus, it is sufficient to select the data symbol that maximizes the numerator of (3.38). Let us focus on the conditional probability density function $p(\mathbf{y}|\mathbf{s}_i)$. We have $\mathbf{n} = \mathbf{y} - \mathbf{s}_i$ from (3.35), therefore

$$p(\mathbf{y}|\mathbf{s}_i) = p(\mathbf{y} - \mathbf{s}_i) = p(\mathbf{n}) = \frac{\exp\left[-\frac{1}{2}(\mathbf{y} - \mathbf{s}_i)R_n^{-1}(\mathbf{y} - \mathbf{s}_i)^T\right]}{(2\pi)^{K/2}|R_n|^{1/2}} \tag{3.39}$$

Expression (3.39) is the multidimensional Gaussian distribution for the noise vector $\mathbf{n}$. Matrix R_n is a $[K \times K]$ noise autocorrelation matrix whose elements are $R_{j,k} = R_n(t_j - t_k)$, where $R_n(\tau)$ is the autocorrelation function of noise $n(t)$. Calculating the logarithm of both sides of (3.38), we obtain

$$\ln P(i|\mathbf{y}) = \ln p(\mathbf{y}|\mathbf{s}_i) + \ln P(\mathbf{s}_i) - \ln P(\mathbf{y}) \tag{3.40}$$

Taking into account the right-hand side of (3.39) in (3.40) and including all the components that do not depend on i in constant C, we obtain

$$\ln P(i|\mathbf{y}) = C + \ln P_i - \frac{1}{2}(\mathbf{y} - \mathbf{s}_i)R_n^{-1}(\mathbf{y} - \mathbf{s}_i)^T \tag{3.41}$$

If we expand the third term in (3.41) we have

$$(\mathbf{y} - \mathbf{s}_i)R_n^{-1}(\mathbf{y} - \mathbf{s}_i)^T = \mathbf{y}R_n^{-1}\mathbf{y}^T - \mathbf{y}R_n^{-1}\mathbf{s}_i^T - \mathbf{s}_iR_n^{-1}\mathbf{y}^T + \mathbf{s}_iR_n^{-1}\mathbf{s}_i^T \tag{3.42}$$

Then if we use this expression in (3.41) and add the first term of the right-hand side of (3.42) to constant C (denoted after this modification as C_1), we obtain

$$\ln P(i|\mathbf{y}) = C_1 + \ln P_i - \frac{1}{2}\mathbf{s}_iR_n^{-1}\mathbf{s}_i^T + \mathbf{y}R_n^{-1}\mathbf{s}_i^T \tag{3.43}$$

Define vector $\mathbf{q}_i$ by the following equation

$$\mathbf{q}_i \Delta t = \mathbf{s}_i R_n^{-1}, \text{ i.e. } \mathbf{q}_i^T \Delta t = R_n^{-1} \mathbf{s}_i^T \tag{3.44}$$

so

$$\mathbf{q}_i \Delta t R_n = \mathbf{s}_i \tag{3.45}$$

For a single element $s_i(t_k)$ of vector $\mathbf{s}_i$ we have, from (3.45)

$$s_i(t_k) = \sum_{m=1}^{K} R_n(t_k - t_m) q_i(t_m) \Delta t \tag{3.46}$$

Let us represent $\mathbf{s}_i R_n^{-1}$ by $\mathbf{q}_i \Delta t$ and use formula (3.46) in (3.43). Then the latter evolves to the form

$$\ln P(i|\mathbf{y}) = C_1 + \ln P_i + \mathbf{y}\mathbf{q}_i^T \Delta t - \frac{1}{2}\mathbf{s}_i \mathbf{q}_i^T \Delta t$$

$$= C_1 + \ln P_i + \sum_{k=1}^{K} y(t_k) q(t_k) \Delta t - \frac{1}{2} \sum_{k=1}^{K} s_i(t_k) q_i(t_k) \Delta t \tag{3.47}$$

Now let the time interval Δt between the samples tend to zero and the number of samples K tend to infinity in such a way that $K \Delta t = T$. If the time interval Δt is reduced to an infinitely small increment dt in the modulation period T, the sums in expression (3.47) evolve to appropriate integrals and formula (3.47) becomes

$$\ln P(i|\mathbf{y}) = C_1 + \ln P_i + \int_0^T y(t) q_i(t) dt - \frac{1}{2} \int_0^T s_i(t) q_i(t) dt \tag{3.48}$$

whereas

$$s_i(t) = \int_0^T R_n(t - \lambda) q_i(\lambda) d\lambda \tag{3.49}$$

Let D_i^2 denote the following expression

$$D_i^2 = \int_0^T s_i(t) q_i(t) dt, \quad i = 0, 1 \tag{3.50}$$

After observation of signal $y(t)$, the receiver making a decision based on the MAP criterion decides that signal $s_0(t)$ has been transmitted if

$$P[0|y(t)] > P[1|y(t)]$$

or, equivalently, if

$$\int_0^T y(t)q_0(t)\mathrm{d}t - \frac{D_0^2}{2} + \ln P_0 > \int_0^T y(t)q_1(t)\mathrm{d}t - \frac{D_1^2}{2} + \ln P_1 \qquad (3.51)$$

When the reverse inequality is true, the receiver decides that signal $s_1(t)$ has been transmitted. Figure 3.12 presents the receiver scheme that results directly from expression (3.51). The receiver is equipped with two correlators and each of them correlates the received signal with the appropriate reference signal $q_0(t)$ or $q_1(t)$ in time interval $[0, T]$. Let us note that the reference signals depend on elementary signal $s_0(t)$ or $s_1(t)$, respectively, and on the noise correlation properties. This observation is a clear result of (3.49), which links both variables with each other. Derivation of the reference signals from (3.49) seems to be a difficult task. However, we can solve this problem if we consider an equivalent form of the optimum MAP receiver in which *matched filters* (MF) are used instead of correlators.

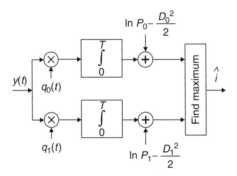

Figure 3.12 Optimal MAP receiver based on correlators

It turns out that the result of correlation of the received signal $y(t)$ with the reference signals, i.e.

$$\int_0^T y(t)q_0(t)\mathrm{d}t \quad \text{and} \quad \int_0^T y(t)q_1(t)\mathrm{d}t \qquad (3.52)$$

can be achieved when the received signal $y(t)$ is fed to the inputs of appropriately selected filters and their outputs are sampled at the moment $t = T$. We will show now that if the filter impulse response is given by expression

$$h_{i,\mathrm{MF}}(t) = q_i(T - t), \quad i = 0, 1 \qquad (3.53)$$

then the results of correlations (3.52) and sampling of the filter outputs at the moment $t = T$ are identical. The reference signal, as with the transmitted one, does not last longer than T seconds so the impulse response (3.53) does not exceed this time interval either.

Therefore the signal at the output of the filter in the ith receiver branch can be described by the convolutional integral

$$z_i(t) = \int_{-\infty}^{\infty} y(\tau) h_{i,\mathrm{MF}}(t-\tau)\mathrm{d}\tau = \int_{0}^{T} y(\tau) q_i\big[T-(t-\tau)\big]\mathrm{d}\tau \tag{3.54}$$

So at $t = T$ the filter output signal is equal to

$$z_i(T) = \int_{0}^{T} y(\tau) q_i\big[T-(T-\tau)\big]\mathrm{d}\tau = \int_{0}^{T} y(\tau) q_i(\tau)\mathrm{d}\tau \tag{3.55}$$

As we see, both receiver configurations yield the same samples at the moment $t = T$, so they are equivalent. The filter with impulse response (3.53) is called a *matched filter* (MF). Figure 3.13 presents the equivalent optimum MAP receiver structure in which the matched filters have been applied.

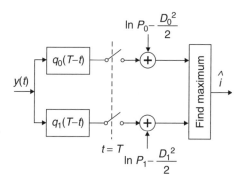

Figure 3.13 Optimal MAP receiver with matched filters

Consider now the transfer function of the matched filter. In order to derive this, let us again analyze formula (3.49). Knowing that reference signal $q_i(t)$ is zero outside the interval $[0, T]$ we can extend the integral limits from $[0, T]$ to $(-\infty, +\infty)$. Therefore

$$s_i(t) = \int_{-\infty}^{\infty} R_n(t-\lambda) q_i(\lambda)\mathrm{d}\lambda \tag{3.56}$$

so the elementary signal can be interpreted as a convolution of the reference signal with the additive noise autocorrelation function. As a result, in the frequency domain the following equation holds

$$S_i(f) = G_n(f) Q_i(f) \tag{3.57}$$

where $G_n(f) = \mathcal{F}[R_n(\tau)]$ is the power density spectrum of the additive noise. The spectral density of the reference signal is therefore

$$Q_i(f) = \frac{S_i(f)}{G_n(f)} \tag{3.58}$$

The impulse response of the matched filter is given by formula (3.53), so taking advantage of (3.58) we get

$$H_{i,\mathrm{MF}}(f) = \mathcal{F}[q_i(T-t)] = \exp(-j2\pi f T)\, Q^*(f) = \frac{\exp(-j2\pi f T) S_i^*(f)}{G_n(f)} \tag{3.59}$$

As we see, the matched filter transfer function depends on the spectral density of elementary signal $s_i(t)$ and the power density spectrum of additive noise.

Consider the particular case of the matched filter when the additive Gaussian noise is white and its power density spectrum is $N_0/2$. Thus

$$H_{i,\mathrm{MF}}(f) = \frac{2}{N_0} S_i^*(f) \exp(-j2\pi f T), \quad \text{so} \quad h_{i,\mathrm{MF}}(t) = \frac{2}{N_0} s_i(T-t) \tag{3.60}$$

In this case the matched filter impulse response is proportional to the mirrored reflection of the elementary signal with respect to the vertical axis. Quite often the elementary signal is symmetric with respect to its maximum. Then the impulse response of the matched filter has a shape that is identical to the elementary signal waveform.

Consider now several particular cases of the optimum MAP receivers. First, let the elementary signals have equal energies, i.e.

$$\int_0^T s_0^2(t)\mathrm{d}t = \int_0^T s_1^2(t)\mathrm{d}t \tag{3.61}$$

Let the additive Gaussian noise be white, so its autocorrelation function is described by formula

$$R_n(\tau) = \frac{N_0}{2}\delta(\tau) \tag{3.62}$$

As a result, the elementary signals and the corresponding reference signals resulting from equation (3.49) are linked by expression

$$s_i(t) = \int_0^T R_n(t-\lambda)q_i(\lambda)\mathrm{d}\lambda = \frac{N_0}{2}\int_0^T \delta(t-\lambda)q_i(\lambda)\mathrm{d}\lambda = \frac{N_0}{2}q_i(t) \tag{3.63}$$

At the same time

$$D_0^2 = \int_0^T s_0(t)q_0(t)\mathrm{d}t = \frac{2}{N_0}\int_0^T s_0^2(t)\mathrm{d}t = \frac{2}{N_0}\int_0^T s_1^2(t)\mathrm{d}t = D_1^2 \tag{3.64}$$

As the terms $D_0^2/2$ and $D_1^2/2$ are equal they do not affect the result of comparison of both sides in inequality (3.51) and they can be omitted in the receiver structure. Thus, the receiver scheme looks like that shown in Figure 3.14.

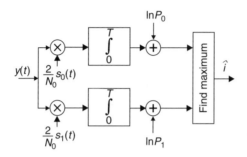

Figure 3.14   Optimal MAP receiver for equal energy signals corrupted by additive white Gaussian noise

If we additionally assume that both signals are transmitted with the same probability, then the terms $\ln P_0$ and $\ln P_1$ do not affect the result of comparison either and they can be removed. Consequently, the scaling factor $2/N_0$ in the reference signals can be omitted too, so the optimum MAP receiver scheme can take the form shown in Figure 3.15, and its decisions result from comparison of the left- and right-hand side of the expression

$$\int_0^T y(t)s_0(t)\mathrm{d}t \lessgtr \int_0^T y(t)s_1(t)\mathrm{d}t \tag{3.65}$$

If the elementary signals are bipolar, i.e. $s_0(t) = -s_1(t)$, then for the same assumptions the receiver is further simplified. For this case the signals at the output of each receiver branch (see Figure 3.15) differ only in sign, so a single branch and a decision circuit with the zero threshold is sufficient. This particular receiver is shown in Figure 3.16a.

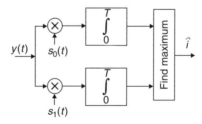

Figure 3.15   Optimal MAP receiver for equiprobable, equal energy signals corrupted by additive white Gaussian noise

So far we have considered the MAP reception of an isolated elementary signal. In practice, a whole sequence of data pulses is transmitted. If the duration of elementary

(a)

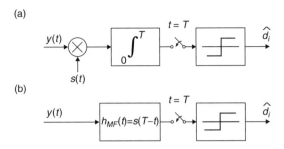

(b)

Figure 3.16 Optimal MAP receiver for equiprobable bipolar signals corrupted by additive white Gaussian noise with application of the correlator (a) and the matched filter (b)

signals does not exceed the modulation period T, as we have assumed so far, the reception processes are disjoint and our analysis still holds. However, we have shown earlier that good spectral properties of the modulated signal are obtained if elementary signals last longer than the modulation period T. As we remember, such signals can be received if the signal at the input of the sampler is a linear combination of pulses having periodical zeros except for the zero moment in which their maximum occurs. Consequently, the assumption about the duration of elementary signals not exceeding the modulation period T can be withdrawn if we apply the matched filter with the sampler sampling the filter output every T seconds and if we use such a waveform of the elementary pulse that the joint impulse response of the transmit and matched filters features zeros every T seconds except for the moment in which its maximum occurs. Therefore, very often the transmit filter has the square root raised cosine characteristics and the matched filter in the receiver has the same characteristics. As a result, their joint characteristics have the raised cosine shape so they conform to the Nyquist criterion. Figure 3.17 presents a scheme of the transmitter with the transmit filter with the impulse response $s(t)$, the additive white Gaussian noise (AWGN) channel and the receiver with the matched filter impulse response $h_{\mathrm{MF}}(t) = s(T_0 - t)$ followed by a sampler and a decision circuit. Figure 3.18 shows the transmit filter impulse response $s(t)$ with the square root raised cosine characteristics, the impulse response of the filter matched to the transmit filter and their joint impulse response. Time T_0 is the effective length of the transmit filter impulse response and it lasts for a few modulation periods T.

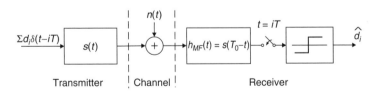

Figure 3.17 Basic scheme of digital PAM transmission with optimum MAP receiver

3.4.2 Optimal Receiver for Multilevel Signals

Generalization of the optimal synchronous MAP receiver onto the application of multilevel signals is very simple. The receiver performs the same operations as those done in each

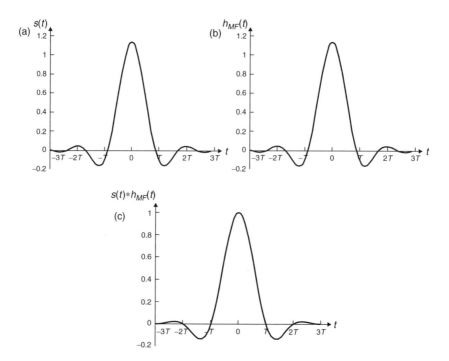

Figure 3.18 Impulse responses for transmit filter (a), matched filter (b) and cascade connection of transmit filter and matched filter (c)

branch of the receiver shown in Figure 3.12, followed by finding the maximum signal among the output signals of each branch. In a general case, the receiver looks like that shown in Figure 3.19. In the decision process according to the MAP criterion the receiver performs the following operation

$$\widehat{d}_i = \underset{d_i}{\arg\max} \left(\int_0^T y(t) q_i(t) dt - \frac{D_i^2}{2} + \ln P_i \right), \quad i = 1, 2, \ldots, M \qquad (3.66)$$

Let us analyze the case of multilevel signals, which often occurs in practical applications. Consider the signals with M-level PAM modulation. Assume that all elementary signals are equiprobable. Assume also that the additive noise is Gaussian and white and its power density spectrum is $N_0/2$. The elementary signals have the following form

$$s_i(t) = d_i s(t), \quad d_i = \pm 1, \pm 3, \ldots, \pm(M - 1) \quad \text{where } M = 2^k \qquad (3.67)$$

For this particular case

$$q_i(t) = \frac{2}{N_0} d_i s(t) \quad \text{and} \quad D_i^2 = \frac{2}{N_0} d_i^2 \int_0^T s^2(t) dt = \frac{2}{N_0} d_i^2 E_s \qquad (3.68)$$

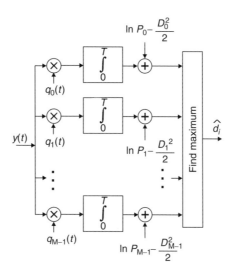

Figure 3.19 Optimal MAP synchronous receiver for M-level transmission

where E_s is the energy of the elementary pulse $s(t)$. The selection criterion applied in the decision device is then reduced to the form

$$\widehat{d_i} = \arg\max_{d_i}\left[d_i \int_0^T y(t)s(t)\mathrm{d}t - \frac{d_i^2}{2}E_s \right], \quad i = 1, 2, \ldots, M \qquad (3.69)$$

Let us note that the receiver needs to perform only one correlation of the received signal $y(t)$ with the elementary pulse $s(t)$. After scaling the correlation result with data symbol d_i and applying the appropriate shift by $-d_i^2 E_s/2$, the receiver searches for the maximum value. Denote the correlator output as U. Thus, selection of data symbol d_i fulfilling the MAP criterion reduces to finding that symbol $\widehat{d_i}$ among all possible data symbols that maximizes the convex quadratic function

$$f(U, d_i) = d_i U - d_i^2 \frac{E_s}{2} \qquad (3.70)$$

It is easy to show that when the received signal has the ideal form $y(t) = d_k s(t)$, function $f(U, d_i)$ achieves its exact maximum for data symbol $\widehat{d_i} = d_k$. Now the following question arises: How much can the received signal U differ from the ideal value at the correlator output equal to

$$d_k \int_0^T s^2(t)\mathrm{d}t = d_k E_s$$

so that the choice $\widehat{d_i} = d_k$ is still optimal? Let data symbol d_k be one of the "internal" data symbols, i.e. $d_k \neq \pm(M-1)$. Knowing that function $f(U, d_i)$ is a quadratic convex function, it is sufficient to determine the lower bound U_1 and upper bound U_2 of this interval. For these bounds the value of function $f(U, d_i)$ for $d_i = d_k$ is equal to the value of this function for data symbols that are neighboring to d_k, i.e. the following equations are fulfilled for U_1 and U_2:

$$f(U_1, d_k) = f\left[U_1, (d_k - 2)\right] \tag{3.71}$$
$$f(U_2, d_k) = f\left[U_2, (d_k + 2)\right]$$

Applying (3.70) in the solution for both equations we receive the bounds in the form

$$U_1 = (d_k - 1)E_s \quad \text{and} \quad U_2 = (d_k + 1)E_s \tag{3.72}$$

These results indicate that if the result of correlation U of the received signal $y(t)$ with the elementary signal $s(t)$ is found in the interval $[(d_k - 1)E_s, (d_k + 1)E_s]$, then the decision circuit should select symbol $\widehat{d_i} = d_k$. We also conclude that the optimum receiver for M-PAM signals consists of a correlator of the received signal with the elementary reference pulse $s(t)$ and an M-level quantizer. The scheme of this receiver is shown in Figure 3.20, whereas the characteristics of the quantizer for the 4-PAM signal is presented in Figure 3.21. The receiver of M-PAM signals often constitutes the basis for receivers of more complicated signals applying M-PAM modulations of a sinusoidal carrier. Such signals will be considered in the next chapter.

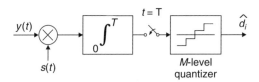

Figure 3.20   Optimal receiver for M-PAM signals

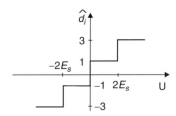

Figure 3.21   Quantizer characteristics for the 4-PAM receiver

3.5   Error Probability at the Output of the Optimal Synchronous Receiver

In this section we will derive the probability of error at the output of the optimal synchronous receiver for the case when binary data are transmitted by means of elementary signals $s_0(t)$ and $s_1(t)$, which are corrupted by additive white Gaussian noise with the power density spectrum equal to $N_0/2$. Assume that both elementary signals, whose duration does not exceed the modulation period equal to T seconds, have equal energy E and the probabilities of generation of elementary signals are equal, i.e. $P[s_0(t)] = P[s_1(t)]$. Thus, condition (3.61) is fulfilled. It is well known that error probability $P(\mathcal{E})$ can be derived by calculation of conditional error probabilities $P[\mathcal{E}|s_i(t)]$ ($i = 0, 1$), and using the formula

$$P(\mathcal{E}) = P[\mathcal{E}|s_0(t)]P[s_0(t)] + P[\mathcal{E}|s_1(t)]P[s_1(t)] \qquad (3.73)$$

Consider the optimal receiver in its basic form, which consists of two correlators and the device selecting the maximum correlator output at moments $t = nT$. First calculate $P[\mathcal{E}|s_0(t)]$. For simplicity let us consider the moment $k = 1$, because in every other modulation period the process is analogous and, due to an assumption about the duration of data symbols not exceeding the modulation period T, the reception processes can be performed separately, one after the other. Assume for the moment that signal $s_0(t)$ has been transmitted in the time interval $[0, T]$. Following our assumption about channel properties, the signal observed at the receiver input is the sum of the elementary signal and noise, i.e.

$$y(t) = s_0(t) + n(t) \qquad (3.74)$$

At the outputs of the correlators shown in Figure 3.15 at $t = T$ we obtain

$$U_0 = \int_0^T \left(s_0(t) + n(t)\right)s_0(t)\mathrm{d}t = E_s + N_{n,0} \qquad (3.75)$$

$$U_1 = \int_0^T \left(s_0(t) + n(t)\right)s_1(t)\mathrm{d}t = \gamma E_s + N_{n,1} \qquad (3.76)$$

where, as previously, E_s is the energy of elementary signals in time interval T and $N_{n,0}$ and $N_{n,1}$ are random variables that result from correlation of the noise $n(t)$ with elementary signals $s_0(t)$ and $s_1(t)$, respectively. Symbol γ denotes the cross-correlation coefficient of both elementary signals, i.e.

$$\gamma = \frac{1}{E_s} \int_0^T s_0(t)s_1(t)\mathrm{d}t \qquad (3.77)$$

The receiver commits an error if signal U_1 at the output of the correlator in which $s_1(t)$ is used as the reference signal is higher than signal U_0 at the output of the correlator applying $s_0(t)$ as the reference signal. Therefore

$$P[\mathcal{E}|s_0(t)] = \Pr\{U_1 > U_0\} = \Pr\{U_0 - U_1 < 0\} \qquad (3.78)$$

Define a random variable $V = U_0 - U_1 = E_s(1 - \gamma) + (N_{n,0} - N_{n,1})$. For the given elementary signals the first component of V is constant. The second component is a difference of two Gaussian random variables. As we remember, the sum (or difference) of two Gaussian random variables remains Gaussian distributed. Therefore random variable V is Gaussian and the error probability conditional on transmitting $s_0(t)$ is equal to the probability that random variable V is lower than zero, i.e.

$$P\big[\mathcal{E}|s_0(t)\big] = \Pr\{V < 0\} \tag{3.79}$$

In order to calculate this probability we have to determine the parameters of Gaussian distribution of V, i.e. its mean and variance. We namely have

$$\mu_V = E[V] = E_s(1 - \gamma) \tag{3.80}$$

and

$$\sigma_V^2 = E[(V - \mu_V)^2] = E[(N_{n,0} - N_{n,1})^2]$$
$$= E[N_{n,0}^2] - 2E[N_{n,0}N_{n,1}] + E[N_{n,1}^2] \tag{3.81}$$

In order to derive the above variance we will consider the first component of the right-hand side of (3.81). Knowing that the additive noise is white, i.e. its autocorrelation function is a Dirac delta of intensity $N_0/2$, we obtain

$$E[N_{n,0}^2] = E\left[\int_0^T s_0(t)n(t)\mathrm{d}t \int_0^T s_0(\tau)n(\tau)\mathrm{d}\tau\right]$$

$$= E\left[\int_0^T\int_0^T s_0(t)s_0(\tau)n(t)n(\tau)\mathrm{d}t\mathrm{d}\tau\right]$$

$$= \int_0^T\int_0^T s_0(t)s_0(\tau)R_n(t - \tau)\mathrm{d}t\mathrm{d}\tau$$

$$= \int_0^T s_0(t)\left[\int_0^T s_0(\tau)\frac{N_0}{2}\delta(t - \tau)\mathrm{d}\tau\right]\mathrm{d}t$$

$$= \frac{N_0}{2}\int_0^T s_0(t)s_0(t)\mathrm{d}t = \frac{N_0}{2}E_s \tag{3.82}$$

It is also easy to show that $E[N_{n,1}^2] = \frac{N_0}{2}E_s$. In a similar way [the reader can perform these calculations by himself/herself by taking (3.82) as a pattern] we can show that

$E[N_{n,0}N_{n,1}] = \frac{N_0}{2}\gamma E_s$. As a result, the variance of the Gaussian distribution of the random variable V is

$$\sigma_V^2 = N_0 E_s (1 - \gamma) \tag{3.83}$$

Knowing that the probability distribution of random variable V is given by the formula

$$p_V(v) = \frac{1}{\sqrt{\pi}\sigma_V} \exp\left[-\frac{(v - \mu_V)^2}{2\sigma_V^2}\right] \tag{3.84}$$

and defining the *complementary error function* (Figure 3.22)

$$\mathrm{erfc}(x) = \frac{2}{\sqrt{\pi}} \int_x^\infty \exp\left(-t^2\right) dt \tag{3.85}$$

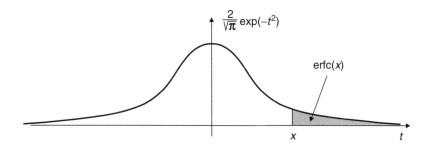

Figure 3.22 Illustration of complementary error function $\mathrm{erfc}(x)$

we can find the probability of error at the receiver output conditional on the transmitted signal $s_0(t)$ by performing the following calculations

$$P(\mathcal{E}|0) = \frac{1}{\sqrt{2\pi}\sigma_V} \int_{-\infty}^{0} \exp\left[-\frac{(v - \mu_V)^2}{2\sigma_V^2}\right] dv = \frac{1}{\sqrt{\pi}} \int_{-\infty}^{\frac{\mu_V}{\sqrt{2}\sigma_V}} \exp\left(-t^2\right) dt$$

$$= \frac{1}{\sqrt{\pi}} \int_{\frac{\mu_V}{\sqrt{2}\sigma_V}}^{\infty} \exp\left(-t^2\right) dt = \frac{1}{2}\,\mathrm{erfc}\left(\frac{\mu_V}{\sqrt{2}\sigma_V}\right)$$

$$= \frac{1}{2}\,\mathrm{erfc}\left[\frac{E_s(1 - \gamma)}{\sqrt{2}\sqrt{E_s N_0(1 - \gamma)}}\right]$$

Therefore

$$P(\mathcal{E}|0) = \frac{1}{2} \, \text{erfc} \left[\sqrt{\frac{E_s}{2N_0}(1-\gamma)} \right] \tag{3.86}$$

Similar calculations lead to the same formula for the probability of error conditional on the transmitted signal $s_1(t)$. Thus, using formula (3.73) we obtain

$$P(\mathcal{E}) = \frac{1}{2} \, \text{erfc} \left[\sqrt{\frac{E_s}{2N_0}(1-\gamma)} \right] \tag{3.87}$$

We conclude from formula (3.87) that the error probability depends on the signal-to-noise ratio (expressed by E_s/N_0) and on the degree of mutual correlation of elementary signals. Selecting $s_0(t)$ and $s_1(t)$ as bipolar signals, we receive the cross-correlation coefficient $\gamma = -1$, which results in the maximum argument of erfc function and, as a result, in minimum error probability. For that case the error probability is

$$P(\mathcal{E}) = \frac{1}{2} \, \text{erfc} \left(\sqrt{\frac{E_s}{N_0}} \right) \tag{3.88}$$

Other values of cross-correlation factor γ are also possible. In particular, elementary signals can be mutually orthogonal. For such signals $\gamma = 0$. In consequence we obtain

$$P(\mathcal{E}) = \frac{1}{2} \, \text{erfc} \left(\sqrt{\frac{E_s}{2N_0}} \right) \tag{3.89}$$

As we have expressed the error probability as a function of E_s/N_0, we may ask how this ratio is associated with the signal-to-noise ratio $\text{SNR} = P_{\text{sig}}/P_{\text{noise}}$. As we remember, owing to spectral shaping of the transmitted signal, its bandwidth moderately exceeds the frequency $1/2T$. In practice it does not exceed $1/T$. Consider the signal-to-noise ratio in this band, assuming that noise is Gaussian and white with the power density spectrum $N_0/2$. Thus, the noise power in the band $(-1/T, 1/T)$ is $P_{\text{noise}} = N_0/2 \times (2/T)$. Therefore

$$\text{SNR} = \frac{P_{\text{sig}}}{P_{\text{noise}}} = \frac{P_{\text{sig}}}{\dfrac{N_0}{2} \times \dfrac{2}{T}} = \frac{P_{\text{sig}} T}{N_0} = \frac{E_s}{N_0} \tag{3.90}$$

where E_s is the mean signal energy per single data symbol.

As we see from (3.89), if we want to achieve the same error probability for the orthogonal signals as for bipolar signals [see (3.88)], the signal-to-noise ratio has to be twice as high, i.e. it must be about 3 dB higher.

3.6   Error Probability in the Optimal Receiver for M-PAM Signals

Knowing the structure of the optimal receiver for M-PAM signals we can relatively easily determine the probability of error at its output. Assume that the receiver consists of a block correlating the received signal with the basic pulse $s(t)$ and a multilevel quantizer with the quantization thresholds given by (3.72) (see Figure 3.20). Assume that white Gaussian noise of power density spectrum $N_0/2$ is an additive distortion and data symbols are equiprobable, i.e. $P(d_i) = 1/M$, where $d_i = \pm 1, \pm 3, \ldots, \pm(M-1)$ and $M = 2^k$. Recall that the received signal is described by the expression

$$y(t) = d_i s(t) + n(t) \tag{3.91}$$

Recalling that we have denoted the energy of the basic pulse $s(t)$ as E_s, we obtain the sample U at the correlator output

$$U = \int_0^T \left[d_i s(t) + n(t) \right] s(t) \mathrm{d}t = d_i E_s + Z_n \tag{3.92}$$

where Z_n is a Gaussian random variable that results from the correlation of noise $n(t)$ and signal $s(t)$. This random variable has zero mean and a variance which is determined on the basis of calculations similar to those performed in (3.82), resulting in the value $\sigma_Z^2 = E_s N_0/2$. Therefore U is also a Gaussian variable with a mean equal to $d_i E_s$ and variance σ_Z^2. In general, symbol error probability can be calculated from the formula

$$P_M(\mathcal{E}) = \sum_{i=1}^M P(d_i) P(\mathcal{E}|d_i \text{ transmitted}) \tag{3.93}$$

In the case of data symbols d_i different from the "outer" ones, i.e. those different from $\pm(M-1)$, taking into account the quantization thresholds (3.72) we obtain the conditional symbol error probability

$$P(\mathcal{E}|d_i \text{ transmitted}) = \Pr\{|U - d_i E_s| > E_s\} \tag{3.94}$$

This probability is shown as the area under both tails of a Gaussian curve. For the "outer" symbols $d_i = \pm(M-1)$ this probability is equal to $\Pr\{[U - (M-1)E_s] < -E_s\}$ for $d_i = M-1$ and $\Pr\{[U + (M-1)E_s] > E_s\}$ for $d_i = -(M-1)$, respectively, because the area of correct decision is limited only on one side. Figure 3.23 illustrates our considerations for 4-PAM modulation. The grey areas indicate conditional error probabilities.

Using (3.93) and (3.94) we can write

$$P_M(\mathcal{E}) = \frac{M-1}{M} \Pr\{|U - E_s| > E_s\} = \frac{M-1}{M} \frac{2}{\sqrt{2\pi}\sigma_Z} \int_{E_s}^\infty \exp\left(-\frac{u^2}{2\sigma_Z^2}\right) \mathrm{d}u \tag{3.95}$$

The denominator of the fraction $(M-1)/M$ is determined by the data symbol probabilities $P(d_i) = 1/M$, whereas the numerator $(M-1)$ reflects the fact that for M-level

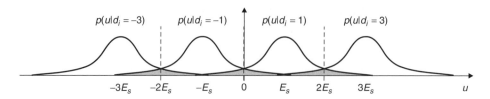

Figure 3.23 Conditional symbol error probability density functions at the correlator output of the optimal receiver for 4-PAM modulated signals

modulation the number of decision areas limited on both sides is $M - 2$, and two decision areas are limited on one side only, which is jointly equivalent to $M - 1$ decision areas limited on both sides. Using again the standard form of the erfc function, we obtain

$$P_M(\mathcal{E}) = \frac{M-1}{M} \frac{2}{\sqrt{\pi}} \int\limits_{\frac{E_s}{\sqrt{2}\sigma_Z}}^{\infty} \exp\left(-\lambda^2\right) d\lambda \tag{3.96}$$

so

$$P_M(\mathcal{E}) = \frac{M-1}{M} \operatorname{erfc}\left(\frac{E_s}{\sqrt{2}\sqrt{N_0 E_s/2}}\right) = \frac{M-1}{M} \operatorname{erfc}\left(\sqrt{\frac{E_s}{N_0}}\right) \tag{3.97}$$

We have derived the formula for symbol error probability for M-PAM signals as a function of the ratio of the basic pulse energy E_s to the noise power density spectrum N_0. An expression for symbol error probability as a function of the mean power of the signal will be more useful if the error probability is presented as a function of the ratio of mean signal energy per data symbol or per data bit to the power density spectrum. Thus, let us determine the mean power of the transmitted signal. We proceed in the following way

$$P_{\text{mean}} = \frac{1}{T} E\left\{\int\limits_0^T [d_i s(t)]^2 \, dt\right\} = \frac{E[d_i^2]}{T} \int\limits_0^T s^2(t) dt = \frac{E[d_i^2]}{T} E_s \tag{3.98}$$

For the considered data symbol set, assuming that all data symbols are equiprobable, we obtain

$$E[d_i^2] = E[(2i - M)^2] = \frac{2}{M} \sum_{i=1}^{M/2} (2i - 1)^2 = \frac{M^2 - 1}{3} \tag{3.99}$$

Therefore

$$P_{\text{mean}} = \frac{M^2 - 1}{3T} E_s \tag{3.100}$$

and we conclude that

$$E_s = \frac{3 P_{\text{mean}} T}{M^2 - 1} = \frac{3 E_{\text{mean}}}{M^2 - 1} = \frac{3 k E_b}{M^2 - 1} \tag{3.101}$$

where $E_{\text{mean}} = P_{\text{mean}} T$ is the mean energy of a single data pulse. The number of modulation levels can be expressed in the form $M = 2^k$, where k is the number of bits mapped onto a single data symbol. Denote mean signal energy per bit as E_b and notice that $E_{\text{mean}} = k E_b$. Now we can express the symbol error probability for M-PAM by the formula

$$P_M(\varepsilon) = \frac{M - 1}{M} \operatorname{erfc} \left(\sqrt{\frac{3k}{M^2 - 1} \frac{E_b}{N_0}} \right) \tag{3.102}$$

If we wish to have a fair comparison of error rates for modulations with different levels, we should plot the error probability curves as functions of E_b/N_0. Figure 3.11 presents several error curves drawn according to (3.102) for different modulation levels M.

3.7 Case Study: Baseband Transmission in Basic Access ISDN Systems

Let us consider a practical example that illustrates our considerations on digital baseband transmission. In the 1980s, along with the progress in digitalization of telecommunication networks affecting mainly switching centers and links between them, it was found that extending the digital network directly to the end users (subscribers) would result in substantial enrichment of the services offered and would finalize digitalization of the whole telecommunication network. In this way the idea of Integrated Services Digital Network (ISDN) systems was established.

As we know, in a classical fixed telephone network each user is connected to the closest switching center via a twisted copper wire pair. The properties of such a channel are considered in Chapter 5. In the so-called *Plain Old Telephone Service* (POTS) a twisted pair is applied only to transmit analog signals representing a voice waveform or a voiceband modem signal. Both are contained in the band limited approximately to 4 kHz. However, a twisted pair offers a much wider channel bandwidth that can be utilized for digital transmission. Thus, the heavy investment in subscriber lines made by telecom operators with the expected return after a few tens of years had to be further exploited. In the first phase of development of digital transmission over subscriber loops, two bearer (B) channels of the rate 64 kbit/s each plus one data (D) channel of the rate 16 kbit/s were offered in the duplex mode. This means that 144 kbit/s was offered to a user for simultaneous transmission in both directions. This transmission system is described as *Basic Access ISDN* (ITU-T G.961 1993). Later the data rate was extended to 2.048 Mbit/s in Europe, equivalent to the primary rate in PCM systems and 1.544 Mbit/s in the USA. Such a transmission technique is called a *High-data-rate Digital Subscriber Line* (HDSL) (ITU-T G.991.1 1998) and was basically performed in parallel over two or three twisted wire pairs. Later, owing to the progress in transmission technology, this method was improved to enable a primary rate transmission on a single twisted wire pair.

A subscriber line with appropriate transceivers which enables transmission of the latter is called a *Single-pair High-speed Digital Subscriber Line* (SHDSL) (ITU-T G.991.2 2003).

In our case study we will present the recommended line codes and pulse shapes mainly for Basic Access ISDN digital transmission.

Let us mention that much faster digital transmission over subscriber loops is currently possible, enabling wideband Internet access using ADSL (*Asymmetric Digital Subscriber Line*), ADSL2, VDSL (*Very high rate Digital Subscriber Line*) or VDSL2 technologies. However, the corresponding methods of digital transmission have not been described yet, as they require modulations of sinusoidal carriers. We will leave them for future consideration.

Before we describe the line codes and appropriate pulse shapes in Basic Access ISDN, let us explain how duplex transmission is possible over a single twisted wire pair. According to the nomenclature adopted for the description of ISDN systems, a transceiver on the user side is part of *Network Termination* (NT), whereas a transceiver on the switching center side is part of *Line Termination* (LT). Although there are various possible techniques of duplex transmission, two techniques have been recommended for Basic Access ISDN and HDLC or SHDLC systems. These are *Echo Cancellation* (ECH) and *Time Compression Multiplex* (TCM).[2] In the ECH method (see Figure 3.24) signals in both directions, i.e. from NT to LT and back, are transmitted simultaneously in the same range of frequencies. The signals are split by a hybrid (which is described in Chapter 5). If the hybrid worked ideally, there would be no signal at the input of the receiver that is

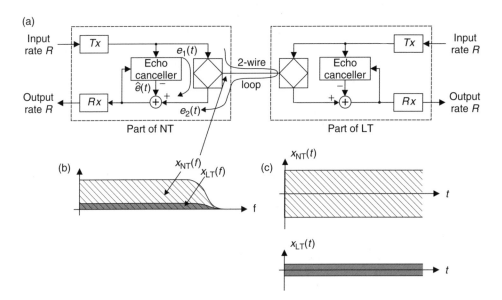

Figure 3.24 Echo cancellation method used in duplex ISDN transmission: (a) basic transmission configuration, (b) signal spectra at the input/output of the NT hybrid, (c) signals at the input/output of the NT hybrid

[2] A more popular explanation for the TCM acronym is *Trellis-Coded Modulation*, which will be the subject of our considerations in the next chapter.

generated by the transmitter located in the same transceiver. The only received signal would be the signal generated by the remote transmitter and sent over the subscriber loop channel. In reality, hybrid ability to attenuate the locally generated signal is limited. Additionally, the signal from the remote transmitter is often attenuated to such a degree that the unwanted local signal component received through a nonideal hybrid is much stronger than the desired signal. Thus, an adaptive filter called an *echo canceller* attempts to generate two unwanted components: $e_1(t)$, the signal leaking through the hybrid; and $e_2(t)$, the signal that arrives at the local receiver as a reflection signal due to nonideal impedance matching in the subscriber loop (see Chapter 5). This signal is then subtracted from the received sum of the desired remote signal and echoes, resulting in a sufficient signal-to-noise ratio to allow for reliable signal detection. As shown in the chapter on channel models (Chapter 5), the attenuation of a twisted wire pair channel substantially increases with frequency, so transmitting in both directions in the same band allows the required bandwidth to be minimized, leading to a higher transmission range.

The second method, TCM, uses the channel in time division mode (see Figure 3.25). On the NT and LT sides there are two switches that, when appropriately synchronized, establish a transmission link by alternating the transmission directions. As we see, no hybrids are required and no echoes distort the received signal. However, the disadvantage

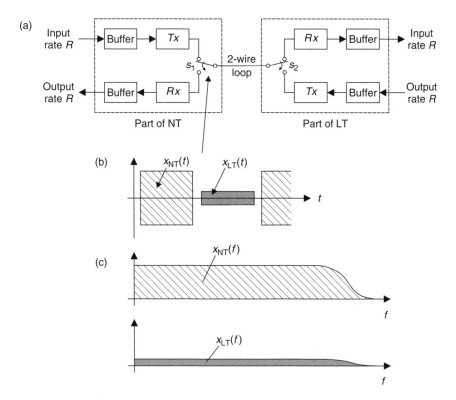

Figure 3.25 The Time Compression Multiplex (TCM) method used in duplex ISDN transmission: (a) basic transmission configuration, (b) time domain signal, (c) signal spectra on the input/output of NT

of this transmission method is the necessity to compress the signal in time more than twice. Half of a time period is devoted to transmission in one direction, but one has to take into account the delay resulting from signal propagation along the subscriber loop channel and a finite time of reversing the transmission direction resulting from the switching processes and rise time of transmit amplifiers. In consequence, the occupied bandwidth is more than twice as large as that used in the ECH method.

Three line codes are presented in the Appendix to ITU-T G.961 Recommendation (ITU-T G.961 1993) as allowable alternatives. The first one is called a Modified Monitoring State Code (MMS 43) and it maps 4-bit blocks into blocks of three ternary symbols with levels +1, 0 and −1 denoted as +, 0 or −. It is sometimes called the 4B3T code. The MMS 43 line code applies four alphabets, S1, S2, S3 and S4. The choice of the alphabet is determined by the current 4-bit block on the encoder input and the previously used alphabet. Table 3.2 shows mapping applied in the MMS 43 line code. Each block of three ternary symbols is accompanied by the number of the alphabet that has to be used in the next application of the line code mapping table.

Table 3.2   MMS 43 line code mapping table

Input block	S1		S2		S3		S4	
0001	0−+	1	0−+	2	0−+	3	0−+	4
0111	−0+	1	−0+	2	−0+	3	−0+	4
0100	−+0	1	−+0	2	−+0	3	−+0	4
0010	+−0	1	+−0	2	+−0	3	+−0	4
1011	+0−	1	+0−	2	+0−	3	+0−	4
1110	0+−	1	0+−	2	0+−	3	0+−	4
1001	+−+	2	+−+	3	+−+	4	−−−	1
0011	00+	2	00+	3	00+	4	−−0	2
1101	0+0	2	0+0	3	0+0	4	−0−	2
1000	+00	2	+00	3	+00	4	0−−	2
0110	−++	2	−++	3	−−+	2	−−+	3
1010	++−	2	++−	3	+−−	2	+−−	3
1111	++0	3	00−	1	00−	2	00−	3
0000	+0+	3	0−0	1	0−0	2	0−0	3
0101	0++	3	−00	1	−00	2	−00	3
1100	+++	4	−+−	1	−+−	2	−+−	3

For example, if alphabet S1 has been used at the previous moment, the block 0000 fed to the line encoder input at the current moment results in the ternary output symbol block +0+ and alphabet S3 will be applied to send the next block. As we can see, the MMS 43 encoder is a device with memory, so it introduces correlation between subsequent ternary data symbols. We should also note that not all possible ternary codewords can be applied. At each moment of the encoding process the encoder is able to select one of 16 ternary codewords (in a single column of the mapping table, appropriately to the selected alphabet) out of 27 combinations of three ternary symbols. If the receiver detects a ternary block that is not an allowable line code sequence, quality control alert can be triggered.

ITU-T Recommendation G.961 does not provide an exact formula for applied pulses; however, it includes a pulse mask, as shown in Figure 3.26. Any pulse shape applied in the NT or LT transmitter has to be contained within this mask.

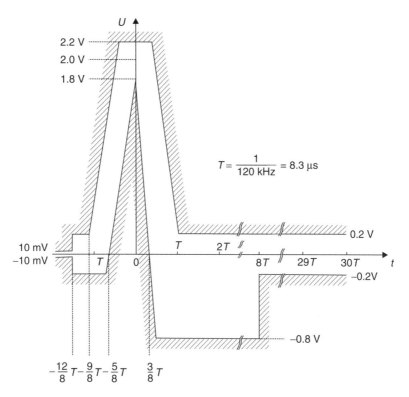

Figure 3.26 Pulse mask for the elementary signal shape applied in ISDN transmission with MMS 43 line code. Reproduced with the kind permission of ITU

Transmission in both directions is organized in frames of 120 ternary symbols. In each frame, besides two bearer channels and the data channel (2B+D) sent in symbols 1–84 and 86–109, the maintenance channel of rate 1 kbit/s is transmitted in symbol 85 and the frame synchronization word consisting of 11 ternary symbols is sent in the block of frame symbols 110–120. The data symbol rate is 120 ksymb/s. Assuming a square root raised cosine shaping filter with roll-off factor of the order of 25–35% applied in the transmitter, the signal bandwidth of around 75–80 kHz is needed for Basic Access ISDN transmission in each direction when the echo cancellation method is applied, although a 3 dB bandwidth of the power density spectrum is about 55 kHz.

Another possible line code listed in the Appendix to ITU-T G.961 Recommendation (ITU-T G.961 1993) is the so-called 2B1Q code. This is a regular 4-PAM modulation. The mapping rule of the 2B1Q code is shown in Table 3.3. The code does not contain any redundancy that could be used for quality monitoring. The frame consists of 120 quaternary symbols transmitted during 1.5 ms. It starts from 9 symbols of the frame synchronization word, followed by 108 symbols carrying bearer and data channels

Table 3.3 2B1Q line code mapping table

First bit (sign)	Second bit (magnitude)	Quaternary symbol
1	0	+3
1	1	+1
0	1	−1
0	0	−3

(2B+D). The frame ends with the maintenance channel requiring 3 quaternary symbols. The required symbol rate is 80 ksymb/s, resulting in a narrower bandwidth compared with the system using the MMS 43 line code. The echo cancellation method is used in duplex transmission. An adequate shape mask can be found in ITU-T Recommendation G.961 (ITU-T G.961 1993).

The third possible line code described in the Appendix to ITU-T G.961 Recommendation (ITU-T G.961 1993) is the so-called SU 32 (Substitutional 3B2T) code. This line code is used to support duplex transmission of 2B+D channels supplemented with an auxiliary channel supporting data CRC, control, supervisory and maintenance functions. The echo cancellation method is applied to ensure duplex transmission. The ternary data symbols of SU 32 codewords are again denoted as +, 0 and −, depending on the polarity of the previous nonzero symbol. Mapping of binary blocks onto ternary codewords is shown in Table 3.4.

Table 3.4 SU 32 (3B2T) line code mapping table

Input block	Output block	Input block	Output block
000	--	100	0-
001	-0	101	+-
010	-+	110	+0
011	0+	111	++

Due to the assumed frame configuration, the required data rates of 2B+D and maintenance channels and the applied SU 32 line code, the symbol rate of a ternary data stream is 108 ksymb/s. The pulse shape signal mask is presented in the above-mentioned ITU-T recommendation.

Although less efficient than the ECH method and probably very rarely applied, the TCM method of duplex transmission is one of the alternatives described in Annex III of ITU-T G.961 Recommendation (ITU-T G.961 1993). For the TCM system configuration a regular AMI line code has been foreseen. A binary "0" is represented by the zero line signal whereas binary "1" is transmitted as a pulse alternately changing polarity. The consequence of time compression is a very high symbol rate, equal to 320 ksymb/s. The positive and negative pulses are rectangular pulses shaped by the low-pass filter with cut-off frequency equal to 640 kHz and an appropriately selected roll-off.

So far we have considered line codes, symbol rates and elementary pulse shapes for Basic Access ISDN transmission. The duplex data rate offered to the user is 144 kbit/s. In the current state of access to the ISDN network it is not a very satisfactory data rate.

The next step of development of data transmission over subscriber loops was HDSL transmission. As we have already mentioned, its aim was to offer data transmission at the rate of the order of 2048 kbit/s on a triple, double or single wire pair. In the last case a more complicated transmission method than those described above was proposed in ITU-T Recommendation G.991.2 in 2003. In fact, the cited recommendation standardizes selected data rates between 192 kbit/s and 2312 kbit/s (in the form of $n \times 64 + i \times 8$ kbit/s) using a Trellis-Coded Pulse Amplitude Modulation (TCPAM). Detailed descriptions of trellis-coded modulations will be presented in Chapter 4.

***

In this chapter we have presented basic information on the modulation of signals in the baseband. We have shown that the spectral properties of the modulated signal strongly depend on the selected shape of the elementary pulse and on the correlation properties of data symbols. We have also derived the structure of the optimum receiver. We have noticed, in particular, that a band-limited channel can be used effectively for data transmission if we modulate elementary pulses that last longer than the modulation period; however, these signals should have specific spectral or time properties expressed by the Nyquist theorem.

In our overview we have concentrated on several forms of pulse amplitude modulation. There are other baseband modulations such as *Pulse Width Modulation* (PWM) or *Pulse Position Modulation* (PPM) but these are mostly used in measurement systems and very specific telecommunication applications, therefore they will not be considered in this book.

3.8 Appendix: Power Spectral Density of Pulse Sequence

Let us derive the formula for power spectral density of the pulse sequence described by (3.2). Let us quote this expression in a slightly modified form

$$x(t) = \sum_{i=-\infty}^{\infty} d_i s(t - iT - T_0) \qquad (3.103)$$

where, as before, d_i denotes data symbol and $s(t)$ describes the shape of elementary data pulses. Random variable T_0 denotes a reference moment on the time axis of an external observer of signal (3.103). It is usually assumed that T_0 has a uniform probability distribution in the interval $[0, T]$. Let us note that formula (3.103) describes a random process in which two variables are random and mutually independent: the data symbol sequence d_i and random initial moment T_0. We will show that the power density spectrum of signal $x(t)$ given by formula (3.103) is expressed by (3.3).

Derivation of the power density spectrum can be performed via calculation of the auto-correlation function of process (3.103), followed by calculation of its Fourier transform. In fact, we use the Wiener-Khinchine theorem on the relationship between the autocorrelation function and power spectral density for a stationary stochastic process. It is worth mentioning that if the random initial moment T_0 is not taken into account, the sequence of data pulses (3.2) is a nonstationary process. For such a case calculation of the autocorrelation function for signal (3.2) would result in a function that would depend not only on the time shift τ but also on the current time moment t. The dependence of the autocorrelation function on time t is periodic. In such a case we speak about the *cyclostationarity* of process (3.2). The property of cyclostationarity can be used in timing recovery performed in the receiver of pulse sequences.

Let us start from derivation of the autocorrelation function of signal (3.103). Using the basic definition of the autocorrelation function we have

$$R_x(t + \tau, t) = E\left[\sum_{i=-\infty}^{+\infty} d_i s(t + \tau - iT - T_0)\sum_{j=-\infty}^{+\infty} d_j s(t - jT - T_0)\right] \quad (3.104)$$

where $E[.]$ denotes ensemble average with respect to both random data symbols and a random initial moment T_0. As we have mentioned, data symbols and the initial moment T_0 are mutually statistically independent. Knowing that the probability distribution of random variable T_0 is described by expression

$$p_{T_0}(t_0) = \begin{cases} \dfrac{1}{T} & \text{for } 0 \leq T_0 \leq T \\[2mm] 0 & \text{otherwise} \end{cases} \quad (3.105)$$

we can present formula (3.104) in the form given below by (3.106). Additionally, we take into account the fact that the data sequence does not change its statistical properties in time so its autocorrelation function $E\left[d_i d_{i+n}\right]$ denoted as $R_d(n)$ does not depend on the current moment i but is a function of time shift n only. Thus, we have

$$R_x(t + \tau, t) = E\left[\sum_{i=-\infty}^{+\infty}\sum_{n=-\infty}^{+\infty} d_i s(t + \tau - iT - t_0)d_{i+n}s(t - iT - nT - t_0)\right]$$

$$= \sum_{n=-\infty}^{+\infty} E\left[d_i d_{i+n}\right]\sum_{i=-\infty}^{+\infty}\int_0^T s(t + \tau - iT - t_0)s(t - iT - nT - t_0)p_{T_0}(t_0)dt_0$$

$$= \sum_{n=-\infty}^{+\infty} E\left[d_i d_{i+n}\right]\sum_{i=-\infty}^{+\infty}\int_0^T s(t + \tau - iT - t_0)s(t - iT - nT - t_0)\frac{1}{T}dt_0$$

$$= \sum_{n=-\infty}^{+\infty} R_d(n)\int_{-\infty}^{\infty} s(t + \tau - \lambda)s(t - nT - \lambda)\frac{1}{T}d\lambda \quad (3.106)$$

The last row in (3.106) is obtained owing to substitution $\lambda = t_0 + iT$ and to replacement of the infinite sum of integrals over period T by a single integral in the infinite interval from minus to plus infinity. In turn, if we apply the substitution $u = t + \tau - \lambda$, we obtain

$$R_x(\tau) = \frac{1}{T}\sum_{n=-\infty}^{+\infty} R_d(n)\int_{-\infty}^{\infty} s(u)s(u - nT - \tau)du \quad (3.107)$$

Calculation of the power density spectrum for signal (3.103) in fact reduces to performing the Fourier transform of formula (3.107). Therefore we have

$$G_x(f) = \mathcal{F}[R_x(\tau)] = \frac{1}{T} \sum_{n=-\infty}^{+\infty} R_d(n) \, \mathcal{F}\left[\int_{-\infty}^{\infty} s(u)s(u - nT - \tau)du \right]$$

$$= \frac{1}{T} \sum_{n=-\infty}^{+\infty} R_d(n) S(f) S^*(f) \exp\left(j2\pi f n T \right) \tag{3.108}$$

where $S(f) = \mathcal{F}[s(t)]$ is the spectral density of pulse $s(t)$. The property

$$\mathcal{F}\left[\int_{-\infty}^{\infty} s(u)s(u - nT - \tau)du \right] = S(f)S^*(f) \exp\left(j2\pi f n T \right) \tag{3.109}$$

can be easily proved by applying basic formulae associated with the Fourier transform and its properties. Thus, the power density spectrum for signal (3.103) can be expressed by formula (3.110), which agrees with expression (3.3)

$$G_x(f) = \frac{1}{T} |S(f)|^2 G_d(f) \tag{3.110}$$

In (3.110) $G_d(f)$ is the power density spectrum of data stream d_i described by formula (3.4), which for the reader's convenience is again given below

$$G_d(f) = \sum_{n=-\infty}^{+\infty} R_d(n) \exp\left(-j2\pi f n T \right) \tag{3.111}$$

Therefore, in a general case the power density spectrum is described by the expression

$$G_x(f) = \frac{1}{T} |S(f)|^2 \sum_{n=-\infty}^{+\infty} R_d(n) \exp\left(-j2\pi f n T \right) \tag{3.112}$$

Consider now a specific, very useful case. Let subsequent data symbols be statistically independent. Then

$$R_d(n) = \begin{cases} E\left[d_i^2\right] = \sigma_d^2 + \mu_d^2 & \text{for } n = 0 \\[2mm] E\left[d_i d_{i+n}\right] = E\left[d_i\right] E\left[d_{i+n}\right] = \mu_d^2 & \text{for } n \neq 0 \end{cases} \tag{3.113}$$

and as a result

$$G_x(f) = \frac{\sigma_d^2}{T} |S(f)|^2 + \frac{\mu_d^2}{T} |S(f)|^2 \sum_{n=-\infty}^{+\infty} \exp\left(-j2\pi f n T \right) \tag{3.114}$$

Taking into account the Poisson formula, i.e.

$$\sum_{n=-\infty}^{+\infty} \exp\left(-j2\pi f n T\right) = \frac{1}{T} \sum_{n=-\infty}^{+\infty} \delta\left(f - \frac{n}{T}\right) \tag{3.115}$$

we obtain the formula for the power density spectrum of signal $x(t)$ in the form

$$G_x(f) = \frac{\sigma_d^2}{T} |S(f)|^2 + \frac{\mu_d^2}{T^2} |S(f)|^2 \sum_{n=-\infty}^{+\infty} \delta\left(f - \frac{n}{T}\right)$$

$$= \frac{\sigma_d^2}{T} |S(f)|^2 + \frac{\mu_d^2}{T^2} \sum_{n=-\infty}^{+\infty} \left|S\left(\frac{n}{T}\right)\right|^2 \delta\left(f - \frac{n}{T}\right) \tag{3.116}$$

A wider interpretation of expression (3.116) is contained in the main course of the chapter.

Problems

Problem 3.1 *Consider unipolar RZ (Return to Zero) line coding in binary transmission at the rate $R = 1/T$, in which the following signal is applied to transmit data symbols d_n*

$$s(t) = A\text{rect}\left(\frac{t - T/4}{T/2}\right) \quad 0 \le t \le T$$

Assume that the transmitted data are equiprobable and statistically independent. Calculate the power spectral density of a random data sequence when the RZ line code is applied. Compare it with the results for unipolar NRZ, bipolar NRZ and Manchester line coding.

Problem 3.2 *Assume bipolar transmission of statistically independent and equiprobable binary data in which the following elementary pulse is applied*

$$s(t) = \begin{cases} \sin\left(\frac{\pi t}{T}\right) & 0 \le t \le T \\ \\ 0 & \text{otherwise} \end{cases}$$

1. *Plot the pulse sequence $x(t)$ given by formula (3.2) if the following data bits a_n are transmitted*

$$011010010001110101011$$

2. *Calculate the power spectral density of $x(t)$ and compare it with the relevant power spectral densities when a rectangular pulse of length T and a raised cosine pulse of length $2T$ are applied.*

Problem 3.3 *Draw an eye diagram for the random bipolar data sequence $x(t)$ described by (3.2) if the applied elementary signal $s(t)$ has the form of a raised cosine pulse of length $2T$, given by formula (3.16).*

Problem 3.4 *Consider a simple discrete model of binary transmission described by the equation*

$$y_n = x_n + v_n$$

where $x_n = \pm A$ and n is the time index. Denote $\Pr\{x_n = -A\} = P_0$ and $\Pr\{x_n = A\} = P_1$. The sample v_n represents a Gaussian noise with zero mean and variance σ^2. At the receiver a simple threshold is applied to decide which symbol, A or $-A$, has been transmitted. Give the formula describing the error probability $P(\mathcal{E}) = \Pr\{\mathrm{dec}(y_n) \neq x_n\}$. Find the optimum threshold that minimizes $P(\mathcal{E})$.

Problem 3.5 *Write a computer program that calculates the values of the impulse response $s(t)$ (3.25) of the filter with the raised cosine characteristics with a given value of the roll-off factor α and subsequently plots them for $\alpha = 0.25$. Perform calculations and the plot for the time interval $(-6T, 6T)$.*

1. *Calculate the amplitude characteristics of the filter.*
2. *Consider the filter impulse response in the time interval $(-3T, 3T)$. Calculate the amplitude characteristics for this case. What are the consequences of shortening the impulse response of the filter? Plot the characteristics on the decibel scale.*
3. *Integrate numerically the product of two pulses – the regular $s(t)$ pulse and the pulse shifted by a multiple of the signaling period T. Draw the conclusions from the result of this integration.*

Problem 3.6 *Solve Problem 3.5 for the square-root raised cosine characteristics of the filter, typically used as transmit and matched filters. It can be shown that the impulse response of this filter is given by the formula*

$$s(t) = \frac{\sin[\pi(1-\alpha)t/T] + (4\alpha t/T)\cos[\pi(1+\alpha)t/T]}{(\pi t/T)[1 - (4\alpha t/T)^2]}$$

Problem 3.7 *Consider the pulse shaping filter characteristics shown in Figure 3.27. Calculate the impulse response of this filter and determine its properties in the context of Theorem 3.2.1.*

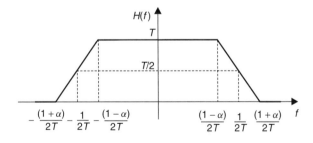

Figure 3.27 Pulse shaping filter characteristics considered in Problem 3.7

Problem 3.8 *Assume the application of data pulses of a rectangular shape in 8-PAM transmission. Data symbols d_n are selected from the alphabet $\{\pm 1, \pm 3, \pm 5, \pm 7\}$.*

1. *Propose a mapping of binary blocks onto data symbols d_n for natural and Gray encoding.*
2. *Draw a plot of PAM signals when natural (a) and Gray (b) encoding is applied and the following data sequence is transmitted*

$$001\ 011\ 101\ 111\ 110\ 100\ 111\ 101\ 001\ 000$$

Problem 3.9 *Apply any pulse from Figure 3.28 as the shaping pulse in binary bipolar transmission over an AWGN channel. Draw the plot of the signal seen at the output of the correlator and at the output of the matched filter applied in an optimum synchronouos receiver for the noise-free case.*

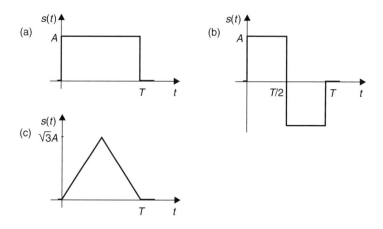

Figure 3.28   Plots of three pulses analyzed in Problem 3.9

Problem 3.10 *Assume that in binary transmission over the AWGN channel the data symbol "0" is transmitted using the pulse shown in Figure 3.28a and the pulse shown in Figure 3.28b is applied for transmission of the data symbol "1". Assume that both data symbols are equiprobable and statistically independent. The power spectral density of the AWGN noise is equal to $N_0/2$.*

1. *Calculate the energy of both pulses in the signaling period T.*
2. *Draw a block diagram of the optimal synchronous receivers for this transmission in which correlators (a) and matched filters (b) are applied.*
3. *Calculate the error probability at the output of this receiver.*

Problem 3.11 *Consider binary bipolar transmission systems in which the pulses from Figure 3.28 are applied. Compare the power spectral density for all the pulses if data symbols "0" and "1" are equiprobable and statistically independent. Draw respective plots of power spectral density.*

Problem 3.12 *Consider a quaternary transmission system in which the elementary pulses have the form $\pm s_1(t), \pm s_2(t)$ for $0 \leq t \leq T$ and are zero outside this time interval. The pulses $s_1(t)$ and $s_2(t)$ have equal energy and are orthogonal. Derive the optimal* *synchronous receiver assuming that additive noise corrupting the transmitted pulses is Gaussian and white with the power spectral density $N_0/2$, and that all data symbols are equiprobable. Derive the decision rule for this receiver and calculate the error probability at its output.*

Problem 3.13 *Consider a ternary transmission system in which equiprobable ternary data symbols $1, 0, -1$ are represented by elementary pulses that have the form $s_1(t) = s(t)$, $s_0(t) = 0$, $s_{-1}(t) = -s(t)$, respectively. The energy of the pulse $s(t)$ in the time interval T is equal to E. Find the scheme of an optimal synchronous receiver when these signals are transmitted through the AWGN channel. Calculate the probability of error at the receiver output.*

Problem 3.14 *Consider the system known in the literature as* Partial Response Signaling *in the version called the* duobinary system. *Its block diagram is shown in Figure 3.29. Assume that binary input data symbols $\{a_n\}$ are equiprobable and statistically independent. Binary data symbols are converted into a bipolar stream $\{b_n\}$ that is fed to the duobinary encoder. The resulting data symbol sequence $\{d_n\}$ is subsequently shaped by an ideal lowpass filter $S(f)$ with bandwidth equal to $1/2T$, where T is the signaling period.*

1. *Find the power spectral density of the signal at the output of the shaping filter $S(f)$.*
2. *Determine the equivalent form of the system that can be described by the equation*

$$x(t) = \sum_{i=-\infty}^{\infty} b_n s_{\mathrm{mod}}(t - iT)$$

   Find a formula for the modified pulse $s_{\mathrm{mod}}(t)$.
3. *Design the receiver that is able to decode the considered signal assuming that spectral shaping is equally divided between the transmitter and receiver, as it is in the case of the transmit and matched filter pair. Find the appropriate thresholds of the decision device and formulate the respective decision rule.*
4. *For the receiver designed in the previous point, show that a decision error can trigger the error propagation effect.*
5. *Assuming perfect decisions in the preceding timing instants, calculate the probability of error for this system. What is the performance difference compared with the regular bipolar system?*
6. *Demonstrate operation of the considered system feeding the exemplary data stream to its input if the system is now supplemented with the so-called differential encoder*

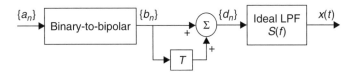

Figure 3.29 Block diagram of the duobinary partial response system transmitter

operating according to the formula

$$a'_n = a_n \oplus a'_{n-1}$$

where $\oplus$ denotes the modulo-2 addition and instead of data bits a_n the data bits a'_n are now converted to bipolar form. Find a new decision device taking into account differential encoding at the transmitter. Is error propagation still dangerous?

Problem 3.15   *Solve Problem 3.14 for the system called* Class-4 Partial Response *or* Modified Duobinary, *in which the data symbols d_n applied to the shaping filter $S(f)$ are received* via *the following operation*

$$d_n = b_n - b_{n-2}$$

4

Digital Modulations of the Sinusoidal Carrier

4.1 Introduction

Most transmission channels are of the bandpass type, i.e. the frequencies of the signal provided to their input are strongly attenuated outside the frequency range limited by certain minimum and maximum frequencies. The difference between the maximum and minimum frequencies is the channel bandwidth. The bandwidth is determined by natural properties of the transmission channel, although it can be a result of administrative regulations aiming at reasonably sharing the electromagnetic spectrum between different systems and their users. The regulations are particularly important for radio systems.

In order to use a bandpass channel in digital transmission, we apply a sinusoidal carrier of the frequency contained in the channel band and manipulate one or more parameters of this signal, depending on the data stream to be carried on this carrier. If we present the sinusoidal signal in the form

$$x(t) = A\cos(2\pi f t + \varphi) \tag{4.1}$$

we see that the subject of manipulation can be amplitude A, frequency f or phase φ. In the simplest case, the change of one of these parameters can occur in a stepwise manner, so digital modulations of a sinusoidal carrier are often described as ASK – *Amplitude Shift Keying*, FSK – *Frequency Shift Keying* or PSK – *Phase Shift Keying*. Figure 4.1 presents typical waveforms for ASK, PSK and FSK. It is assumed here that single bits of a data stream modulate the carrier.

Consider the basic digital modulations mentioned above. Figure 4.1 illustrates a particular case of ASK, FSK and PSK modulations. Let us note that for all waveforms shown in Figure 4.1 a single signaling period contains an integer number of sinusoidal periods. Additionally, all possible changes of elementary signals always occur at the moments of zero crossings of the sinusoidal carrier. Obviously, this is a rare case in practice, but it allows a clear presentation of the basic properties of modulations to be given.

Let us start with the ASK modulation. According to the rule applied in Figure 4.1a, a sinusoidal signal represents message "1", whereas absence of the signal represents

---

Introduction to Digital Communication Systems   Krzysztof Wesołowski
© 2009 John Wiley & Sons, Ltd

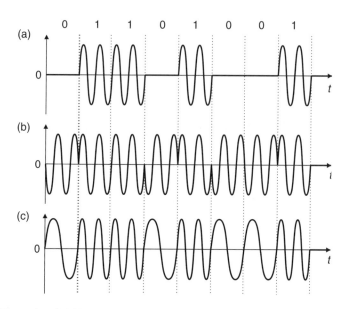

Figure 4.1 Example of ASK (a), PSK (b) and FSK (c) waveforms with a binary modulating sequence

message "0". Therefore, the ASK modulation constitutes a choice of one of two elementary signals

$$s_0(t) = 0, \quad \text{or} \quad s_1(t) = A\cos(2\pi f_c t), \quad \text{for } t \in [0, T] \tag{4.2}$$

As we see, the ASK-modulated signal can be generated by switching on and off a sinusoidal generator of frequency f_c for a duration of a single bit T. As in the case of baseband modulations, the time interval between the moments in which subsequent elementary signals are generated is called a *modulation period*.

In turn, in the case of FSK modulation two elementary signals have the form

$$s_0(t) = A\cos(2\pi f_0 t), \quad s_1(t) = A\cos(2\pi f_1 t), \quad \text{for } t \in [0, T] \tag{4.3}$$

where f_0 and f_1 are the so-called FSK *nominal frequencies*. If the nominal frequencies are not selected in such a way that a multiple number of periods of $s_0(t)$ and $s_1(t)$ is contained in the modulation period T, then two cases are possible:

- The FSK-modulated signal is obtained by passing a signal from one of two free-running generators of sinusoidal carriers of frequencies f_0 and f_1 to the modulator output – as a result, a phase discontinuity at the ends of the modulation periods occurs, which has a substantial influence on the signal spectral properties,
- The FSK-modulated signal is obtained by changing one of the parameters of a single sine wave oscillator. Consequently, phase continuity is preserved in the moments of the start of new modulation periods. Unfortunately, analysis of FSK-modulated signals generated in this way is much more difficult than in the first case.

Finally, the PSK-modulated signal can be represented by the following two elementary signals

$$s_0(t) = A\cos(2\pi f_c t), \qquad s_1(t) = A\cos(2\pi f_c t + \pi) = -A\cos(2\pi f_c t) \qquad (4.4)$$

Let us note that signal $s_1(t)$ features the opposite polarization with respect to signal $s_0(t)$, so PSK modulation can be treated not only as the phase keying but also as a particular case of DSB-SC modulation,[1] in which a binary rectangular random signal is a modulating signal.

In telecommunication practice more than one signal parameter is very often used in manipulation of a digitally modulated sinusoidal waveform. Typically, amplitude and phase are manipulated simultaneously and the transmitted elementary signals are determined not by single message bits but by whole binary blocks. The application of elementary trigonometric formulas allows amplitude-and-phase-modulated signals to be treated as a sum of two amplitude-modulated waveforms of the same frequency that are phase-shifted by 90° with respect to each other. Since the carrier phase-shifted by 90° with respect to the reference carrier is called a quadrature carrier, digital modulation in which amplitude modulation of two orthogonal carriers is applied is called *Quadrature Amplitude Modulation* (QAM).

Consider now a very general model of operations performed by a modulator. This model covers practically all digital modulations of a sinusoidal carrier and is described by the formula

$$s(t) = x^I(t)\cos 2\pi f_c t - x^Q(t)\sin 2\pi f_c t = \mathrm{Re}\{x(t)\exp(j2\pi f_c t)\} \qquad (4.5)$$

where Re{.} is the real part of a complex variable, and $x(t) = x^I(t) + jx^Q(t)$. Functions $x^I(t)$ and $x^Q(t)$ describe the signal modulating the cosinusoidal and sinusoidal carrier of frequency f_c, respectively. These signals are called the *in-phase* and *quadrature components*. We are able to represent any modulation by an appropriate choice of signals $x^I(t)$ and $x^Q(t)$. By applying $x(t)$ in the description of a given modulation we are able to consider each modulation on the complex plane, showing the points that are characteristic for the applied modulation and the trajectories along which the signal passes between these points. Signal $x(t)$ is called the *baseband equivalent signal*.

In general, digital modulations can be divided into linear and nonlinear ones. Linear modulations are those for which signals $x^I(t)$ and $x^Q(t)$ may be generated by linear circuits. ASK, QAM and PSK as well as their variations may be generated using such circuits. It is not always possible in the case of FSK and its derivatives, so generally this type of modulation is considered to be nonlinear.

A digitally modulated signal may be processed using different kinds of receivers, which in turn ensure different levels of error probability. Generally, digitally modulated signals can be the subject of *synchronous (coherent)* or *asynchronous (noncoherent)* reception. For the first type of reception the receiver needs to know the elementary signals and their starting and ending points on the time axis, as well as the exact value of the carrier phase. If the knowledge about the carrier phase is not used in the receiver or it cannot be

---

[1] Let us recall that DSB-SC (Double SideBand - Supressed Carrier) denotes an amplitude modulation of the form $x(t) = m(t)\cos 2\pi f_c t$, where $m(t)$ is a zero-mean modulating signal.

acquired due to technical difficulties or cost reasons, we talk about noncoherent reception. Thus, we intuitively feel that the performance of a noncoherent receiver is worse than that of a coherent one. In the course of this chapter we will consider both types of receivers and we will study receiver schemes for basic types of digital modulations.

4.2 Optimal Synchronous Receiver

In our considerations on digital modulations and on performance evaluation of different receiver types let us assume that the message bits are equiprobable. Let us also assume that the only disturbance in the channel is a white Gaussian noise of power spectral density equal to $N_0/2$. Channels that introduce other kinds of disturbances, e.g. they unequally attenuate the signal frequency components, will be considered in Chapters 5 and 6.

The derivation of the synchronous receiver structure for a digital modulation in which M elementary signals are applied is analogous to the derivation of the receiver for M-ary baseband modulations that was presented in the previous chapter, therefore we will not analyze it in detail. However, we will show the scheme of the optimal receiver for an M-ary modulation, which will be further applied for several digital modulations of a sinusoidal carrier. Let us denote elementary signals used by the modulator as $s_i(t)$ $(i = 0, \ldots, M-1)$. Their energy within a modulation period is given by the formula

$$E_i = \int_0^T s_i^2(t)\mathrm{d}t \tag{4.6}$$

Since we consider modulations of a sinusoidal carrier, the elementary signals have the form in which a sinusoidal signal is contained. However, we may treat these signals in the same way as baseband signals. The only difference is the mathematical form that describes them. As we already know the basic structure of the optimal synchronous receiver, we easily conclude that the scheme of the optimal receiver looks like that shown in Figure 4.2. The receiver is *synchronous*, because in the correlation with the received signal $y(t)$ the receiver applies the reference signals in the full form of the elementary signals, including their carrier phase. Thus, the receiver consists of M correlators that apply the appropriate elementary signals as the reference ones, the circuits that shift the signal levels according to possibly different values of the elementary signal energy E_i and a decision block that selects the highest value on the outputs of all the receiver branches. As we already know, the set of correlators can be replaced by the block of matched filters followed by the circuits, which sample the output of each matched filter at the end of each modulation period.

The generic scheme shown in Figure 4.2 may be the basis for a series of simpler structures of optimal receivers for several digital modulations.

First, let us consider a binary modulation of a sinusoidal carrier. The analysis of this modulation does not differ from the one performed in Chapter 3. Let us recall that in this modulation two elementary signals $s_0(t)$ and $s_1(t)$ are applied. If the energies of both signals are equal to E_s, if the additive Gaussian noise is white and if binary modulating symbols are equally probable, then we can apply the formula describing the probability

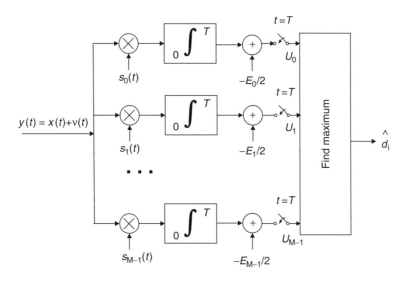

Figure 4.2   Scheme of the optimal receiver for M-ary modulation assuming equiprobable elementary signals transmitted over the additive white Gaussian noise channel

of error on the output of the optimal receiver derived in the previous chapter, i.e.

$$P(\mathcal{E}) = \frac{1}{2}\,\mathrm{erfc}\left[\sqrt{\frac{E_s}{2N_0}(1-\gamma)}\,\right] \tag{4.7}$$

where, as we remember, γ is the cross-correlation coefficient of the elementary signals. Inspection of formula (4.7) allows us to note that the error probability depends on the signal-to-noise ratio (due to E_s/N_0) and the degree of cross-correlation of the elementary signals. If we select bipolar elementary signals then, as in the case of baseband transmission, we receive $\gamma = -1$, which in turn ensures the maximum argument of the erfc function and the minimum error probability, which is then equal to

$$P(\mathcal{E}) = \frac{1}{2}\,\mathrm{erfc}\left(\sqrt{\frac{E_s}{N_0}}\right) \tag{4.8}$$

In turn, if the elementary signals are mutually orthogonal ($\gamma = 0$), then

$$P(\mathcal{E}) = \frac{1}{2}\,\mathrm{erfc}\left(\sqrt{\frac{E_s}{2N_0}}\right) \tag{4.9}$$

Both bipolar and orthogonal signals are often used in modulations of a sinusoidal carrier and we will study them further. In cases in which not all the assumptions set above are valid, e.g. those related to the probabilities of message symbols, equal energy of elementary signals or whiteness of the additive noise, the probability of error should be determined individually for each case on the basis of the specific properties of the analyzed modulation.

4.3   Optimal Asynchronous Receiver

Let us consider the optimal asynchronous reception of signals representing a binary sequence. Let bit "0" be represented by elementary signal $s_0(t)$, and let bit "1" be represented by signal $s_1(t)$. In general, let the signals have the form

$$s_0(t) = r_0(t) \cos(2\pi f_{c,0} t + \theta_0)$$

$$s_1(t) = r_1(t) \cos(2\pi f_{c,1} t + \theta_1) \qquad (4.10)$$

where $r_0(t)$ and $r_1(t)$ are any signals taking nonzero values in the interval $0 \le t \le T$ and equal to zero outside of it. Let us note that formula (4.10) is general enough to be valid for three modulations, i.e. ASK, PSK and FSK, if appropriate forms of $r_0(t)$ and $r_1(t)$, appropriate frequencies $f_{c,0}$ and $f_{c,1}$ and phases θ_0 and θ_1 are selected. Let the signal observed on the receiver input have the form

$$y(t) = s_i(t) + \nu(t) \qquad (4.11)$$

Let us assume that the optimal asynchronous receiver is not able to recover the signal phase θ_i, so θ_i is treated as a random variable. As we remember, for the optimal synchronous receiver the minimization of error probability resulting from application of the *Maximum a Posteriori* (MAP) criterion is equivalent to searching for such a value $\widehat{i}$ $(i = 0, 1)$ for which $p(\mathbf{y}|\mathbf{s}_i) P_i$ is maximized, i.e.

$$\widehat{i} = \arg \max_i \{ p(\mathbf{y}|\mathbf{s}_i) P_i \} \qquad (4.12)$$

where, as before

$$\mathbf{y} = \mathbf{s}_i + \boldsymbol{\nu}$$

$$\mathbf{y} = [y(t_1), y(t_2), \ldots, y(t_K)]$$

$$\mathbf{s}_i = [s_i(t_1), s_i(t_2), \ldots, s_i(t_K)]$$

$$\boldsymbol{\nu} = [\nu(t_1), \nu(t_2), \ldots, \nu(t_K)]$$

and we denote the probability of transmission of elementary signal $s_i(t)$ as $P_i = P[s_i(t)]$. The modulation period T is divided into K intervals of width $\Delta t = T/K$. If we take into account the phase θ_i, criterion (4.12) takes the form

$$\widehat{i} = \arg \max_i \{ p(\mathbf{y}|\mathbf{s}_i, \theta_i) P_i \} \qquad (4.13)$$

If we assume that subsequent noise samples contained in the signal $\mathbf{y}$ are Gaussian and statistically independent, we get

$$p(\mathbf{y}|\mathbf{s}_i, \theta_i) = \prod_{k=1}^{K} p_\nu \left[y(t_k) - s_i(t_k) \right]$$

$$= (2\pi\sigma)^{-K/2} \exp \left\{ -\frac{1}{2\sigma^2} \sum_{k=1}^{K} \left[y(t_k) - s_i(t_k) \right]^2 \right\} \qquad (4.14)$$

We will now determine the noise variance σ^2 that appears in (4.14).

White noise can be treated as an asymptotic case of the noise that has a uniform power spectral density equal to $N_0/2$ and is band-limited by an ideal low-pass filter of width W if W tends to infinity. Since the power spectral density of this noise is given by the formula

$$G_v(f) = \frac{N_0}{2} \, \text{rect} \left(\frac{f}{2W} \right)$$

the noise autocorrelation function is described by the expression

$$R_v(\tau) = N_0 W \, \text{sinc}(2\pi \tau W)$$

As a result, if the noise is sampled at time instants that are distant by $\Delta t = 1/2W$ seconds, then the samples are uncorrelated, because except for the moment $\tau = 0$ the noise autocorrelation function $R_v(\tau)$ is equal to zero at multiples of time distance $\Delta t = 1/2W$. As the noise samples are Gaussian and uncorrelated, they are statistically independent and their variance is equal to

$$\sigma^2 = R_v(0) = N_0 W = \frac{N_0}{2\Delta t} \tag{4.15}$$

If we extend the number K of signal samples taken within modulation period T, assuming that $K \Delta t = T$, we observe that the sum in formula (4.14) transforms into the integral, the sequences of signal samples change into signals that are continuous functions of time and time interval Δt tends to an infinitely small increment dt. Thus

$$p\left[y(t)|s_i(t), \theta_i\right] = \lim_{K \to \infty} p(\mathbf{y}|\mathbf{s}_i, \theta_i)$$

$$= C \exp \left[-\frac{1}{N_0} \int_0^T y^2(t)dt - \frac{1}{N_0} \int_0^T s_i^2(t)dt + \frac{2}{N_0} \int_0^T y(t)s_i(t)dt \right]$$

$$\tag{4.16}$$

The value of C is determined by the formula

$$C = \lim_{K \to \infty} \left(\frac{\Delta t}{\pi N_0} \right)^{K/2} \tag{4.17}$$

and it does not have any influence on the decision-making by the receiver, because it is the same for all i.

Since the phases θ_0 and θ_1 are unknown to the receiver, the decision criterion takes the form

$$\hat{i} = \text{arg} \left\{ \max_i E_{\theta_i} \left[p\big(y(t)|s_i(t), \theta_i\big) \right] P_i \right\} \tag{4.18}$$

where $E_{\theta_i}[.]$ denotes the ensemble average with respect to phase θ_i. As we see, the optimal asynchronous receiver makes the decisions on the basis of conditional probability density functions of signal $y(t)$, which are averaged with respect to phase θ_i. It is usually assumed that the values of phase θ_i are uniformly distributed in the interval $[0, 2\pi]$. If we take into account the specific form of $s_0(t)$ and $s_1(t)$ determined by (4.10) in the components of the exponent in (4.16), and if we assume that the bandwidth of $r_0(t)$ and $r_1(t)$ is small compared with the carrier frequencies $f_{c,0}$ and $f_{c,1}$, then we have

$$\frac{1}{N_0} \int_0^T s_i^2(t) dt = \frac{1}{N_0} \int_0^T r_i^2(t) \cos^2(2\pi f_{c,i} t + \theta_i) dt$$

$$= \frac{1}{2N_0} \int_0^T r_i^2(t) dt + \frac{1}{2N_0} \int_0^T r_i^2(t) \cos(4\pi f_{c,i} t + 2\theta_i) dt \approx \frac{E_i}{N_0} \qquad (4.19)$$

where

$$E_i = \frac{1}{2} \int_0^T r_i^2(t) dt \qquad (4.20)$$

If we use the expression

$$s_i(t) = r_i(t) \cos(2\pi f_{c,i} t + \theta_i)$$

$$= r_i(t) \cos 2\pi f_{c,i} t \cos \theta_i - r_i(t) \sin 2\pi f_{c,i} t \sin \theta_i \qquad (4.21)$$

then for the last component of (4.16) we receive

$$\frac{2}{N_0} \int_0^T y(t) s_i(t) dt = \frac{2}{N_0} \left(y_i^I \cos \theta_i - y_i^Q \sin \theta_i \right) = \frac{2}{N_0} r_i \cos(\theta_i + \psi_i) \qquad (4.22)$$

where

$$y_i^I = \int_0^T y(t) r_i(t) \cos 2\pi f_{c,i} t \, dt, \quad y_i^Q = \int_0^T y(t) r_i(t) \sin 2\pi f_{c,i} t \, dt \qquad (4.23)$$

and

$$r_i = \sqrt{\left(y_i^I\right)^2 + \left(y_i^Q\right)^2}, \quad \psi_i = \text{arctg} \frac{y_i^Q}{y_i^I}$$

Let us insert (4.19) and (4.22) into (4.16) and calculate the ensemble average with respect to phase θ_i. Let us recall our assumption that the values of phase θ_i are uniformly distributed in the interval $[0, 2\pi]$. According to (4.18) we obtain

$$E_{\theta_i}[p(y(t)|s_i(t), \theta_i)]$$

$$= CE_{\theta_i}\left[\exp\left(-\frac{1}{N_0}\int_0^T y^2(t)dt - \frac{E_i}{N_0}\right)\exp\left(\frac{2}{N_0}r_i\cos(\theta_i + \psi_i)\right)\right]$$

$$= C\exp\left(-\frac{1}{N_0}\int_0^T y^2(t)dt - \frac{E_i}{N_0}\right)\frac{1}{2\pi}\int_0^{2\pi}\exp\left(\frac{2}{N_0}r_i\cos(\theta_i + \psi_i)\right)d\theta_i \qquad (4.24)$$

Let us take advantage of the formula that describes the modified Bessel function of the first kind of zero order

$$I_0(z) = \frac{1}{2\pi}\int_0^{2\pi}\exp(z\cos\alpha)d\alpha \qquad (4.25)$$

On the basis of (4.24) we get

$$E_{\theta_i}[p(y(t)|s_i(t), \theta_i)] = C\exp\left(-\frac{1}{N_0}\int_0^T y^2(t)dt\right)\exp\left(-\frac{E_i}{N_0}\right)I_0\left(\frac{2r_i}{N_0}\right) \qquad (4.26)$$

Since constant C and factor $\exp[(-1/N_0)\int_0^T y^2(t)dt]$ are the same for each i, the decision criterion reduces to the form

$$\widehat{i} = \arg\left\{\max_i\left[\exp\left(-\frac{E_i}{N_0}\right)I_0\left(\frac{2r_i}{N_0}\right)P_i\right]\right\}, \quad i = 0, 1 \qquad (4.27)$$

The general form of the optimal asynchronous receiver results directly from formula (4.27) and is shown in Figure 4.3. It contains correlators that are used to calculate the in-phase and quadrature signal components, which are subsequently used to compute the envelope sample r_i. As we already know from the previous chapter, we can obtain the same samples if we apply the filters matched to the signals $r_0(t)\cos 2\pi f_{c,0}t$ and $r_1(t)\cos 2\pi f_{c,1}t$, respectively. As a result, we can draw an equivalent scheme of the receiver from Figure 4.3. This scheme is shown in Figure 4.4. In general, the optimal receiver has to calculate the value of the modified Bessel function, which can be a problem. However, if the energies of elementary signals and elementary signal probabilities are equal, then it is no longer necessary to multiply the values of the Bessel function by the factor $P_i\exp(-E_i/N_0)$ $(i = 0, 1)$, so calculations of the Bessel function are not needed. In order to make a decision about the transmitted data i, it is sufficient to check which argument of the modified Bessel function is higher, as this function is monotonic (see

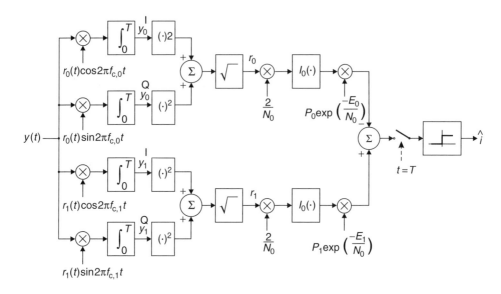

Figure 4.3 General scheme of the optimal asynchronous receiver for binary modulated signals

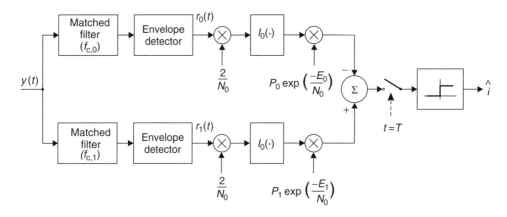

Figure 4.4 Equivalent form of the optimal asynchronous receiver using matched filters

Figure 4.5). As it is not necessary to multiply the signals on both branches by $2/N_0$, the optimal asynchronous receiver reduces to the form shown in Figure 4.6. It consists of two parallel branches composed of filters matched to the appropriate elementary signals and of envelope detectors. The branch output signals are subtracted and their difference determines the decision $\hat{i}$.

We will return to specific forms of the optimal asynchronous receiver again when considering particular digital modulations.

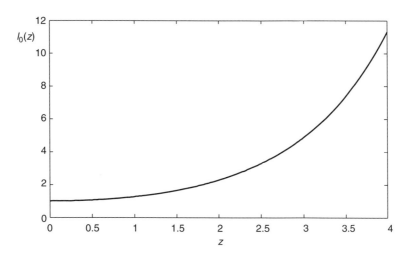

Figure 4.5 Plot of the modified Bessel function of order zero $I_0(z)$

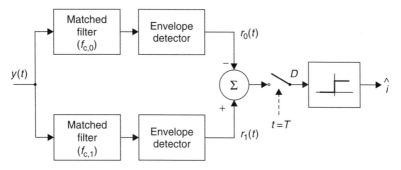

Figure 4.6 Optimal asynchronous receiver for binary modulated signals of equal energies and probabilities of the elementary signals

4.4 ASK Modulation

First let us consider ASK modulation. We will present both synchronous and asynchronous receivers for ASK-modulated signals. Nowadays ASK modulation is not applied in its pure form. Instead, more advanced binary modulations are used that feature, for example, a constant envelope. Despite this we will devote our attention to ASK modulation because the results of our considerations will be useful in the analysis of other digital modulations.

4.4.1 Synchronous Receiver for ASK-Modulated Signals

First let us focus on the optimal synchronous receiver scheme. Although elementary signals $s_0(t)$ and $s_1(t)$ described by formula (4.2) are mutually orthogonal, their energies

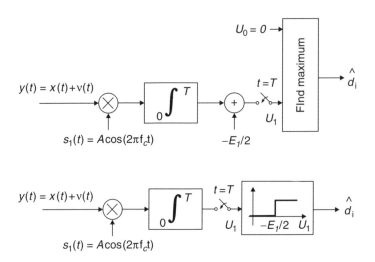

Figure 4.7 Two equivalent forms of the synchronous receiver for ASK-modulated signals

are not equal, as $s_0(t) = 0$ and consequently $E_0 = 0$. As a result, the synchronous receiver shown in Figure 4.2 reduces to the form presented in the upper part of Figure 4.7.

The bottom receiver shown in Figure 4.7 applies a single correlator with a nonlinear decision circuit. Its decision threshold is half the energy value of elementary signal $s_1(t)$. Relatively simple calculations, similar to those shown in the previous chapter for the optimal baseband receiver, lead to the formula

$$P\{\mathcal{E}|s_0(t) \text{ was transmitted}\} = P\{\mathcal{E}|s_1(t) \text{ was transmitted}\} = \frac{1}{2} \text{erfc}\left(\sqrt{\frac{E_1}{4N_0}}\right) \quad (4.28)$$

Thus, taking advantage of the formula

$$P(\mathcal{E}) = P\{\mathcal{E}|s_0(t) \text{ was transmitted}\} P\big[s_0(t)\big] + P\{\mathcal{E}|s_1(t) \text{ was transmitted}\} P\big[s_1(t)\big]$$

we receive the following expression for error probability on the output of the synchronous receiver of ASK-modulated signals

$$P(\mathcal{E}) = \frac{1}{2} \text{erfc}\left(\sqrt{\frac{E_1}{4N_0}}\right) \quad (4.29)$$

Let us note that energy E_1 is the energy of its nonzero elementary signal $s_1(t)$ and not the mean energy of the ASK. If we assume equal probabilities of both elementary signals then the mean energy of the ASK signal is $E_1/2$, so the result obtained is the same as for the orthogonal elementary signals of energy E_s. Figure 4.8 illustrates the calculation of the error probability for a synchronous receiver of ASK signals.

Formulas describing spectral properties of baseband digitally-modulated signals, which are already known to us, can be directly applied to the case of passband signals as well.

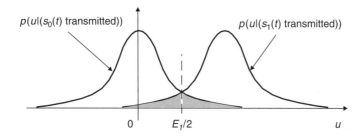

Figure 4.8 Conditional error distributions for a synchronous receiver of ASK-modulated signals

If we take into account that the Fourier transform of elementary signal $s_1(t)$ is

$$F[s_1(t)] = F\left[\text{rect}\left(\frac{1}{T}\right) A \cos(2\pi f_c t)\right]$$

$$= \frac{AT}{2}\{ \text{sinc}[\pi(f - f_c)T] + \text{sinc}[\pi(f + f_c)T]\} \tag{4.30}$$

we obtain the power spectral density of the ASK signal in the form

$$G_{\text{ASK}}(f) = \frac{A^2 T}{8}\{ \text{sinc}^2[\pi(f - f_c)T] + \text{sinc}^2[\pi(f + f_c)T]\} \tag{4.31}$$

As we can see, the shape of the power density spectrum is the square of the sinc function shifted by $\pm f_c$. However, formula (4.31) is valid if the modulation period is much longer than the period of the carrier frequency f_c, or equivalently if the carrier frequency f_c is much higher than the modulation rate $1/T$. In such a case two spectral components in (4.30) centered at f_c and $-f_c$ do not overlap. Let us note that the power density spectrum of the ASK signal has the same shape as that for rectangular signals, but the spectrum is shifted in frequency by $\pm f_c$ and is appropriately scaled.

4.4.2 Asynchronous Reception of ASK-Modulated Signals

Spectral properties of ASK-modulated signals are not impressive. Moreover, it can be difficult to maintain synchronism of the reference signal in the correlator with the received signal, since there are time periods in which a zero signal is transmitted. In fact, ASK modulation was used in the early days of digital communications in which synchronous reception was rather difficult to implement. Therefore, asynchronous reception, similar to AM envelope signal detection, was usually applied.

Let us consider an optimal asynchronous receiver for ASK-modulated signals, which can be treated as a particular case of the optimal asynchronous receiver considered previously. Let us apply a general form of elementary signals (4.10). We see that for ASK modulation $r_0(t) = 0$, $r_1(t) = A$, and the carrier frequency is $f_{c,1} = f_c$. In that particular case the scheme of the optimal asynchronous ASK receiver reduces to the form shown in Figure 4.9a. However, such a receiver also requires calculation of the modified Bessel function for the given value of $r_1(t)\frac{2}{N_0}$. On the output of the sampling block the receiver checks if the signal is higher or lower than zero. Such a receiver scheme results from

(a)

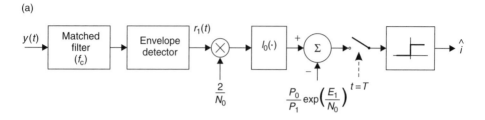

(b)

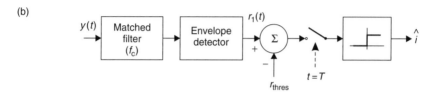

Figure 4.9 Optimal asynchronous receiver for ASK-modulated signals: (a) in a full version; (b) without calculation of the modified Bessel function when $P_0 = P_1$

the fact that the second branch of the optimum receiver would give the zero signal on its output, so it does not need to be implemented. Fortunately, we can further substantially simplify the optimum asynchronous ASK receiver, taking advantage of the following argumentation.

Let us assume that the transmitted binary symbols are equiprobable, i.e. $P_0 = P_1$. We wish to find a threshold value r_{thres} on the output of the envelope detector for which the sample on the input of the sampling block shown in Figure 4.9a is equal to zero at the moment $t = T$. This envelope value can be determined from the equation

$$I_0\left(\frac{2r_{\text{thres}}}{N_0}\right) - \exp\left(\frac{E_1}{N_0}\right) = 0 \tag{4.32}$$

Let us write equation (4.32) in the form

$$I_0\left(r_{\text{thres}}\sqrt{\frac{2}{E_1 N_0}}\sqrt{\frac{2E_1}{N_0}}\right) = \exp\left(\frac{E_1}{N_0}\right) \tag{4.33}$$

which is similar to the equation $I_0(b_0\sqrt{2\gamma}) = \exp(\gamma)$ known from mathematical literature. The latter equation has a solution b_0 that depends on parameter γ. The form of the solution is

$$b_0 = \sqrt{2 + \gamma/2}$$

In our case we have

$$b_0 = r_{\text{thres}}\sqrt{\frac{2}{E_1 N_0}} \quad \text{and} \quad \gamma = \frac{E_1}{N_0}$$

We conclude that the optimal threshold value is

$$r_{\text{thres}} = \sqrt{\frac{E_1 N_0}{2}} \sqrt{2 + \frac{E_1}{2N_0}} = \frac{E_1}{2} \sqrt{1 + \frac{4}{E_1/N_0}} \qquad (4.34)$$

As we can see, the optimal threshold depends on the energy E_1 of elementary signal $s_1(t)$ and on the noise level. If the signal-to-noise ratio is high, then the optimal threshold value is approximately equal to $r_{\text{thres}} = E_1/2$, because the argument of the square root in (4.34) is close to unity. Figure 4.9b shows the scheme of the receiver based on the derivation presented above.

The receiver consists of a bandpass filter matched to signal $s_1(t)$, an envelope detector of the matched filter output, a sampling circuit and a decision device with the decision threshold r_{thres}. If a zero signal is transmitted, then the noise passes through the bandpass filter and its envelope is determined in the envelope detector. If $s_1(t)$ is transmitted, the envelope detector determines the envelope of the sum of responses of the filter matched to $s_1(t)$ and Gaussian noise.

Figure 4.10 presents another equivalent version of the ASK asynchronous receiver. The bandpass filter extracts the band in which signal $s_1(t)$ is transmitted. The envelope detector determines the envelope of the signal passing through the filter, i.e. signal $s_1(t)$ with additive noise, or noise only. The filter matched to the envelope is used to maximize the signal-to-noise ratio. For each modulation period a sample from the filter output is acquired. On its basis a decision about the transmitted data symbol is made. Let us note that if the envelope has a rectangular shape, then the filter matched to it is in fact an integrator over the modulation period, which is subsequently reset after its output value is read. Figure 4.11 shows an example of the signal waveforms in several receiver points.

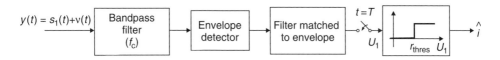

Figure 4.10   Equivalent form of the asynchronous receiver for ASK-modulated signals

4.4.3   Error Probability on the Output of the Asynchronous ASK Receiver

Let us consider the error probability in the case of transmitting equiprobable data symbols "0" and "1", i.e. when $P_0 = P_1$. It is sufficient to find the conditional probabilities of error $P(\varepsilon|\text{symbol "0" is transmitted})$ and $P(\varepsilon|\text{symbol "1" is transmitted})$. We assume that white Gaussian noise of power density $N_0/2$ is an additive disturbance. To make our calculations easier, let us consider the receiver in Figure 4.12. This receiver is equivalent to that shown in Figure 4.9b, in which correlators are used instead of the matched filter. We also assume that the carrier frequency is much higher than the modulation rate, i.e. $f_c \gg 1/T$.

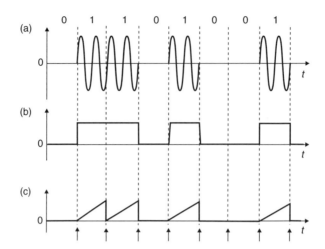

Figure 4.11   Example of idealized waveforms in several points of the receiver from Figure 4.10: (a) on the output of the bandpass filter, (b) on the output of the envelope detector, (c) on the output of the filter matched to the envelope (sampling moments are indicated by arrows)

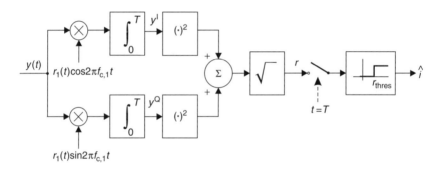

Figure 4.12   Optimal asynchronous receiver for ASK signals with the application of correlators if $P_0 = P_1$

The signal appearing on the receiver input during modulation period T is described by the formula

$$y(t) = \begin{cases} v(t) & \text{if symbol "0" is transmitted} \\ r_1(t)\cos(2\pi f_c t + \theta) + v(t) & \text{if symbol "1" is transmitted} \end{cases} \qquad (4.35)$$

As previously, phase θ is unknown to the receiver. The noise $v(t)$ can be presented as a combination of baseband components $n_c(t)$ and $n_s(t)$, using the formula

$$v(t) = n_c(t)\cos(2\pi f_c t + \theta) - n_s(t)\sin(2\pi f_c t + \theta) \qquad (4.36)$$

First, let us consider the case when a zero signal is transmitted. Then only the noise $v(t)$ is observed on the receiver input. It is correlated in the in-phase and quadrature branches with appropriate reference signals, resulting in samples y^I and y^Q, respectively. Since the correlation of the received signal with the reference signal is equivalent to the linear filtration, samples y^I and y^Q are Gaussian random variables. Simple calculations that are similar to those performed for the optimal baseband receiver result in the following values of the variance of samples y^I and y^Q

$$\sigma_I^2 = \sigma_Q^2 = \sigma^2 \approx \frac{N_0}{2} E_1 \tag{4.37}$$

The sign of approximate equality in (4.37) is a result of the fact that the integrals of the form $\int_0^T r_1^2(t) \cos 4\pi f_{c,1} t \, dt$ are approximately equal to zero for $f_{c,1} \gg 1/T$. We will use this fact a few times during error probability calculations.

Let us determine the signals on the output of the in-phase and quadrature correlators shown in Figure 4.12. These signals are

$$y^I = \int_0^T v(t) r_1(t) \cos 2\pi f_c t \, dt \quad \text{and} \quad y^Q = \int_0^T v(t) r_1(t) \sin 2\pi f_c t \, dt \tag{4.38}$$

Using formula (4.36) for noise $v(t)$ in (4.38) we obtain

$$y^I = u_c \cos \theta - u_s \sin \theta \quad \text{and} \quad y^Q = -u_c \sin \theta - u_s \sin \theta \tag{4.39}$$

where

$$u_c = \frac{1}{2} \int_0^T n_c(t) r_1(t) \, dt \quad \text{and} \quad u_s = \frac{1}{2} \int_0^T n_s(t) r_1(t) \, dt \tag{4.40}$$

Samples y^I and y^Q on the output of both correlators can be treated as a pair of orthogonal signals that can be represented in polar coordinates by the amplitude (envelope) r and phase ψ, i.e.

$$r = \sqrt{\left(y^I\right)^2 + \left(y^Q\right)^2}, \quad \psi = \text{arctg}\left(\frac{y^Q}{y^I}\right) \tag{4.41}$$

Putting (4.39) in (4.41), we obtain $r = \sqrt{u_c^2 + u_s^2}$.

In order to calculate the conditional probability of error given that the symbol "0" was transmitted, we have to know the probability density function of the envelope r, because the decision is made on the basis of r. In general, taking into account the joint probability distribution of the signal pair y^I and y^Q, the value of the incremental probability is the same whichever system of coordinates (rectangular or polar) is applied. Therefore the following equation holds

$$p_{y^I, y^Q}(y^I, y^Q) dy^I dy^Q = p_{r,\psi}(r, \psi) dr d\psi \tag{4.42}$$

Due to statistical independence of samples y^I and y^Q and their Gaussian distribution, the joint probability distribution $p_{y^I,y^Q}(y^I, y^Q)$ can be presented in the form

$$p_{y^I,y^Q}(y^I, y^Q) = \frac{1}{2\pi\sigma^2} \exp\left[-\frac{(y^I)^2}{2\sigma^2}\right] \exp\left[-\frac{(y^Q)^2}{2\sigma^2}\right]$$

$$= \frac{1}{2\pi\sigma^2} \exp\left(-\frac{r^2}{2\sigma^2}\right) \tag{4.43}$$

Since we also know from the mathematical analysis that $dy^I dy^Q = rdrd\psi$, we have

$$p_{y^I,y^Q}(y^I, y^Q)dy^I dy^Q = \frac{r}{2\pi\sigma^2} \exp\left(-\frac{r^2}{2\sigma^2}\right) drd\psi$$

We conclude from the above expression and from equation (4.42) that

$$p_{r,\psi}(r, \psi) = \frac{r}{2\pi\sigma^2} \exp\left(-\frac{r^2}{2\sigma^2}\right) \tag{4.44}$$

On the basis of (4.44) we finally obtain the envelope probability distribution in the form

$$p_0(r) = \int_0^{2\pi} p_{r,\psi}(r, \psi)d\psi = \frac{r}{\sigma^2} \exp\left(-\frac{r^2}{2\sigma^2}\right), r \geq 0 \tag{4.45}$$

where subscript "0" denotes the envelope probability distribution for data symbol "0".

Let us recall that $\sigma^2 = N_0 E_1/2$. Formula (4.45) describes the *Rayleigh distribution*. As a result, we can express the probability of error under the condition that symbol "0" was transmitted in the following form

$$P(\mathcal{E}|\text{symbol "0" was transmitted}) = \int_{r_{thres}}^{\infty} p_0(r)dr \tag{4.46}$$

where r_{thres} is the appropriately selected threshold value of the envelope r.

Let us consider the second possible situation, i.e. transmission of symbol "1". The signal observed on the receiver input is then the sum of the signal carrying the data symbol and of the noise: $y(t) = r_1(t)\cos(2\pi f_c t + \theta) + v(t)$. Let us calculate the signal on the output of both correlators as we have done for the first case. We then get the formulae

$$y^I = (E_1 + u_c)\cos\theta - u_s \sin\theta$$

$$y^Q = -[(E_1 + u_c)\sin\theta + u_s \cos\theta] \tag{4.47}$$

Let us note that every pair of orthogonal samples (y^I, y^Q) is equivalent to the respective pair $[(E_1 + u_c), u_s]$, because they differ only due to rotation by angle θ. If we denote $u'_c = E_1 + u_c$, then we can write

$$p_{r,\psi}(r, \psi)\mathrm{d}r\mathrm{d}\psi = p_{y^I, y^Q}(y^I, y^Q)\mathrm{d}y^I\mathrm{d}y^Q = p_{u'_c, u_s}(u'_c, u_s)\mathrm{d}u'_c u_s \qquad (4.48)$$

Random variable u'_c is Gaussian with the mean E_1 and variance given by formula (4.37). As in the previous case, both random variables, u'_c and u_s, are statistically independent, therefore

$$p_{u'_c, u_s}(u'_c, u_s)\mathrm{d}u'_c u_s = \frac{1}{2\pi\sigma^2}\exp\left[-\frac{(u'_c - E_1)^2 + u_s^2}{2\sigma^2}\right]\mathrm{d}u'_c u_s \qquad (4.49)$$

If we put (4.49) in (4.48), apply the equality $\mathrm{d}u'_c \mathrm{d}u_s = r\mathrm{d}r\mathrm{d}\psi$ and the knowledge resulting from Figure 4.13, after a few simple derivation steps we get

$$p_{r,\psi}(r, \psi)\mathrm{d}r\mathrm{d}\psi = \exp\left(-\frac{E_1^2}{2\sigma^2}\right)\frac{r}{2\pi\sigma^2}\exp\left(-\frac{r^2 - 2E_1 r\cos\psi}{2\sigma^2}\right)\mathrm{d}r\mathrm{d}\psi \qquad (4.50)$$

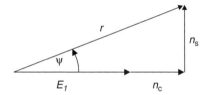

Figure 4.13 Envelope r as a function of the orthogonal components of the signal and noise on the output of the correlators in the asynchronous receiver

Derivation of the probability density function of envelope r for the case when symbol "1" is transmitted results in the formula

$$p_1(r) = \int_0^{2\pi} p_{r,\psi}(r, \psi)\mathrm{d}r\mathrm{d}\psi$$

$$= \exp\left(-\frac{E_1^2}{2\sigma^2}\right)\exp\left(-\frac{r^2}{2\sigma^2}\right)\frac{r}{2\pi\sigma^2}\int_0^{2\pi}\exp\left(-\frac{E_1}{\sigma^2}r\cos\psi\right)\mathrm{d}r\mathrm{d}\psi \qquad (4.51)$$

If we apply expression (4.25) in part of (4.51), we conclude that the probability density function $p_1(r)$ is determined by the formula

$$p_1(r) = \frac{r}{\sigma^2}\exp\left(-\frac{E_1^2 + r^2}{2\sigma^2}\right)I_0\left(\frac{E_1 r}{\sigma^2}\right) \qquad (4.52)$$

This probability density function is called the *Rice distribution*. This time the conditional probability of error given that the symbol "1" was transmitted is

$$P(\mathcal{E}|\text{symbol "1" wastransmitted}) = \int_0^{r_{\text{thres}}} p_1(r)dr \qquad (4.53)$$

Probability density functions $p_0(r)$ and $p_1(r)$ of envelope r conditioned on the transmitted symbol "0" or "1", respectively, are shown in Figure 4.14. The grey area corresponds to the conditional probability of error. Finally, the probability of error on the output of the optimal asynchronous receiver for ASK-modulated signals is given by the following formula

$$P(\mathcal{E}) = \frac{1}{2} \int_{r_{\text{thres}}}^{\infty} p_0(r)dr + \frac{1}{2} \int_0^{r_{\text{thres}}} p_1(r)dr \qquad (4.54)$$

The value of the error probability for a given value of signal-to-noise ratio can be found only numerically. For large values of signal-to-noise ratio the error probability can be approximated by the formula

$$P(\mathcal{E}) \approx \frac{1}{2}\left(1 + \sqrt{\frac{1}{2\pi\rho}}\right)\exp\left(-\frac{\rho}{2}\right) \qquad (4.55)$$

where $\rho = \dfrac{E_1}{2N_0}$.

The ASK modulation is hardly ever applied in current systems, however the main concept of an asynchronous receiver for digital signals can be well explained using this modulation scheme. In subsequent paragraphs we will present other digital modulations, using the results obtained so far.

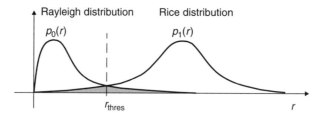

Figure 4.14 Plots of the conditional probability density functions of the envelope of an ASK-modulated signal at the sampling moment

4.5   FSK Modulation

Let us consider a binary FSK modulation with elementary signals described by formula (4.3). As we have mentioned before, an FSK signal can be generated by switching between

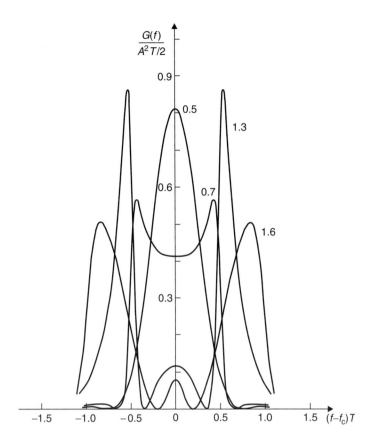

Figure 4.15 Power density spectrum of CPFSK signals for several values of modulation index h (Peebles 1986)

two oscillators that independently generate sinusoidal waveforms of frequency f_0 and f_1, respectively. Another way of generating an FSK signal relies on modification of parameters of a single sinusoidal oscillator. The latter method preserves phase continuity of the generated FSK signal when the signal is switched from one nominal frequency to the other. This kind of modulation is usually called *Continuous Phase* FSK – CPFSK). Phase continuity atrongly affects the shape of the power density spectrum of the FSK signal. However, calculation of the power density spectrum is beyond the scope of this book. In Figure 4.15, quoted after (Peebles 1986), we plot the power density spectrum for several values of the *modulation index* $h = 2\Delta f T$, where $\Delta f = f_c - f_0 = f_1 - f_c$, with f_c being the carrier frequency.

As we can see, the shape of the power density spectrum highly depends on the values of modulation index h. If it is equal to $1/2$, a substantial part of the power is located around the carrier frequency. The higher the value of h, the wider the power spread on the frequency axis and the less power contained between nominal frequencies f_0 and f_1.

In the next sections we will consider synchronous and asynchronous reception of FSK signals.

4.5.1 Discussion of Synchronous Reception of FSK Signal

Let us note that both elementary signals of the FSK-modulated signal have equal energy per modulation period. Thus, assuming additive white Gaussian noise and equal probability of data symbols, we conclude that the error probability on the output of an FSK synchronous receiver can be calculated using formula (4.7). To apply this formula, we need to determine the cross-correlation coefficient γ for FSK elementary signals. Let us additionally assume that the period of the carrier signals of frequency f_0 and f_1 is much smaller than the modulation period T, i.e. $f_0, f_1 \gg 1/T$. Under these assumptions calculation of the cross-correlation coefficient γ between elementary signals $s_0(t)$ and $s_1(t)$ results in formula (4.56), which is illustrated in Figure 4.16.

$$\gamma = \frac{\sin\left[2\pi(f_1 - f_0)T\right]}{2\pi(f_1 - f_0)T} \tag{4.56}$$

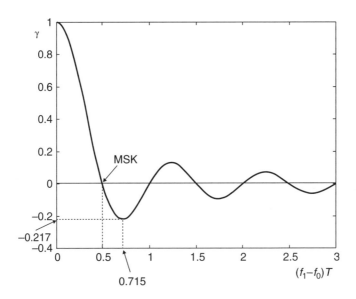

Figure 4.16 Cross-correlation coefficient of the elementary signals in FSK modulation as a function of the normalized difference of nominal frequencies f_0 and f_1

We can easily deduce that we can receive different values of the probability of error (4.7) depending on the difference of nominal frequencies $f_1 - f_0$ normalized with respect to modulation period T. The smallest value of the cross-correlation coefficient appears for $(f_1 - f_0)T = 0.715$ and is equal to -0.217. This cross-correlation coefficient guarantees the minimum of probability of error on the output of the synchronous receiver. As we already know, such a synchronous receiver consists of two correlators correlating the received signal with the elementary signals $s_0(t)$ and $s_1(t)$, respectively, and the decision device that selects the information symbol that is related to the higher correlator output.

Let us put $\gamma = -0.217$ in (4.7). We see from the resulting formula that if we want to ensure a given level of error probability for different modulations, the signal-to-noise ratio must be over $2\,\mathrm{dB}$ higher for FSK modulation than for a bipolar (e.g. BPSK) modulation. For $\gamma = 0$ the elementary signals become mutually orthogonal. It occurs for example if $h = (f_1 - f_0)T = 0.5$. This is the smallest value of the modulation index h for which elementary signals are orthogonal, therefore the FSK modulation with $h = 0.5$ is called MSK (*Minimum Shift Keying*). We will discuss MSK in the section devoted to constant envelope modulations.

4.5.2 *Asynchronous Reception of FSK Signals*

FSK-modulated signals are mostly received in an asynchronous manner. The optimal structure of an asynchronous FSK receiver is similar to the structure of the receiver derived for ASK-modulated signals. Let us note that an FSK signal can be treated as a concatenation of two complementary ASK signals: one for which the modulator sends a zero signal if data symbol "0" is transmitted and a carrier signal of frequency f_1 if data symbol "1" is transmitted; and the other for which the modulator sends a carrier signal of frequency f_0 if data symbol "0" is transmitted and a zero signal if data symbol "1" is transmitted. Therefore, the asynchronous receiver for FSK signals has two parallel branches consisting of the receivers for ASK signals. Such a single ASK receiver has already been shown in Figure 4.4. The equivalent form of the FSK receiver is presented in Figure 4.17.

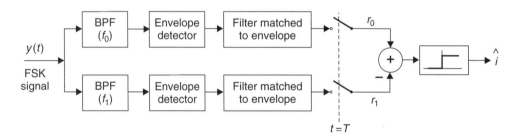

Figure 4.17 Asynchronous optimal receiver for FSK signals

The FSK signal corrupted by additive noise undergoes passband filtration performed by two filters of center frequencies $f_c \pm \Delta f$ equal to the nominal frequencies f_0 and f_1, respectively, which in turn are related to data symbols "0" and "1". In a given modulation period in which one of the nominal frequencies is transmitted we observe a noise signal on the output of one filter, whereas on the output of the second filter we observe the carrier signal plus noise. Subsequently, envelope detectors determine the envelopes in both branches. In order to maximize the signal-to-noise ratio, the envelope detector output signals are fed to the matched filters. The matched filters act in the same way as in a synchronous receiver. The outputs of the matched filters are sampled after each modulation period and the resulting samples are compared. The larger sample indicates the data symbol that was more probably transmitted.

4.5.3 Error probability for Asynchronous FSK Receiver

In this section we will calculate the error probability for the optimal asynchronous receiver for FSK signals, assuming that data symbols "0" and "1" are equiprobable. Let us recall that energies of both elementary signals are the same and equal to E_s. In our calculations we will use the results previously obtained for ASK modulation.

Let us assume that data symbol "1" has been transmitted, i.e. the transmitter has sent signal $s_1(t)$. As a result, a signal of frequency f_1 appears in the lower branch of the receiver shown in Figure 4.17. Envelope sample r_1 in the lower branch is Rice distributed, whereas on the output of the upper branch filter we observe only the Gaussian noise. This means that envelope sample r_0 is Rayleigh distributed. The receiver commits an error if $r_0 > r_1$. The probability that this inequality holds true can be derived by calculation of the probability that $r_0 > r_1$ for a given value r_1, and averaging this result along the whole range of possible values of r_1 taking into account the probability distribution of r_1. This means that the conditional probability of error given that the data symbol "1" was transmitted can be calculated from the formula

$$P(\mathcal{E}|\text{symbol "1" was transmitted}) = \int_{r_1=0}^{\infty} p_1(r_1) \left[\int_{r_0=r_1}^{\infty} p_0(r_0) dr_0 \right] dr_1 \qquad (4.57)$$

In (4.57) we assume that envelope samples in both receiver branches are statistically independent. One can prove that this assumption is fulfilled if both the center frequency $f_c = (f_0 + f_1)/2$ and the difference frequency $\Delta f = f_1 - f_c = f_c - f_0$ are multiples of $1/4T$. Consequently, the elementary signals applied in FSK modulation are mutually orthogonal. Calculation of the integral within the square brackets in (4.57) gives the result

$$\int_{r_1}^{\infty} p_0(r_0) dr_0 = \int_{r_1}^{\infty} \frac{r_0}{\sigma^2} \exp\left(-\frac{r_0^2}{2\sigma^2}\right) dr_0 = \exp\left(-\frac{r_1^2}{2\sigma^2}\right) \qquad (4.58)$$

Putting this in (4.57) and using the formula for the Rice distribution (4.52) we have

$$P(\mathcal{E}|\text{symbol "1" was transmitted}) = \int_{r_1=0}^{\infty} \frac{r_1}{\sigma^2} \exp\left(-\frac{E_s^2 + r_1^2}{2\sigma^2}\right) \exp\left(-\frac{r_1^2}{2\sigma^2}\right) I_0\left(\frac{E_s r_1}{\sigma^2}\right) dr_1$$

$$= \int_{r_1=0}^{\infty} \frac{r_1}{\sigma^2} \exp\left(-\frac{E_s^2 + 2r_1^2}{2\sigma^2}\right) I_0\left(\frac{E_s r_1}{\sigma^2}\right) dr_1 \qquad (4.59)$$

Substituting $\sqrt{2}r_1$ for u in integral (4.59) and shifting the expression $\exp(-E_s^2/4\sigma^2)$ in front of the integral, we receive

$$P(\mathcal{E}|\text{symbol '1' was transmitted}) = \frac{1}{2}\exp\left(-\frac{E_s^2}{4\sigma^2}\right)\int\limits_{u=0}^{\infty}\frac{u}{\sigma^2}\exp\left[-\frac{\left(\frac{E_s}{\sqrt{2}}\right)^2 + u^2}{2\sigma^2}\right]I_0\left(\frac{\frac{E_s}{\sqrt{2}}u}{\sigma^2}\right)du$$

(4.60)

Let us note that the integration contained in formula (4.60) is performed on the Rice distribution in the whole range of its argument, therefore its result is equal to unity. Thus, if we take into account that $\sigma^2 = E_s N_0/2$ [see formula (4.37)], we obtain a simple result

$$P(\mathcal{E}|\text{symbol "1" was transmitted}) = \frac{1}{2}\exp\left(-\frac{E_s}{2N_0}\right)$$

(4.61)

Calculation of the conditional error probability given that the symbol "0" was transmitted leads to an identical result because both receiver branches are symmetric to each other. Since we have assumed that data symbols "0" and "1" are equiprobable, the final formula for the error probability achieves the form

$$P(\mathcal{E}) = \frac{1}{2}\exp\left(-\frac{E_s}{2N_0}\right)$$

(4.62)

In Figure 4.31 we compare the error probability for several binary modulations and several types of receiver, including the asynchronous FSK receiver discussed above. We can see that this receiver features the worst quality, although for high signal-to-noise ratios the difference between the synchronous and asynchronous receiver is relatively small.

4.5.4 Suboptimal FSK Reception with a Frequency Discriminator

Application of a frequency discriminator in an FSK receiver yields another method of asynchronous FSK reception. In this type of receiver an FSK signal is treated as an FM signal for which the modulating signal is a sequence of bipolar pulses representing the binary sequence. Thus, the frequency discriminator converts an instantaneous frequency into a signal with the amplitude proportional to this frequency. Further operations are performed in the baseband. A phase-locked loop (PLL) is a typical circuit performing frequency discrimination in the receivers. Figure 4.18 presents a general scheme of such a receiver.

The appropriate choice of the input bandpass filter in the FSK receiver is a relevant issue for the FSK receiver with a frequency discriminator. For a given value of the modulation index h, which is equivalent to a given distance between nominal

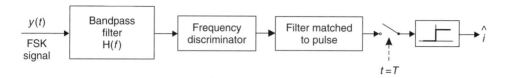

Figure 4.18   An FSK receiver based on a frequency discriminator

frequencies f_0 and f_1 with respect to the modulation rate $1/T$, we can find the optimal filter bandwidth that will ensure the minimum probability of error for each input filter characteristics. Let us consider the formula for the probability of error for the receiver with a frequency discriminator and with an input bandpass filter of a given characteristic $H(f)$.

Let us denote the FSK signal amplitude at the input of the bandpass filter $H(f)$ as A. Let $H_B(f)$ be the characteristic of the baseband equivalent filter that is associated with the bandpass filter characteristic by the formula

$$H(f) = \frac{1}{2}\left[H_B(f + f_c) + H_B(f - f_c)\right] \tag{4.63}$$

Figure 4.19 shows this relation in graphical form. Let $\Delta f = (f_1 - f_0)/2$. Let us introduce the parameter $\beta = \Delta f T$. We can prove that the error probability for the FSK receiver with a frequency discriminator and matched filter is described by the formula

$$P(\mathcal{E}) = \frac{1}{2}\,\text{erfc}\left[\sqrt{3\beta^2 \frac{A^2 T}{2N_0}\,|H_B(\Delta f)|^2}\,\right] \tag{4.64}$$

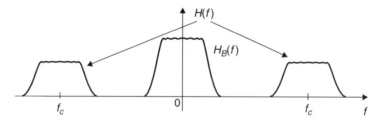

Figure 4.19   Input bandpass filter characteristic $H(f)$ and the characteristic $H_B(f)$ of its baseband equivalent filter

In order to minimize the error probability we have to maximize the argument of the erfc function. To do this, we can increase parameter β by increasing the distance between nominal frequencies f_0 and f_1. This leads to an increase of the argument of the erfc function until Δf becomes so large that $|H_B(\Delta f)|^2$ starts to decrease. This finally causes a decrease in the argument of the erfc function. Thus, for the given filter characteristic $H_B(f)$, or equivalently $H(f)$, there is an optimal value of β. Table 4.1 lists optimal

Table 4.1 Optimal values of index β and the distance between nominal frequencies for an FSK receiver with a frequency discriminator

Type of filter $\|H_B(f)\|^2$	Value of index β	Optimum value of distance $f_1 - f_0$
Gaussian $\exp[-f^2/(2B_{3dB})^2]$	$0.31 B_{3dB} T$	$0.7 B_{3dB}$
Butterworth, 5th order $1/[1 + (f/B_{3dB})^{10}]$	$0.40 B_{3dB} T$	$0.81 B_{3dB}$
Ideal $\|H_B(f)\|^2 = 1 \quad \|f\| \leq B_{3dB}$ $\|H_B(f)\|^2 = 0 \quad$ otherwise	$0.5 B_{3dB} T$	B_{3dB}

values of parameter β and the distance between nominal frequencies f_1 and f_0 for a few types of the input filter.

In conclusion, let us stress once more that the performance of various types of FSK receivers differs substantially, which manifests itself in a different immunity to noise. The synchronous receiver that uses the largest knowledge about the signal features the lowest probability of erronous detection. The asynchronous receiver utilizing envelope detection is slightly worse, whereas the FSK receiver with the frequency discriminator features the lowest detection quality. Despite this, the latter receiver is often applied due to its simple implementation.

4.6 PSK Modulation

Let us recall that elementary signals applied in digital phase modulation are given by formula (4.4). As we have already mentioned during analysis of this formula, the elementary signals are bipolar, i.e. $s_1(t)$ has the opposite polarity with respect to $s_0(t)$. Therefore, it is easy to show that elementary signals $s_0(t)$ and $s_1(t)$ have equal energy and their correlation coefficient is $\gamma = -1$. In consequence, for synchronous reception we obtain the probability of error described by formula (4.8). Let us note that in the case of a standard *Phase Shift Keying* (PSK) modulation the elementary signals last for exactly T seconds, i.e. the modulation period. Signal components $x^I(t)$ and $x^Q(t)$ appearing in the general formula (4.5) have, for PSK, the following form for the time interval $iT \leq t \leq (i+1)T$

$$x^I(t) = d_i A \, \text{rect}\left(\frac{t - T/2 - iT}{T}\right), \qquad \text{where} \quad d_i = \pm 1$$

$$x^Q(t) = 0 \tag{4.65}$$

As we can see, the gate function rect (.) determines the spectral properties of the PSK signal defined by formula (4.4). After simple calculations, assuming again that $f_c \gg 1/T$, we get

$$G_{\text{PSK}}(f) = \frac{A^2 T}{4} \left\{ \text{sinc}^2\left[\pi(f - f_c)T\right] + \text{sinc}^2\left[\pi(f + f_c)T\right] \right\} \tag{4.66}$$

Due to the application of bipolar signals, the optimal synchronous receiver is simplified to a single branch consisting of a correlator that performs correlation of the received signal with the elementary signal $s_0(t) = A \cos(2\pi f_c t)$, a sampling block and a decision device in the form of a comparator with zero threshold. The receiver is shown in Figure 4.20.

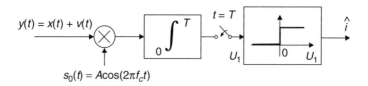

Figure 4.20   Block diagram of a synchronous PSK receiver

We can see from (4.65) that the spectral properties of the PSK modulation are not very good, similar to those for ASK modulation. They can be improved by replacing the gate function applied in (4.65), which describes a modulating pulse in the baseband, with another function, e.g. one of the functions presented in Chapter 3. The pulses with the raised cosine or square root raised cosine characteristics are particularly advantageous for improving the spectral properties of PSK modulation.

4.7   Linear Approach to Digital Modulations – M-PSK Modulation

So far we have considered binary modulations. However, if a higher data rate compared with that obtained with a binary modulation is required in transmission over a band-limited channel, then the solution is to increase the number of elementary signals of a given modulation. This is also true for PSK modulation. Figure 4.21 presents the so-called signal constellations for 2-, 4- and 8-PSK modulations. A signal constellation is a set of points determined by the in-phase I and quadrature Q components of the data symbols. Thus, the coordinates of signal constellation points are the data symbol pairs that modulate the baseband pulses of the in-phase $x^I(t)$ and quadrature $x^Q(t)$ components, respectively. As two mutually orthogonal signal components are modulated, such

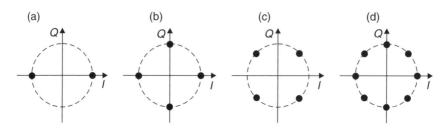

Figure 4.21   PSK signal constellations for binary PSK (BPSK) (a), 4-PSK (b), 4-PSK in the QPSK version (c) and 8-PSK (d)

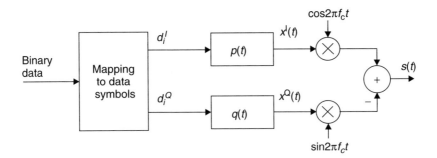

Figure 4.22 Linear modulator for a two-dimensional digital modulation

modulation is considered as two-dimensional. If we consider PSK modulation as linear, the modulation process can be presented in a linear form, as shown for the general case in Figure 4.22.

The elements of a binary data sequence in the form of single bits or bit blocks are mapped onto pairs of elementary symbols d_i^I and d_i^Q, which constitute an excitation for the transmit baseband filters characterized by the impulse responses $p(t)$ and $q(t)$, respectively. Thus, the signals modulating the in-phase and quadrature carriers are described by the formulae

$$x^I(t) = \sum_{i=-\infty}^{\infty} d_i^I p(t - iT) \quad x^Q(t) = \sum_{i=-\infty}^{\infty} d_i^Q q(t - iT) \tag{4.67}$$

We can map bit blocks onto pairs of elementary symbols d_i^I and d_i^Q in several ways. The so-called *Gray code* is most frequently applied. For this code neighboring constellation points on the in-phase and quadrature plane are assigned binary blocks that differ in the smallest possible number (mostly one) of positions. Figure 4.23 illustrates how binary blocks are assigned to 8-PSK constellation points according to the Gray code.

The main reason for application of the Gray code is the observation that the most frequent errors are due to selection of the neighboring constellation point with respect to the one that was actually transmitted. If binary blocks assigned to the neighboring points differ only in one position, then a single binary error will be made. Other binary block assignments would lead to an increased number of binary errors. There are, however, situations for which other forms of mapping are more advantageous for transmission quality.

Several linear modulations can be presented using formula (4.67). If we assume $q(t) = 0$, $d_i^I = \pm1$ and $p(t) = \mathrm{rect}((t - T/2)/T)$, we obtain BPSK – *Binary Phase Shift Keying* – Figure 4.21a. In turn, if we select $d_i^I = d_i^Q = \pm1$ and $p(t) = q(t) = \mathrm{rect}((t - T/2)/T)$, we obtain QPSK – *Quadrature Phase Shift Keying* – shown in Figure 4.21c. 8-PSK modulation is obtained assuming data symbols d_i^I and d_i^Q in the form of $\cos(i\pi/4)$ and $\sin(i\pi/4)$ ($i = 0, \ldots, 7$), respectively.

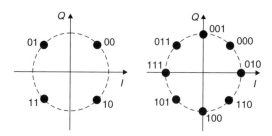

Figure 4.23 Mapping of the binary blocks onto the elementary signals for QPSK and 8-PSK modulations according to the Gray code

As we have already mentioned, the signal spectrum can be shaped by appropriate selection of the transmit filter impulse responses $p(t)$ and $q(t)$. A typical impulse response of the transmit filter applied in digital communication systems is a time function $p(t)$ whose frequency characteristic has the shape of a square root raised cosine. In the receiver, the same receive filter characteristic is applied. The spectrum shaped in this way ensures a low level of spectral sidelobes and high energy in the mainlobe. However, let us note that the transmit filter response to a data symbol lasts for a few modulation periods, so in the case of transmission of a whole data symbol sequence the responses of the transmit filter to particular data symbols overlap.

So far we have considered signal constellations for some PSK modulations. They are formed out of the points on the (I, Q) plane, which represent data symbols modulating the in-phase and quadrature components. Such a constellation does not illustrate how the modulator changes elementary signals during the modulation period. We can obtain a plot showing this process if we draw both components $x^I(t)$ and $x^Q(t)$ along mutually orthogonal axes. The resulting trajectories show the instantaneous amplitude and phase of the modulated signal as a function of time. Strictly speaking, the distance of a trajectory point to the origin of the coordinate system is the instantaneous value of the envelope of the modulated signal, whereas the angle between the line connecting the trajectory point with the origin of the coordinate system and the I axis is the instantaneous signal phase (not taking into account the phase resulting from the carrier signal). In some communication systems, e.g. in cellular radio, it is crucial that the signal envelope is constant or fluctuates only within a small range. This need results from the application of power amplifiers in their nonlinear range, which is usually caused by limited energy resources, in particular in mobile terminals. Thus, the applied modulation should be robust to nonlinear distortions. In Section 4.14 we consider the influence of the nonlinear characteristics of the power amplifier on the transmitted signal.

Figure 4.24a presents envelope trajectories for QPSK modulation in which a transmit filter with the square root raised cosine characteristics is applied. Recall that for each time instant the current envelope value is the distance between the trajectory point and the coordinate system origin. Let us note that at some moments the envelope value is close to zero. Therefore, despite good spectral properties achieved due to filtering, the modulated signal is not robust against nonlinear distortions. We can obtain a visible improvement if we shift the symbol timing clock of one signal component with respect to the other, i.e. if $q(t) = p(t - T/2)$. In such a case we talk about

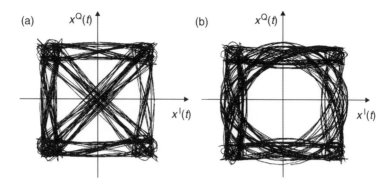

Figure 4.24 Envelope trajectories for QPSK (a) and OQPSK (b) modulations using the filter with square root raised cosine characteristics

OQPSK – *Offset Quaternary Phase Shift Keying*. This type of modulation is illustrated in Figure 4.24b.

Another means of decreasing the envelope dynamics, leading to higher robustness against nonlinear distortions, relies on adding an extra phase shift of $\pi/4$ to the modulated signal in each modulation period. Applying the phase difference between two consequtive modulation periods and the additional phase shift mentioned above results in the so-called $\pi/4$-DQPSK modulation that appears in some mobile communication systems.

In a typical $\pi/4$-DQPSK modulation a binary stream is divided into dibits, which determine the phase shift with respect to the phase in the previous modulation period. Possible phase shifts are $\pm\pi/4$ and $\pm3\pi/4$. Figure 4.25 presents envelope trajectories for any possible dibit combinations. The envelope is not constant but its fluctuations are moderate. We see that the envelope is never equal to zero, which positively influences the signal robustness against nonlinear distortions. The signal shown in Figure 4.25 was generated by application of the transmit filter with the square root raised cosine characteristics and with the roll-off factor equal to 0.35.

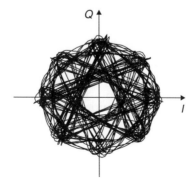

Figure 4.25 Envelope trajectories for $\pi/4$-DQPSK-modulated signal

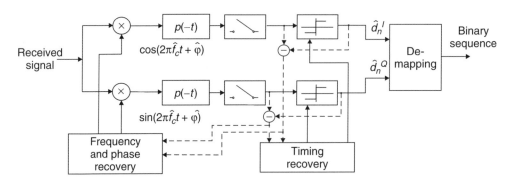

Figure 4.26 Optimal receiver for QPSK signals with carrier and timing recovery blocks

Figure 4.26 presents a typical synchronous receiver scheme for a QPSK signal [we assume that the transmit filters in the transmitter shown in Figure 4.22 are identical, i.e. $q(t) = p(t)$].

The received signal (possibly after down-conversion to the intermediate frequency) is given to the inputs of two synchronous demodulators that consist of multipliers and low-pass filters. These filters are simultaneously matched to the transmit filters applied in the transmitter. Let us recall that for a nondistorting channel with additive white Gaussian noise the impulse response of the receive filter should be a mirrored reflection of the transmit filter impulse response. For a transmit filter with the characteristics given by formula (4.20), for which the impulse response is symmetric with respect to its center, the receive filter has an impulse response that is identical to that of the transmit filter. Therefore both filters constitute a pair. The frequency and phase recovery block shown in Figure 4.26 makes use of the difference between the decision generated in the decision device for the in-phase and quadrature components and the signal components given to the input of the decision device. This difference is a measure of the phase difference. The rate of change of the phase difference carries information about frequency offset with respect to the received signal.

The above frequency and phase recovery block with feedback is one of several possible solutions. There are other phase and frequency recovery circuits that are based on nonlinear processing of the received signal, resulting in creation of a discrete spectral component at the multiple of the carrier frequency. After extracting this spectral component from the received signal and dividing it in a frequency divisor, we obtain the carrier frequency that can be applied by the reference carrier. One possible way to adjust the symbol timing circuit is also based on the signal difference between the signal on the decision device input and the signal on its output.

A negative consequence of using the carrier phase recovery circuits is the so-called *phase ambiguity*. Due to the performed mathematical operations these circuits recover the received signal phase modulo-180° (for BPSK modulation) or modulo-90° (for 4-PSK or other modulations with constellations that are symmetric with respect to both coordinate axes).

4.8 Differential Phase Shift Keying (DPSK)

4.8.1 PSK Modulation with Differential Coding and Synchronous Detection

Phase ambiguity, as we said above, is a negative consequence of the carrier phase recovery that takes place in a synchronous receiver. To remove phase ambiguity we must apply differential coding before feeding the data symbols to the modulator input. The simplest case of the BPSK modulation is shown in Figure 4.27.

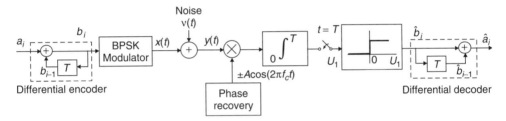

Figure 4.27 Transmitter and synchronous receiver for BPSK-modulated signals with differential coding

Let us use Figure 4.27 for explanation of the differential encoding. This operation is based on modulo-2 addition of the input bit to the bit resulting from differential encoding in the previous timing instant. Therefore, the BPSK input signal b_i is given by the equation

$$b_i = a_i \oplus b_{i-1} \tag{4.68}$$

Symbols b_i are recovered in the receiver on the output of the BPSK demodulator. First, let us assume that the phase recovery circuit operating with 180° phase ambiguity has recovered the received signal phase correctly. Let the decided symbol $\widehat{b}_i$ be the same as the transmitted symbol b_i. Then, differential decoding leads to the following result

$$\widehat{a}_i = \widehat{b}_i \oplus \widehat{b}_{i-1} = (a_i \oplus b_{i-1}) \oplus b_{i-1} = a_i \tag{4.69}$$

Naturally, the receiver operates correctly if the decision device makes correct decisions. If the decision is incorrect, it will appear in the differential decoder in two succesive timing instants due to the applied memory cell. Thus, a single decision error will trigger two binary errors. This is the price paid for removing the influence of phase ambiguity caused by the phase recovery circuit applied at the receiver.

Now let us assume that the phase recovery circuit generates the reference signal shifted by 180° with respect to the received signal. Thus, the correlator output sample is generated with inverted polarity with respect to the output sample that would appear if the phase were recovered properly. In consequence, assuming the correct decision of the decision device, its output symbol is $\widehat{b}_i = \overline{b}_i = 1 \oplus b_i$. Therefore, we get the following signal on the output of the differential decoder

$$\widehat{a}_i = \widehat{b}_i \oplus \widehat{b}_{i-1} = (1 \oplus b_i) \oplus (1 \oplus b_{i-1})$$

$$= b_i \oplus b_{i-1} = a_i \oplus b_{i-1} \oplus b_{i-1} = a_i \qquad (4.70)$$

As we can see, the final decision related to symbol a_i remains correct independently of the presence or absence of the 180° phase shift caused by the carrier recovery circuit. Let us stress once more that the condition for correct operation of the whole system is the lack of errors on the output of the decision device. Let us also note that the receiver considered so far is synchronous, since it recovers the carrier phase (although it performs this task with 180° or 90° ambiguity in some other cases).

We can easily calculate the error probability for the receiver of differentially encoded BPSK if we apply the results for a synchronous BPSK receiver. As we have already noticed, in order to achieve the correct signal from the differential decoder, correct decisions from the decision device have to be made in the current and preceding timing instants. The output decision will also be correct if the decision device commits an error in the current and preceding timing instants. Knowing the probability of an incorrect decision (4.8), we receive the following formula for the probability P_C of the correct data symbol decision

$$P_C = \left[1 - \frac{1}{2} \operatorname{erfc}\left(\sqrt{\frac{E_s}{N_0}} \right) \right]^2 + \left[\frac{1}{2} \operatorname{erfc}\left(\sqrt{\frac{E_s}{N_0}} \right) \right]^2 \qquad (4.71)$$

so the symbol error probability is

$$P(\mathcal{E}) = 1 - P_C = 1 - \left[1 - \frac{1}{2} \operatorname{erfc}\left(\sqrt{\frac{E_s}{N_0}} \right) \right]^2 - \left[\frac{1}{2} \operatorname{erfc}\left(\sqrt{\frac{E_s}{N_0}} \right) \right]^2$$

$$= \operatorname{erfc}\left(\sqrt{\frac{E_s}{N_0}} \right) - \frac{1}{2} \operatorname{erfc}^2\left(\sqrt{\frac{E_s}{N_0}} \right) \qquad (4.72)$$

If we compare (4.72) with (4.8), we see that the error probability is higher for the synchronous receiver of differentially encoded BPSK signals than for the receiver of the pure BPSK. The value of the first component in (4.72) is twice as high as the value of the single component in (4.8), whereas the value of the second, negative component is substantially smaller than that in (4.8).

4.8.2 Asynchronous DPSK Receivers

Now let us treat the BPSK modulator and differential encoder jointly. Let us assume that the elementary signal $s_0(t)$ [see formula (4.4)] is generated for $b_i = 0$, whereas data symbol $b_i = 1$ implies generation of the elementary signal $s_1(t)$, as it occurs in a typical BPSK modulator. Let $b_0 = 0$, i.e. the encoder memory cell contains a zero in the zero timing instant. From the respective waveform in Figure 4.28 we see that the 180° phase shift with respect to the current phase is triggered by the data symbol $a_i = 1$, whereas there is no phase shift if the data symbol $a_i = 0$ appears on the modulator input. Such a modulation is called *Differential Phase Shift Keying* (DPSK) and it can be achieved also through differential encoding and regular BPSK modulation.

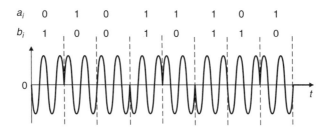

Figure 4.28   Example of DPSK-modulated signal waveform

In real channels distortions are introduced by multipath propagation, flat and selective fading, intersymbol interference and phase jitter. As a result, the complexity of the synchronous receiver grows and sometimes implementation of the synchronous receiver is impossible, particularly if the phase and frequency recovery blocks generate estimates of the carrier frequency and phase of insufficient quality. In such cases, instead of regular PSK modulation a DPSK modulation with asynchronous reception can be applied. Let us consider the DQPSK signal, i.e. the signal with four-phase differential shift keying in the version analogous to that shown in Figure 4.21c. The signal is described by formula (4.5), and the in-phase and quadrature baseband components of signal (4.5) are determined by the expressions

$$x^I(t) = \sum_{n=-\infty}^{\infty} d_n^I p(t - nT) \quad x^Q(t) = \sum_{n=-\infty}^{\infty} d_n^Q p(t - nT) \tag{4.73}$$

This time, the two-bit information is carried neither by the data symbols d_n^I and d_n^Q themselves nor by their argument $\phi_n = \arg(d_n^I + j d_n^Q)$. It is contained in the phase difference between successive modulation periods, $\Delta\phi_n = \phi_n - \phi_{n-1}$. The scheme of such a receiver is shown in Figure 4.29 (Okunev 1979).

Signals on the output of both receive filters are sampled once per modulation period and are proportional to the cosine and sine of the angle $\phi_n + \theta$, respectively, where θ represents the carrier phase difference between the received signal and the demodulating

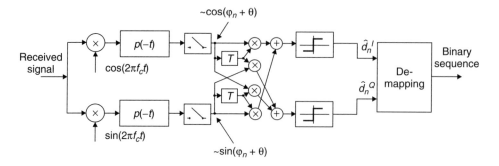

Figure 4.29   Asynchronous receiver for DQPSK signals

signals. The receiver does not track the carrier phase of the received signal. The samples that are proportional to cosines and sines of the angles $\phi_n + \theta$ are stored in the memory cells denoted by T, so both current and previous samples are at the disposal of the receiver. Let us note that in order to make a decision it is sufficient to know in which quarter of the coordinate system the angle $\Delta\phi_n$ is located. In order to detect it, it is sufficient to determine the sign of functions $\sin \Delta\phi_n$ and $\cos \Delta\phi_n$. They can be calculated on the basis of well-known trigonometric formulas

$$\cos(\Delta\phi_n) = \cos\left[(\phi_n + \theta) - (\phi_{n-1} + \theta)\right]$$

$$= \cos(\phi_n + \theta)\cos(\phi_{n-1} + \theta) + \sin(\phi_n + \theta)\sin(\phi_{n-1} + \theta)$$

$$\sin(\Delta\phi_n) = \sin\left[(\phi_n + \theta) - (\phi_{n-1} + \theta)\right]$$

$$= \sin(\phi_n + \theta)\cos(\phi_{n-1} + \theta) - \cos(\phi_n + \theta)\sin(\phi_{n-1} + \theta) \qquad (4.74)$$

Let us note that the system of multipliers and adders in Figure 4.29 exactly follows formula (4.74). The whole receiver operates correctly under the assumption that the changes of angle θ between succeeding modulation periods are insignificant, i.e. the phase jitter is slow as compared with the duration of a single modulation period. If the impulse response of the baseband filter $p(t)$ lasted for the modulation period T and $p(t)$ had a rectangular shape, the optimum receive filter would be an integrator, so it would realize a correlation. Thus, such DQPSK receiver is called a *correlative receiver*.

For DPSK modulation for which information is carried by the phase difference between the current and previous modulation periods, besides a correlative receiver another type of asynchronous receiver is also possible. This receiver is called an *autocorrelative receiver*, because correlation is performed between the currently received signal and the signal received during the previous modulation period and stored in the delay line. Two versions of a DPSK autocorrelative receiver are shown in Figure 4.30.

Let us consider the operation of both versions. Transmitted data are contained in the phase difference $\Delta\phi = 0$ or π. Let the phase of the received signal be θ. Let the following assumptions be fulfilled:

- a multiple number of periods of a carrier signal are contained in the modulation period;
- the carrier frequency is much higher than the modulation rate $1/T$.

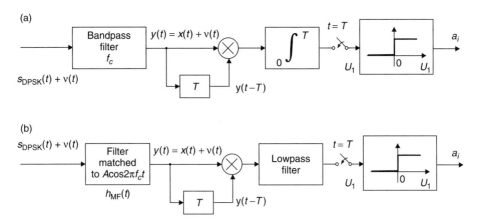

Figure 4.30 Autocorrelative receiver for binary DPSK signals: (a) suboptimal receiver; (b) optimal receiver with a matched filter

Let us consider the suboptimal receiver shown in Figure 4.30a. Neglecting noise for a while, which is approximately equivalent to the assumption of a high signal-to-noise ratio, we can describe the sample on the output of the correlator as a random variable U_1:

$$U_1 = \int_0^T A\cos(2\pi f_c t + \varphi_n + \theta)A\cos[2\pi f_c(t-T) + \varphi_{n-1} + \theta]\,dt$$

$$= \frac{A^2}{2}\int_0^T \left[\cos(\varphi_n - \varphi_{n-1}) + \cos(4\pi f_c t - 2\pi f_c T + 2\theta + \varphi_n + \varphi_{n-1})\right]dt$$

$$= \frac{A^2 T}{2}\cos(\Delta\varphi_n) \tag{4.75}$$

As we see, the signal on the input of the decision device is proportional to the cosine of the angle $\Delta\varphi_n$. In order to check if this angle is equal to 0 or π, it is sufficient to check the sign of its cosine. The scheme is rather simple but its performance is inferior to the other receiver types, because in reality the noise from the current and previous modulation periods disturbs the correlation process. Let us note that the correlator is preceded by a bandpass filter, which narrows the signal spectrum to a necessary minimum in order to ensure the maximum signal-to-noise ratio without distorting the useful signal. Denoting the bandpass filter bandwidth as B, one can show (see Park 1978) that for high signal-to-noise ratio the error probability on the receiver output can be approximated by the following formula

$$P_{\text{DPSK}}(\mathcal{E}) = \frac{1}{2}\,\text{erfc}\left(\sqrt{\frac{1}{1 + \dfrac{BT}{2}\dfrac{N_0}{E_s}}\dfrac{E_s}{2N_0}}\right) \tag{4.76}$$

For a typical filter bandwidth $B = 3/T$ and for $E_s/N_0 = 10\,\mathrm{dB}$, formula (4.76) can be approximated by the expression (Couch 1987)

$$P_{\mathrm{DPSK}}(\mathcal{E}) = \frac{1}{2}\,\mathrm{erfc}\left(\sqrt{\frac{E_s}{2N_0}}\right) \tag{4.77}$$

so the performance of the considered receiver is comparable to the performance of a synchronous receiver for orthogonal elementary signals.

Now let us consider the second type of receiver. It can be proved that this is an optimal asynchronous receiver for DPSK signals. The impulse response of the matched filter at the receiver input is given by the formula

$$h_{\mathrm{MF}}(t) = \frac{1}{T}\,\mathrm{rect}\left(\frac{t - T/2}{T}\right)\cos 2\pi f_c t \tag{4.78}$$

so the initial phase θ of the received signal is not taken into account. Let the bipolar data sequence, which reflects the transmitted binary sequence a_i, be denoted as d_i, and let the bipolar sequence c_i represent the binary sequence b_i on the output of the differential encoder. The data symbols of sequences c_i and d_i take the values ± 1. We can easily note that differential coding can be described not only by formula (4.68), but also by the expression

$$c_i = d_i c_{i-1} \tag{4.79}$$

As a result, the DPSK signal generated by the transmitter can be described by

$$s(t) = \left\{\sum_{i=-\infty}^{\infty} c_i\,\mathrm{rect}\left[\frac{t - (i - 0,5)T}{T}\right]\right\} A\cos(2\pi f_c t + \theta) \tag{4.80}$$

Signal $y(t)$ observed on the receiver input is the sum of signal $s(t)$ and white additive Gaussian noise $v(t)$ of power spectral density $N_0/2$. On the output of the matched filter we obtain the component resulting from the DPSK signal and the noise component. The first component, under the assumption that $f_c \gg 1/T$, is

$$x(t) = \left\{\sum_{i=-\infty}^{\infty} c_i \Lambda\left[\frac{t - (i + 1)T}{T}\right]\right\} A\cos(2\pi f_c t + \theta) \tag{4.81}$$

where $\Lambda(t)$ is a triangular function described by the formula

$$\Lambda(t) = \begin{cases} 1 - |t| & \mathrm{dla}\ |t| < 1 \\ 0 & \mathrm{dla}\ |t| > 1 \end{cases} \tag{4.82}$$

The noise component on the matched filter output is a Gaussian passband noise of the form

$$n(t) = n_c(t)\cos(2\pi f_c t + \theta) - n_s(t)\sin(2\pi f_c t + \theta)$$

with the power spectral density and autocorrelation function equal to, respectively

$$G_n(f) = |H_{MF}(f)|^2 \frac{N_0}{2} \quad \text{and} \quad R_n(\tau) = \frac{N_0}{4T} \Lambda\left(\frac{\tau}{T}\right) \cos(2\pi f_c \tau) \qquad (4.83)$$

where $H_{MF}(f) = \mathcal{F}[h_{MF}(t)]$ is the matched filter transfer function, and the noise on the output of the matched filter has the variance $\sigma^2 = N_0/4T$ (we propose that interested readers calculate the noise variance in a similar way as we have done for asynchronous reception of ASK signals). Let us note that, under the assumption that $f_c = k/T$, the processes $n(t)$ and $n(t-T)$, where $n(t-T)$ appears on the output of the element delaying the signal by modulation period T, are mutually uncorrelated [cf. formula (4.83)]. Because they are Gaussian they are also statistically independent. Denoting

$$n(t-T) = n_c(t-T) \cos\left[2\pi f_c(t-T) + \theta\right]$$

$$- n_s(t-T) \sin\left[2\pi f_c(t-T) + \theta\right]$$

$$= n_{c,T}(t) \cos(2\pi f_c t + \theta) - n_{s,T}(t) \sin(2\pi f_c t + \theta) \qquad (4.84)$$

we obtain the following expressions for the samples multiplied at the moment $t = kT$

$$y(kT) = \left[\frac{c_k A}{2} + n_c(kT)\right] \cos(2\pi f_c kT + \theta) - n_s(kT) \sin(2\pi f_c kT + \theta)$$

$$y\big((k-1)T\big) = \left[\frac{c_{k-1} A}{2} + n_{c,T}(kT)\right] \cos(2\pi f_c kT + \theta)$$

$$- n_{s,T}(kT) \sin(2\pi f_c kT + \theta) \qquad (4.85)$$

The signal that is a product of the signals $y(t)$ and $y(t-T)$ is filtered in the low-pass filter so the components around $2f_c$ are attenuated. Assume again that the signal-to-noise ratio is so high that the noise components can be neglected for a while. Then the product of the signals on the output of the lowpass filter at the time instant $t = kT$ is

$$y(kT)y\big[(k-1)T\big] \approx \frac{c_k c_{k-1} A^2}{8} = \frac{c_{k-1}^2 d_k A^2}{8} = \frac{d_k A^2}{8} \qquad (4.86)$$

As we can see, this result is proportional to the bipolar data symbol d_k, so it is sufficient to determine the polarization of the product of the signals $y(t)$ and $y(t-T)$ at the moments $t = kT$.

4.8.3 Discussion on the Error Probability of the Optimal Asynchronous DPSK Receiver

Considerations that lead to the structure of the optimal asynchronous receiver for DPSK signals can be based on the following argumentation.

Let us note that in DPSK modulation, as for BPSK modulation, the elementary signals used are $s_0(t) = s(t)$ and $s_1(t) = -s(t)$. In the modulation period T each signal has energy E_s. However, for DPSK a phase shift with respect to the current phase occurs if the data symbol $a_i = 1$. This means that the data symbol $a_i = 1$ is mapped onto a pair of elementary signals, $[s(t), -s(t)]$ or $[-s(t), s(t)]$. Similarly, the data symbol $a_i = 0$, for which the phase shift does not take place, is mapped onto a pair of elementary signals, $[s(t), s(t)]$ or $[-s(t), -s(t)]$. In each case a transmitted bit is represented by a pair of elementary signals of energy $2E_s$. Let us note that any signal pair assigned to the data symbol $a_i = 0$ is orthogonal to any signal pair assigned to the data symbol $a_i = 1$. Thus, the orthogonal signals of equal energy $2E_s$ are used for transmission. The error probability for such a case has already been the subject of our considerations and has resulted in formula (4.62). This time the signal energy is twice as high as in the case of orthogonal signals of energy E_s for which formula (4.62) is valid, so we get

$$P_{\text{DPSK}}(\mathcal{E}) = \frac{1}{2} \exp\left(-\frac{E_s}{N_0}\right) \tag{4.87}$$

If we compare the above formula with formula (4.62) describing the error probability for the optimal asynchronous receiver for FSK orthogonal signals, we see that the argument of the function $\exp(-x)$ contained in both formulas is twice as high for DPSK compared with FSK. From this observation we conclude that for DPSK the error probability curve is shifted by 3 dB to the left with respect to the curve for orthogonal FSK signals. Therefore, in order to obtain the same probability of error, the DPSK system with optimal asynchronous reception requires a signal-to-noise ratio that is twice as small compared with the FSK system.

4.8.4 Comparison of Binary Modulations

In the previous paragraphs we considered different binary modulations and several receivers. Our considerations usually resulted in formulas for error probability for respective modulations and associated receivers. Now we will use these results for plotting the curves of error probability versus signal-to-noise ratio expressed in dB (Figure 4.31). We draw the following conclusions from analysis of the plots:

1. Error probability decreases monotonically with an increase of E_s/N_0.
2. For any value of E_s/N_0, PSK modulation with a synchronous receiver ensures lower probability of error than any other binary modulation. One can show that the PSK system with a synchronous receiver is the optimal solution for the additive white Gaussian noise channel.
3. PSK and DPSK modulations with synchronous receivers require an E_s/N_0 ratio that is 3 dB lower compared with FSK modulation with synchronous and nonsynchronous receivers, respectively.
4. For high E_s/N_0 DPSK modulation with a nonsynchronous receiver and FSK modulation with a nonsynchronous receiver are only slightly worse (by less than 1 dB) than PSK and FSK modulations with synchronous receivers, respectively.

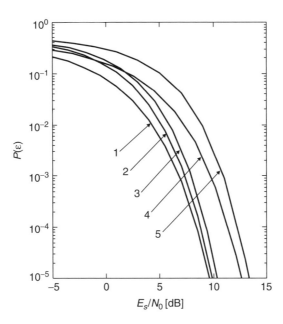

Figure 4.31 Error probability for several binary modulations and several types of reception: synchronous PSK reception (1), synchronous DPSK reception (2), nonsynchronous DPSK reception (3), synchronous FSK reception (4), nonsynchronous FSK reception (5)

4.9   Digital Amplitude and Phase Modulations – QAM

4.9.1   General Remarks

So far we have limited our considerations to the modulations in which only a single parameter of the carrier, such as amplitude, frequency or phase, is manipulated. As we have seen, in order to achieve a high throughput in the transmission over a band-limited channel, we cannot only apply M-ary PSK (or its differential version), but we can use digital modulations in which both amplitude and phase of the carrier are manipulated simultaneously. Such modulations are often denoted as AM/PM (*Amplitude Modulation/Phase Modulation*). As for an M-ary PSK, these modulations can be interpreted as amplitude modulation of two mutually orthogonal carriers and are often described as QAM (*Quadrature Amplitude Modulation*). Figure 4.32 presents examples of QAM constellations.

The first modulation shown in Figure 4.32 is actually an AM/PM. Constellation points are located on four circles reflecting four amplitude values. The signals characterized by a given amplitude can have one of four possible phases. Let us note that the phases of constellation points on neighboring amplitude circles are mutually shifted by $45°$. Thus the distance between constellation points on the I–Q plane is maximized and they become more distinguishable from each other. As a result, the error probability is minimized. The remaining modulations shown in Figure 4.32 are QAM: 16-QAM of a square shape

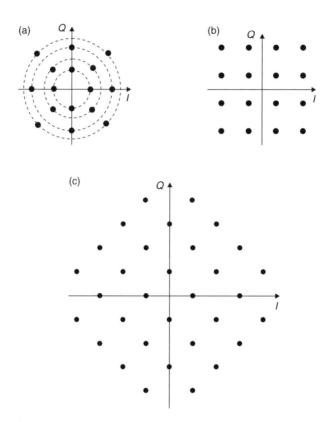

Figure 4.32 Examples of signal constellations for AM/PM and QAM modulations: (a) 16-AM/PM, (b) 16-QAM, (c) 32-QAM cross-constellation

and 32-QAM of a cross shape. All three modulations are applied in many communication systems in which a high binary throughput per spectrum unit is required and the signal-to-noise ratio is sufficiently high to allow for application of multilevel modulations.

QAM is best described by formulas (4.5) and (4.67). The signals that modulate in-phase and quadrature carriers are baseband signals with PAM. The block diagram of a QAM modulator is shown in Figure 4.33 and the synchronous receiver matched to it is presented in Figure 4.34.

Similarly, as for an M-ary PSK, mapping of binary data blocks onto QAM elementary signals has to be performed. Gray coding is applicable in this case as well, however this kind of mapping requires more careful analysis. This time Gray coding is not applied for the constellation points located on a circle, but it is applied simultaneously in two dimensions. In real communication systems differential encoding is typically applied as well, since the carrier phase synchronization block recovers the received signal phase modulo-90°. In this context we wish our QAM to be *90° phase invariant*, i.e. we have to apply such a differential encoding and binary block mapping that if the reference carrier phase is recovered in the receiver with the phase shift of $k\pi/2$ with respect to the original one, the data blocks on the output of the differential decoder are correct. Because the invariance with respect to the reference phase should be preserved modulo-90°, two bits

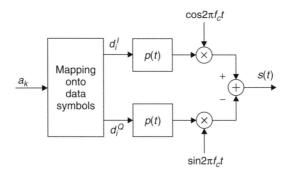

Figure 4.33   Block diagram of a QAM modulator with spectrum shaping performed in the base-band

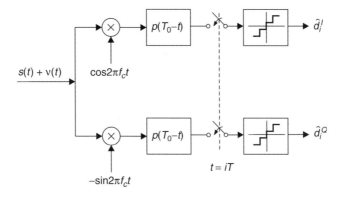

Figure 4.34   Block diagram of the synchronous receiver for QAM signals with spectral shaping in the transmitter

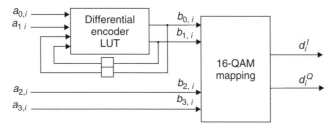

Figure 4.35   Scheme of a differential encoder with a look-up table (LUT) and a 16-QAM mapping block

are differentially encoded in each binary block. Figure 4.35 presents a general scheme of the differential encoder and QAM symbol mapper for a 16-QAM. An example of mapping of 4-bit blocks onto 16-QAM constellation points and the table of the differential encoder are shown in Figure 4.36. Let us note that for constellation points equidistant from the origin of the coordinate system and differing between each other by a multiple of $90°$,

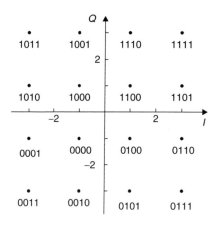

Input bits		Previous output bits		Current output bits	
$a_{0,i}$	$a_{1,i}$	$b_{0,i-1}$	$b_{1,i-1}$	$b_{0,i}$	$b_{1,i}$
0	0	0	0	0	1
0	0	0	1	1	1
0	0	1	0	0	0
0	0	1	1	1	0
0	1	0	0	0	0
0	1	0	1	0	1
0	1	1	0	1	0
0	1	1	1	1	1
1	0	0	0	1	1
1	0	0	1	1	0
1	0	1	0	0	1
1	0	1	1	0	0
1	1	0	0	1	0
1	1	0	1	0	0
1	1	1	0	1	1
1	1	1	1	0	1

Binary block mapping: $b_{0,i}, b_{1,i}, b_{2,i}, b_{3,i}$

Figure 4.36 Binary block mapping onto 16-QAM constellation points and the differential encoding table (in accordance with ITU-T V.32 Recommendation 1993)

two last bits of the binary block mapped to them are identical. Moreover, all constellation points located in a given quarter have an identical first two bits of the block mapped to them. Such mapping ensures modulo-90° phase invariance.

4.9.2 Error Probability for QAM Synchronous Receiver

Probability of error for QAM with an optimal receiver can be easily derived if the results for a baseband multilevel PAM are applied. Let us recall that the error probability for M-ary PAM is given by the formula

$$P_{M\text{-PAM}}(\mathcal{E}) = \frac{M-1}{M} \operatorname{erfc}\left(\sqrt{\frac{3k}{M^2-1}\frac{E_b}{N_0}}\right) \qquad (4.88)$$

where k is the number of bits assigned to a single PAM symbol. Let us note that an M-ary QAM signal can be treated as a superposition of two independent $\sqrt{M}$-ary PAM signals placed on two orthogonal sinusoidal signals with the same carrier frequency. This is exactly the case if $M = 2^{2m}$. Let us consider this case first. Note that the investigated QAM has 4, 16, 64, 256, etc. constellation points, so the signal constellation has a quadratic shape.

We will derive the error probability by calculating the probability P_C of the correct reception first. The symbol decision will be correct if there is no erroneous decision in either of the two dimensions of the modulation, so we can write

$$P_C = [1 - P_{\sqrt{M}\text{-PAM}}(\mathcal{E})]^2$$

Therefore

$$P_{M\text{-QAM}}(\mathcal{E}) = 1 - P_C = 1 - [1 - P_{\sqrt{M}\text{-PAM}}(\mathcal{E})]^2 \qquad (4.89)$$

In consequence

$$P_{M\text{-QAM}}(\mathcal{E}) = 2P_{\sqrt{M}\text{-PAM}}(\mathcal{E}) - [P_{\sqrt{M}\text{-PAM}}(\mathcal{E})]^2 \qquad (4.90)$$

Application of formula (4.88) in (4.89) is not straightforward. Let us note that the mean energy per modulation period for a single modulation dimension is half of the energy $E_s = kE_b$ of a two-dimensional QAM signal (E_b is the mean energy in the modulation period per single data bit). Similar observation is valid for the noise component too. For high values of E_b/N_0 the squared error probability $P_{\sqrt{M}\text{-PAM}}(\mathcal{E})$ is small compared with the first component, therefore the second term of (4.90) can be neglected. Finally, we obtain the following approximate formula for probability of error for an M-ary QAM modulation

$$P_{M-\text{QAM}}(\mathcal{E}) \approx 2\left(1 - \frac{1}{\sqrt{M}}\right)\text{erfc}\left[\sqrt{\frac{3kE_b}{2(M-1)N_0}}\right] \qquad (4.91)$$

When the number k of bits in a block mapped onto a constellation point is odd ($M = 32, 128, 512$), the QAM signal cannot be decomposed into two independent mutually orthogonal one-dimensional modulations. Exact calculation of error probability should take this fact into account. However, for high signal-to-noise ratios approximation (4.91) can still be used.

In communication systems maximum power of the signal generated by a transmitter is often one of the main limitations. One system must not interfere with the functioning of other systems that use the neighboring transmission lines or neighboring spectral bands. Sometimes the maximum allowable power is determined by the maximum value of the signal that can be fed to the amplifier input without saturating it. The considered limitation is often defined as a maximum allowable mean power or a maximum peak power. Taking into account one of these limitations we can find a QAM or AM/PM that ensures the lowest error probability on the output of a synchronous receiver if the signal is transmitted over an additive white Gaussian noise channel. In the general case, finding the best modulation can be a difficult task. This task can be simplified if we assume a high signal-to-noise ratio on the receiver input. In this case, the most probable errors are caused by selection of the signal constellation points that are the neighboring constellation points with respect to those signals that have actually been transmitted. Thus, our aim can be formulated in the reversed form: Find the signal constellation featuring the lowest mean power among possible constellation sets with the same number of constellation points and the same minimum Euclidean distance among constellation points. Another possible criterion for the constellation selection is the sensitivity of the receiver performance to the phase recovery inaccuracies.

4.9.3 Multidimensional Modulations

Development of new digital transmission systems, microprocessors and digital signal processing techniques paved the way to new concepts in digital modulations (Forney *et al.* 1984). In order to maximize the minimum distance between constellation points, the number of modulation dimensions can be increased. QAM considered so far are two-dimensional (cf. those shown in Figure 4.32). The dimensions are determined by carriers that are orthogonal to each other. Since in each modulation period a pair of data symbols (d_i^I, d_i^Q) is transmitted, these symbols can be interpreted as coordinates of a particular constellation point. However, if we treat two subsequent modulation periods jointly, i.e. we consider the vector $(d_i^I, d_i^Q, d_{i+1}^I, d_{i+1}^Q)$ as coordinates of a single constellation point in four-dimensional space, then we achieve additional freedom in selection of the constellation points. We choose these points in such a way that the minimum Euclidean distance between neighboring points is maximized. In general, the coordinates of the first two dimensions (d_i^I, d_i^Q) do not need to be selected from the same constellation set as the coordinates (d_{i+1}^I, d_{i+1}^Q) in the next modulation period. This approach results in a decrease in error probability compared to two-dimensional QAM. The above idea can be further expanded to a higher number of dimensions.

The disadvantage of getting the modulation more robust against noise by expanding its number N of dimensions is a serious increase in complexity of a decision device that makes joint decisions in N-dimensional space on the whole data symbol block. For example, for $N = 4$, decisions are made jointly on the blocks $(d_i^I, d_i^Q, d_{i+1}^I, d_{i+1}^Q)$. Although the achieved gain is finally quite moderate, the idea of increase in modulation dimensions found an application (together with other methods not considered in this section) in telephone modems, e.g. in the ITU-T V.34 modem (ITU-T V34 Recommendation 1997).

4.10   Constant Envelope Modulations – Continuous Phase Modulation (CPM)

A constant envelope of a transmitted signal is a highly desired feature in some wireless communication systems. The reason for this is the necessity to use the whole range of amplifier characteristics, including its nonlinear part. Thus, high power efficiency can be achieved, which is particularly important for battery powered devices.

A constant envelope signal can be achieved using the phase (PM) or frequency (FM) modulation if the modulating signal is a continuous signal $m(t)$ whose amplitude does not exceed a given maximum value $m_{\max}$. In the general case, the FM is described by the expression

$$x(t) = x^I(t) \cos 2\pi f_c t - x^Q(t) \sin 2\pi f_c t \tag{4.92}$$

where

$$x^I(t) = r \cos \varphi(t) \quad \text{and} \quad x^Q(t) = r \sin \varphi(t) \tag{4.93}$$

and

$$\varphi(t) = 2\pi k_{\text{FM}} \int_{-\infty}^{t} m(\tau)\mathrm{d}\tau \quad \text{where} \quad |m(t)| \le m_{\max} \tag{4.94}$$

The coefficient k_{FM} is the *FM modulation index*, $k_{\text{FM}} = \Delta f / m_{\max}$, and Δf is the frequency deviation, i.e. maximum possible difference between the instantaneous signal frequency and the carrier frequency. FM was used for voice transmission in the first generation analog mobile cellular radio and cordless telephony. It is also used in FM radio broadcasting.

In the case of digital transmission using Continuous Phase Modulation (CPM), the equation describing the function of phase in time, $\varphi(t)$, used in (4.93) in which data symbols are contained is

$$\varphi(t) = 2\pi h \sum_{i=-\infty}^{n} a_i \int_{-\infty}^{t} g(\tau - iT)\mathrm{d}\tau \qquad \text{for} \qquad nT \leq t < (n+1)T \qquad (4.95)$$

Formula (4.95) describes the instantaneous phase determined by the data symbol sequence $\{a_i\}$ ($i = -\infty, ..., n$). Parameter $h = 2\Delta f T$ is the so-called digital modulation index, whereas Δf is, as previously, the frequency deviation, and T is the modulation period. Data symbols are usually bipolar (i.e. $a_i = \pm 1$), although in some cases the number of possible data symbols is higher. Function $g(t)$ is called a *frequency impulse* and it describes how the instantaneous frequency changes in time due to a single data symbol $a_i = 1$. Instantaneous frequency is determined with respect to the carrier frequency f_c. Let us note that the instantaneous frequency can be found from the formula

$$f(t) = \frac{1}{2\pi} \frac{\mathrm{d}\varphi(t)}{\mathrm{d}t} = h \sum_{i=-\infty}^{n} a_i g(\tau - iT) \qquad \text{where} \qquad nT \leq t < (n+1)T \qquad (4.96)$$

In turn

$$q(t) = \int_{-\infty}^{t} g(\tau)\mathrm{d}\tau \qquad (4.97)$$

is the so-called *phase impulse* and describes how the instantaneous phase changes in time due to a single data symbol $a_i = 1$. In the simplest case the frequency impulse $g(t)$ has a rectangular shape of height $1/2T$ and it lasts for a modulation period T. Thus, the instantaneous frequency with respect to the carrier frequency is equal to $\pm \Delta f$. The frequency and phase impulses for this case are shown in Figure 4.37.

Let us notice that the signal described by equations (4.95) and (4.97) features a continuous phase function for any integrable shape of the phase impulse $g(t)$. Such modulations

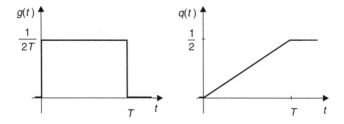

Figure 4.37   Frequency and phase impulses for continuous phase FSK modulation

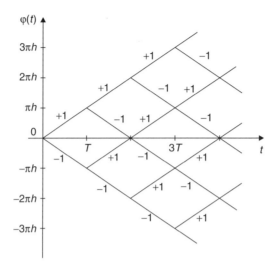

Figure 4.38   Phase tree for continuous phase FSK modulation with modulation index h

are often called *Continuous Phase Modulation* – CPM. This property has a fundamental meaning for spectral properties of a modulated signal. In practice, among several possible FSK modulations a continuous phase FSK signal (CPFSK) is most frequently applied.

Figure 4.38 presents possible phase paths $\varphi(t)$ in a few modulation periods starting from the zero time instant, in which the initial phase is equal to zero. The plot obtained is the so-called *phase tree* and it characterizes a given modulation. Phase trees can differ owing to the choice of the modulation index h and selection of the applied phase impulse $q(t)$. Supplying the modulator with the data symbol $+1$ implies the walk on the phase tree in the upper direction, whereas the data symbol -1 causes the walk to be in the lower direction, starting from the phase appearing at the beginning of the current modulation period. As we see, a given data sequence is in fact equivalent to the corresponding path on the phase tree. Thus, we could say that the aim of the demodulator is to find the best path on the phase tree on the basis of the received signal. Futher inspection of the phase tree shows another important feature of CPM. As we have already mentioned, the phase walk at a given modulation period depends not only on the supplied data symbol but also on the operation of the modulator in the past and the phase achieved up to the current moment. Therefore we can say that CPM is a modulation with memory. Thus, the optimum reception should not be limited to a single modulation period only, because the modulator memory should be used. In the case shown in Figure 4.38 the memory is equal to one modulation period, because the same phase value can appear again at the earliest after two modulation periods. Therefore the optimum receiver should analyze the signals from the current and preceding modulation periods. One can show that such a receiver should select the data symbol for which the generated FSK signal should have the lowest Euclidean distance from the received signal.

It was found quite early that both the shape of the frequency (or, equivalently, phase) impulse and the value of the modulation index h have a fundamental influence on the spectral properties of the CPM signal. For the frequency impulse shown in Figure 4.37

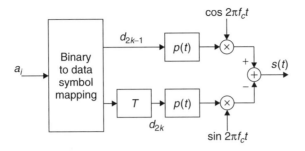

Figure 4.39 Linear MSK modulator

and for the modulation index $h = 1/2$ we obtain a special case of the FSK modulation, the so-called *Minimum Shift Keying* (MSK), i.e. the FSK with the minimum deviation that ensures orthogonality of the elementary signals. The scheme of the modulator is shown in Figure 4.39.

Contrary to FSK with values of the modulation index $h \neq 1/2$, MSK can be interpreted as a linear modulation for which the superposition rule with respect to the modulating signal holds true. Therefore the MSK-modulated signal can be described by formula (4.5). The baseband shaping filter impulse response $p(t)$ is shown in Figure 4.40a, and the impulse response of the similar filter $q(t)$ is shown in Figure 4.40b.

For MSK signals the baseband shaping filter $p(t)$ applied in the in-phase branch of the modulator is described by the formula

$$p(t) = \text{rect}\left(\frac{t}{2T}\right) \cos\left(\frac{\pi}{2T}t\right) \tag{4.98}$$

whereas the baseband shaping filter $q(t)$ used in the quadrature branch of the modulator has the impulse response

$$q(t) = p(t - T) = \text{rect}\left(\frac{t - T}{2T}\right) \sin\left(\frac{\pi}{2T}t\right) \tag{4.99}$$

Finally, the MSK-modulated signal can be expressed in the linear form using the following formula

$$x(t) = \sum_{k=0}^{n/2} \left[d_{2k-1} p(t - 2kT) \cos(2\pi f_c t) - d_{2k} p\big(t - (2k + 1)T\big) \sin(2\pi f_c t)\right] \tag{4.100}$$

Figure 4.40 presents the baseband shaping filter impulse responses and shows the example of the in-phase and quadrature baseband signal components that finally result in a particular phase function of time.

The constant value of the FSK, MSK or other signal envelope can be shown graphically on a complex plane if we plot the position of the points as a function of time, where the real coordinate is the amplitude of the in-phase component $x^I(t)$ and the imaginary

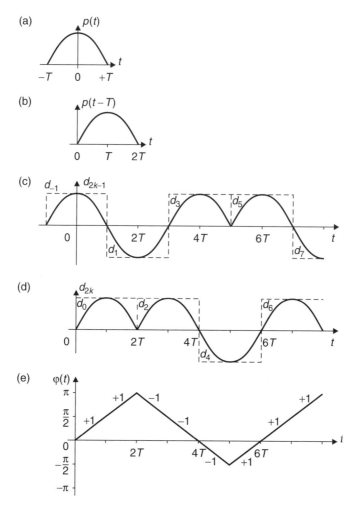

Figure 4.40 Impulse response of the in-phase filter (a) and quadrature filter (b). Example of the in-phase baseband signal (c), the quadrature baseband signal (d) and the resulting phase function of time (e) for the MSK signal

coordinate is the quadrature component $x^Q(t)$. From (4.93) we conclude that the envelope is given by the formula

$$r(t) = \sqrt{\left(x^I(t)\right)^2 + \left(x^Q(t)\right)^2} = r\sqrt{\cos^2\varphi(t) + \sin^2\varphi(t)} = r = \text{const} \qquad (4.101)$$

The constant envelope is independent of the form of variability of the phase function $\varphi(t)$. Figure 4.41 presents the plot of the envelope and phase functions of time for MSK modulation ($h = 0.5$). Let us note that in this case the angle $\varphi(t)$ changes by $\pm\pi/2$ during a single modulation period T. The phase at the end of this period depends on the current data symbol and also on the phase achieved at the end of the previous modulation period. Therefore we conclude from this plot that MSK modulation has memory.

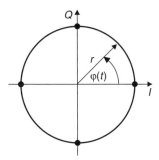

Figure 4.41 Envelope and phase for MSK signal shown on the in-phase and quadrature plane

Several frequency impulses were investigated in research on CPM. The influence of these impulses on the signal spectral properties and detection capabilities was also evaluated. For an appropriately designed modulation the sequence of elementary signals generated by the modulator for different data sequences should differ from each other as much as possible. We can achieve this feature by lengthening the frequency impulse response beyond the modulation period T. This leads to increased complexity of the receiver, as the appropriately modified detection algorithm (e.g. the Viterbi algorithm) has to be applied.

One of the most successful solutions in terms of the signal spectral properties is application of a frequency impulse of the form

$$g(t) = \frac{1}{\sqrt{2\pi}\sigma T} \exp\left(\frac{-t^2}{2\sigma^2 T^2}\right) * \text{rect}\left(\frac{t}{T}\right) \qquad (4.102)$$

where $*$ denotes convolution, and $\text{rect}(t/T)$ describes a rectangular pulse of unit height that lasts from $-1/2T$ to $1/2T$. Such a modulation is known as GMSK (*Gaussian Minimum Shift Keying*). This is a modulation with a minimum frequency shift ($h = 1/2$). Its frequency impulse is a Gaussian-filtered rectangular pulse. For GMSK modulation $\sigma = \sqrt{\ln 2}/(2\pi BT)$, where B is a 3-dB bandwidth of the Gaussian filter. The impulse $g(t)$ usually spans a few modulation periods so subsequent pulses that are the response to the succeeding data symbols interfere with each other. As already mentioned, the Viterbi algorithm is useful in such signal detection.

GMSK modulation is applied in mobile cellular GSM systems and some cordless telephony systems. Particularly good spectral properties of GMSK-modulated signals are the reason for GMSK application. These properties are expressed as a narrow mainlobe and low levels of spectral sidelobes of the modulated signal. A simple scheme of the GMSK modulator is shown in Figure 4.42.

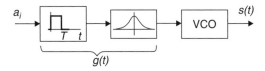

Figure 4.42 Block diagram of GMSK modulator

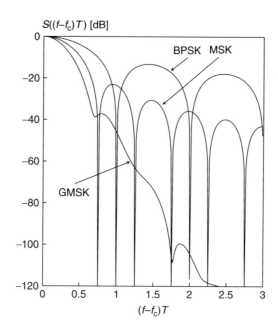

Figure 4.43   Power spectral density of GMSK in comparison with MSK and BPSK signals

The GMSK modulator operates as follows. First, the binary data stream is formed into the NRZ (*Non-Return to Zero*) bipolar sequence of pulses lasting T seconds each. The resulting sequence is given to the input of the lowpass filter of Gaussian characteristics (both in time and frequency domains), so the joint response to a single data symbol is given by formula (4.102). The effective duration of this impulse response is around $(4 \div 5)T$ for $BT = 0.3$. The filter output signal controls the voltage-controlled oscillator (VCO).

Figure 4.43 presents the power density spectrum of the signal on the output of the ideal GMSK modulator. The plot is drawn around the carrier frequency f_c. Power spectral densities for BPSK and MSK with the same modulation period T are shown for comparison. We observe an impressive improvement in the spectral properties of the GMSK signal compared with the other modulations. GMSK is currently one of the best modulations with a constant envelope with respect to its spectral efficiency.

As we have already mentioned, the increased receiver complexity and intersymbol interference, which is meant as overlapping of the modulator responses to subsequent data symbols, is the price paid for excellent spectral properties of GMSK signals. In practice, GSMK signals are demodulated using a synchronous receiver with detection of a whole sequence of data symbols (as it is typically applied in a GSM system). An alternative demodulator solution is a simplified asynchronous receiver that is based on a frequency discriminator (e.g. applied in DECT cordless telephony).

It was found (Laurent 1986) that GMSK, in particular for $BT = 0.3$ characteristic for GSM and DECT systems, can be effectively approximated by a linear modulation described by formulae (4.5) and (4.67) with the appropriately selected baseband shaping pulse $p(t)$. The optimum pulse $p(t)$ has a shape similar, but not identical, to the Gaussian

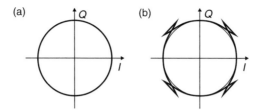

Figure 4.44 Envelopes of an ideal GMSK signal (a) and its linear approximation (b)

shape. For a linear modulator using the pulse $p(t)$ the signal envelope is not fully constant (see Figure 4.44).

Several other CPM have been investigated, however regular CPFSK, MSK and GMSK are the most popular so we have concentrated our attention on them.

4.11 Trellis-Coded Modulations

4.11.1 Description of Trellis-Coded Signals

Performance of data transmission over many communication channels is not satisfactory. Therefore, there have been attempts to find modulation and error correction codes that would ensure the required transmission quality. As we already know, application of an error correction code leads to the necessity for transmission of additional parity bits that allow the error pattern to be identified and corrected. In order to maintain an unchanged information data bit rate, we can take one of the following approaches:

- Increase the data symbol rate while preserving the number of constellation points in the applied modulation – this approach leads to spectrum broadening, which may not be acceptable for band-limited channels.
- Increase the number of constellation points in the modulation while keeping the data symbol rate constant – assuming constant mean signal power, this approach leads to an increase in the symbol error rate that has to be compensated for by the applied error correction code.

It turns out that the first method cannot be applied due to channel bandwidth limitations, whereas for transmission over band-limited channels the results of the second solution are at most moderate, leading to a substantial increase in receiver complexity. The reason for rather poor results in searching for a good error correction code was in treating modulation and error correction as separate processes.

In 1982 Ungerboeck published a paper (Ungerboeck 1982) in which he proposed to treat modulation and coding jointly. This approach made a breakthrough in digital communication systems of the 1980s. The criterion of the code and modulation optimization selected by Ungerboeck was the maximization of the minimum Euclidean distance between the sequences of constellation points that are allowed due to the coding process and are set through the mapping of the binary blocks received from the error correcting code encoder onto the signal constellation points. In this joint approach to coding and modulation mostly convolutional codes are applied. Convolutional codes that are combined

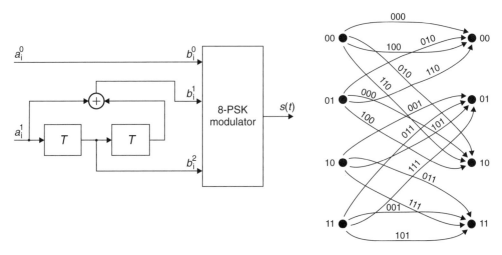

Figure 4.45   Trellis encoder with the parameters $k = 2$ and $n = 3$ and the associated trellis diagram

with appropriately selected modulations are often called *Trellis-Codes*, and the appropriate modulations are called *Trellis-Coded Modulations* (TCM). In general, the trellis encoder is a finite state machine stimulated by a sequence of k-bit blocks $(a_i^0, a_i^1, \ldots, a_i^{k-1})$ and generating a sequence of n-bit blocks $(b_i^0, b_i^1, \ldots, b_i^{n-1})$, where index i denotes subsequent modulation periods. The operation of this finite state machine is described by a trellis diagram. Figure 4.45 shows a four-state trellis encoder combined with an 8-PSK modulator and the trellis diagram associated with the encoder.

The number of encoder states results from the number of possible data combinations contained in the encoder memory cells. The trellis diagram presents all paths from each state in the ith moment to possible states in the $(i + 1)$st moment. Each path is associated with the appropriate binary encoder output block. For example, in Figure 4.45 the binary block 100 is associated with the path between the states $u_1 = 00$ and $u_1 = 00$ as well as between $u_2 = 01$ and $u_3 = 10$. Let us note that the trellis diagram shown in Figure 4.45 features parallel paths. They result from the uncoded bit a_i^0, which directly appears on the encoder output in position b_i^0. In general, the number of parallel paths between states is equal to 2^{n-k}. Let us denote the trellis encoder output signal by s_j. Let this signal be mapped to the binary block $(b_i^0, b_i^1, \ldots, b_i^{n-1})$ that represents the number j in the binary form. For the trellis diagram shown in Figure 4.45 there exist the signals $s_0, \ldots, s_7$, which correspond to each of eight phases of the 8-PSK modulation.

The task of the decoder is to select the best sequence among all the sequences that can appear due to coding, i.e. the one that is the closest in the Euclidean distance sense to the received sequence of samples. As we remember, this criterion of the signal selection is called the *Maximum Likelihood* (ML) criterion. In the decoding process errors can arise. In the so-called *error event*, which occurs in the decoder, the decoded path diverges from the correct one on the trellis diagram, and merges with it again after some modulation periods (decoder timing instants). Figure 4.46 presents two out of many possible error events for the trellis code shown in Figure 4.45 on the assumption that the correct path on the trellis is associated with the signal sequence $s_0, s_0, \ldots, s_0$, which in turn represents

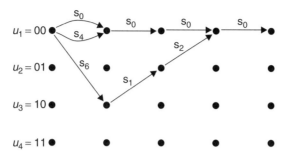

Figure 4.46 Example of error events on the trellis diagram for the code presented in Figure 4.45

the blocks 000 on the encoder output. Let us note that if parallel paths on the trellis diagram exist then the error event may be as short as a single modulation period. The example of such an event is shown in Figure 4.46. The error event (s_6, s_1, s_2) shown in this figure is one of eight possible signal sequences associated with the marked path between states u_1, u_3, u_2, u_1. For high values of the signal-to-noise ratio, error events that have the lowest Euclidean distance d_{min} to the correct trellis path dominate, therefore they practically determine the achievable error rate. The number of error events characterized by the lowest Euclidean distance influences the error rate too. For that reason the mapping of binary blocks produced by the convolutional encoder onto the elementary signals or constellation points of the modulator has a crucial meaning for the overall trellis code performance.

Ungerboeck formulated the following rules for the mapping of binary blocks onto elementary signals:

1. Signals associated with all parallel paths on the trellis diagram should be characterized by the highest Euclidean distance between them on the constellation diagram.
2. Signals associated with all paths diverging from or merging to a given state should be characterized by the second highest Euclidean distance between them.

The procedure of assigning binary blocks to elementary signals is called *set partitioning* and will be explained by the example of a trellis-coded signal based on 8-PSK shown in Figure 4.47, i.e. based on the modulation presented in Figure 4.45.

In the first step of 8-PSK set partitioning, when the minimum Euclidean distance between 8-PSK constellation points is $d_0 = \sqrt{2 - \sqrt{2}}r$, the constellation points are divided into two four-point constellations for which the minimum distance is $d_1 = \sqrt{2}r$. In the second step each of these constellations is further divided into two two-point constellations according to the same rule. The minimum distance for them is $d_2 = 2r$. As a result we obtain four constellations, denoted by symbols A, B, C and D, that are the subsets of the primary constellation. The subsets can be assigned the following dibits

$$A \longleftarrow 00, \quad B \longleftarrow 10, \quad C \longleftarrow 01, \quad D \longleftarrow 11$$

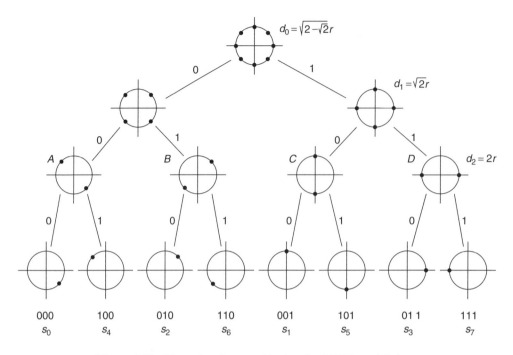

$$d_0 = \sqrt{2-\sqrt{2}}\,r$$

$$d_1 = \sqrt{2}\,r$$

$$d_2 = 2r$$

000	100	010	110	001	101	011	111
s_0	s_4	s_2	s_6	s_1	s_5	s_3	s_7

Figure 4.47   Example of set partitioning for 8-PSK modulation

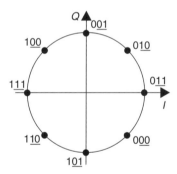

Figure 4.48   Binary block assignment to elementary signals (constellation points) of 8-PSK modulation according to the set partitioning rule

In the last set partitioning step, single constellation points are selected from each subset A, B, C and D. The final result of the binary block assignment to 8-PSK constellation points is shown in Figure 4.48.

According to the set partitioning rule, parallel transitions on the trellis diagram are characterized by the signal constellation points related to the binary blocks $(0A, 1A)$, $(0B, 1B)$, $(0C, 1C)$ or $(0D, 1D)$. Taking into account the second rule of set partitioning we conclude that the signals associated with the paths diverging from the same state or merging into the same state should belong to the pairs of constellation subsets (A, B) or (C, D). The trellis encoder drawn in Figure 4.45 is one of the encoders found by Ungerboeck that fulfills the above stated set partitioning rules.

The minimum distance between constellation points for pure QPSK modulation is $d_{min} = \sqrt{2}r$. Therefore, in order to obtain the same performance for high signal-to-noise ratios, as in the case of 8-PSK trellis-coded modulation, the signal-to-noise ratio value for uncoded QPSK must be

$$G = \frac{d^2_{TCM, \, min}}{d^2_{QPSK, \, min}} = 2 \qquad (4.103)$$

times higher. We find that the *asymptotic gain* G, which is achieved owing to trellis coding, in our case is equal to 3 dB. Let us stress that this performance improvement has been achieved without bandwidth expansion.

The same set partitioning rule can be applied for 16-QAM. It is shown in Figure 4.49 down to the stage in which the subsets A, B, C, D, E, F, G and H are determined. The reader is kindly asked to analyze this figure by himself/herself.

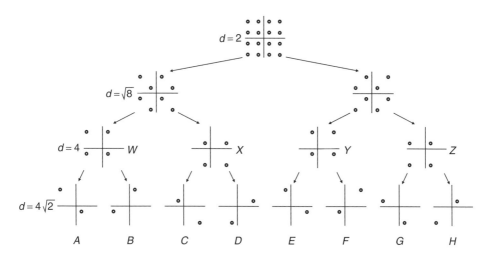

Figure 4.49   Set partitioning for 16-QAM

The joint modulation and coding approach has been applied e.g. in telephone modems (ITU-T V.32, V.33 and V.34 Recommendations). Figure 4.50 presents the scheme of the trellis encoder and 32-QAM signal constellation associated with it, designed for duplex transmission at the data rate of 9600 bit/s. The additional advantage of the presented trellis code and binary block mapping to the constellation points is modulo-90° phase invariance of the signal constellation. Another system in which TCM is applied is SHDSL – Single-pair High-speed Digital Subscriber Line. In this system TCM coding is associated with PAM, resulting in serial transmission at rates of 192–2312 kbit/s (see ITU-T G.991.2 2003).

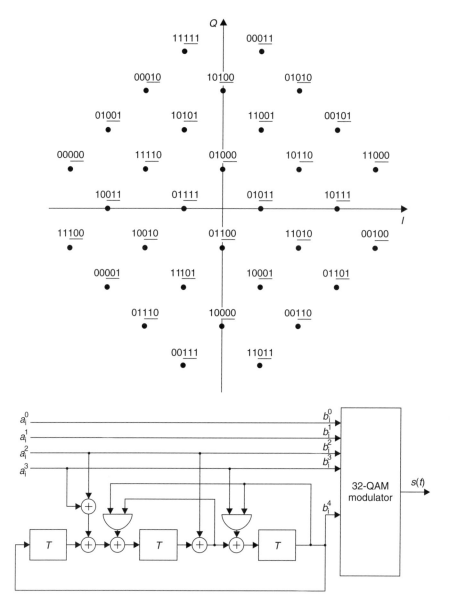

Figure 4.50 Trellis encoder and signal constellation for 32-TCM according to ITU-T V.32 Recommendation (1993)

In the literature we can find tables of useful trellis codes (Ungerboeck 1987a, 1987b) with asymptotic gains, achievable with the associated PAM, PSK and QAM. Complexity of the tabularized codes expressed as the number of encoder states ranges between 4 and 512. The achievable coding gains reach 6 dB, which is an impressive achievement. However, in order to reach it, a relatively computationally demanding decoding algorithm of the trellis-coded signals is required, such as the Viterbi algorithm.

4.11.2 Decoding of the Trellis-Coded Signals

Let us consider the operation of the receiver for trellis-coded signals with the example of the signal constellation and trellis code shown in Figure 4.51. The trellis diagram and binary block mapping onto signal constellation points have been repeated for convenience again and have been supplemented with notation of the signal subsets A, B, C and D containing pairs of signals with opposite polarization.

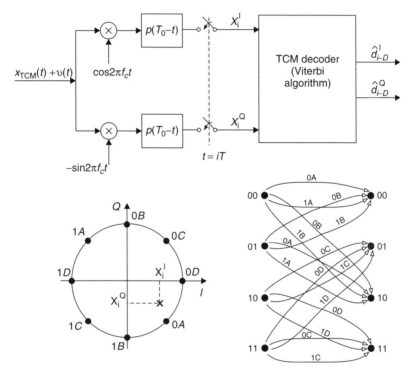

Figure 4.51 Scheme of the synchronous receiver for trellis-coded signals corrupted by additive noise and the example of the signal mapping and the trellis diagram of the applied trellis code

Let us assume that the trellis-coded 8-PSK signal corrupted by the additive white Gaussian noise of power density $N_0/2$ reaches the receiver. First, the received signal is synchronously demodulated: it is multiplied by the reference in-phase and quadrature carrier signals and the results are filtered by a pair of lowpass filters that are at the same time matched to the baseband signal pulses applied in the modulator. A typical receive filter has a square root raised cosine characteristic and it is matched to the transmit filter of the same shape. Therefore, the additive noise power density spectrum on the outputs of both receive filters has a raised cosine shape, and the noise autocorrelation function has periodical zeros in the modulation period intervals (except for the zero argument). Therefore the noise samples picked up in both in-phase and quadrature branches at the modulation period intervals are uncorrelated. Because the noise samples on the filter outputs are Gaussian, they are also statistically independent. This important conclusion will be applied in the reception procedure.

As we know from Chapter 1, in making the decisions the decoder very often searches for a data symbol sequence $\mathbf{d}_n$, which ensures a maximum conditional probability density function for the sequence of the received samples $\mathbf{x}_n$ when the data sequence $\mathbf{d}_n$ has been transmitted. This means that the decoder selects the data sequence $\widehat{\mathbf{d}}_n$ for which the following expression holds

$$\widehat{\mathbf{d}}_n = \arg\max_{\mathbf{d}_n} p(\mathbf{x}_n | \mathbf{d}_n) \qquad (4.104)$$

where $\mathbf{d}_n = (d_1, d_2, \ldots, d_n)$ and $\mathbf{x}_n = (x_1, x_2, \ldots, x_n)$. Note that $d_i = d_i^I + j d_i^Q$ is a complex data symbol that describes coordinates of the signal constellation point transmitted in the ith modulation period, while $x_i = x_i^I + j x_i^Q$ is a complex representation of the samples taken in the ith moment from the outputs of the matched filters of the synchronous detector. Let us note that the complex sample of the received signal x_i is a sum of the data symbol d_i and the complex sample of the additive noise $n_i = n_i^I + j n_i^Q$. In-phase and quadrature noise components are statistically independent not only along the time axis but also with respect to each other. Then, knowing that noise samples are Gaussian distributed and statistically independent, we can write the conditional probability density function $p(\mathbf{x}_n | \mathbf{d}_n)$ in the following form

$$p(\mathbf{x}_n | \mathbf{d}_n) = \prod_{i=1}^{n} p(x_i | d_i) = \prod_{i=1}^{n} p(x_i^I | d_i^I) p(x_i^Q | d_i^Q)$$

$$= \prod_{i=1}^{n} \frac{1}{\sqrt{2\pi}\sigma} \exp\left[-\frac{(x_i^I - d_i^I)^2}{2\sigma^2} \right] \frac{1}{\sqrt{2\pi}\sigma} \exp\left[-\frac{(x_i^Q - d_i^Q)^2}{2\sigma^2} \right]$$

$$= \left(\frac{1}{2\pi\sigma^2} \right)^n \exp\left(-\frac{1}{2\sigma^2} \sum_{i=1}^{n} |x_i - d_i|^2 \right) \qquad (4.105)$$

Searching for the maximum likelihood data sequence $\mathbf{d}_n$ will be much easier if we calculate the natural logarithm of both sides of formula (4.105), i.e.

$$\ln p(\mathbf{x}_n | \mathbf{d}_n) = n \ln \frac{1}{2\pi\sigma^2} - \frac{1}{2\sigma^2} \sum_{i=1}^{n} |x_i - d_i|^2 \qquad (4.106)$$

Applying criterion (4.104), we should select that data sequence $\widehat{\mathbf{d}}_n$ for which the Euclidean distance of the data sequence constellation points to all points characterizing the received in-phase and quadrature samples on the (I, Q) plane is minimized. Thus, criterion (4.104) reduces to the form

$$\widehat{\mathbf{d}}_n = \arg\min_{\mathbf{d}_n} \sum_{i=1}^{n} |x_i - d_i|^2 \qquad (4.107)$$

As we remember from our considerations on convolutional code decoding, searching for the maximum likelihood data sequence according to criterion (4.107) is effectively realized by the Viterbi algorithm. Its operation differs from soft-decision convolutional code decoding in the way in which path metrics are calculated. First of all, the calculations are performed on complex samples. Typically, the trellis diagram has parallel transitions. These transitions imply that when the sample $x_i = x_i^I + jx_i^Q$ is received, the closest (in the Euclidean sense) constellation points representing each subset A, B, C and D are selected. For example, for sample x_i in Figure 4.51 the following signals from each subset are selected: $0A$, $1B$, $0C$ and $0D$. These signals take part in determination of the minimum cost of reaching each trellis state in the ith moment. This is equivalent to finding the best route on the trellis diagram, the so-called *survivor*, to each state. Let us denote the survivor costs for the trellis states 00, 01, 10 and 11 in the ith moments as $S_{0,i}$, $S_{1,i}$, $S_{2,i}$ and $S_{3,i}$, respectively. Calculation of the cost of reaching each state in the $(i + 1)$st moment is performed by finding for each state m ($m = 0, 1, 2, 3$) the minimum

$$S_{m,(i+1)} = \min\left[\left(S_{k,i} + |x_i - d_{i,X}|^2\right), \left(S_{j,i} + |x_i - d_{i,Y}|^2\right)\right] \tag{4.108}$$

where at the given moment the mth state can be reached from the kth or jth state. Transitions from the kth to mth state and from the jth to mth state are associated with the choice of the data symbols in the ith moment from X and Y subsets, respectively ($X, Y \in \{A, B, C, D\}$). For example, if we analyze the trellis diagram in Figure 4.51 we find that reaching the first state in the given moment is possible from the first or second state from the previous moment and the data signals associated with these paths belong to the subsets $X = A$ and $Y = B$, respectively.

As for the decoding of a convolutional code using the Viterbi algorithm, it has been observed that in the trellis code decoding the survivors reaching each state merge into a single path on the trellis diagram with a probability close to unity about $(3 \div 5)L$ timing instants back, where L is the trellis code constraint length. Therefore, it is possible to issue final decisions upon the transmitted data symbols d_{i-D} with the delay $D = (3 \div 5)L$. This is also the size of memory needed by the processor implementing trellis decoding using the Viterbi algorithm. In such memory not only transitions between states but also the signals associated with them have to be remembered. Let us recall that these signals have to be stored because more than one signal is associated with each transition between states due to parallel transitions on the trellis diagram.

The above description is related to the reception of trellis-coded signals in the presence of additive noise. The situation becomes more complicated if the channel distorts the spectrum of the transmitted signal. Thus, the Viterbi TCM decoder is typically preceded by the block that mainly corrects the channel characteristics. This leads to the situation in which the signal samples on the input of the TCM decoder are TCM signal samples corrupted by the additive noise, thus our system model considered above remains valid.

4.12 Multitone Modulations

So far we have considered digital modulations of a sinusoidal carrier where the binary signal is transmitted bit by bit (for binary modulations) or binary block after binary block (for multilevel modulations). However, most real communication channels substantially differ from the channel model in which additive white Gaussian noise is the only impairment. Real channels very often feature strict spectral limitations and introduce amplitude and phase distortions. When data symbol rate (measured in symbols per second) is getting closer and closer to the channnel bandwidth expressed in Hz, the phenomenon of inter-symbol interference is getting more and more visible. The length of the channel impulse response increases to such an extent that the channel responses to particular data symbols partially overlap. As a result, serious signal distortions occur and signal constellations on the channel output may completely lose their shape. We can also observe the eye pattern with the eye partially or even fully closed. This is an indication that the decision process may yield unreliable results.

There are two solutions to the problem of fast data transmission over band-limited channels:

- application of a channel equalizer that compensates for the channel amplitude and phase distortions; the equalizer often works in an adaptive mode, i.e., it automatically adjusts its parameters to the channel characteristics, providing signal samples with negligible intersymbol interference at the sampling moments;
- division of the data stream to be transmitted into a set of substreams that are transmitted in parallel on many subcarriers; each subcarrier carries data symbols at such a low data rate that intersymbol interference practically does not occur.

The first solution will be described in a separate chapter. The second solution leads to the introduction of multitone modulations and we will consider it in this section.

The idea of multitone modulations first appeared in the late 1950s. Multitone modu-lations were successfully applied for data transmission rates above 2400 bit/s in military data modems operating on high frequency (HF) channels of the acoustic bandwidth. Achieving such a data rate was a real success considering the early stages of electronics development at that time. Multitone modulation was rediscovered in the late 1980s when the work on digital radio broadcasting began. Currently multitone modulation is the basis of *Digital Audio Broadcasting* (DAB), the terrestrial and handheld segment of *Digital Video Broadcasting* (DVB-T and DVB-H), fast digital transmission in digital subscriber loops (*Asymmetric Digital Subscriber Line* – ADSL; *Very High Bitrate Digital Subscriber Line* – VDSL; and their modifications), fast wireless LAN (WLAN) links, and broadband wireless access to digital networks known as WiMAX.

Let us assume that the binary data stream is equal to R bit/s. If a single carrier with 16-QAM were applied, the symbol data rate would be $R/4$ symb/s. Instead, let us apply N subcarriers that are appropriately located in the channel band. Each of them should carry the substream of R/N bit/s. If the 4-PSK modulation is applied on each subcarrier, the symbol data rate on each of them is equal to $R/(2N)$ symb/s.

Let us consider data transmission at the rate of 2400 bit/s over an HF channel. This is an old traditional application of multitone modulation[2]. Application of a 16-QAM single-carrier transmission is relatively difficult due to nonlinear distortions introduced by the power amplifier, time variablity of the channel characteristics and intersymbol inter-ference introduced by the channel. If the 4-PSK (or QPSK) modulation is considered, the symbol data rate is 1200 symb/s. The modulation period is then equal to 0.8333 ms. Unfortunately, due to the property of the HF channel known as multipath propagation,

[2] Current military standards describe a single-carrier transmission with QAM for which an advanced receive algorithm has to be applied.

intersymbol interference arises because the channel impulse response can span the time period up to 1.5 and sometimes even 5 ms. Additionally, channel characteristics can vary so quickly in comparison with the symbol data rate that tracking them requires a complicated channel estimation algorithm. A simple solution is the application of multitone transmission. Let us assume $N = 24$ QPSK-modulated subcarriers in parallel. Each subcarrier carries the substream of 100 bit/s, which is equivalent to 50 symb/s. The modulation period on each subcarrier lasts for 20 ms. Thus, in the worst case intersymbol interference distorts 1/4 of the modulation period. Typically, however, it does not exceed 7.5% of the modulation period. The subcarrier frequencies should be selected in such a way that the subcarrier signals are mutually orthogonal in the part of the modulation period in which the influence of multipath propagation and intersymbol interference is negligible. Figure 4.52 presents the multipath channel response to a rectangular pulse of duration equal to T. In its initial part we observe some ripples that appear when the transmitted signal components arrive at the receiver at intervals resulting from multipath propagation. After a time interval lasting for a duration of the channel impulse response, the channel response to the data pulse achieves its steady state. From this moment until the end of the modulation period we can consider the signals transmitted on different subcarriers as mutually orthogonal if their subcarrier frequecies are properly selected. The modulation period is divided into two parts: the so-called *guard time* T_g, in which the channel response to the data symbol is still unstable; and the *orthogonality period* T_{ort}. Typically T_g does not exceed 20% of the modulation period.

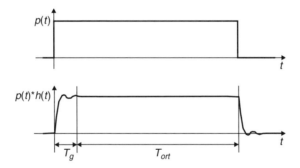

Figure 4.52 Example of the multipath channel response to the rectangular baseband equivalent signal with the denoted period of the guard time T_g and the orthogonality period T_{ort}

Two sinusoidal signals with arbitrary initial phases and amplitudes are mutually orthogonal if their frequencies are different multiples of the reciprocal of the orthogonality period, i.e. if

$$f_{c,i} = \frac{i}{T_{ort}}, \quad f_{c,j} = \frac{j}{T_{ort}}, \quad i \neq j \tag{4.109}$$

Then we have

$$\int_0^{T_{ort}} \cos\left(2\pi \frac{i}{T_{ort}}t + \varphi_i\right) \cos\left(2\pi \frac{j}{T_{ort}}t + \varphi_j\right) dt = 0 \quad \text{for} \quad i \neq j \tag{4.110}$$

It is easy to verify (4.110) using elementary trigonometric formulas. Similarly we have

$$\int_0^{T_{ort}} \cos\left(2\pi \frac{i}{T_{ort}}t + \varphi_i\right) \sin\left(2\pi \frac{j}{T_{ort}}t + \varphi_j\right) dt = 0 \quad \text{for any } i, j \tag{4.111}$$

Knowing, that (4.110) and (4.111) hold true, we can state that the following signals $s_i(t)$ and $s_j(t)$ determined in the time interval $nT < t < (n+1)T$ are mutually orthogonal in the orthogonality period T_{ort}:

$$s_i(t) = a_{i,n} \cos\left(2\pi \frac{i}{T_{ort}}t\right) - b_{i,n} \sin\left(2\pi \frac{i}{T_{ort}}t\right) \quad \text{and}$$

$$s_j(t) = a_{j,n} \cos\left(2\pi \frac{j}{T_{ort}}t\right) - b_{j,n} \sin\left(2\pi \frac{j}{T_{ort}}t\right) \tag{4.112}$$

where $a_{i,n}$, $b_{i,n}$ and $a_{j,n}$, $b_{j,n}$ are the data symbols modulating the ith and jth subcarriers in the nth modulation period, respectively. Application of the data symbols $a_{i,n}$, $b_{i,n}$ and $a_{j,n}$, $b_{j,n}$ implies that both PSK and QAM can be applied on particular subcarriers.

Mutual orthogonality of the signals using different subcarrier frequencies is necessary for the correct demodulation process. We will explain it below. Due to mutual orthogonality of signals on different subcarriers, this type of modulation is often denoted as OFDM (*Orthogonal Frequency Division Multiplexing*).

Let the multitone signal be described in the time period $[nT, (n+1)T]$ by the formula

$$x(t) = \sum_{i=1}^{N} \left[a_{i,n} \cos\left(2\pi \frac{i}{T_{ort}}t\right) - b_{i,n} \sin\left(2\pi \frac{i}{T_{ort}}t\right) \right] \tag{4.113}$$

Considering the ith subcarrier of frequency i/T_{ort} we see that it is a QAM signal with the baseband pulse shaping filter $p(t)$ in the form of a rectangular pulse of the unit amplitude. The shape of the power density spectrum of such a signal is a square of the sinc function. Figure 4.53 presents an example of a power density spectrum of signal $x(t)$ centered around a given carrier frequency. The spectrum is plotted in the decibel scale with respect to this center frequency. This is the power density spectrum of a terrestrial digital video broadcasting (DVB-T) signal, in which 1705 or 6817 OFDM subcarriers are applied. If 1705 subcarriers are used, the system is often referred to as a $2k$ system, whereas if 6817 subcarriers are applied, the DVB-T system works in the so-called $8k$ mode. Let us note that the more subcarriers in a given bandwidth, the faster the power density spectrum falls outside the mainlobe.

We also see that the power density spectrum is almost flat in the frequency range in which subcarrier frequencies are generated. Power density spectra of neighboring subcarrier signals partially overlap. Thus, the following problem arises: how to extract the signals of the particular jth subcarrier in order to enable the detection of information symbols $a_{j,n}$ and $b_{j,n}$. Mutual orthogonality of subcarrier signals turns out to be crucial for solving this problem. In order to extract the data symbols on a particular subcarrier it

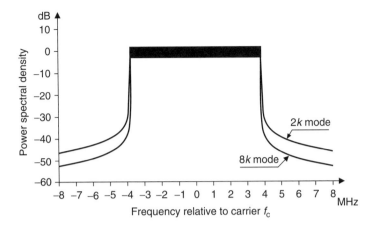

Figure 4.53 Power spectral density of the OFDM DVB-T signal $x(t)$ for $2k$ and $8k$ subcarriers © European Telecommunications Standards Institute 2004. Further use, modification, redistribution is strictly prohibited. ETSI standards are available from http://pda.etsi.org/pda

is sufficient to correlate the composite signal $x(t)$ with the selected subcarrier signal. We have

$$
U_{j,n}^{I} = \int_{T_g}^{T} \left\{ \sum_{i=1}^{N} \left[a_{i,n} \cos\left(2\pi \frac{i}{T_{ort}} t\right) - b_{i,n} \sin\left(2\pi \frac{i}{T_{ort}} t\right) \right] \right\} \cos\left(2\pi \frac{j}{T_{ort}} t + \theta_j\right) dt
$$

$$
= \frac{a_{j,n} T_{ort}}{2} \cos \theta_j \tag{4.114}
$$

and

$$
U_{j,n}^{Q} = \int_{T_g}^{T} \left\{ \sum_{i=1}^{N} \left[a_{i,n} \cos\left(2\pi \frac{i}{T_{ort}} t\right) - b_{i,n} \sin\left(2\pi \frac{i}{T_{ort}} t\right) \right] \right\} \sin\left(2\pi \frac{j}{T_{ort}} t + \theta_j\right) dt
$$

$$
= \frac{b_{j,n} T_{ort}}{2} \cos \theta_j \tag{4.115}
$$

where θ_j denotes the phase difference of the jth reference subcarrier in comparison with the phase of the jth received subcarrier. If synchronous reception on each subcarrier is applied, then the phase $\theta_j = 0$, or, what is more often, before making a decision on data symbols $(a_{j,n}, b_{j,n})$ on the basis of the correlators outputs $(U_{j,n}^{I}, U_{j,n}^{Q})$, the phase shift is estimated and subsequently cancelled. However, if the channel introduces various phase shifts on different subcarriers or the phase shifts vary in time so quickly that they cannot be tracked by the receiver, DPSK modulation on each subcarrier can be used and the correlation reception similar to that shown in Figure 4.29 can be applied. We can imagine a few correlators operating in parallel in the receiver for different subcarriers.

The same number of modulators is needed in the transmitter. However, if the number of subcarriers grows, implementation of such a receiver becomes unreasonable. Fortunately, achievements in digital signal processing are the remedy for the increased complexity of both the transmitter and the receiver. Let us note that if the OFDM signal is placed around a certain carrier frequency f_c, formula (4.113) can be presented for the time period $[nT, (n + 1)T]$ in the form

$$x(t) = \text{Re}\left[\exp(j2\pi f_c t) \sum_{i=-N/2}^{N/2-1} (a_{i,n} + jb_{i,n}) \exp\left(j2\pi \frac{i}{T_{ort}} t \right) \right] \quad (4.116)$$

or, after renumerating the data symbols and subcarriers from 0 to $N - 1$

$$x(t) = \text{Re}\left[\exp\left(j2\pi (f_c - \frac{N/2}{T_{ort}}) t \right) \sum_{i=0}^{N-1} (a_{i,n} + jb_{i,n}) \exp\left(j2\pi \frac{i}{T_{ort}} t \right) \right]$$

The first component in (4.116) symbolizes the frequency shift to the carrier frequency f_c and it can be omitted in further analysis if the baseband equivalent OFDM signal is considered. Denoting the data symbols in the baseband equivalent part $x_B(t)$ of signal $x(t)$ as $d_{i,n} = a_{i,n} + jb_{i,n}$ we obtain

$$x_B(t) = \text{Re}\left[\sum_{i=0}^{N-1} d_{i,n} \exp\left(j2\pi \frac{i}{T_{ort}} t \right) \right] \quad \text{for } t \in [nT, (n + 1)T] \quad (4.117)$$

Let us calculate the samples of the baseband equivalent signal $x_B(t)$ at the time instants $t = kT_{ort}/N$. We obtain

$$x_B\left(\frac{k}{N} T_{ort} \right) = \text{Re}\left[\sum_{i=0}^{N-1} X_B(i) \exp\left(j2\pi \frac{ik}{N} \right) \right] \quad (4.118)$$

where $X_B(i) = d_{i,n}$ for $k \in [0, N - 1]$. Inspection of formula (4.118) allows us to note that the samples of signal $x_B(t)$ can be received as a real part of the discrete inverse Fourier transform of the samples in the frequency domain, which are given as $X_B(i) = d_{i,n}$. We can determine the sample block of signal $x_B(t)$ effectively by implementing formula (4.118) with the Inverse Fast Fourier Transform (IFFT). Similarly, we can show that discrete correlation with respect to the in-phase and quadrature subcarriers (see Figure 4.29) performed in parallel for all subcarriers can be implemented using the Fast Fourier Transform (FFT). According to this transformation, a signal sample on frequency i/T_{ort} can be derived from the formula [for comparison, see expressions (4.114) and (4.115)]

$$X_B(i) = \sum_{k=1}^{N} x_B\left(\frac{kT_{ort}}{N} \right) \exp\left(-j2\pi \frac{ik}{N} \right) \quad (4.119)$$

The real part of the right-hand side of (4.119) is a discrete implementation of the correlator with respect to the cosinusoidal reference signal of frequency i/T_{ort}, whereas its imaginary part is a discrete implementation of the correlation of signal $x_B(t)$ with respect to the sinusoidal reference signal of the same frequency. Performing these calculations in parallel with the use of the FFT results in a substantial reduction of the number of necessary calculations. Let us note that the samples generated using IFFT/FFT span the orthogonality period only. In order to protect the signal against the influence of the channel impulse response resulting from multipath propagation, the transmitter applies the so-called *cyclic prefix*. The appropriately selected signal samples fill the whole guard time T_g. These samples are copies of the signal from the last part of the orthogonality period. In the receiver the cyclic prefix is removed before correlation starts.

The final, simplified scheme of the discrete transmitter and receiver for OFDM signals is shown in Figure 4.54. Let us note that prior to the decisions made by the receiver on subcarrier symbols, the FFT output samples representing the discrete correlators outputs and denoted by $Y(i)$, $(i = 0 \ldots, N-1)$ are multiplied by complex coefficients $C(i)$. These coefficients correct the received correlator samples by compensating for the attenuation and phase shifts that are introduced by the transmission channel on each subcarrier. Very often these coefficients are set adaptively.

Let us consider how the shift of the correlation period influences the signals on the correlator outputs when the cyclic prefix is applied. Let us neglect the impact of additive

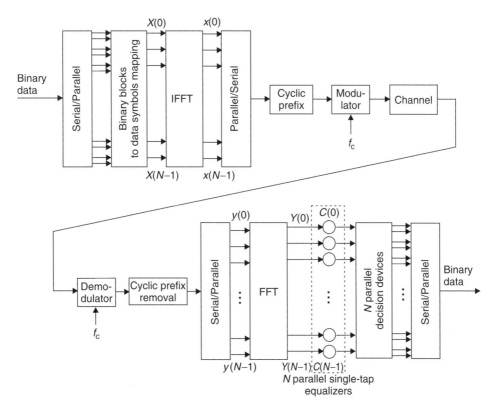

Figure 4.54   Block diagram of the OFDM transmitter and receiver in which IFFT/FFT algorithms are applied

noise on the correlators output signals for a while. Let the samples acquired within the modulation period have the form

$$\underbrace{x_B(N-L), x_B(N-L+1), \ldots, x_B(N-1)}_{\text{cyclic prefix}}, x_B(0), x_B(1), \ldots, x_B(N-1) \qquad (4.120)$$

where $x_B(k) = x_B(kT_{\text{ort}}/N)$. Let us note that the first L samples constitute a cyclic prefix and they are the copies of the last L samples from the modulation period. In the discrete correlation process formula (4.119) is applied to the samples $x_B(0), x_B(1), \ldots, x_B(N-1)$. Let us suppose that due to the synchronization block operation the samples $x_B(N-1), x_B(0), x_B(1), \ldots, x_B(N-2)$ are applied in the correlation process. Let us calculate the result of the correlation

$$X'_B(i) = x_B(N-1) + \sum_{k=1}^{N-1} x_B(k-1) \exp\left(j2\pi \frac{ki}{N}\right)$$

$$= x_B(N-1) + \sum_{l=0}^{N-2} x_B(l) \exp\left[j2\pi \frac{(l+1)i}{N}\right]$$

$$= x_B(N-1) \exp\left(j2\pi \frac{i}{N}\right) \exp\left[j2\pi \frac{i(N-1)}{N}\right]$$

$$+ \exp\left(j2\pi \frac{i}{N}\right) \sum_{l=0}^{N-2} x_B(l) \exp\left(j2\pi \frac{li}{N}\right)$$

$$= \exp\left(j2\pi \frac{i}{N}\right) \left\{ x_B(N-1) \exp\left[j2\pi \frac{i(N-1)}{N}\right] \right.$$

$$\left. + \sum_{l=0}^{N-2} x_B(l) \exp\left(j2\pi \frac{li}{N}\right) \right\} = X_B(i) \exp\left(j2\pi \frac{i}{N}\right) \qquad (4.121)$$

As we see, the ith sample of the spectrum $X'_B(i)$ differs from the correlation result based on the sample set $x_B(0), x_B(1), \ldots, x_B(N-1)$ resulting in the frequency domain sample $X_B(i)$ only in the phase rotation by the angle $2\pi i/N$. However, if the cyclic prefix were not used and the guard time were filled with zeros, the correlation realized on the sample sequence $0, x_B(0), x_B(1), \ldots, x_B(N-2)$ would result in the loss of mutual orthogonality of particular subcarrier signals. Due to the zero value of the first sample applied in the correlation process, the correlation is practically performed in the time period that is shorter than the orthogonality period.

The cyclic prefix plays another important role. It can be applied in maintaining OFDM symbol timing synchronization, i.e. in determination of the start of a correlation period, or finding the possible frequency offset between the received signal and the reference signal used in the receiver. If there were no additive noise on the input of OFDM receiver and the channel were not distorting the OFDM signal, then the cyclic prefix and the last L received signal samples used in the correlation period would be identical. This fact can be used for detection of the cyclic prefix in the received signal by performing correlation of the received samples in the window of length equal to L samples with the samples that are distant by N samples. At the moment in which the prefix samples are exactly correlated with the end part of the samples from the modulation period, the correlator output is maximum. Appearance of noise and distortions due to the echoes from the previous OFDM symbol decreases the level of this maximum. Despite this fact, this method is applied in maintaining the timing synchronization in OFDM transmission. Thus, the cyclic prefix length should significantly exceed the length of the channel impulse response. If the carrier frequencies of the received and reference signals differ by Δf, then the phase shift between the signal samples distant by N samples accumulates to the value of $2\pi \Delta f N T_s$, where T_s is the sampling period. This observation can be applied in determination of the frequency shift by finding the phase argument of the maximum of the correlation function between signals that are distant by the period of $N T_s$ seconds (N samples). The block scheme of detection of the beginning of the orthogonality period and determination of the frequency offset is shown in Figure 4.55.

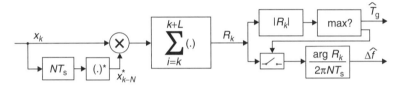

Figure 4.55 Scheme of timing synchronization and frequency offset detection based on the correlation of the cyclic prefix with the end part of the modulation period sample block

A cyclic prefix usually lasts no longer than 20–25% of the OFDM modulation period. Thus, treating it as a base for synchronization acquisition would be unreliable. Alternatively, this process would be very slow. In order to make the synchronization process fast and reliable the OFDM transmission is often organized in frames. The first few OFDM symbols in a frame are specially selected and known to the receiver. Thus, the full length of these symbols can be applied for timing and carrier synchronization purposes.

OFDM modulation allows for highly flexible use of transmission properties of a channel. If a feedback channel from the receiver to the transmitter can be established, then the results of the channel transfer function measurements performed in the receiver can be transmitted back to the transmitter. As a result, the channel measurements can be applied by the transmitter in the optimization of the OFDM transmission link. The measure of transmission quality on a particular OFDM subcarrier is the magnitude of this subcarrier or, more generally, the signal-to-noise ratio measured on it. Thus, for sufficiently high signal-to-noise ratio the required error probability can be achieved for a sufficiently high modulation level. If the amplitude of a given subcarrier is lower, a lower level modulation should be applied on it. In conclusion we can state that the transmission rate, which is controlled through selection of the modulation level, can be modified on the basis of information on channel properties received through the feedback channel.

4.13 Case Study: OFDM Transmission in DVB-T System

Let us illustrate the problems that have been considered in this chapter with an example of OFDM transmission in the DVB-T system. This will allow us to confront our theoretical knowledge with some important details of a real system. As the topic of our current chapter is modulation, we will focus on modulations applied in DVB-T and on some issues related to them.

The ETSI Standard EN 300 744 (ETSI 2004) describes the framing structure, channel coding and modulations applied in the DVB-T system. As we have mentioned earlier, OFDM has been selected as the main modulation method. Two modes of transmission are applicable in DVB-T systems, based on $2k$ and $8k$ IFFT/FFT processing. In the $2k$ mode $K = 1705$ subcarriers are used, whereas in $8k$ mode $K = 6817$ subcarriers are applied for information transfer and link maintenance. Besides the $2k$ and $8k$ modes applied in DVB-T, there is also the $4k$ mode applicable in the mobile DVB-H system for handheld digital TV receivers that has been described in the same standard. In our analysis we will concentrate on DVB-T only.

Digital transmission in the DVB-T system is organized in OFDM frames. Each frame consists of 68 OFDM symbols. Four frames constitute a superframe. The symbols in the OFDM frame are numbered from 0 to 67. The OFDM symbols carry data and reference information. Out of $K = 1705$ or $K = 6817$ subcarriers applied in $2k$ or $8k$ modes, respectively, part of them contains data symbols and the other part, consisting of selected subcarriers, carries the scattered and continual reference signals, which are called *pilots*. The pilots are used for frame, frequency and timing synchronization, channel estimation and transmission mode identification. The remaining subcarriers are applied to transport the *Transmission Parameter Signaling* (TPS) block, which describes the mode (FFT size), constellation of the modulation, error correction coding and the OFDM guard interval parameters that will be applied in the next frame.

The DVB-T system is to replace analog TV, so it occupies channels of the same width as those assigned to analog TV. In many countries, e.g. in European countries, the bandwidth of a single analog TV channel is equal to 8 MHz; however, there are also 6 and 7 MHz TV channels. The ETSI Standard EN 300 744 describes OFDM transmission parameters for all the above channel bandwidths. We will focus on the 8 MHz channel as it gives enough information about the system parameters. Transmission parameters for 6 and 7 MHz channels differ in detail and are not dealt with in this book. The interested reader is referred to the original ETSI standard. Table 4.2 presents OFDM parameters for 8 MHz channels for the $2k$ and $8k$ modes.

Tables 4.3 and 4.4 show further OFDM parameters for 8 MHz channels that are mostly related to the sampling period T_s and different lengths of the cyclic prefix. As we see, OFDM transmission in the DVB-T system can be set in a very flexible way depending on the current propagation environment.

Table 4.2 OFDM signal parameters for the $8k$ and $2k$ modes for 8 MHz channels

Parameter	$8k$ mode	$2k$ mode
Number of subcarriers K	6817	1705
Value of subcarrier number K_{min}	0	0
Value of subcarrier number K_{max}	6816	1704
Duration of the orthogonality period T_{ort}	896 μs	224 μs
Subcarrier spacing $1/T_{ort}$ (approx.)	1116 Hz	4464 Hz
Spacing between subcarriers K_{min} and K_{max} $(K-1)/T_{ort}$	7.61 MHz	7.61 MHz

Table 4.3 Duration of guard and orthogonality periods in the 8k mode for 8 MHz channels

Mode	8k mode			
Guard interval T_g/T_{ort}	1/4	1/8	1/16	1/32
Duration of orthogonality interval		$8192T_s$ $896\,\mu s$		
Duration of guard time interval T_g	$2048 \times T_s$ $224\,\mu s$	$1024 \times T_s$ $112\,\mu s$	$512 \times T_s$ $56\,\mu s$	$256 \times T_s$ $28\,\mu s$
Symbol duration $T = T_g + T_{ort}$	$10240 \times T_s$ $1120\,\mu s$	$9216 \times T_s$ $1008\,\mu s$	$8704 \times T_s$ $952\,\mu s$	$8448 \times T_s$ $924\,\mu s$

Table 4.4 Duration of guard and orthogonality periods in the 2k mode for 8 MHz channels

Mode	2k mode			
Guard interval T_g/T_{ort}	1/4	1/8	1/16	c1/32
Duration of orthogonality interval		$2048T_s$ $224\,\mu s$		
Duration of guard time interval T_g	$512 \times T_s$ $56\,\mu s$	$256 \times T_s$ $28\,\mu s$	$128 \times T_s$ $14\,\mu s$	$64 \times T_s$ $7\,\mu s$
Symbol duration $T = T_g + T_{ort}$	$2560 \times T_s$ $280\,\mu s$	$2304 \times T_s$ $252\,\mu s$	$2176 \times T_s$ $238\,\mu s$	$2112 \times T_s$ $231\,\mu s$

In general, the DVB-T OFDM signal is described by the formula (ETSI 2004)

$$x(t) = \mathrm{Re}\left\{ e^{j2\pi f_c t} \sum_{m=0}^{\infty} \sum_{l=0}^{67} \sum_{k=K_{\min}}^{K_{\max}} c_{m,l,k} \psi_{m,l,k}(t) \right\} \tag{4.122}$$

where

$$\psi_{m,l,k}(t) = \begin{cases} \exp\left[j2\pi \frac{k'}{T_{ort}} \left(t - T_g - lT - 68mT \right) \right] & (l+68m)T \le t \le (l+68m+1)T \\ 0 & \text{otherwise} \end{cases} \tag{4.123}$$

As we see, $\psi_{m,l,k}(t)$ is a complex kth subcarrier signal that is generated in the mth OFDM frame during the lth OFDM symbol, where k' is the subcarrier index with respect to the center frequency, i.e. $k' = k - (K_{\max} - K_{\min})/2$. The complex data symbol carried on the kth subcarrier that is generated in the mth OFDM frame during the lth OFDM symbol is denoted by $c_{m,l,k}$. Recall that T denotes the OFDM symbol modulation period.

Data carrying subcarrier symbols $c_{m,l,k}$ are selected according to the QPSK, 16-QAM or 64-QAM signal constellations shown in Figures 4.56, 4.57 and 4.58. Figure 4.56 presents standard uniform QPSK and QAM constellations. Figures 4.57 and 4.58 show nonuniform QAM constellations for which each constellation point is obtained by adding the complex number $\pm(\alpha - 1) \pm j(\alpha - 1)$ to the original constellation symbol z from Figure 4.56

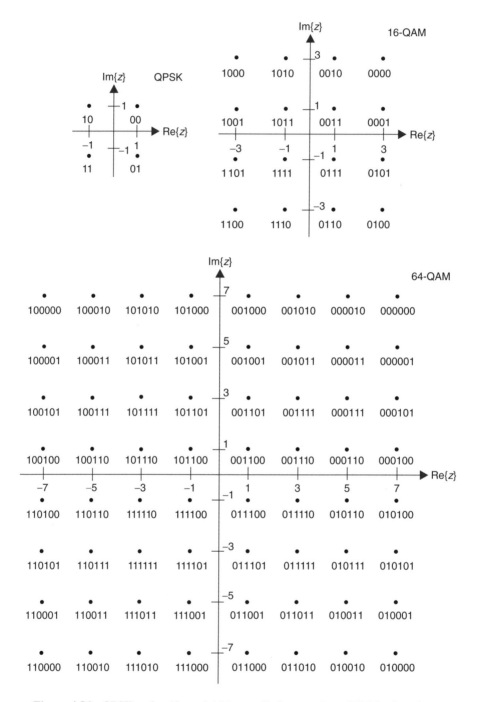

Figure 4.56　QPSK and uniform QAM constellations used on OFDM subcarriers

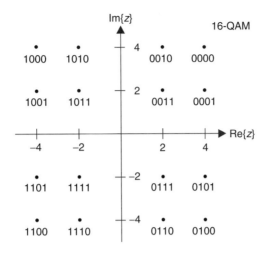

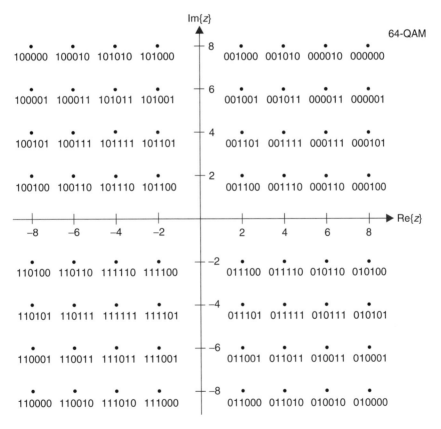

Figure 4.57 Nonuniform QAM constellations ($\alpha = 2$) used on OFDM subcarriers

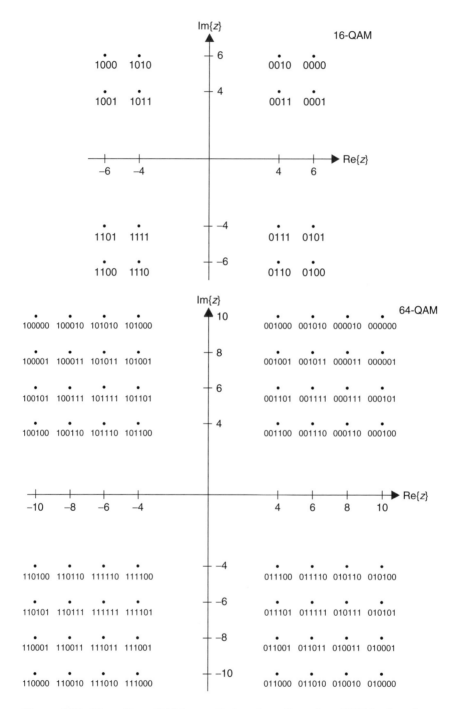

Figure 4.58 Nonuniform QAM constellations ($\alpha = 4$) used on OFDM subcarriers

(with signs of α depending on the quarter of the coordinate system in which a particular constellation point z is located). Shifting constellation points further outside the origin allows less sophisticated receivers to treat the constellation clusters in each quarter of the coordinate system as single points and consider the received signal as the signal in which applied modulation has fewer constellation points than the original one. This is possible due to appropriate binary block assignment to each constellation point. The data symbols $c_{m,l,k}$ are normalized in the sense that $E[|c|^2] = 1$. This means that the mean power of the data symbols is equal to unity. Therefore, each constellation point z is normalized depending on the modulation level and the value of α. In Table 4.5 the normalization factors are listed for all uniform and nonuniform modulations.

Table 4.5 Normalization factors for different modulations

Modulation scheme		Normalization factor
QPSK		$c = z/\sqrt{2}$
16-QAM	$\alpha = 1$	$c = z/\sqrt{10}$
	$\alpha = 2$	$c = z/\sqrt{20}$
	$\alpha = 4$	$c = z/\sqrt{52}$
64-QAM	$\alpha = 1$	$c = z/\sqrt{42}$
	$\alpha = 2$	$c = z/\sqrt{60}$
	$\alpha = 4$	$c = z/\sqrt{108}$

As we have already mentioned, subcarriers transmit both data symbols and reference symbols. Figure 4.59 presents the location of scattered pilot symbols along frequency and time axes. The scattered pilot symbols feature the increased power and are BPSK modulated for more reliable channel estimation and synchronization. Thus, their symbols

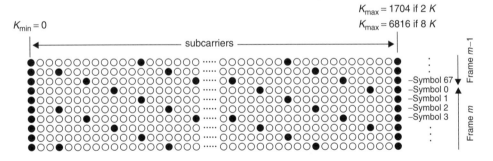

TPS pilots and continual pilots not shown

● pilot symbol
○ data symbol

Figure 4.59 Pilot and data symbols located on OFDM subcarriers in successive OFDM symbols in the DVB-T OFDM frame © European Telecommunications Standards Institute 2004. Further use, modification, redistribution is strictly prohibited. ETSI standards are available from http://pda.etsi.org/pda

are described by formula

$$\text{Re}\{c_{m,l,k}\} = \frac{4}{3} \cdot 2 \left(\frac{1}{2} - w_k \right)$$ (4.124)

$$\text{Im}\{c_{m,l,k}\} = 0$$

where w_k is the output of the pseudorandom binary sequence (PRBS) generator described by the polynomial $p(x) = x^{11} + x^2 + 1$. The power increase is due to the factor $4/3$ contained in (4.124). As we have mentioned before, both scattered pilots and continual pilots are transmitted on appropriate subcarriers. Thus, some selected subcarriers are exclusively devoted to the transmission of the reference symbols. There are 45 such subcarriers in the $2k$ mode and 177 subcarriers in the $8k$ mode. The continual pilots create an irregular pattern of boosted subcarrier symbols along the frequency axis. The pilot symbols are also generated by the binary PRBS and are given by formula (4.124).

Finally, the remaining subcarriers are used to transmit the TPS block. Its correct detection is crucial for operation of the receiver in the next frame, therefore the TPS block is transmitted along the OFDM frame in parallel on 17 subcarriers in the $2k$ mode and 68 subcarriers in the $8k$ mode. The contents of the TPS block have already been described. DBPSK (differential BPSK) modulation is applied on each subcarrier carrying the TPS block. The first symbol of the TPS block constitutes a reference symbol for the DBPSK modulation and its value is also determined by the same PRBS generator.

Summarizing, out of $K = 1705$ subcarriers used in the $2k$ mode of DVB-T OFDM transmission 1512 subcarriers are used for data transmission, whereas the remaining ones carry scattered or continual pilot symbols or carry transmission parameter sequence information. Similarly, for $K = 6817$ subcarriers in the $8k$ mode 6048 are used for data transmission. The remaining ones are used as pilots or to carry TPS blocks.

The above short sketch of digital modulation schemes applied in the DVB-T system shows us how advanced modulation techniques can be applied in modern communication systems. A similar modulation system also operates in DVB-H transmission for which receivers should be handheld devices.

4.14 Influence of Nonlinearity on Signal Properties

Nonlinear distortions introduced by a given block, e.g. by a power amplifier placed in a transmitter or a satellite transponder, are often characterized by the so-called AM-AM and AM-PM characteristics. The former characteristics show how the amplitude of the input signal influences the amplitude of the output signal, whereas the latter describe how the nonlinear block shifts the signal phase depending on the input signal amplitude.

Let $r_{\text{out}} = A(r_{\text{in}})$ be the AM-AM characteristics and $\varphi_{\text{out}} = \Phi(r_{\text{in}})$ be the AM-PM characteristics of a nonlinear block. Figure 4.60 presents AM-AM and AM-PM characteristics for a typical model of nonlinearity that well characterizes the properties of the so-called Travelling Wave Tube (TWT). The TWT is often the main amplifying element of the satellite transponder. Let us consider an input signal of the form

$$s(t) = r(t) \cos \left[2\pi f_c t + \theta(t) \right]$$ (4.125)

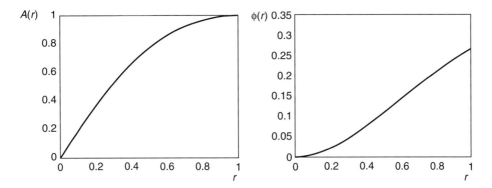

Figure 4.60 AM-AM and AM-PM characteristics of the model of the nonlinear amplifier as a function of the normalized input signal envelope r

where $r(t)$ denotes the signal envelope and $\theta(t)$ is the phase function of time. Let us recall that by using the signal pair $[r(t), \theta(t)]$ we can characterize an arbitrary modulation, including a digital one. Recalling formula (4.5) and using the in-phase and quadrature components we can easily show that $r(t) = \sqrt{[x^I(t)]^2 + [x^Q(t)]^2}$, and $\theta(t) = \mathrm{arctg}[x^Q(t)/x^I(t)]$. Thus, the signal on the output of a nonlinear block can be described by the expression

$$x(t) = A\big[r(t)\big] \cos\Big\{2\pi f_c t + \theta(t) + \Phi\big[r(t)\big]\Big\} \qquad (4.126)$$

First, let us assume that the signal envelope is constant, i.e. $r(t) = r = const$, and the transmitted data signal is fully represented by the signal $\theta(t)$. If we consider the signal with a constant envelope and with phase or frequency modulation, then

$$x(t) = A(r) \cos\big[2\pi f_c t + \theta(t) + \Phi(r)\big] \qquad (4.127)$$

As a result, the signal on the output of the nonlinear block is characterized by a constant amplitude and an additional but constant phase shift. Thus, the spectral properties of the transmitted signal remain unchanged (only scaling due to amplification and a constant phase shift occur). The information carrying phase function $\theta(t)$ is not distorted either. We conclude that the nonlinarity is harmless for constant envelope signals.

The situation becomes much more complicated if the input signal envelope varies in time. For simplicity let us consider the input signal of the form

$$x(t) = r(t) \cos 2\pi f_c t \qquad (4.128)$$

where, as previously, $r(t)$ is the signal envelope $[r(t) \geq 0]$. In order to illustrate the influence of nonlinearity on signal (4.128) let us represent the AM-AM characteristics in the form of the polynomial expansion

$$r_{\text{out}} = A_1 r_{\text{in}} + A_2 r_{\text{in}}^2 + A_3 r_{\text{in}}^3 + \cdots \qquad (4.129)$$

Let us consider an even simpler case in which only two first expansion components in (4.129) are taken into account, i.e.

$$r_{\text{out}} = A_1 r_{\text{in}} + A_2 r_{\text{in}}^2 \tag{4.130}$$

and the nonlinear block does not introduce AM-PM distortions, as is often the case for solid-state amplifiers. Then the signal on the output of a nonlinear block is

$$x(t) = \left[A_1 r(t) + A_2 r^2(t) \right] \cos 2\pi f_c t \tag{4.131}$$

and the spectrum of $x(t)$ is expressed by the formula

$$X(f) = \frac{1}{2} \left[M(f - f_c) + M(f + f_c) \right], \tag{4.132}$$

where $M(f) = \mathcal{F}\left[A_1 r(t) + A_2 r^2(t) \right]$

Let us denote the spectral density of signal $r(t)$ as $R(f)$. The spectral density of the signal envelope on the output of the nonlinear block is then equal to

$$M(f) = A_1 R(f) + A_2 R(f) * R(f) \tag{4.133}$$

As we see, new spectral components appear in the signal on the output of the nonlinear block and they distort the original signal spectrum. While the spectral width of the nonzero values of the convolution $R(f) * R(f)$ is doubled compared with the original spectrum $R(f)$, we see that nonlinearity results not only in spectral distortions but also in spectrum re-growth. Thus, nonlinearity appearing in the transmitter is particularly disadvantageous in radio transmission because it causes crosstalk to the neighboring radio channels. Figure 4.61 presents spectral re-growth and spectral distortion of the signal transmitted through the nonlinear block.

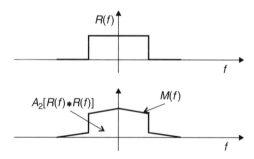

Figure 4.61 Illustration of the spectral distortion resulting from transmission of a signal through the block featuring second-order nonlinearity

Problems

Problem 4.1 *Derive the formula for the cross-correlation coefficient γ given by (4.56) for two elementary signals used in FSK modulation described by (4.3). Assume that the nominal frequencies f_0 and f_1 of the elementary signals are much higher than the signaling rate $1/T$. Show that the optimal cross-correlation coefficient γ for two elementary signals used in FSK modulation is equal to -0.217 when a synchronous FSK receiver is applied. Calculate the exact value of the loss in E_s/N_0 for this receiver compared with an optimal synchronous receiver of BPSK signals with the same energy.*

Problem 4.2 *Assume that for an FSK signal the nominal frequency f_1 is higher than the nominal frequency f_0. Plot the FSK signal characterizing the data sequence 0110011100 and for the asynchronous optimal FSK receiver shown in Figure 4.17 draw the idealized waveforms at the outputs of bandpass filters, envelope detectors and filters matched to the envelope.*

Problem 4.3 *Consider a suboptimal FSK receiver with a frequency discriminator. The receiver is shown in Figure 4.18 and the probability of error at its output is given by formula (4.64). Assume that the square root raised cosine filter with a given roll-off factor α is a baseband equivalent filter of the bandpass filter applied in the receiver. Find the optimal parameter $\beta = (f_1 - f_0)T/2$ and the distance between the nominal frequencies f_0 and f_1 that minimize the probability of error (4.64). Give the quantitative results for $\alpha = 0.1, 0.25$ and 0.5. Compare them with the results for other filter characteristics shown in Table 4.1.*

Problem 4.4 *Consider PSK transmission with synchronous reception. Assume that the signal energy over the modulation period T is equal to E_s, the PSK signal is corrupted by an additive white Gaussian noise of power spectral density equal to $N_0/2$, and data symbols are statistically independent and equiprobable. Calculate the probability of error at the output of the synchronous receiver when the phase of the reference carrier signal differs by θ from the ideal one. Draw the plot of the error probability versus E_s/N_0 for a few values of the phase error θ. What is the allowable phase error θ if the loss equivalent to $0.5\,dB$ in E_s/N_0 compared with ideal reception can be tolerated? In drawing the plots apply the tables of the erfc function contained in the Appendix or use Matlab.*

Problem 4.5 *Consider PSK transmission with synchronous reception. Assume that the signal energy over the modulation period T is equal to E_s, the PSK signal is corrupted by an additive white Gaussian noise of power spectral density equal to $N_0/2$, and data symbols are statistically independent and equiprobable. Calculate the probability of error at the output of the synchronous receiver when the timing recovery block produces a timing clock that differs by τ seconds from the ideal timing moments. Draw the plot of the error probability versus E_s/N_0 for a few values of the timing error τ. What is the allowable timing error τ if the loss equivalent to $0.5\,dB$ in E_s/N_0 compared with the ideal reception can be tolerated? In drawing the plots apply the tables of the erfc function contained in the Appendix or use Matlab.*

Problem 4.6 *For high values of the signal-to-noise ratio (SNR) in M-QAM or M-PSK transmission a predominating error event in a synchronous receiver is due to selection of a constellation point that is closest to the transmitted one. Thus, the minimum Euclidean distance between constellation points is an important factor influencing the error probability value. Assume that 16-PSK and 16-QAM feature the same minimum distance d. What is the difference in E_b/N_0 between both modulations if, for high SNR, they have the same probability of error at the synchronous receiver output?*

Problem 4.7 *Consider synchronous reception of QPSK and 8-PSK signals in the presence of an additive white Gaussian noise of power spectral density equal to $N_0/2$. Let the mean energy over the modulation period be the same for both modulations. Show that the difference in the required E_b/N_0 expressed in dB between both modulations needed to achieve the same probability of error approaches the value of $10\log_{10}\left(3d_{8PSK}^2/2d_{QPSK}^2\right)$ when the SNR increases. Calculate this value in dB.*

Problem 4.8 *Consider QPSK transmission with differential encoding that ensures robustness to phase ambiguity of multiples of 90°. Design the mapping of the binary blocks onto QPSK signal constellation points and calculate the contents of the look-up table of the differential encoder shown in Figure 4.62.*

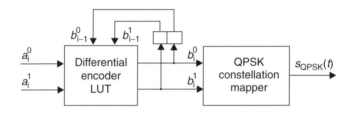

Figure 4.62 Block diagram of the DQPSK modulator

Problem 4.9 *Check, by considering the autocorrelation properties of the differentially encoded QPSK data symbols, if the process of differential encoding has an impact on the spectral properties of the QPSK signal compared with the QPSK signal without differential encoding. Apply the results of your considerations in spectral properties of the random data sequence derived in Chapter 3.*

Problem 4.10 *Solve Problem 4.8 for 32-QAM.*

Problem 4.11 *Draw the block diagram of the correlative receiver for a DBPSK signal similar to that shown for DQPSK in Figure 4.29. Assume for simplicity that instead of the matched filters $p(-t)$ shown in Figure 4.29 two correlators are applied. Give the formulas for signals at the outputs of the correlators and the input to the decision device when the DBPSK signal corrupted by additive noise $v(t)$ is found at the receiver input.*

Problem 4.12 *Compare the formulas for the probability of error for synchronous reception of BPSK and QPSK signals. Show that when the SNR increases, the probability of error approaches the same value for both modulations.*

Problem 4.13 *Probability of error at the output of a synchronous receiver for a differentially encoded PSK signal is given by formula (4.72). Derive a similar formula for the probability of error at the output of a synchronous receiver for DQPSK signals. We assume that data symbols are statistically independent and equiprobable and an additive noise is Gaussian and white with the power spectral density $N_0/2$.*

Problem 4.14 *As we know, the MSK modulation can be considered as a linear modulation. An MSK signal can be generated by a linear transmitter shown in Figure 4.39 for which the pulse shaping filters in the in-phase and quadrature branches are selected according to formulas (4.98) and (4.99). Design the optimal linear receiver for MSK signals.*

Problem 4.15 *Draw the phase tree for a continuous phase FSK signal featuring the modulation index $h = 0.7$. Show the path of the signal on this tree if the data sequence at the modulator input is 011010001011. Assume that binary zero is represented by the data symbol $a_i = -1$ and binary one by the data symbol $a_i = 1$.*

Problem 4.16 *Let the frequency pulse $g(t)$ in a continuous phase FSK signal have the raised cosine shape lasting for the modulation period T, i.e. let it be given by the formula*

$$g(t) = \frac{A}{2}\left[1 + \cos\frac{2\pi}{T}\left(t - \frac{T}{2}\right)\right] \; for \; 0 \le t \le T$$

What is the appropriate value of the signal amplitude A? Draw the phase impulse $q(t)$ for the calculated amplitude A and subsequently sketch the phase tree.

Problem 4.17 *Let us assume a continuous phase FSK signal with modulation index $h = 0.5$ and frequency pulse $g(t)$ lasting no longer than the modulation period T. Besides the phase tree the equivalent way to characterize the continuous phase FSK signal is the trellis diagram, in which the number of states is determined by the number of possible phases achievable after each modulation period. Determine the trellis diagram showing possible transitions between the trellis states from the nth to the $(n + 1)$st moment. Denote the paths traversed due to the data symbol $a_n = -1$ by solid lines and the paths traversed due to the data symbol $a_n = 1$ by dashed lines. Demonstrate the path on the trellis diagram for the continuous phase FSK signal starting at the zero moment from the zero phase when the continuous phase FSK signal is used to transmit the following data sequence: $\mathbf{a} = [1, -1, 1, -1, -1, 1, 1, 1, 1, -1]$.*

Problem 4.18 *Solve Problem 4.17 if the continuous phase FSK signal has the modulation index $h = 3/4$ but other parameters retain their previous values. How many states does the trellis have?*

Problem 4.19 *Consider the trellis diagram developed in Problem 4.17. How could the Viterbi algorithm be applied at the receiver to determine the best path on the trellis diagram and the data sequence associated with this path starting from the zero moment up to the current moment n? Assume that the Viterbi algorithm measures the "distance" between the signal $x(t)$ received during the nth modulating period $((n - 1)T \le t \le T)$ and signal*

$s(t, \theta_{n-1}, a_n)$ *according to the formula*

$$\lambda(\theta_{n-1}, a_n) = \int_{(n-1)T}^{nT} |x(t) - s(t, \theta_{n-1}, a_n)|^2 \, dt$$

The Viterbi algorithm attempts to find the best path for which the sum of such distances is minimized. Recall that the reference signal is given by formula $s(t, \theta_{n-1}, a_n) = \cos(2\pi f_c t + \theta_{n-1} + 2\pi h a_n q(t - (n-1)T))$, *with* $(n-1)T \leq t \leq nT$. *In our case* $h = 1/2$ *and the phase impulse* $q(t) = \frac{1}{2T} t$ *for* $0 \leq t \leq T$.

Problem 4.20 *Consider TCM in which 16-QAM and the trellis encoder shown in Figure 4.63 are applied. Assume that the binary block to 16-QAM constellation mapping is done as in Figure 4.49. Determine the trellis diagram for the TCM configuration. How many parallel transitions between the states exist for the trellis diagram? What is the minimum distance for the considered TCM?*

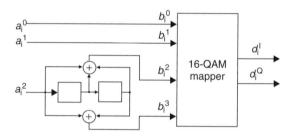

Figure 4.63   TCM encoder with 16-QAM mapper

Problem 4.21 *Consider the TCM with 8-PSK for which the encoder is shown in Figure 4.45. Assume that the TCM signal is corrupted by a white Gaussian noise. The TCM signal is processed by the receiver shown in Figure 4.51. Let the* (x_i^I, x_i^Q) *pairs received at the outputs of the samplers be*

$$(0.9, 0.1), (-0.78, 0.85), (-0.1, 1.1), (0.5, 0.3), (-0.3, -0.9),$$
$$(1.3, -0.5), (-0.1, -1.1), (-0.15, 1.2)$$

Assuming that the trellis encoder starts from the state $(0, 0)$, *demonstrate the operation of the TCM Viterbi detector. Determine the sequence of TCM constellation points indicated by the Viterbi detector and give the binary sequence decided by the detector.*

Problem 4.22 *Let us consider an OFDM transmission system for indoor radio applications such as WLANs. In order to incorporate the channel impulse response within the guard period and to retain sufficient synchronization abilities, it is often assumed that the the guard interval is four times longer than the average length of the channel impulse response. Design the parameters of the OFDM system, in particular the modulation period, the guard interval, the size of digital modulation applied on each subcarrier, the subcarrier frequency separation and the number of applied subcarriers if the system requirements are* *the following: the system binary throughput for the user is 20 Mbit/s, the assumed channel impulse reponse length is 200 ns, the signal bandwidth is 15 MHz, the FEC coding rate R = 1/2.*

Problem 4.23 *Consider OFDM transmission over a channel with the transfer function H(f). The system configuration is shown in Figure 4.64. An additive white Gaussian noise with the power spectral density $N_0/2$ is added at the receiver input. Let the channel transfer function be*

$$H(f) = 0.7107 \cdot \left\{1 - 0.99 \exp\left[-j2\pi(f - f_0)\tau\right]\right\}$$

The channel transfer function describes a two-path channel with the second path delayed by τ with respect to the first one. Let us assume that the cyclic prefix applied in the OFDM signal is longer than the delay τ. In such a situation the signal sample at each FFT output can be described by the formula

$$Y(i) = H(i)X(i) + N_v(i), \quad i = 1, \dots, N$$

where X(i) is the 16-QAM data symbol of the mean energy per bit E_b transmitted on the ith subcarrier, $H(i) = H(f_i)$ is the channel transfer function coefficient on the ith subcarrier of frequency f_i and $N_v(i)$ is the noise component resulting from the correlation of the noise from the receiver input performed by the FFT demodulator. Let the equalizer applied at the FFT output invert the channel transfer function, thus $C(i) = H^{-1}(i)$. The assumed system parameters are: $N = 8$, $f_1 = 300\,Hz$, subcarrier separation $\Delta f = 400\,Hz$, energy per transmitted bit $E_b = 0.001\,W\cdot s$ and the FFT demodulator preserves the level of the signal power. The channel parameters are: $N_0 = 3.2: 10^{-6}\,W/Hz$, notch frequency $f_0 = 1500\,Hz$ and second path delay $\tau = 0.3\,ms$. Calculate the probability of error for each subcarrier and find the probability of error for the whole OFDM signal. Is there any subcarrier that has a decisive influence on the overall system performance?

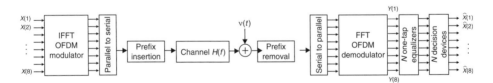

Figure 4.64   Block diagram of the OFDM system considered in Problem 4.23

5

Properties of Communication Channels

5.1 Introduction

The communication channel is a key element in the whole chain of blocks constituting a communication system, because the structure and operation of many remaining blocks strictly depend on its physical properties. For example, the channel passband implies the range of applicable carrier frequencies and the bandwidth of the applied signal. A given noise level allows for application of a modulation featuring the appropriate number of levels, whereas fluctuations of the amplitude characteristics and nonlinearity of the phase characteristics determine the necessity of application of respective equalization structures at the receiver.

In this chapter we will describe basic properties of several kinds of communication channels and present the methods of modeling these channels. Construction of a channel model allows us to conduct both theoretical and simulation analysis of the investigated communication system. Hardware and software channel simulators are useful tools in simulation investigations and laboratory tests.

We will start the description of channel properties with the introduction of the concept of the baseband equivalent channel, which is very useful in channel characterization and modeling. Then we will describe the most frequently used transmission channel, i.e. telephone channel. We will mention ambiguity of the telephone channel term. We will also present properties of a subscriber loop, i.e. a twisted copper wire pair connecting a subscriber's telephone or modem with line equipment in the telephone exchange. We will further show properties of a twisted copper wire pair channel that imply much wider possibilities of its use, as compared with a typical application for transmission of an acoustic signal in the range of frequencies between 300 and 3400 Hz. Next we will consider basic properties of selected radio channels, i.e. the channel used in cellular radio, the indoor channel typical for wireless local area networks, the line-of-sight horizontal microwave channel and the HF channel. Finally, we will present basic features of an optical fiber channel.

Introduction to Digital Communication Systems   Krzysztof Wesołowski
© 2009 John Wiley & Sons, Ltd

5.2   Baseband Equivalent Channel

Figure 5.1 presents a typical model of the digital transmission system in which modulation of both the in-phase and quadrature carriers is applied. The in-phase and quadrature data symbols, d_n^I and d_n^Q, are fed to the inputs of the baseband pulse-shaping filters. Very often the pulse-shaping filter has the square root raised cosine characteristics. Thus, the same characteristics are selected for the filters applied in synchronous demodulators, which also function as the matched filters. The signals at the outputs of the pulse-shaping filters are given by the formula

$$u^I(t) = \sum_{n=-\infty}^{\infty} d_n^I p(t - nT)$$

$$u^Q(t) = \sum_{n=-\infty}^{\infty} d_n^Q p(t - nT) \qquad (5.1)$$

The signals are shifted to the channel passband by the modulators in which sinusoidal and cosinusoidal carriers of frequency f_c are applied. The sum of both signals, denoted in Figure 5.1 as $s(t)$, is transmitted through the channel with the characteristics $G_P(f)$. Equivalently, the channel is characterized in time domain by the impulse response $g_P(t)$. At the receiver synchronous demodulation and matched filtering are performed. We assume that the demodulator uses the ideal reference phase of the locally generated carriers. If the transmission channel with the characteristics $G_P(f)$ were undistorting, then the sampling of the matched filters outputs at the appropriate time phase would result in signals proportional to the data symbols d_n^I and d_n^Q. Unfortunately, in reality the channel distorts the transmitted signal and the samples at the matched filters outputs depend not only on the data symbols d_n^I and d_n^Q but also on some preceding and following data symbols.

In order to characterize this phenomenon more precisely, let us derive the transfer function between two data inputs denoted in Figure 5.1 by the dotted line A and two receiver outputs denoted by the second dotted line B. In our derivation we apply the

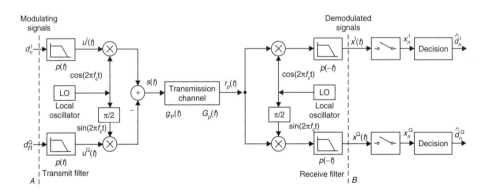

Figure 5.1   Model of a digital transmission system with two-dimensional modulation and shaping of the spectrum in the baseband

symbolic notation in which the in-phase signals are represented by the real part of the complex waveform, whereas the quadrature signals are represented by the imaginary part of the complex waveform. Therefore, we can introduce the baseband complex signal $u(t)$, which has the form

$$u(t) = u^I(t) + ju^Q(t) = \sum_{n=-\infty}^{\infty} (d_n^I + jd_n^Q)p(t - nT) \tag{5.2}$$

Denoting $d_n = d_n^I + jd_n^Q$ as the complex data symbol, we can express the signal $u(t)$ as

$$u(t) = \sum_{n=-\infty}^{\infty} d_n p(t - nT) = p(t) * \sum_{n=-\infty}^{+\infty} d_n \delta(t - nT) \tag{5.3}$$

The transmitted signal $s(t)$ is then given by the formula

$$s(t) = u^I(t) \cos 2\pi f_c t - u^Q(t) \sin 2\pi f_c t = \text{Re}\{u(t) \exp(j2\pi f_c t)\} \tag{5.4}$$

or, equivalently

$$s(t) = \frac{1}{2} \left[u(t) \exp(j2\pi f_c t) + u^*(t) \exp(-j2\pi f_c t) \right] \tag{5.5}$$

where $(.)^*$ denotes a complex conjugate. In turn, the spectrum of the transmitted signal is given by the expression

$$S(f) = \mathcal{F}[s(t)] = \frac{1}{2} \left[U(f - f_c) + U^*(-f - f_c) \right] \tag{5.6}$$

Since the bandpass channel with the characteristics $G_P(f)$ is a physical channel, its impulse response is a real function of time. Thus, taking into account the Fourier transform properties the characteristics fulfill the following equality

$$G_P^*(-f) = G_P(f) \tag{5.7}$$

Let $G_B(f)$ be the part of the channel characteristics $G_P(f)$ located on the positive part of the frequency axis after shifting it to the coordinate system origin by f_c Hz, i.e.

$$G_B(f - f_c) = \begin{cases} G_P(f) & \text{for} \quad f > 0 \\ 0 & \text{for} \quad f \le 0 \end{cases} \tag{5.8}$$

Consequently

$$G_B^*(-f - f_c) = \begin{cases} 0 & \text{for} \quad f \le 0 \\ G_P^*(-f) & \text{for} \quad f > 0 \end{cases} \tag{5.9}$$

and

$$G_P(f) = G_B(f - f_c) + G_B^*(-f - f_c) \tag{5.10}$$

From (5.10) we can write the expression describing the channel impulse response as

$$
\begin{aligned}
g_P(t) &= \mathcal{F}^{-1}\left[G_P(f)\right] = g_B(t)\exp\left(j2\pi f_c t\right) + g_B^*(t)\exp\left(-j2\pi f_c t\right)\\
&= 2\mathrm{Re}\left\{g_B(t)\exp\left(j2\pi f_c t\right)\right\}
\end{aligned}
\tag{5.11}
$$

Let us denote the signal at the channel output as $r_P(t)$, so

$$r_P(t) = 2\mathrm{Re}\left\{r(t)\exp\left(j2\pi f_c t\right)\right\} \tag{5.12}$$

Its spectrum depends on the input signal spectrum and the channel transfer function, as in the following formula

$$R_P(f) = G_P(f)S(f) \tag{5.13}$$

On the basis of (5.6) and (5.10) we have

$$R_P(f) = \frac{1}{2}\left[U(f - f_c) + U^*(-f - f_c)\right]\left[G_B(f - f_c)G_B^*(-f - f_c)\right] \tag{5.14}$$

Since the channel and the input signal are bandpass, we have

$$
\begin{aligned}
U(f - f_c) &= 0, \quad G_B(f - f_c) = 0 \quad \text{for} \quad f < 0\\
U(f + f_c) &= 0, \quad G_B(f + f_c) = 0 \quad \text{for} \quad f > 0
\end{aligned}
\tag{5.15}
$$

Therefore, taking into account (5.15) we obtain the following formula from (5.14)

$$
\begin{aligned}
R_P(f) &= \frac{1}{2}\left[U(f - f_c)G_B(f - f_c) + U^*(-f - f_c)G_B^*(-f - f_c)\right]\\
&= \frac{1}{2}\left[R(f - f_c) + R^*(-f - f_c)\right]
\end{aligned}
\tag{5.16}
$$

We conclude from (5.16) that

$$R(f) = U(f)G_B(f) \tag{5.17}$$

which is equivalent in the time domain to the expression

$$r(t) = r^I(t) + jr^Q(t) = u(t) * g_B(t) \tag{5.18}$$

The impulse response $g_B(t)$ characterizes the so-called *baseband equivalent channel*. Figure 5.2 shows the characteristics of the baseband equivalent channel in comparison with the bandpass channel. In the general case the impulse response $g_B(t)$ is a complex

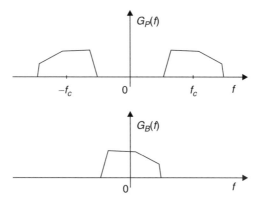

Figure 5.2   Example of the bandpass channel characteristics and the characteristics of the baseband equivalent channel

function of time. Let us recall that it would be a real function of time if the characteristics $|G_B(f)|$ were even-symmetric with respect to zero frequency and $\arg[G_B(f)]$ were an odd-symmetric function of frequency, i.e. if the following expressions were fulfilled

$$|G_B(f)| = |G_B(f)| \ \text{ and } \ \arg[G_B(f)] = -\arg[G_B(-f)] \tag{5.19}$$

These features of $G_B(f)$ are equivalent to the even symmetry of the bandpass channel amplitude characteristics $|G_P(f)|$ with respect of the carrier frequency f_c and the odd symmetry of the bandpass channel phase characteristics with respect to the carrier frequency. Typically they are not preserved in real transmission channels.

The receiver aims at recovery of the data symbols d_n^I and d_n^Q on the basis of the received signal $r_P(t) = \mathcal{F}^{-1}\left[R_P(f)\right]$. Since from (5.12) we have

$$r_P(t) = 2\left[r^I(t)\cos 2\pi f_c t - r^Q(t)\sin 2\pi f_c t\right] \tag{5.20}$$

we can derive the signals at the outputs of the synchronous demodulators. For the in-phase demodulator we have

$$x^I(t) = \left[r_P(t)\cos 2\pi f_c t\right] * p(-t) \tag{5.21}$$
$$= 2\left[r^I(t)\cos^2 2\pi f_c t - r^Q(t)\cos 2\pi f_c t \sin 2\pi f_c t\right] * p(-t)$$
$$= r^I(t) * p(-t)$$

The remaining components in (5.21) are concentrated around the doubled carrier frequency so they are attenuated by the lowpass filter that is matched to the transmitted pulses. Similar derivation performed for the quadrature demodulator results in the formula

$$x^Q(t) = r^Q(t) * p(-t) \tag{5.22}$$

Therefore, in the complex notation the signal $x(t)$ at the demodulator output is the following

$$x(t) = x^I(t) + jx^Q(t) = r(t) * p(-t) \tag{5.23}$$

Using (5.18) and (5.3), we get

$$x(t) = u(t) * g_B(t) * p(-t) \tag{5.24}$$

$$= \left[\sum_{n=-\infty}^{+\infty} d_n \delta(t - nT) \right] * p(t) * g_B(t) * p(-t)$$

$$= \sum_{n=-\infty}^{+\infty} d_n \delta(t - nT) * h(t)$$

So, finally

$$x(t) = \sum_{n=-\infty}^{+\infty} d_n h(t - nT) \tag{5.25}$$

where

$$h(t) = p(t) * g_B(t) * p(-t) \tag{5.26}$$

Formulae (5.25) and (5.26) indicate that the whole system between the data symbol inputs in the transmitter and the matched filters outputs in the receiver can be replaced by a single filter characterized by the complex impulse response $h(t) = h_{re}(t) + jh_{im}(t)$. This impulse response is a convolution of the transmit filter impulse response, the impulse response of the filter matched to the transmit filter and the impulse response of the baseband equivalent channel. The system is presented in Figure 5.3 and it is supplemented with the in-phase and quadrature noise sources.

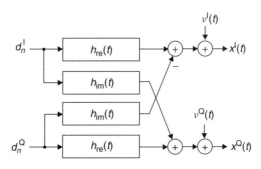

Figure 5.3 Simple baseband model of a digital transmission system with in-phase and quadrature components

As we conclude from Figure 5.3, the output signals $x^I(t)$ and $x^Q(t)$ not only contain the components resulting from the appropriate data symbols, d_n^I snd d_n^Q, repectively, but also contain distorting components that are the result of signal transmission through the cross-connecting filters with the impulse reponses $h_{\text{im}}(t)$.

Summarizing, the system shown in Figure 5.1 can be equivalently described by the formula

$$x(t) = \sum_{n=-\infty}^{+\infty} d_n h(t - nT) + v(t) \qquad (5.27)$$

In further considerations we will often use formula (5.27) for digital transmission system modeling.

5.3 Telephone Channel

5.3.1 Basic Elements of the Telephone Network Structure

Originally, the telephone network was designed to transmit analog signals representing human voice in the range 300–3400 Hz. However, owing to technological progress it is more and more often used for transmission of digital data. Due to heavy investment costs of its deployment and despite new alternative transmission channels and systems, such as wireless or cellular systems, the telephone channel will be further exploited for decades to come. Additionally, new applications are offered on subscriber channels to retain their attractiveness. Therefore it seems valuable to present properties of particular channel types applied in a telephone network as a whole.

Some subscribers still use their telephone lines for transmitting digital signals from and to their computers via modems. A telephone modem generates analog signals featuring the bandwidth similar to that of human voice transmitted over a telephone channel. This is necessary because some transmission devices applied in the transmission chain limit the signal bandwidth. Consider some basic configurations of a telephone link. Let us start from a configuration typical for an "analog communications era", i.e. the configuration that is almost obsolete in many countries.

Figure 5.4 presents a block diagram of a telephone link between subscriber A and B, implemented in analog technology. The subscriber equipment, i.e. a telephone or a modem, is connected with a twisted copper wire pair to so-called *Line Equipment* (LE) in the front-end part of the switching exchange. The length of a twisted wire pair ranges from a few tens of meters up to a few kilometers. Its properties will be descibed in detail in one of the following sections. A twisted copper wire pair can be used in a much more effective way than for transmission of human voice or voiceband acoustic modem signals. The line equipment connects a subscriber with a telephone network, matches the received and transmitted signal levels, separates galvanically the subscriber loop and the switching exchange, etc. In the switching exchange a link to another switching exchange is established. The signal arrives at the destination switching exchange to which subscriber B is connected through his/her twisted wire pair. On the way to subscriber B the signal passes the chain of transmission lines and subsequent switching exchanges. In the switching exchanges the process of concentration of many telephone links takes place

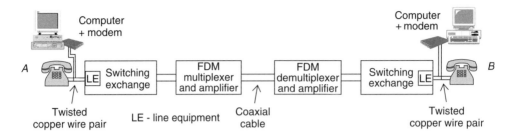

Figure 5.4   Simplified scheme of the analog telephone link

and some of them are further transferred to a particular switching exchange. Thus, the links between switching exchanges have to be able to transmit many telephone signals at the same time. They are typically made of wideband coaxial cables. Due to analog transmission using a coaxial cable as a common transmission medium, a multiple access method based on frequency division, i.e. FDMA (*Frequency Division Multiple Access*) is applied. The common cable signal is formed in the *Frequency Division Multiplexer* (FDM), which is part of the transmitter of the carrier telephony. In order to ensure a high cable throughput, each component link signal is band-limited to the range 300–3400 Hz with a passband filter. In the days of design and deployment of such systems human voice was the only transmitted signal, so these filters were designed by taking into account its properties and the properties of the human ear. Consequently, because fluctuations of the group delay characteristics of the order of a millisecond are not detectable for a human being, they were not taken into account in the filter design. Such fluctuations are, however, very important for data transmission. The filter output signals modulate carriers spaced every 4 kHz using *Single Sideband Modulation* (SSB).[1] In the receive part of the carrier telephony system, i.e. in the FDM demultiplexer, the signals on the appropriate carriers are demodulated and down-converted so that particular links are extracted. Filtering is also an important part of this process. As a result of SSB modulation and demodulation *frequency offset* often arises, because the carrier frequencies in the modulator and demodulator are not identical. The difference between them can reach a few Hz. There can be a few FDM multiplexers and demultiplexers along a particular link. Characteristics of the telephone channel seen between subscriber A and B can have the shape shown in Figure 5.5.

Figure 5.6 shows a typical configuration of a modern telephone network in which mostly digital technology is applied. An analog signal generated by a subscriber or his/her modem is transmitted over a twisted wire pair to the *concentrator* of the switching exchange. A concentrator is often an external part of the switching exchange. The signals from each subscriber are turned, in the line equipment of the concentrator, into a binary stream. On the subscriber side the concentrator is connected with tens or hundreds of subscribers. Statistically, about one-tenth of the lines on the switching exchange side are sufficient to serve the traffic offered by subscribers. Their signals converted to binary streams are further transmitted using the *Time Division Multiplexing* rule. Transmission between a

---

[1] Recall that SSB relies on amplitude modulation of the carrier by the passband signal. As a result a double-side band signal is received, which is subsequently filtered retaining one sideband only. Alternatively, a single sideband-modulated signal can be obtained using a special transmitter structure. The reader is asked to consult Couch (1987) for more information, if needed.

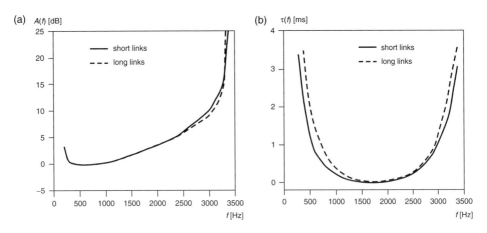

Figure 5.5   Average attenuation characteristics (a) and group delay characteristics (b) for a medium or long link including subscriber loops (Carrey *et al.* 1984)

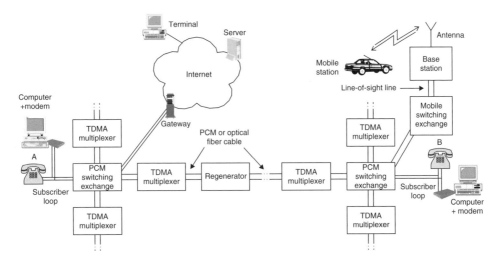

Figure 5.6   Example of a telephone network configuration based on a PCM system including connections to other networks (such as Internet or cellular radio) Reproduced with the kind permission of ITU

concentrator and its switching exchange is digital. The key elements in concentrators are PCM (*Pulse Code Modulation*) encoders and decoders with filters, known as *PCM* codecs (Figure 5.7). A *hybrid* is a circuit placed on the subscriber input of the codec that splits transmission directions from and to the subscriber because in a subscriber loop the signals flow in both directions concurrently. Figure 5.7 presents a possible hybrid scheme and applies the principle of a balanced bridge. A hybrid operates correctly if the impedance Z_L of the subscriber loop seen at points $A-B$ is fully balanced by impedance $Z = Z_L$ in the bottom right branch of the bridge. This impedance is called a *balance impedance*. The resistances in upper branches of the bridge are identical. Impedance Z should be equal

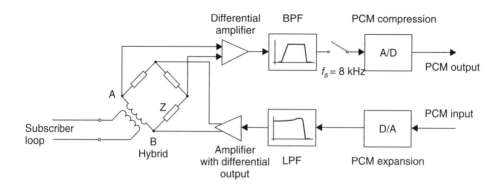

Figure 5.7 Functional scheme of a PCM codec (BPF, bandpass filter; LPF, lowpass filter)

to the loop impedance Z_L in the whole range of the used frequency band. However, this condition cannot be precisely fulfilled. Typically, a simple RC circuit with parameters selected on a statistical basis is applied. Let us note that owing to the transformer, the subscriber loop is galvanically separated from the main PCM codec circuit.

On the side of the analog-to-digital (A/D) conversion there is a lowpass or bandpass filter that narrows the signal bandwidth to the range 300–3400 Hz. The upper band limitation results from the necessity to avoid spectrum overlapping due to the applied 8 kHz sampling frequency. Lower band limitation results from desired attenuation of a power line frequency (50 or 60 Hz) and its harmonics induced in a twisted wire pair. In turn, on the digital-to-analog (D/A) conversion side the lowpass filter smoothes a staircase form of the signal produced by the D/A converter. Figure 5.8 shows the characteristics of the filter preceding the A/D conversion and the characteristics of the filter on the D/A converter output of the M5913 PCM codec produced by SGS-Thomson Microelectronics (1993). As we know, the characteristics of the applied analog-to-digital and digital-to-analog converters are nonlinear in order to make the quantization noise the least perceptible to a listener. The characteristics of both converters are reciprocal to each other, so their nonlinearity mutually compensates. In accordance with ITU-T recommendations, the nonlinear characteristics of the analog-to-digital conversion, called *compression characteristics*, are

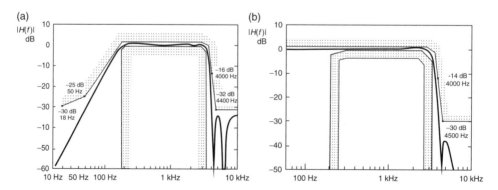

Figure 5.8 Characteristics of the M5913 PCM codec filters: (a) the filter preceding the A/D converter and (b) the filter following the D/A converter (SGS-Thomson Microelectronics 1993)

given by the formula

$$
f(x) = \begin{cases} \dfrac{A}{1 + \ln A} x & \text{for} \quad 0 \le |x| \le \dfrac{1}{A} \\[3ex] \dfrac{\text{sgn}(x)}{1 + \ln A} \left[1 + \ln(A|x|) \right] & \text{for} \quad \dfrac{1}{A} \le |x| \le 1 \end{cases} \tag{5.28}
$$

These are the so-called A-law characteristics and a typical value of A is 87.6. The A-law characteristics are used in European PCM systems. In the USA and Japan different compression characteristics are applied, known as μ-law, given by the formula

$$
f(x) = \frac{\text{sgn}(x)}{\ln(1 + \mu)} \ln(1 + \mu|x|) \tag{5.29}
$$

In the receive part of the PCM codec, in which the digital PCM stream is converted back into analog form, *expansion characteristics* are applied. The data stream resulting from A/D processing has the binary rate of 64 kbit/s and it represents signal samples acquired at the frequency of 8 kHz. Data streams in both transmission directions are the object of electronic switching in the switching exchange and they are subsequently time multiplexed according to the TDM principle. The binary streams organized in a hierarchical manner in the form of structures of higher orders are subsequently transmitted between switching exchanges mostly over optical fiber links. In more traditional implementations coaxial cables are applied. The coaxial or optical fiber cable sections are separated by devices called *regenerators*, which receive the signal attenuated by the transmission line, recover the timing clock for the data signal, amplify the data signal and recover a primary shape that is adequate for transmission in the next cable segment.

A telephone network is often used for communication with other communication networks, such as the Internet and several public land mobile networks. In particular in cellular networks the internal PCM links are often implemented using line-of-sight terrestrial radio links. Line-of-sight terrestrial radio links are also an alternative for wireline links, particularly in environments in which deployment of wireline systems can be very difficult or expensive.

Summarizing, we see that channel properties that are crucial for digital transmission highly depend on the particular structure of the telephone network and, within it, on the applied filters and properties of subscriber loops.

5.3.2 Telephone Channel Properties

A typical telephone channel features various disturbances and distortions. Due to the applied filters and separating transformers, the telephone channel is essentially passband. Unfortunately, the channel amplitude characteristics are not flat and the group delay characteristics are not linear, so we observe amplitude and delay distortions.

Amplitude distortions are introduced by filters that have been designed for voice signal transmission. Let the channel transfer function be denoted as $H_c(f)$. We apply the mathematical description of the channel by assuming its linearity (we neglect a small nonlinear effect caused by amplifiers applied in the transmission chain). Recall that the channel amplitude characteristics are $|H_c(f)|$, whereas the phase characteristics are given by the formula $\varphi(f) = \arg[H_c(f)]$. Instead of amplitude and phase characteristics, a channel is often described by its *attenuation characteristics* and *delay or group delay characteristics*, denoted by $A(f)$ and $\tau(f)$, respectively. They are given by the expressions

$$A(f) = 20\log_{10}\frac{|H_c(f)|}{|H_c(f_0)|} \qquad \tau(f) = \tau(f_0) - \frac{1}{2\pi}\frac{\mathrm{d}\{\arg[H_c(f)]\}}{\mathrm{d}f} \qquad (5.30)$$

where f_0 is a selected frequency for which attenuation and delay are references for measurements at other frequencies. ITU-T Recommendations M.1020, M.1025 and M.1040 (ITU-T Series M 1997) determine the limits within which the channel characteristics have to be fitted to fulfill the defined requirements. An example of the limits for attenuation and delay characteristics is shown in Figure 5.9. These limits are defined for a leased telephone line of special quality, useful for international links. An example of channel characteristics fulfilling these requirements is also shown in this figure. The leased line is a link that has been set up by the telecom operator for a group of subscribers. It is a high quality but expensive solution. In times of wide access to the Internet and satellite and wireless communications, the importance of such links has diminished.

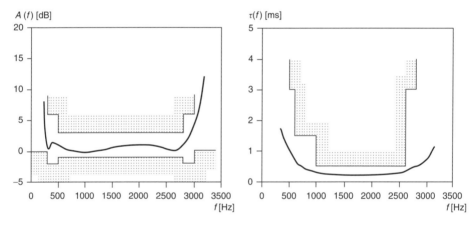

Figure 5.9 Limits of attenuation and delay characteristics determined by ITU-T M.1020 Recommendation and an example of channel characteristics. Reproduced with the kind permission of ITU

In 1984 the measurement results for US telephone links were published (Carrey *et al.* 1984), taking into account the potential use of telephone links in data transmission. Figure 5.10 presents statistical measurement results of attenuation and delay characteristics for links of medium (as for the USA) length between 550 km and 1700 km. Although these data are relatively old, they sufficiently characterize possible attenuation and delay characteristics of a telephone channel. Well-designed equipment should work reliably for the channel characteristics shown by curve (c).

Besides linear distortions described by attenuation and delay characteristics of a telephone channel, one can expect several kinds of noise. *Thermal noise* is caused by various electronic devices operating in a transmission link, in particular amplifiers and receivers.

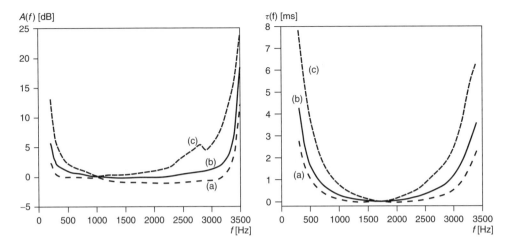

Figure 5.10 Amplitude and delay characteristic measurement results for medium length links in the USA: (a) lower limit of 1% of the best links, (b) average characteristics, (c) lower limit of 99% of links (Carrey *et al.* 1984)

Such noise is modeled as a Gaussian process with a flat power density spectrum. *Quantization noise* results from converting analog samples of the processed signal into 8-bit representations in the PCM codec according to one of the characteristics (5.28) or (5.29). The source of *impulse noise* can be crosstalk from neighboring telephone links placed in a common cable when impulse dialing is applied at a given moment. However, this source of impulse noise is already rare due to massive application of DTMF signaling.[2] Impulse noise can also be caused by switching processes in switching exchanges or induced in a subscriber loop by external devices. It is observed as a sequence of pulses significantly exceeding a typical noise level.

Another already mentioned distortion is *frequency offset* resulting from the application of carrier telephony in old telephone networks. If it is not compensated for, it can be a reason for errors in data transmission. Telephone modems standardized in ITU-T Recommendations Series V (1997) should be able to cope with the frequency offset of up to 7 Hz.

The reasons for *phase jitter* are the power supplies of electronic devices applied in a telephone link and the power line harmonics induced in subscriber loops. Phase jitter is revealed as an undesired phase modulation in the transmitted signal. Phase jitter can be measured as a spread of zero crossings with respect to the reference tone. The measurements reported in (Carrey *et al.* 1984) indicate that the peak value of the phase jitter does not exceed 20°.

The level of *nonlinear distortions* in telephone channels is low. For example, when a sinusoidal measurement tone is applied, its second and third harmonics are about 30 dB below the first harmonics. Nonlinear distortions are caused by nonideal compression and expansion characteristics in PCM codecs or nonlinearity of power amplifiers in analog

---

[2] DTMF (Dual-Tone Multi-Frequency) is a method of signaling the dialed number by selection of one of sixteen combinations of two tones, each taken from a separate set of four frequencies.

systems. Nonlinear distortions are usually neglected in system analysis. Their influence is taken into account only in very advanced receivers.

The last distortion to be described is *echo*. Figure 5.11 shows a situation typical for both analog and digital transmission in which the phenomenon of echo shows itself. The main source (although not the only one) of echo is nonideal operation of hybrids (see Figure 5.7). *Near-end echo* arises due to nonideal match of hybrid and loop impedances. If the hybrid worked ideally, then the signal from subscriber *A* would pass through the loop and would be directed exclusively to the upper branch of the link; actually, however, part of the signal is reflected and returns back to the receiver of subscriber *A*, creating a distortion. Recall that a similar hybrid is located in the subscriber device where two transmission directions over a subscriber loop have to be split. Nonideal operation of the local hybrid also causes near-end echo; this time however, the echo signal is not attenuated by the subscriber loop (see Figure 5.12). *Far-end echo* also has its origin in nonideal operation of the hybrid. However, it is caused by the crosstalk of the signal from the input to output branch of the hybrid on the right-hand side of Figure 5.11. This signal returns (together with the signal from subscriber *B*) to subscriber *A*, creating a disturbance. Besides nonideal operation of a hybrid the reasons for echos are impedance mismatches occurring in different places of a telephone link, which cause signal reflections.

Figure 5.12 illustrates the phenomenon of echo arising due to hybrid and subscriber loop mismatch, in which the hybrid is part of the subscriber's DCE – *Data Communication Equipment*. As we have mentioned above, the echo arising due to the crosstalk from

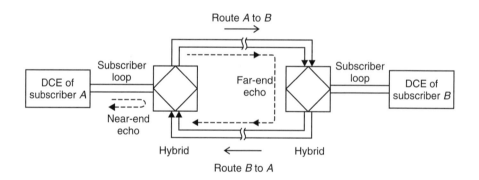

Figure 5.11   Phenomenon of near-end and far-end echo

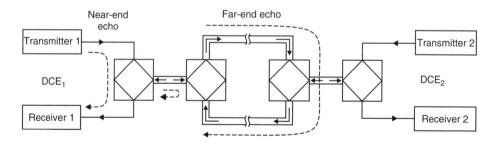

Figure 5.12   Creation of echoes in systems with hybrids in the end-user devices

a local transmitter to a local receiver is particularly harmful because its level can be tens of dB higher than the level of a desired signal received from the remote transmitter (Transmitter 2 in Figure 5.12).

The phenomenon of echo has a crucial influence on a data transmission system if the system works in duplex mode, i.e. transmission is performed in two directions simultaneously in a two-wire link when both directions occupy the same frequency band.

5.4 Properties of a Subscriber Loop Channel

Properties of a channel connecting a single subscriber to a telephone switching exchange became particularly important when full network digitalization took place. In many countries electromechanical switching exchanges were replaced by electronic ones and transmitted signals starting from concentrators are already in a digital form. Only "the last mile" connection, i.e. the connection between a switching exchange concentrator and a subscriber's equipment, was analog. As we have learnt, the band limitation to the range between 300 and 3400 Hz has nothing to do with the properties of the twisted wire pair. Digitalization of this last part of a communication link was possible owing to the introduction of ISDN – *Integrated Services Digital Network*. As a result, a subscriber loop channel could be used much more efficiently. The next steps were the introduction of HDSL (*High-Rate Digital Subscriber Line*), ADSL (*Asymmetric Digital Subscriber Line*) and VDSL (*Very High Bitrate Digital Subscriber Line*) and their upgrades.

As we have already mentioned in the case study in Chapter 3, basic access to ISDN networks is performed by a duplex transmission in a subscriber loop at the rate of 144 ($2 \times 64 + 16$) kbit/s. The methods of duplex transmission in such channel were intensively investigated in the 1980s (Lechleider 1986; Modestine *et al.* 1986). In the 1990s HDSL transmission at the rate up to 2 Mbit/s was investigated. In ADSL multicarrier transmission (known as OFDM; see Chapter 4 on modulations of sinusoidal carriers for details) was introduced. In particular, it is applied to provide wideband access to the Internet. In all those cases the transmission medium is a twisted copper wire pair connecting a subscriber with the closest switching exchange.

In typical situations a subscriber loop consists of serial connection of a number of sections of copper pairs which are part of the communication cable. These sections differ in wire diameter, length and insulation material. Currently polyethylene is mostly applied as an insulation material. In some countries the number of sections is in practice limited to two.

Taking into account the wide bandwidth used by HDSL and VDSL systems, each section of the copper wire pair can be treated as a transmission line. Thus, the transmission line theory can be applied to find a subscriber channel transfer function. On the basis of this we can find a channel impulse response that is particularly important in the description of digital transmission.

Let us treat a single twisted wire pair section as a two-port with input and output voltages and currents denoted as U_{in}, I_{in}, U_{out}, I_{out}, respectively. In the steady state the two-port can be characterized by the following matrix equation

$$\begin{bmatrix} U_{in} \\ I_{in} \end{bmatrix} = \begin{bmatrix} A & B \\ C & D \end{bmatrix} \begin{bmatrix} U_{out} \\ I_{out} \end{bmatrix} \qquad (5.31)$$

where A, B, C and D are complex functions of frequency that characterize the two-port properties. One can prove that serial connection of a number of two-ports is characterized by the product of matrices $ABCD$ describing each of the two-ports separately.

Let us focus on a single homogeneous section of a subscriber loop of incremental length dx. According to transmission line theory, a section of copper wire pair of length dx can be represented by the lumped circuit shown in Figure 5.13. Symbols G', R', C' and L' denote conductance, resistance, capacity and inductance per unit length, respectively. These values are basic parameters of the subscriber loop section. Unfortunately, most of them depend on frequency. Due to the *skin effect*, resistance R' increases with frequency. This feature is a result of the fact that with increasing frequency the current flows within a thinner and thinner external layer of the conductor. Inductance L' consists of two components: external and internal. The first component depends on geometrical properties of the loop and magnetic properties of the applied materials. Because the copper wire does not possess ferromagnetic properties, this component does not depend on frequency. Internal inductance is associated with the electromagnetic field generated by the current flowing in the conductor. Due to the skin effect, the current inside the conductor and, consequently, the electromagnetic field decrease with an increase of frequency. Therefore internal inductance gradually decreases to zero with a frequency increase. Capacity C' is the only parameter that does not depend on frequency and is a function of geometrical properties of the subscriber loop. Conductance G' is associated with capacity C' through the formula $G' = \text{tg}\delta \cdot 2\pi f C'$, where $\text{tg}\delta$ is the so-called *loss coefficient*. It depends on the insulation material applied in the cable and is relatively constant in the relevant range of frequencies, except for paper insulation (Schmid 1976). As we can see, the cable properties significantly depend on frequency. Table 5.1 presents typical values of basic parameters of a twisted copper wire pair for a few representative types of cables (Schmid 1976).

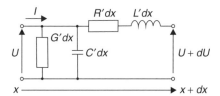

Figure 5.13 Representation of the incremental section of the transmission line using a two-port with lumped elements

For clarity of applied notation we will not show explicitly the dependence of copper wire pair parameters on frequency, however we will take this dependence into account during calculation of the loop transfer function. It is well known from the transmission line theory (Davidson 1978; Schmid 1976) that for sinusoidal excitation parameters A, B, C and D of the line matrix are given by the following formulae

$$A = D = \cosh \gamma l$$

$$B = Z_0 \sinh \gamma l \qquad\qquad (5.32)$$

$$C = Z_0^{-1} \sinh \gamma l$$

Table 5.1 Basic parameters of symmetric copper pairs (Schmid 1976)

Diameter [mm]	Max. resistance [Ω/km] (20°C)	Capacity [nF/km]	Inductance [mH/km]	Conductance [μS/km]
Local cable, paper insulation				
0.3	530	50	0.7	0.5
0.4	300	36	0.7	0.5
0.6	130	42	0.7	0.5
0.8	73.2	42	0.7	0.6
Local cable, polyethylene insulation				
0.3	530	50	0.7	0.1
0.4	300	50	0.7	0.1
0.6	130	50	0.7	0.1
0.8	73.2	55	0.7	0.1

where l is the line length. *Characteristic impedance* Z_0 and *propagation coefficient* γ are described by expressions

$$Z_0(f) = \sqrt{\frac{R' + j2\pi f L'}{G' + j2\pi f C'}} \tag{5.33}$$

$$\gamma(f) = \alpha(f) + j\beta(f) = \sqrt{(R' + j2\pi f L')(G' + j2\pi f C')} \tag{5.34}$$

Consider now a single subscriber loop section of length l with load impedance Z_L connected to its output. Let the signal at the input of the section be a sinusoid of frequency f denoted as $U(f)$. Figure 5.14 illustrates the considered case. Calculate the input impedance $Z_{in}(f) = U_{in}(f)/I_{in}(f)$ and the transfer function $H(f) = U_{out}(f)/U_{in}(f)$. Knowing that $U_{out}(f) = I_{out}(f)Z_L$, from formula (5.31) we obtain the following expression

$$\begin{bmatrix} U_{in}(f) \\ I_{in}(f) \end{bmatrix} = \begin{bmatrix} A & B \\ C & D \end{bmatrix} \begin{bmatrix} Z_L I_{out}(f) \\ I_{out}(f) \end{bmatrix} \tag{5.35}$$

from which we conclude that

$$Z_{in}(f) = \frac{U_{in}(f)}{I_{in}(f)} = \frac{AZ_L + B}{CZ_L + D} \tag{5.36}$$

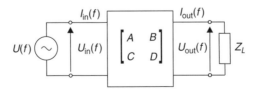

Figure 5.14 Representation of a single cable section in the form of a two-port with load impedance Z_L

One can easily verify that if $Z_L = Z_0$ then $Z_{in}(f) = Z_0$. Similarly, if the line remains open then

$$Z_{in}(f) = Z_0 \frac{\cosh \gamma l}{\sinh \gamma l} \qquad (5.37)$$

In turn, the transfer function of the considered section can be derived from the first equation of equation system (5.31). Knowing that $I_{out}(f) = U_{out}(f)/Z_L$, we get

$$H(f) = \frac{U_{out}(f)}{U_{in}(f)} = \frac{1}{A + BZ_L^{-1}} \qquad (5.38)$$

On the basis of formuale (5.31), (5.32) and (5.38) we are able to calculate the transfer function of any serial connection of particular sections of twisted copper wire pair. Let us stress once more that a serial connection of several loop sections is represented by the product of $ABCD$ matrices characterizing each section.

In some countries during deployment of new subscriber loops, installment of unloaded loop taps is allowable. In this way higher flexibility of installation of possible telecommunication equipment is achieved. Such taps do not have a big influence on voice signal transmission, however they are meaningful if a subscriber loop is used for digital transmission, as happens in the case of ISDN. Since these taps remain unloaded, signals reflect from their ends, which in turn causes echoes. Unloaded taps can be represented in the form of a two-port and the $ABCD$ matrix, shown in Figure 5.15.

Let us present an example, cited after Werner (1991).

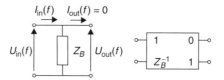

Figure 5.15   Representation of an unloaded line tap

Example 5.4.1 *Consider a subscriber loop consisting of two sections of lengths l_1 and l_2 featuring characteristic impedances Z_{01} and Z_{02}, respectively. In the junction point of two sections the third section is connected in the form of the unloaded tap. A sinusoidal voltage from the source with internal impedance Z_S is fed to the loop input. Figure 5.16 presents the considered loop configuration. Based on our considerations we can formulate the following equation system*

$$\begin{bmatrix} U_{in} \\ I_{in} \end{bmatrix} = \begin{bmatrix} 1 & Z_S \\ 0 & 1 \end{bmatrix} \begin{bmatrix} A_1 & B_1 \\ C_1 & D_1 \end{bmatrix} \begin{bmatrix} A_3 & B_3 \\ C_3 & D_3 \end{bmatrix} \begin{bmatrix} A_2 & B_2 \\ C_2 & D_2 \end{bmatrix} \begin{bmatrix} U_{out} \\ I_{out} \end{bmatrix} \qquad (5.39)$$

where for $i = 1, 2$ we have

$$A_i = D_i = \cosh \gamma_i l_i$$

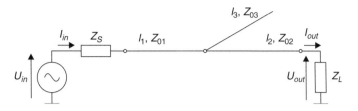

Figure 5.16   Configuration of the subscriber loop with unloaded tap considered in Example 5.4.1

$$B_i = \sinh \gamma_i l_i$$

$$C_i = Z_{0i}^{-1} \sinh \gamma_i l_i$$

and

$$A_3 = D_3 = 1$$

$$B_3 = 0$$

$$C_3 = Z_B^{-1}$$

Recall that both γ_i, Z_{0i} ($i = 1, 2$) and Z_B depend on frequency. Equation system (5.39) can be replaced by a simple equation system similar to (5.31), for which the ABCD matrix is determined by expression

$$\begin{bmatrix} A & B \\ C & D \end{bmatrix} = \begin{bmatrix} 1 & Z_S \\ 0 & 1 \end{bmatrix} \begin{bmatrix} A_1 & B_1 \\ C_1 & D_1 \end{bmatrix} \begin{bmatrix} A_3 & B_3 \\ C_3 & D_3 \end{bmatrix} \begin{bmatrix} A_2 & B_2 \\ C_2 & D_2 \end{bmatrix} \qquad (5.40)$$

So based on (5.40) we can derive the transfer function $H(f)$ of the loop, taking into account formula (5.38). Calculating the inverse Fourier transform of $H(f)$, we can determine the channel impulse response $h(t)$.

Concluding our presentation of a subscriber loop channel, we present a typical form of the channel transfer function and corresponding channel impulse response. Figure 5.17 shows the results of calculations similar to those sketched in Example 5.4.1. Calculations were performed for a subscriber loop of length $l = 3.3$ km constructed from a single section of a twisted copper wire pair of diameter 0.4 mm. The copper pair had the following parameters: $R' = 300 \,\Omega/\text{km}$, $C' = 42 \,\text{nF/km}$, $L' = 1 \,\text{mH/km}$, $G' = 1 \,\mu\text{S/km}$, $Z_S = Z_L = 300 \,\Omega$. Analysis of Figure 5.17 allows us to formulate the following observations. Due to the loop length of 3.3 km the signal reaching the end of the loop is substantially attenuated. This attenuation is seen in the plot of transfer function $H(f)$ around zero frequency. In the range of a few tens of kHz the attenuation grows relatively quickly with frequency. Starting from about 50 kHz, attenuation increase is significantly slower, with attenuation achieving the value of 80 dB at a frequency of around 2 MHz. When the subscriber loop is used in the acoustic range, e.g. for analog voice signal transmission, only a small fraction of the available band is occupied. If the signal features much wider bandwidth, attenuation of this signal increases with frequency. In order for

(a)

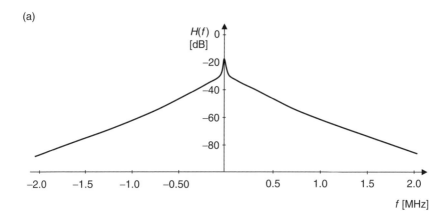

(b)

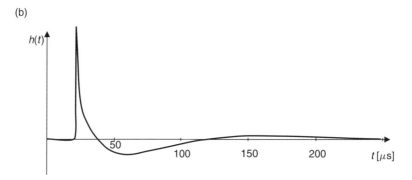

Figure 5.17   Characteristics of a subscriber loop channel of length 3.3 km and parameters described in the text: (a) amplitude characteristics, (b) channel impulse response

transmission to be as robust as possible, the signal spectrum should be concentrated in the lower part of the channel band.

   Interesting conclusions also can be drawn from analysis of the channel impulse response. The channel has a relatively wide bandwidth, so its impulse response has the form of a sharp impulse, which, however, features a long decaying time; in our example it exceeds 200 μs. The time duration of the impulse response has a crucial meaning for application of appropriate transmission methods and receiver structures. In practice, a long "tail" of the channel impulse response causes the channel responses to subsequent input pulses to overlap. This phenomenon is known as *intersymbol interference*. Intersymbol interference is considered in Chapter 6.

   So far we have focused on properties of the subscriber loop channel transfer function and the corresponding channel impulse response. However, besides linear distortions causing intersymbol interference, some other disturbances are present in a subscriber loop channel. The most important disturbance is *crosstalk*. As we have mentioned, a subscriber loop is constructed of sections that, at each level of telephone network distribution, are physically placed close to many other twisted copper wire pairs connecting other subscribers. In particular, there are a number of wire pairs close to the switching

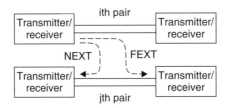

Figure 5.18 Scheme of creation of NEXT and FEXT

exchange. Communication cables are specially constructed so as to minimize crosstalk and differential transmission is performed; however, minor imperfections cause the signals propagating in different copper pairs placed in the same cable electromagnetically to interfere with each other. Such interference is described as NEXT (*Near-End Crosstalk*) and FEXT (*Far-End Crosstalk*). Both kinds of crosstalk are illustrated in Figure 5.18. The NEXT generated in the jth pair by signal transmission in the ith pair is interference resulting from the operation of a transmitter of the ith pair placed at the same end of the copper wire pair as the receiver of the jth pair. Figure 5.18 shows the creation of a crosstalk between two selected copper wire pairs. In reality, the receiver operating at the output of the jth pair receives crosstalk signals appearing in the jth pair due to ongoing transmission in many other pairs (so-called *direct crosstalk*), the signals being *indirect crosstalk*. The receiver also receives signals that are the result of a chain of crosstalk between subsequent pairs, leading to the final crosstalk to the jth pair. Such a situation often occurs at the end devices on the switching exchange side. In turn, the FEXT is interference that originates in the transmitter located on the remote loop end. In this case the interfering signal is attenuated across the whole line length, so it appears at the receiver input as a disturbance on a much lower level than the level of NEXT. One can prove (see (Gibbs and Addie 1979; Werner 1991)) that if the subscriber loop is sufficiently long and the considered frequency range is sufficiently wide, e.g. appropriate for HDSL transmission, then the mean square of the amplitude characteristics of the NEXT is proportional to $f^{3/2}$, i.e.

$$|H_{NEXT}(f)|^2 \approx K_{NEXT} \cdot |f|^{3/2} \qquad (5.41)$$

whereas the mean square of the amplitude characteristics of FEXT can be described by the formula

$$|H_{FEXT}(f)|^2 \approx K_{FEXT} \cdot f^2 \cdot l \cdot \exp[-2\alpha(f)l] \qquad (5.42)$$

where $\alpha(f)$ is the attenuation coefficient determined by equation (5.34), and l is the line length. The form of crosstalk appearing at the receiver input depends on the form of transmission in a given cable. In some copper pairs analog voice transmission takes place, however some others can be used for data transmission at the same rate as the rate that is applied in the reference twisted copper wire pair. If all the crosstalk sources can be considered as statistically independent and none of them dominates, then the joint crosstalk signal can be modeled as white Gaussian noise. In a general case the crosstalk signal observed at the receiver input can be modeled as colored Gaussian noise with the power density spectrum described by the formula

$$G_{NEXT}(f) = |H_{NEXT}(f)|^2 G_{in}(f)$$

$$G_{FEXT}(f) = |H_{FEXT}(f)|^2 G_{in}(f) \tag{5.43}$$

where $G_{in}(f)$ is the power density spectrum of the crosstalk source. This model is not correct if transmission in all pairs within the same cable is performed at the same rate in a synchronous way.

Apart from both kinds of crosstalk, there are other additive disturbances in subscriber loops. As in a typical telephone channel, impulse noise can be observed. According to Werner (1990) impulse noise is one of the main limitations in the application of subscriber loops in data transmission. It is difficult to describe impulsive noise in a statistical sense. The measurements indicate that its properties depend on the time of day, location and kind of switching exchange, as well as other factors. According to Werner (1990) distorting pulses occur about one to five times per minute and their peak values are in the range between 5 and 20 mV, whereas their duration is contained in the interval 30–150 μs. Most of the impulse noise energy is concentrated below 40 kHz, so the influence of this kind of noise can be limited if the spectrum of the transmitted signal is placed above this frequency. This can be achieved by appropriate line coding or modulation.

Thermal noise is another source of additive disturbance and is typically modeled as white Gaussian noise. Its level is about 45–60 dB lower than the level of the used signal.

Echo is also a problem in data transmission over subscriber loops. It is particularly significant if a subscriber loop is used in duplex transmission in the same frequency band. The main echo component is the result of a mismatch between a subscriber loop and the hybrid to which a local transmitter and receiver are connected. Other causes of echo are small inaccuracies in the matching of particular loop segments, and, in some countries, parallel unloaded taps.

Summarizing, as in the case of other transmission channels, the choice of modulation type and duplex transmission method resulting from the channel properties determines the transmission range in the subscriber loop. The task of a transmission system designer having at his/her disposal the telephone network with given statistics of the loop lengths and loop parameters is to design a system that will be accessible for the highest possible percentage of subscribers residing in the area served by the telephone network.

As we have already mentioned, the twisted copper wire pair is expected to be used for communication with subscribers for many years, although expansion of radio access techniques and enhancing their service offer will change the importance of a wireline subscriber loop. More and more often "the last mile" connection will be realized in the form of a wireless subscriber loop or a wideband radio access to a digital network. Cellular systems and their data transmission capabilities are growing rapidly, successfully competing with fast subscriber loop access systems. Another observed tendency is the approach of optical fiber links closer and closer to end users, offering data transmission rates that will be difficult to achieve in radio access systems due to the spectral limitations of the latter.

5.5 Line-of-Sight Radio Channel

Line-of-sight radio links are an alternative to wireline links applied to carry digital streams generated by several systems. They can be the streams transmitted between switching exchanges, between a base station and a base station controller in a cellular radio system, or they can be a digital stream representing TV or audio signals applied in the distribution part of digital broadcasting systems.

A channel model of line-of-sight radio characterizes a specific transmission scenario. Transmission takes place between two strongly directive mutually visible antennas. These antennas are placed on high masts or towers in order to ensure a wide transmission range and to minimize defraction by several terrain obstacles. Figure 5.19 symbolically presents such a situation. Radio transmission takes place in several microwave bands. Table 5.2 shows the frequency ranges applied in different line-of-sight systems, quoted after Manning (1999).

In typical situations antennas are deployed on towers of height 50–100 m. The distance between the towers depends on the applied carrier frequency. Frequencies above 10 GHz are subject to strong attenuation, which increases significantly during rainfalls. Table 5.3 presents typical distances between antennas, depending on the applied band.

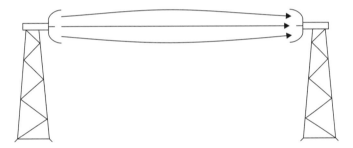

Figure 5.19   Channel of a line-of-sight radio link

Table 5.2   Frequency ranges applied in line-of-sight radio links

Band [GHz]	Range [GHz]	Comment
2	1.7–2.7	Currently applied by DECT and PCS
4	3.8–4.2	High throughput; the band used by a public operator
6	5.9–7.1	High throughput; the band used by a public operator
7–8	7.1–8.5	Long distance links; medium or high throughput
11	10.7–11.7	High throughput; the band used by a public operator
13	12.7–13.3	Small or medium throughput
15	14.4–15.4	All throughputs
18	17.7–19.7	The band used by a public operator; small or medium throughput
23	21.2–23.6	All throughputs
26	24.5–26.5	All throughputs
38	37–39.5	All throughputs

Table 5.3 Allowable distances between antennas
depending on the applied frequency band

Band	Maximum distance between antennas
7 GHz	>39 km
13/15/18 GHz	15–30 km
23/26 GHz	5–15 km
38 GHz	<5 km

Most of the time a microwave line-of-sight channel is well described by the model of an additive white Gaussian noise channel. The signal emitted by a transmit antenna reaches the receive antenna over a specular path, because a typical antenna emits a signal beam of angular width about $1°$. The channel introduces attenuation because only part of the energy reaches the receiver and the remaining part dissipates in the troposphere. Unfortunately, in a small fraction of time the signal propagation between the transmit and receive antennas becomes multipath due to bad meteorological conditions. Such *multipath propagation* has been symbolically sketched in Figure 5.19. We observe multipath propagation, because atypical change of the refraction coefficient as a function of height of the atmosphere layers takes place and part of the emitted energy that would normally be dissipated is reflected from those layers and reaches the receive antenna. Several propagation paths differ in length, so they introduce different delays. In consequence, the received signal is a sum of many delayed and attenuated replicas of the transmitted signal. The impulse response of the channel in the state of multipath propagation can be characterized by the following formula

$$h(t) = \sum_{k=1}^{N} A_k \delta(t - \tau_k) \qquad (5.44)$$

where A_k and τ_k are the gain coefficient and delay of the kth path, respectively. The corresponding channel transfer function is then equal to

$$H(f) = \sum_{k=1}^{N} A_k \exp(-j2\pi f \tau_k) \qquad (5.45)$$

As in the case of a telephone channel, the line-of-sight radio channel can be described by its attenuation characteristics $A(f) = -20 \log_{10} |H(f)|$ and the delay characteristics $\tau(f) = -\frac{1}{2\pi} \frac{d[\arg H(f)]}{df}$. For some values of coefficient A_k and delay τ_k the channel introduces strong attenuation as a result of the fact that some frequency components arrive at the receiver in opposite phases but with similar levels of amplitudes. In turn, some other frequency components arrive at the receiver along different paths and add to each other constructively. Thus, we observe deep attenuation in some frequency ranges and much smaller attenuation in other ranges. This phenomenon is dynamic in time. If attenuation changes in time practically equally over the whole range of signal frequencies, we talk about *flat fading*. When the attenuation variations strongly depend on frequency, we observe *selective fading*. Both phenomena have a substantial influence on the binary error rate and if the latter exceeds the assumed level, e.g. 10^{-3}, we mean that transmission is no longer possible and the system status is declared as a *system outage*. Usually the variations of channel characteristics in time are much slower than the applied signaling rate of the transmitted signal. In such a situation a channel appears to be quasi-static. As we see, relative channel time variability is of primary importance for digital transmission.

Measurements indicate that when a line-of-sight channel of bandwidth 50 MHz or less is in the multipath propagation mode, typically a single fade occurs inside the passband. It has the form of strong attenuation at a given frequency or the form of attenuation increasing in the direction of one of the ends of the channel passband. The measurements reported in Martin *et al.* (1983) show that in extreme cases the fading rate achieves up to 100 dB/s, whereas the fade can move along the frequency axis at the rate of about 10–30 MHz/s, achieving a rate up to 100 MHz/s. For the carrier frequency of a few GHz a channel of bandwidth of the order of 50 MHz can be considered as narrowband. Therefore, it is not possible to extract individual components of the signal reaching the receiver along individual paths characterized by the pairs A_k and τ_k. Thus, these parameters cannot be estimated. In practical situations a single selective fade appears within the band of the channel in the multipath mode, thus a few mathematical models have been worked out to characterize line-of-sight channels in this mode. The most popular one is the Rummler model (Rummler 1979; Rummler 1981). The channel transfer function used in this model is described by the formula

$$H(f) = a\{1 - b \exp[-j2\pi(f - f_0)\tau]\} \tag{5.46}$$

Coefficient a describes attenuation introduced by the channel, coefficient b is relative attenuation of the path delayed by τ with respect to the first path and f_0 determines the frequency for which the maximum attenuation occurs. Figure 5.20 presents an example of the amplitude characteristics of concatenation of the Rummler channel model using the

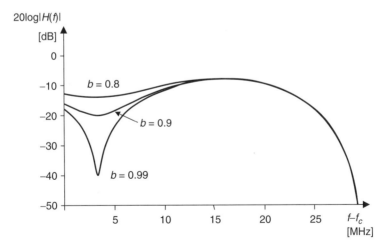

Figure 5.20 Amplitude characteristics of a serial connection of the Rummler channel model ($a = 1$; $b = 0.8, 0.9, 0.99$; $\tau = 6.3$ ns; $f_0 = 3$ MHz) and the filter with the raised cosine characteristics (roll-off factor $\alpha = 0.5$)

filter with raised cosine characteristics. The latter filter jointly describes the transmit and receive filters applied in a line-of-sight radio system. The channel amplitude characteristics distorted by selective fading are a source of intersymbol interference.

Summarizing, let us stress that most of the time a channel of line-of-sight radio is well characterized by a band-limited channel with additive white Gaussian noise. In the multipath mode occurring during the remaining small fraction of time, the channel characteristics are well defined by formula (5.46).

5.6 Mobile Radio Channel

Mobile communication systems have been developing intensively since the 1980s. Mobile phones have become a common good. Transmission between a mobile phone and a fixed part of the radio system has a digital form. Thus, it would be advantageous to characterize mobile communication channel with respect to digital transmission. A more detailed description of the cellular radio channel can be found in Pätzold (2002). At this point we will present only essential information related to this topic.

In a cellular mobile system the system coverage area is divided into subareas called *cells*. Very often the cells are further divided into *sectors*. *Mobile stations* communicate with a *base station* that is installed in each cell or sector. A mobile station can change its location with respect to the base station with which it currently communicates or move to the area covered by another base station. Base stations emit the signal in wide angle resulting from the number of sectors, e.g. 120° for three sectors or 360° in the case of an omnidirectional antenna. A mobile station is also equipped with an omnidirectional antenna to ensure approximately the same link quality at different positions of a mobile station user with respect to the base station. The kinds of applied antennas, place of their installment and possibility of moving the mobile stations with respect to the base station determine the properties of a mobile communication channel.

Figure 5.21 presents an example scenario of a signal propagation from a base station to a mobile station located in a moving vehicle. A wide emission angle of the base station antenna and omnidirectionality of the mobile station antenna imply reception of the signal by the mobile station from all possible directions. The signal components appear as a result of reflections from terrain obstacles and refractions on their sharp edges. From our own observations we know very well that communication is also possible when there is no optical visibility between a transmit and receive antenna. This means that a mobile station is able to operate when receiving only the signal components that have been reflected or refracted from the obstacles. In general, a signal travels from a base station to a mobile station (and vice versa) in the form of differently attenuated and delayed replicas of the transmitted signal.

In a typical situation the signal propagates along a few distinguishable paths, and in the direct neighborhood of a mobile station it is locally dispersed on the obstacles located close to it. The propagation scenario is relatively static in a short time interval, although this quasi-stationarity depends on the signaling rate, applied carrier frequency and velocity of the mobile station. In a scenario typical for a popular GSM system that operates in the 900, 1800 and 1900 MHz bands at a signalling rate of 270.833 kbits, we assume that the channel is approximately static in a period of about 150 data symbols even if the mobile

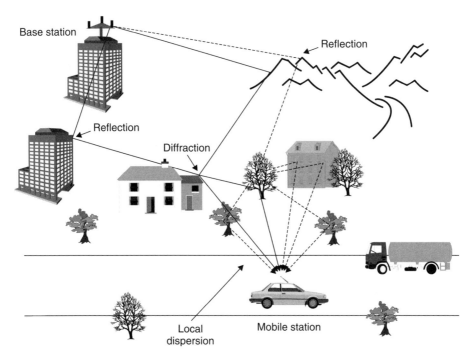

Figure 5.21 An example scenario of radio propagation between base and mobile station (Wesołowski 2002)

station operating in the 900 MHz band moves at a speed up to 250 kmph or at a speed of about 130 kmph if the system operates in the 1800 MHz band.

The *Doppler effect* resulting from movement of the mobile station or elements of the propagation environment is an important phenomenon observed in mobile channels. The Doppler effect causes a frequency shift called the *Doppler frequency*. The Doppler frequency of each signal component depends on the angle φ_i between the direction of the mobile station route and the direction of arrival of the ith received signal component. It also depends on the carrier frequency f_c and the velocity v of the mobile station. The Doppler frequency is described by the formula

$$f_{Di} = f_c \cdot \frac{v}{c} \cos \varphi_i \tag{5.47}$$

where c is the speed of electromagnetic wave propagation. Let us note that the Doppler frequency has its maximum magnitude equal to $f_{D\,\text{max}} = f_c v/c$ if the direction of movement of a mobile station is the same as or opposite to the direction of the electromagnetic wave propagation.

We can show (see (Wesołowski 2002)) that for a given signal propagation scenario we can define a model in which only baseband signals are applied. As we already know, such a channel model is called a baseband equivalent channel model. The model is sketched in Figure 5.22 and consists of a tapped delay line, where the tap coefficients are functions

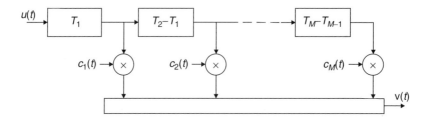

Figure 5.22   Short-term mobile channel model

of time. All the signals and tap coefficients are complex functions of time and their real and imaginary components correpond to the in-phase and quadrature signal components. Values $T_1, T_2, \ldots, T_M$ denote delays at which the signals reach a receiver along M main propagation paths. Signal dispersion in the local vicinity of the mobile station causes minor changes in delays and also results in the signal arriving at the receive antenna from different directions. Movement of the mobile station, or alternatively of the elements of the propagation environment, is taken into account in the time variation of the tap coefficients. The impulse response of the modeled channel is given by the following formula

$$h(t, \tau) = \sum_{k=1}^{M} c_k(t)\delta(\tau - T_k) \tag{5.48}$$

Let us note that there are two arguments representing time. The first of them, i.e. t describes time flow associated with the time variability of the channel characteristics, whereas the second one, i.e. τ is related to a short term channel description in time span of the order of a few lengths of the channel impulse response. If the channel were fully static then the tap coefficients would not depend on time and the channel impulse response would only depend on τ.

It remains to consider which function should be applied to describe the time variability of the weighting coefficients $c_k(t)$, $k = 1, 2, \ldots, M$. This function depends on the Doppler frequency, and on geometrical properties of the dispersing environment around the mobile station. Jakes (1974) proved that in the case of the signal approaching the receive antenna with equal angular power density from all the directions, the power spectral density of signal $c_k(t)$ as a function of the Doppler frequency is given by the formula

$$G_k(f_D) = \begin{cases} \dfrac{\sigma_k^2}{\pi\sqrt{f_{D,\max}^2 - f_D^2}} & \text{for } |f_D| < f_{D,\max} \\ 0 & \text{otherwise} \end{cases} \tag{5.49}$$

In order to prove the formula, see Jakes (1974). However, if both the dispersed signal and the signal along a specular path jointly reach the receiver, the signal component featuring the lowest delay has a power spectral density that is the sum of expression (5.49) and the Dirac delta function placed on the Doppler frequency $f_{D,\text{dir}}$ resulting from the velocity of the mobile station and the angle between the velocity vector and the direction of arrival of the direct, specular wave. The frequency $f_{D,\max}$ appearing in formula (5.49) is a maximum Doppler frequency whereas σ_k^2 symbolizes the mean power of the kth weighting coefficient $c_k(t)$. Figure 5.23 presents the power spectral density described by formula (5.49) supplemented by a Dirac delta pulse of intensity σ_{dir}^2 corresponding to the mean power of the signal from the direct, specular path.

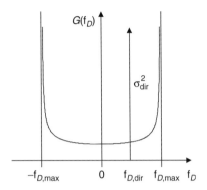

Figure 5.23 Power spectral density of the weighting coefficient of the kth propagation path of model (5.48) describing the channel impulse response

The above considerations allow us to derive another way of characterizing the time properties of a transmission channel in mobile radio systems. Namely, knowing the mean powers $\sigma_1^2, \sigma_2^2, \ldots, \sigma_M^2$ of the signals reaching the receiver over subsequent propagation paths and path delays $T_1, T_2, \ldots, T_M$ we often present a so-called power delay profile $P(\tau)$ describing the mean power distribution in time. The power delay profile is given by the formula

$$P(\tau) = \sum_{k=1}^{M} \sigma_k^2 \delta(\tau - T_k) \tag{5.50}$$

This formula describes a statistical signal spread in time resulting from multipath propagation. An example of such a plot is shown in Figure 5.24. It is a fragment of the channel model description contained in the ETSI standard of radio transmission in GSM 05.05 (2000). It shows a power delay profile describing the spread of the mean power for signal propagation in a typical urban environment. Let us stress again that the model characterized by Figure 5.22 and formulae (5.48) and (5.50) is valid in short time intervals only. If the propagation environment changes sufficiently, delays $T_1, T_2, \ldots, T_M$ and the path mean powers $\sigma_1^2, \sigma_2^2, \ldots, \sigma_M^2$ can change their values and a new channel model has to be used.

Multipath propagation and associated dispersion of the transmitted signal in time is obviously not the only distortion occurring in a cellular mobile radio channel. The immanent feature of the mobile cellular channel is the occurrence of disturbances from use of the same channel frequencies in cells located at a "safe" distance from the reference cell in which the given channel frequencies are applied. This is a so-called *co-channel interference*. Each receiver receives a few signals at the given channel frequency: the

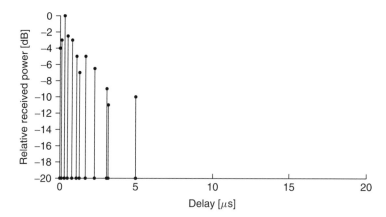

Figure 5.24 Power delay profile for a typical urban environment, according to ETSI GSM 05.05 Standard (2000)

desired one and some others generated in other cells. Their number depends on the system design, in particular on the applied *multiple access* method, i.e. sharing time and frequency resources by many system users, or the number of applied sectors in the cells.

The next kind of disturbances are the signals pervading from the neighboring channels on the frequency axis. They create the so-called *inter-channel interference*. Appropriate selection of frequency channels in each cell allows the influence of this kind of disturbance to be minimized. Another disturbance that occurs in a cellular mobile radio channel is impulsive noise, which mostly originates from human activities or car ignition systems.

Apart from the disturbances described above, the cellular mobile radio channel features some nonlinearities caused by transmit amplifiers. Nonlinearity is particularly visible if the transmitted signal has a high peak-to-average power ratio. For these reasons, if possible, this ratio should be minimized. The most common way to avoid nonlinearities is to apply a signal featuring a constant envelope. This topic is considered in more detail in Chapter 4.

5.7   Examples of Other Radio Channels

5.7.1   Wireless Local Area Network (WLAN) Channel

Development of digital wireless transmission systems opens wide perspectives for new services applications. One of them is wireless access to local computer networks. In wireless local area networks (WLANs) digital transmission is performed over short distances up to a few hundred meters. This distance has a crucial influence on the scale of physical phenomena taking place in WLAN channels. Generally, the properties of a WLAN channel are similar to those of a cellular radio channel. The differences result from the channel bandwidth, the range of applied frequencies and the distance between a mobile terminal and a base station called, in WLAN nomenclature, an *Access Point* (AP). The type of environment in which a signal propagates and channel nonstationarity are of great importance as well.

A mobile terminal is most often a functional part of a portable computer such as a laptop or palmtop. However, the computer often remains in one place while some elements of the propagation environment are moving, e.g. people walking in the immediate vicinity of the WLAN terminal or of the AP. Physical properties of the buildings in which digital transmission is performed also affect the overall channel properties. Generally, a WLAN channel features multipath propagation, although its time spread is much smaller than in a cellular radio channel. This is a result of small distances between a mobile terminal and an AP and of geometrical properties of the room or office space in which transmission is performed. Echoes arising due to reflections from the walls depend on the properties of the construction materials of which the walls have been built. More information on WLAN channel properties can be found in Wesołowski (2002) or Rappaport (1996).

5.7.2   Channel in Satellite Transmission

Satellite links (Figure 5.25) are often used in digital transmission. A good example of satellite link application is the distribution of TV signals in the satellite segment of the Digital Video Broadcasting (DVB) system (DVB-S). In DVB-S the signals are broadcast from the satellite to many users, so transmission is mostly unidirectional.[3] There are two-directional satellite systems as well, such as VSATs (*Very Small Aperture Terminals*). The VSATs are used in data transmission on satellite links with antennas of diameters in the range 1.2–1.8 m, installed directly on the subscriber's premises. We should also mention personal satellite communication systems such as IRIDIUM or GLOBALSTAR, in which a user terminal is similar to a cellular phone.

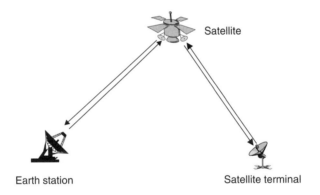

Figure 5.25   Basic scheme of a satellite link (Li 1980) © 1980 IEEE

A characteristic feature of a typical satellite channel is the existence of a specular path between a transmitter and receiver. Because the distance between them is very large (around 35780 km for a geostationary satellite, or 700–800 km for LEO – *Low Earth Orbit* – satellites), strong signal attenuation is observed. This implies relatively low values of the signal-to-noise ratio. Consequently, low-level modulations are mostly applied. If a satellite is placed sufficiently high above the horizon, the propagation is mainly single-path.

---

[3] There are special services that require two-directional transmission as well. An example of such services is the so-called Digital Satellite News Gathering (DSNG).

A satellite *transponder* plays a crucial role in the determination of satellite channel properties. Its function is similar to that of a regenerator in cable transmission. The transponder receives strongly attenuated signals from an earth station or a user terminal, decides upon the received data, shapes data pulses, places the signal in the spectral range appropriate for transmission in a downlink direction and amplifies the signal to such a level that a sufficient signal-to-noise ratio at the receiver is ensured. The system located on the satellite has a limited power supply, therefore a power-efficient amplifier has to be applied. In consequence, the amplifier operates in the range where nonlinearity of its characteristics is already observed. Nonlinearity of the transponder's characteristics is one of the important features of a satellite channel. In order to minimize its influence, digital modulations featuring a low peak-to-average power ratio, e.g. QPSK, are applied.

5.7.3 Short-Wave (HF) Channel

A short-wave or *High-Frequency* (HF) channel has been used in radio communications for many years. Its importance has decreased due to satellite communications, but it is used in special point-to-point applications such as military, maritime or diplomatic. Owing to specific features of wave propagation in the range between 2 and 30 MHz, i.e. the HF range, if the carrier frequency is appropriately selected it is even possible to communicate between continents.

In the frequency range of 2–30 MHz electromagnetic waves emitted along the earth's surface are heavily attenuated. Consequently, the transmission range is relatively small and decreases with a carrier frequency increase. For example, for a carrier frequency of 5 MHz and transmitted signal power of 1 kW, the transmission range is about 100–500 km along the earth's surface and about 1000 km over the sea surface (Wiesner 1984). Fortunately, the transmission range can be considerably higher due to ionospheric reflections.

The ionosphere is a layer of the atmosphere located above the troposphere. It starts at a height of about 60 km and reaches a height of 600 km. It consists of a few layers and, due to solar radiation, ionization of gases contained within it takes place. The intensity of the process depends on the time of day and the season of the year. Ionization causes partial absorption of the HF radiation emitted by a transmitter. However, the other part of the signal is reflected and directed back to the earth's surface (Figure 5.26). The reflecting layer is irregular and its height varies in time, causing the Doppler effect and mulipath propagation. Electromagnetic waves can be also reflected from the earth's surface. Therefore a multiple process of reflections from the earth and the ionosphere is possible, which is shown in Figure 5.27. Taking all this into consideration, we can represent the HF channel in a short time range with a model similar to that shown in Figure 5.22. The duration of the channel impulse response, i.e. the time spread resulting from the multipath propagation, is contained in the range between 0.5 and 5 ms, and typically is about 1.5 ms. The Doppler effect and fluctuations in signal attenuation are basic reasons for time variability of the HF channel and the frequency spectrum shift, which in extreme situations can reach up to 10 Hz. We can conclude on the basis of the literature studies that time-varying tap coefficients of the channel model can be interpreted as sample functions of random processes of bandwidth of the order of a few herz. Let us note that in typical HF data transmission applications the transmission band is of acoustic width due to the fact that the main HF application is voice transmission.

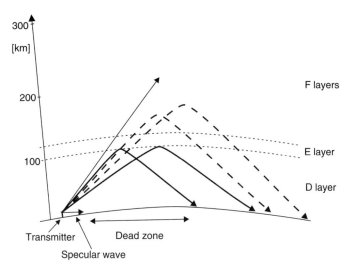

Figure 5.26   Wave propagation in the HF channel with ionospheric reflections

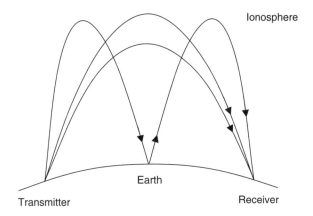

Figure 5.27   Examples of paths with a single reflection and triple reflections in the HF channel

In special applications, e.g. military, the bandwidth of the radio channel for land mobile services is often 25 kHz and the transmission range is a few kilometers for signals in bands between 20 up to 512 MHz. As we see, the frequencies start in the HF band, cover the VHF (*Very High Frequency*: 30–300 MHz) range and reach the lower part of the UHF (*Ultra High Frequency*: 300 MHz–3 GHz) band. For such scenarios, the transmission channel is almost flat fading and can be modeled in the form of a multiplier with a time-varying channel gain coefficient.

5.8   Basic Properties of Optical Fiber Channels

Transmission systems using light pulses propagated along optical fibers were introduced in the 1970s and since then have steadily gained importance. Optical fiber systems and physical phenomena associated with light propagation over optical fibers are usually the topics of separate courses in electronic and communication studies, so in this section we will present only some basic optical fiber channel properties shown from the point of view of digital transmission systems.

Electromagnetic waves applied in optical fiber transmission have wavelengths between 800 and 1600 nm. Let us note that in optical fiber systems wavelengths rather than frequencies are used in system description, unlike many other communication systems. Figure 5.28 presents the attenuation per unit length given in dB/km for a typical optical fiber as a function of wavelength. Three wavelength ranges, the so-called windows, are of practical use: 800–900 nm (*the first transmission window*), 1250–1350 nm (*the second transmission window*) and 1500–1600 nm (*the third transmission window*). The frequency of light emission of wavelength λ is given by the expression

$$f = \frac{v_p}{\lambda} = \frac{c/n}{\lambda} \qquad (5.51)$$

where v_p denotes the light propagation velocity in the given transmission medium measured in m/s, c is the light propagation velocity in a vacuum and n is the light refraction index in a given propagation medium. In the case of glass this index is 1.5, therefore $v_p = 2 \times 10^8$ m/s. Let us note that for the third transmission window, taking advantage of formula (5.51), we receive the upper and lower band limits equal to 1.25×10^{14} and 1.33×10^{14}, respectively, so the bandwidth of the window is equal to 0.08×10^{14} Hz, i.e. it is equal to 8 THz. In radio systems such a bandwidth is difficult to imagine. Theoretically it makes possible extremely fast digital transmission or, equivalently, parallel

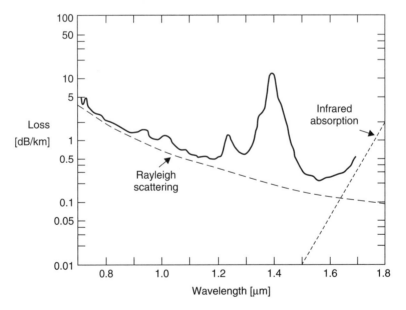

Figure 5.28   Plot of optical fiber attenuation as a function of wavelength. Reproduced with kind permission of IEEE (Li 1980 © IEEE 1980)

transmission of many very fast data streams using different carrier wavelengths. According to the plot in Figure 5.28, in the third transmission window an optical fiber shows attenuation of about 0.3 dB/km only,[4] therefore the sections of optical fibers between regenerators can be very long. In this context optical fiber transmission has a number of advantages: a very wide bandwidth available for data transmission allowing very high data rates, the possibility of deployment of very long links without regenerators, robustness to electromagnetic perturbations such as atmospheric storms or man-made impulse noise, lack of crosstalk between optical fibers placed in the same cable, difficulty in intercepting transmitted signals, galvanic separation of transmitters and receivers, and a very low error rate resulting from a high signal-to-noise ratio.

At the same time an optical fiber channel has some limitations resulting from the construction of optical fibers and light-emitting sources. The fundamental physical law on which optical fiber transmission is based is Snell's law. According to this law the ratio of the sines of the angles of incidence θ_1 and refraction θ_2 between two media is equivalent to the opposite ratio of the indices of refraction n_1 and n_2 of these media. This law is described by the formula

$$\frac{\sin \theta_1}{\sin \theta_2} = \frac{n_2}{n_1} \tag{5.52}$$

and is illustrated in Figure 5.29.

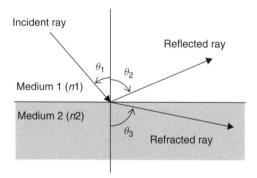

Figure 5.29   Illustration of Snell's law

A typical optical fiber is built of a core, a cladding layer and an external coating called a jacket. The first two elements have a significant impact on wave propagation within the core. The refractive index of the core is larger than that of the cladding, so light propagates exclusively inside the core if the light ray is introduced into the core at the appropriate angle.

Light pulses transmitted through an optical fiber are subject to dispersion. There are a few kinds of dispersion. Generally, dispersion implies spreading of the transmitted pulses, which in turn causes intersymbol interference and constitutes a practical limitation for the

---

[4] Due to technological progress the attenuation level has further decreased and current solutions ensure attenuation below 0.2 dB/km.

transmission medium. The measure of dispersion of the transmitted pulses is the value of $\Delta\tau$ given by formula

$$\Delta\tau = (M + M_g)L\Delta\lambda \qquad (5.53)$$

where M is the coefficient of the *material dispersion*, M_g is the coefficient of *waveguide dispersion* resulting from the optical fiber geometry, L denotes the fiber length in km and $\Delta\lambda$ is the spectral width of the light-emitting source, given in nanometers. The joint dispersion coefficient $(M + M_g)$ is equal to 120, 0 and 15 ps/(nm·km) for wavelengths of 850, 1300 and 1550 nm, respectively.

Material dispersion results from dependence of the refractive index on the light wavelength. The light sources emit pulses that are not fully monochromatic. The bandwidth of emission for a laser transmitter is of the order of 1–3 nm, however for a light-emitting diode it is in the range 30–50 nm. The components of the emitted light featuring different colors diffract on the core boundary, which leads to different path lengths for these color components and eventually causes pulse spreading. Material dispersion can hardly be avoided. However, owing to the appropriate construction of an optical fiber we can avoid the so-called *modal dispersion*.

Construction of an optical fiber is characterized by a specific profile. Three basic profiles are shown in Figure 5.30. Figure 5.30a visualizes a step-index optical fiber. In this type of fiber light rays travel along the fiber in a multimodal way, causing modal dispersion. Energy for different modes propagates at different velocities, resulting in additional spreading of the light pulses. The fastest propagation takes place along the concentric path. Other paths that feature reflections from the core and cladding boundaries are longer and the signal propagating along them reaches the receiver after a longer time. The sum of signals arriving along different paths causes pulse spreading, as in multipath radio propagation. As a result, such fibers are applied if the transmission bandwidth and the length of the transmission line are not large.

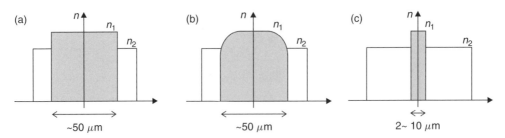

Figure 5.30 Three types of optical fibers: multimode step-index (a), multimode graded-index (b), monomode step-index (c)

Modal dispersion can be significantly decreased when a graded-index fiber is applied (Figure 5.30b), for which the refractive index varies with the distance from the fiber axis in such a way that propagation times of rays along each path are equalized. Finally, by applying a monomode fiber in which the core has a diameter of the order of 2–10 μm, we limit propagation to a single mode only, the one in which propagation is concentric.

Monomode fibers are preferred for applications in which a wide bandwidth is required and transmission lines are very long.

Besides dispersion, which in practice limits the transmission rate, noise is an additional disturbance occurring in optical fibers. Additive noise has a level a few orders lower than in traditional communication systems. However, there also exists multiplicative noise, whose level depends on the generated signal power.

The characteristic properties of optical fiber transmission also reflect the properties of applied light sources that are controlled by electrical signals. Typical sources are *Laser Diodes* (LD) and *Light-Emitting Diodes* (LED). These devices are controlled by current. The power P_{Tx} of an emitted light signal depends on the value of the current i according to the formula

$$P_{Tx} = k_0 + k_1 i \tag{5.54}$$

where k_0 and k_1 are constants. Let us note that this dependence is nonlinear in the sense of the superposition rule. At the optical fiber output a photodetector converts the received light signal back into an electrical current signal. This time current i at the output of a photodetector depends on the received light signal power according to the formula

$$i = \rho P_{Rx} \tag{5.55}$$

where ρ is the *photodetector sensitivity* and $\rho \approx 0.5$ mA/mW. Due to the nonlinear character of the transmitter, only simple modulations may be used in optical fiber links compared with those applied in copper wire or radio links.

5.9 Conclusions

The transmission channel is an extremely important block in the communication system. Its properties determine possible types of digital transmission, achievable data rates, the construction of a transmitter and receiver, and the level of complexity of applied channel coding or data block exchange procedures such as ARQ. Knowledge of the telephone channel properties, and the properties of its part, i.e. the subscriber loop, allows better use of them in digital transmission. Radio channels are also very important. Owing to the fact that we know their properties, and owing to advanced digital signal processing techniques applied in the transmitter and receiver, they have become a popular transmission medium. By performing a rough overview of the properties of an optical fiber channel we have tried to show its specific features and its ability to carry data streams with very high throughputs.

Problems

Problem 5.1 *Let us assume that the signal at the analog-to-digital converter (A/D) has a Gaussian distribution with a zero mean and variance σ^2. The n-bit linear A/D converter spans the range $[-4\sigma, 4\sigma]$. Calculate the signal-to-quantization noise power ratio assuming that the number of quantization levels, 2^n, is so high that we can assume that the quantization noise has uniform distribution.*

Problem 5.2 *Solve Problem 5.1 for the input signal of the same Gaussian distribution and for a nonlinear A/D converter. The A/D converter can be decomposed into a linear quantizer and nonlinear compressor operating according to the μ-law given by formula (5.29).*

Problem 5.3 *Prove formulae (5.32) using the general transmission line theory.*

Problem 5.4 *From the general expression describing the currents and voltages and the related ABCD matrix of the two-port shown in Figure 5.14, derive the values of A, B, C and D of the ABCD matrix for the circuits shown in Figure 5.31.*

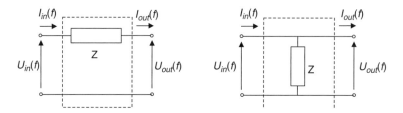

Figure 5.31 Two circuits considered in Problem 5.4

Problem 5.5 *Show that if several line segments are connected in series, then the line ABCD matrix of the whole chain of line segments is a product of the line matrices of each line segment.*

Problem 5.6 *Write a computer program that calculates the channel transfer function and the corresponding channel impulse response of a chain of subscriber loop line segments of given lengths $l_1, l_2, \ldots, l_N$ and line parameters R_i', C_i', L_i', G_i' ($i = 1, 2, \ldots, N$). Take into account the load impedance Z_L and the source impedance Z_S, and a single bridged unloaded tap, as an option. Draw the channel transfer function and the impulse response for the line configuration shown in Figure 5.16 when the first segment, of diameter equal to 0.4 mm, has the length $l_1 = 2$ km and parameters $R' = 300\ \Omega/km$, $C' = 42\ nF/km$, $L' = 1\ mH/km$ and $G' = 1\ \mu S/km$, and the second segment has the length $l_2 = 1$ km and the same parameters. The bridged tap has the length $l_3 = 0.5$ km, the same line parameters and it remains unloaded. The source and load impedances are $Z_S = Z_L = 300\ \Omega$.*

Problem 5.7 *Receivers of GSM signals operating in the 900 MHz band have been designed to cope with the Doppler effect when they move with a speed up to 250 km/h. Calculate the maximum Doppler frequency for this case. What is the maximum speed of the moving GSM receiver if it operates in the 1800 MHz band?*

Problem 5.8 *Consider 16-QAM transmission over a flat fading radio channel. Let us model the channel as in Figure 5.32. The channel attenuates the transmitted signal equally in its whole passband and we assume that the gain G shown in Figure 5.32 has the Rayleigh distribution [see formula (4.45)] with variance $\sigma_G^2 = 1$. The power of the signal at the channel input is normalized to unity. The power of the additive white Gaussian noise in the signal band $\left(-\frac{1}{T}, \frac{1}{T}\right)$ is 30 dB lower than the signal power at the channel input (T is the signaling period). If the binary error rate is 10^{-3} or higher the transmission system is considered to be in the outage state. Assuming that in regular system operation single-bit*

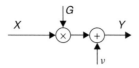

Figure 5.32   Channel model applied in Problem 5.8

errors dominate owing to the applied Gray coding, calculate the probability of the system outage. If needed, compute the probability of system outage numerically.

Problem 5.9   *Consider the baseband equivalent mobile communication channel model described by formula (5.48). Let it be additionally characterized by the power delay profile (5.50) in which $T_k = k \Delta T$ $(\Delta T = 3.69 \, \mu s, \ k = 1, \dots, M)$, the number of signal paths is $M = 6$ and the mean power of the signal reaching the receiver over the kth path is $\sigma_k^2 = \sigma_0^2 \exp(-k/a)$. Let $\sigma_0^2 = 1$ and $a = 2.1714$.*

1. Draw a power delay profile of this channel in the dB scale.
2. Based on the determined power delay profile, and assuming line-of sight (LOS) signal propagation, write a program simulating time variability of the channel impulse response (5.48). In the case of LOS propagation, the channel complex tap coefficients are the sample functions of mutually independent random processes characterized by the Rayleigh distribution of their envelope and the uniform distribution of their phase (check Chapter 4 in order to find that such a process can be synthesized as the sum of two independent in-phase and quadrature Gaussian processes). Apply appropriate filters that shape the time variability of the tap coefficients, resulting in the approximate shape of the Doppler spectrum (5.49). Assume that the maximum Doppler frequency is equal to 200 Hz. Calculate the series of channel transfer functions for the moments $t = n \Delta T (n = 1, 2, \dots)$. In general, for each moment t we have

$$H(f, t) = \int_{-\infty}^{\infty} h(t, \tau) \exp[-j2\pi f \tau] d\tau$$

so apply the adequate and properly scaled FFT algorithm. Assume $\Delta T = 0.5$ ms. Draw the series of transfer functions in the form of a three-dimensional plot along frequency f and time t axes (in Matlab use mesh *or* surf *commands for that purpose).*

6

Digital Transmission on Channels Introducing Intersymbol Interference

6.1 Introduction

Physical channels used in transmission of digital signals can be rarely represented by a non-distorting channel model with additive white Gaussian noise (AWGN) as the only impairment. As we have already mentioned in the previous chapter, the only channel that can be represented most of the time by a band-limited AWGN channel model is a microwave line-of-sight channel. The majority of channels are characterized not only by a limited bandwidth but also by a channel transfer function in which particular frequency components of transmitted signals are unequally attenuated (causing *amplitude distortion*) and unequally delayed (creating *delay distortion*). These effects are the result of the physical properties of the transmission medium and of the imperfect design of transmit and receive filters applied in the transmission system. A good example of the first is the radio channel, in which the transmitted signal reaches the receiver along many different paths through reflections, diffractions and dispersion on the terrain obstacles. As a result, particular signal path components arriving with various attenuations and delays are combined at the receiver. The delayed components can be considered as echoes that cause time dispersion of the transmitted signal. If time dispersion is greater than a substantial fraction of the signaling period, the channel responses to the subsequent data signals overlap. This effect is known as *intersymbol interference* (ISI). Thus, the signal observed at the receiver input contains information on a certain number of data signals simultaneously. In many cases the channel impulse response spans tens of signaling periods and intersymbol interference appears to be a major impairment introduced by the channel.

The destructive influence of intersymbol interference on a digital communication system performance has to be counteracted by special receiver and/or transmitter design. The part of the receiver that counteracts ISI is called the channel *equalizer*. Very often transmission channel characteristics are either not known at the beginning of a data transmission session or they are time variant. Therefore, it is advantageous to make the equalizer

Introduction to Digital Communication Systems   Krzysztof Wesołowski
© 2009 John Wiley & Sons, Ltd

adaptive. The adaptive equalizer is able to adaptively compensate for the distorting channel characteristics and simultaneously track the changes of channel characteristics in time. The latter property is a key feature of equalizers used in digital transmission over nonstationary radio channels.

Since the invention of an equalizer in the early 1960s, hundreds of papers have been devoted to this subject. Adaptive equalization is usually the topic of a separate chapter in leading books on digital communication systems (Proakis 2000; Barry *et al.* 2003; Gitlin *et al.* 1992) and separate books tackle this subject as well (Clark 1985; Ding and Ye 2001). Adaptive equalization is also a well-documented application example in books devoted to adaptive filters (Haykin 2002; Macchi 1995). Interested readers are referred to the bibliography included in such papers as Qureshi (1985) and Taylor *et al.* (1998). The current chapter is partially based on the tutorial published by the author (Wesołowski 2003).

In this chapter we will concentrate on basic structures and algorithms of adaptive equalizers. We will start with a general description of the intersymbol interference channel. Subsequently, we will divide the equalizers into several classes. We will continue our considerations with the basic analysis of adaptation criteria and algorithms for linear and decision feedback equalizers. Then we will concentrate on adaptive algorithms and equalizer structures applying the MAP (*Maximum a Posteriori*) symbol-by-symbol detector and the MLSE (*Maximum Likelihood Sequence Estimation*) detector. We will also describe basic structures and algorithms of adaptive equalization without a training sequence (*blind equalization*). Further, we will tackle turbo-equalizers as a new application of the turbo principle. Finally, we will consider equalizers for MIMO systems. As a case study we will consider the operation of a typical MLSE detector applied in GSM receivers.

6.2 Intersymbol Interference

Let us describe the phenomenon of intersymbol interference using the baseband equivalent channel model (see Chapter 5). As we remember, the output of the baseband equivalent channel is described by the formula

$$x(t) = x^I(t) + jx^Q(t) = \sum_{i=-\infty}^{+\infty} d_i h(t - iT) + v(t) \qquad (6.1)$$

where the complex pulse $h(t)$ is a convolution of the transmit filter response $p(t)$, receive filter response $p(-t)$ and channel impulse response converted to the baseband $g_B(t)$, i.e. $h(t) = p(t) * g_B(t) * p(-t)$. Function $v(t)$ is the noise $n(t)$ filtered by the receive filter, so $v(t) = n(t) * p(-t)$. The baseband equivalent channel model considered in this chapter is shown in Figure 6.1.

The task of the digital system receiver is to find the most probable data symbols d_n^I and d_n^Q on the basis of the signal $x(t)$ observed at its input. Recall that if the channel were non-distorting, then, assuming appropriate shaping of the transmit filter $p(t)$ and the filter $p(-t)$ matched to it, it would be possible to find periodic sampling moments at the outputs of the receive filters such that the samples of signals $x^I(t)$ and $x^Q(t)$ would

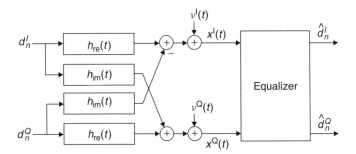

Figure 6.1   Equivalent transmission system model with the equalizer

contain information on single data symbols only. However, the distortion introduced by
the channel makes the finding of such sampling moments impossible. Thus, it is necessary
to apply a special system block denoted in Figure 6.1 as an equalizer, which is able to
detect data symbols on the basis of $x(t)$ [or equivalently $x^I(t)$ and $x^Q(t)$] or its samples.
An equalizer is in fact a kind of receiver that either minimizes the influence of intersymbol
interference or uses it constructively in decisions concerning the transmitted data.

Although the signals $x^I(t)$ and $x^Q(t)$ are represented in Figure 6.1 as continuous time
functions, typically, due to digital implementation, the equalizer accepts their samples
only. Let us temporarily assume that the equalizer input samples are taken with the
symbol period T, with the time offset τ with respect to the zero moment. Then the
equalizer input signal is expressed by the equation

$$x_n = x(nT + \tau) = \sum_{i=-\infty}^{\infty} d_i h(nT + \tau - iT) + v(nT + \tau)$$

$$= \sum_{i=-\infty}^{\infty} d_i h_{n-i} + v_n = \sum_{i=-\infty}^{\infty} h_i d_{n-i} + v_n \qquad (6.2)$$

or

$$x_n = h_0 d_n + \sum_{i=-\infty, \, i\neq 0}^{\infty} h_i d_{n-i} + v_n \qquad (6.3)$$

where $h_{n-i} = h(nT + \tau - iT)$. The first term in (6.3) is proportional to the data symbol
to be decided on. The second term is a linear combination of previous and future data
symbols and expresses intersymbol interference. It should be eliminated or constructively
used by the equalizer. The third term is the additive noise and cannot be eliminated.

6.3   Channel with ISI as a Finite State Machine

Let us consider the baseband equivalent channel model again. The samples taken at the
output of the receive filter with the sampling rate $1/T$ are described by formula (6.2).

This formula can be approximated by the expression

$$x_n = \sum_{i=0}^{L} h_i d_{n-i} + v_n \tag{6.4}$$

in which we have assumed that the channel impulse response effectively has a finite duration equal to $L + 1$. A sample taken from the channel output in the nth time instant is the sum of the unobservable channel output

$$r_n = \sum_{i=0}^{L} h_i d_{n-i} \tag{6.5}$$

and additive noise v_n. In fact, equation (6.5) describes the output sample of a finite impulse response filter that models the channel. Let us assume that data symbols supplied to the channel input (6.4) are selected from a finite alphabet, they are statistically independent and have a given probability distribution. As a result, the output samples $\{r_n\}$ take the values from a finite set of all possible linear combinations of subsequent $L + 1$ data symbols. A particular sequence $\{r_n\}$ is a sample function of the Markov chain. The current state of this chain is determined by the data sequence that is currently contained in the tapped delay line of the finite impulse response filter modeling the channel (6.5). Therefore the channel model can be interpreted as a finite state machine with the states determined by the content of the tapped delay line. Its excitation is the data symbol currently supplied to its input, and the output signal is equal to r_n. Transfer of the finite state machine from a given state in the nth moment to a new state in the $(n + 1)$st moment is determined by the input symbol. Thus, it is clear that a given input data sequence is uniquely associated with the channel state sequence. Let us illustrate this with a simple example.

Example 6.3.1 *Let us consider a channel model with three taps with weighting coefficients h_0, h_1, h_2. Bipolar data symbols $d_n = \pm 1$ are fed to its input. This model is presented in Figure 6.2a. A typical way to illustrate the functioning of a finite state machine is to present its state diagram. The trellis diagram shown in Figure 6.2b is an alternative way to represent a finite state machine. The considered channel has four states, because it has two delay elements and the data stored in them are bipolar. In the general case of M-ary data symbols fed to the channel with $L + 1$ taps, the number of states is equal to M^L. Figure 6.2c presents all possible paths on the trellis diagram starting in the zero moment from the state determined by the data sequence $(1, 1)$. The dotted line denotes the sequence of states through which the channel evolves when a given sequence of data symbols is supplied to its input. These data symbols are shown above the arrows symbolizing the paths between subsequent states. There are M^n possible data sequences starting from the zero moment and ending in the nth timing instant. Recall that in our example $M = 2$. Finding the most likely data sequence in the receiver on the basis of the sequence of samples $\{x_n\}$ seems to be quite a complex task.*

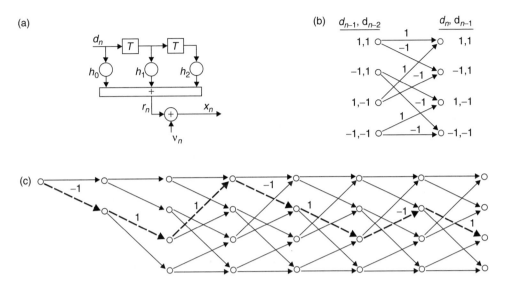

Figure 6.2   Model of the ISI channel with three taps (a), trellis diagram of the channel (b) and possible state sequences (c)

6.4   Classification of Equalizer Structures and Algorithms

Channel equalization can be performed by linear or nonlinear methods. Decisions upon the data symbols can be made by the equalizer on a symbol-by-symbol basis or performed on the whole sequence. Figure 6.3 presents the classification of equalization structures.

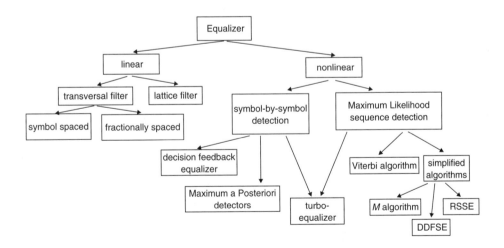

Figure 6.3   Classification of the equalization structures

Within the class of linear receivers the equalizer based on an FIR *transversal filter* is
of great importance. It is implemented using symbol-spaced or fractionally spaced taps.
A lot of attention has also been paid in the literature to the linear equalizer applying a
lattice filter (Satorius and Alexander 1979). The latter, although more complicated than
the transversal filter, assures faster convergence of the adaptation algorithm. Despite this,
due to implementation simplicity, FIR equalizers are most common.

In the case of channels characterized by the occurrence of deep notches, nonlinear
receivers are used. The simplest version of a nonlinear receiver is the *Decision Feedback Equalizer* (DFE) (Monsen 1971). The Maximum Likelihood Sequence Estimation
(MLSE) equalizer, which is more computationally intensive, is applied in GSM receivers
and high-speed telephone modems. It detects a whole sequence of data symbols, usually
using the *Viterbi algorithm* (Forney 1972). Sometimes the MLSE equalizer not only
detects the data symbols but also supplies their likelihoods. This is particularly important if the equalizer is followed by an error correction code decoder that uses soft inputs.
The MLSE equalizer using the modified Viterbi algorithm, which produces soft outputs, is
called a *SOVA equalizer* (cf. the SOVA algorithm for channel code decoding described in
Chapter 2). If the intersymbol interference is caused by a long channel impulse response
or if the data symbol alphabet is large, the MLSE equalizer becomes infeasible due
to excessive computational complexity. Several suboptimal structures and procedures
can be applied instead, e.g. *Reduced State Sequence Estimation* (RSSE; Eyuboglu and
Qureshi 1988), *Delayed Decision Feedback Sequence Estimation* (DDFSE; Duel-Hallen
1992) or the *M* algorithm (Anderson and Mohan 1984). Another approach is a nonlinear
symbol-by-symbol detection using the *Maximum a Posteriori* (MAP) criterion. The algorithm of Abend and Fritchman (1970) is one example of such an approach. The MAP
algorithms are usually computationally complex.

Modern digital communication systems, particularly those operating on radio channels,
are often equipped with a channel code that improves the overall transmission quality.
Treating the channel introducing ISI as an inner convolutional encoder and the channel
encoder applied in the transmitter as an outer code encoder (see the section on concatenated coding in Chapter 2) we are able to apply the turbo principle in the receiver. Thus,
the equalizer (linear, MLSE- or MAP-based) is treated as an inner code decoder and the
whole decoding process is performed in an iterative manner.

The key feature of all equalization structures is their ability to perform initial adaptation
to the channel characteristics (*start-up equalization*) and to track channel characteristics in
time. In order to acquire the initial adaptation, an optimization criterion has to be defined.
Historically, the first criterion was minimization of the maximum value of intersymbol
interference (*Mini-Max criterion*) resulting in the *Zero-Forcing* (ZF) equalizer. The most
popular adaptation criterion is minimization of the *Mean Square Error*, resulting in the
MSE equalizer. In this criterion the expectation of the squared error signal at the equalizer
output is minimized. Finally, the criterion used in the fastest adaptation algorithms relies
on minimization of the *Least Squares* (LS) of errors. The equalizer using the algorithm
based on this criterion is called an *LS* equalizer. The equalizer parameters are selected to
minimize the squared sum of the equalizer output signal errors that would be achieved if
these parameters were used starting from the initial moment of adaptation. Some other cost
functions can be selected if the equalizer coefficients are derived without the knowledge
of the transmitted data symbols.

The equalizer parameters are derived in accordance with a chosen adaptation criterion by an *adaptation algorithm*. Most of the algorithms are *recursive* – the adaptation at a given moment is performed iteratively, taking advantage of the results achieved in the previous adaptation step. In special cases *fast start-up equalization* algorithms are applied, resulting in extremely fast calculation of coarse equalizer parameters that are good enough to start regular data transmission and are later refined. Some of these algorithms are known as *noniterative* and others are *Recursive Least Squares* (RLS) algorithms.

The adaptation process of a typical equalizer can be divided into two phases. In the first phase, the training data sequence known to the receiver is transmitted. The adaptation algorithm uses this sequence as a reference for adjustment of the equalizer coefficients; thus the equalizer is in *training mode*. After achieving the equalizer parameters which result in a sufficiently low probability of errors made by the equalizer decision device, the second phase of adaptation begins in which the equalizer starts to use the derived decisions in its adaptation algorithm. We say that the equalizer is then in the *decision-directed* mode.

In some cases, in particular in point-to-multipoint transmission, sending the training sequence to initiate a particular receiver is not feasible. Thus, the equalizer must be able to adapt without a training sequence. Its algorithm is based exclusively on the general knowledge about the data signal statistics and on the signal reaching the receiver. Such an equalizer is called *blind*. Blind adaptation algorithms are generally either much slower or much more complex than data-trained algorithms.

6.5   Linear Equalizers

The linear equalizer is the simplest structure most frequently used in practice. Let us consider the receiver applying a transversal filter. The scheme of such an equalizer is shown in Figure 6.4. The output signal y_n at the nth moment depends on the input signal samples x_{n-i} and the equalizer coefficients $c_{i,n}$ ($i = -N, \ldots, N$) according to the equation

$$y_n = \sum_{i=-N}^{N} c_{i,n} x_{n-i} \qquad (6.6)$$

The equalizer output signal is a linear combination of $2N + 1$ subsequent samples of the input signal. Indexing the equalizer coefficients from $-N$ to N reflects the fact that the reference tap is located in the middle of the tapped delay line of the equalizer and that, typically, not only previous data symbols with respect to the reference one but also some future symbols influence the current input signal sample.

6.5.1   ZF Equalizers

Historically, the earliest equalizers used the *Mini-Max* adaptation criterion. This resulted in the simplest algorithm, which is sometimes used in the equalizers applied in line-of-sight microwave radio receivers. Let us neglect the additive noise for a while. Taking into account (6.3) and (6.6) we obtain

$$y_n = \sum_{i=-N}^{N} c_{i,n} \sum_{k=-\infty}^{\infty} h_k d_{n-i-k} \qquad (6.7)$$

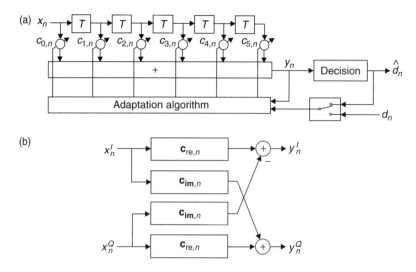

Figure 6.4 Linear adaptive equalizer: (a) basic structure, (b) structure equivalent to the complex filter applying real filters

Substituting $j = i + k$ we get

$$y_n = \sum_{i=-N}^{N} c_{i,n} \sum_{j=-\infty}^{\infty} h_{j-i} d_{n-j} \tag{6.8}$$

or equivalently

$$y_n = \sum_{j=-\infty}^{\infty} g_{j,n} d_{n-j} \quad \text{where} \quad g_{j,n} = \sum_{i=-N}^{N} c_{i,n} h_{j-i} \tag{6.9}$$

and $g_{j,n}$ are the samples of cascade connection of the discrete channel and the equalizer. In the *Mini-Max* criterion the equalizer coefficients $c_{i,n}$ $(i = -N, \ldots, N)$ are adjusted to minimize the expression

$$I = \frac{1}{g_{0,n}} \sum_{j=-\infty, \, j \neq 0}^{\infty} |g_{j,n}| \tag{6.10}$$

Let us note that, because of the finite number of adjustable equalizer coefficients, it is possible to set to zero only part of the ISI samples observed at the output of the equalizer filter. One can show that in order to set the ISI samples to zero, under the assumption that the data symbols are uncorrelated and equiprobable, it suffices to set the equalizer coefficients to force the following equality to be fulfilled,

$$E\left(e_n d_{n-i}^*\right) = 0 \quad \text{for } i = -N, \ldots, N \tag{6.11}$$

where the error e_n in the training mode is given by the expression

$$e_n = y_n - d_n \tag{6.12}$$

or $e_n = y_n - \text{dec}(y_n)$ in the decision-directed mode, and where $(.)^*$ denotes a complex conjugate. In fact, substituting in (6.11) the expression for e_n and y_n we obtain from (6.9)

$$E\left[e_n d_{n-i}^*\right] = E\left[\left(\sum_{j=-\infty}^{\infty} g_{j,n} d_{n-j} - d_n\right) d_{n-i}^*\right] = \begin{cases} 0 \text{ for } i = 0, \text{ if } g_{0,n} = 1 \\ \\ 0 \text{ for } i \neq 0, \ i \in \langle -N, N \rangle, \text{ if } g_{i,n} = 0 \end{cases} \tag{6.13}$$

By forcing condition (6.13), $2N$ intersymbol interference samples can be set to zero. Therefore, such an equalizer is called a *Zero-Forcing equalizer*. If the equalizer was infinitely long it would be able to completely eliminate the ISI at its output. The cascade connection of the channel and equalizer would have a discrete impulse response in the form of a unit pulse. Therefore, the equalizer would ideally inverse the channel frequency characteristics. Such an equalizer could be adjusted iteratively according to the equation

$$c_{i,n+1} = c_{i,n} - \alpha E\left[e_n d_{n-i}^*\right] \quad \text{for } i = -N, \ldots, N \tag{6.14}$$

where α is an appropriately selected small constant, called the adaptation step size. However, replacing the ensemble average with its stochastic estimate, we obtain the following equation for the coefficients' adjustment, which is easily implementable even at a very high symbol rate

$$c_{i,n+1} = c_{i,n} - \alpha e_n d_{n-i} \quad \text{for } i = -N, \ldots, N \tag{6.15}$$

for real equalizers, and

$$c_{i,n+1} = c_{i,n} - \alpha e_n d_{n-i}^* \quad \text{for } i = -N, \ldots, N \tag{6.16}$$

for complex ones. More details on the ZF equalizer can be found in Lucky *et al.* (1968). The ZF equalizer attempting to inverse the channel characteristics amplifies the noise in those frequency regions in which the channel particularly attenuates the signal.

6.5.2 MSE Equalizers

As we have already mentioned, the most frequent adaptation criterion is minimization of the mean square error (MSE), i.e.

$$\min_{\{c_{i,n}, \ i=-N,\ldots,N\}} E\left[|e_n|^2\right] \tag{6.17}$$

where the error is given by equation (6.12). Direct calculations of the mean square error $\mathcal{E}_n^{MSE} = E\left[|e_n|^2\right]$ with respect to the equalizer coefficients $\mathbf{c}_n =$

$\left[c_{-N,n}, \ldots, c_{0,n}, \ldots, c_{N,n} \right]^T$ lead to the following dependence of the MSE on the coefficients for the real equalizer

$$\mathcal{E}_n^{MSE} = E[e_n^2] = E[(\mathbf{c}_n^T \mathbf{x}_n - d_n)(\mathbf{x}_n^T \mathbf{c}_n - d_n)]$$
$$= \mathbf{c}_n^T A \mathbf{c}_n - 2\mathbf{b}^T \mathbf{c}_n + E\left[|d_n|^2 \right] \tag{6.18}$$

where $A = E\left[\mathbf{x}_n \mathbf{x}_n^T \right]$ ($\mathbf{x}_n = \left[x_{n+N}, \ldots, x_n, \ldots, x_{n-N} \right]^T$) is the input signal autocorrelation matrix and $\mathbf{b} = E\left[d_n \mathbf{x}_n \right]$ is the vector of cross-correlation between the current data symbol and the equalizer input samples. The autocorrelation matrix A is positive definite (all its eigenvalues are positive). It is well known from algebra that for such a matrix expression (6.18) has a single and global minimum. The minimum can be found if we set the condition

$$\frac{\partial \mathcal{E}_n^{MSE}}{\partial \mathbf{c}_n} = \begin{bmatrix} \dfrac{\partial \mathcal{E}_n^{\text{MSE}}}{\partial c_{-N,n}} \\ \vdots \\ \dfrac{\partial \mathcal{E}_n^{\text{MSE}}}{\partial c_{N,n}} \end{bmatrix} = 2A\mathbf{c}_n - 2\mathbf{b} = 0 \tag{6.19}$$

The result is the well-known Wiener-Hopf equation for the optimum equalizer coefficients

$$A\mathbf{c}_{\text{opt}} = \mathbf{b} \tag{6.20}$$

An efficient method of achieving the optimum coefficients and the minimum MSE is to update the equalizer coefficients iteratively with adjustments proportional to the negative value of the gradient of $\mathcal{E}_n^{MSE}$ calculated for the current values of the coefficients, i.e.

$$c_{i,n+1} = c_{i,n} - \alpha_n \frac{\partial \mathcal{E}_n^{MSE}}{\partial c_{i,n}} \quad \text{for} \quad i = -N, \ldots, N \tag{6.21}$$

where α_n is a small positive value called the adjustment step size. Generally, it can be time variant, which is expressed by the time index n. Calculation of the gradient $\frac{\partial \mathcal{E}_n^{MSE}}{\partial c_{i,n}}$ leads to the result

$$\frac{\partial \mathcal{E}_n^{MSE}}{\partial c_{i,n}} = \frac{\partial E\left[|e_n|^2 \right]}{\partial c_{i,n}} = 2E\left[e_n \frac{\partial e_n}{\partial c_{i,n}} \right] = 2E\left[e_n x_{n-i} \right] \quad \text{for } i = -N, \ldots, N \tag{6.22}$$

Replacing the gradient calculated in (6.22) by its stochastic estimate $e_n x_{n-i}$ ($i = -N, \ldots, N$) for the real equalizer we receive the stochastic gradient (LMS – *Least Mean Square*) algorithm

$$c_{i,n+1} = c_{i,n} - \gamma_n e_n x_{n-i} \quad \text{for } i = -N, \ldots, N \tag{6.23}$$

where $\gamma_n = 2\alpha_n$. One can show that the analogous equation for the complex equalizer is

$$c_{i,n+1} = c_{i,n} - \gamma_n e_n x^*_{n-i} \text{ for } i = -N, \ldots, N \tag{6.24}$$

Figure 6.5 presents a scheme of the linear transversal equalizer with the tap coefficients adjusted according to algorithm (6.24). The switch changes its position from 1 to 2 after a sufficiently long training mode.

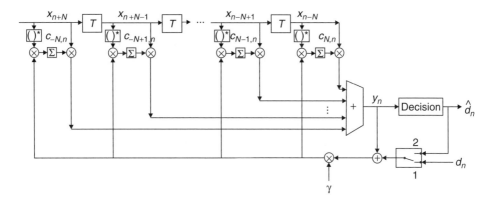

Figure 6.5   Adaptive MSE gradient equalizer

The convergence rate of the LMS algorithm depends on the value of the step size γ_n. This problem has been thoroughly researched. Generally, the value of the step size depends on the eigenvalue distribution of the input signal autocorrelation matrix A (Proakis 2000). G. Ungerboeck (1972) derived a simple "engineering" formula for the step size, which results in fast and stable convergence of the LMS adaptive equalizer. The initial step size is described by the formula

$$\gamma_0 = \frac{1}{(2N + 1)E\left[|x_n|^2\right]} \tag{6.25}$$

where $E\left[|x_n|^2\right]$ is the mean input signal power and is equal to each element of the main diagonal of the autocorrelation matrix A. When the equalizer taps are close to their optimum values, the step size should be decreased in order to prevent a too high level of the residual mean square error (e.g. $\gamma_\infty = 0.2\gamma_0$).

Algorithm (6.24) is very simple and it is often applied in digital transmission systems for which the channel characteristics are stationary or change very slowly in time with respect to the symbol rate. Thus, there are no strict time limits on the training sequence applied in the start-up phase. However, time limits for the start-up phase must be introduced if transmission has the form of short data packets starting with a preamble that plays the role of a training sequence. Obviously the preamble should be much shorter than the transmitted packets and the equalizer start-up procedure should be fast.

6.5.3 LS Equalizers

Particularly fast initial equalizer convergence is achieved if the *Least Squares* adaptation criterion is applied. The coefficients of a linear equalizer are set in order to minimize the following cost function with respect to the filter coefficient vector $\mathbf{c}_n$

$$\mathcal{E}_n^{LS} = \sum_{i=0}^{n} \lambda^{n-i} \left| \mathbf{c}_n^T \mathbf{x}_i - d_i \right|^2 \tag{6.26}$$

For each moment n, the algorithm minimizes the weighted summed squared error starting from the initial moment up to the current moment n, which would be achieved if the current coefficient vector calculated on the basis of the whole signal sequence up to the nth moment were applied in the equalizer from the initial moment. The window coefficient λ^{n-i} ($\lambda \leq 1$) causes gradual forgetting of past errors and is applied for nonstationary channels to follow the changes in the channel characteristics. The calculation of (6.26) leads to equations similar to (6.18) and (6.20)

$$\mathcal{E}_n^{LS} = \mathbf{c}_n^T R_n \mathbf{c}_n - 2\mathbf{c}_n^T \mathbf{q}_n + \sum_{i=0}^{n} \lambda^{n-i} |d_i|^2 \tag{6.27}$$

$$R_n \mathbf{c}_{n,\text{opt}} = \mathbf{q}_n \tag{6.28}$$

where

$$R_n = \sum_{i=0}^{n} \lambda^{n-i} \mathbf{x}_i^T \mathbf{x}_i = \lambda R_{n-1} + \mathbf{x}_n^T \mathbf{x}_n \text{ and } \mathbf{q}_n = \sum_{i=0}^{n} \lambda^{n-i} d_i \mathbf{x}_i \tag{6.29}$$

Instead of solving the set of linear equations (6.28) at each subsequent moment, we can find the optimum coefficients iteratively using the results derived at the previous time instant. Below we list the equations of the standard *Recursive Least Squares* (Kalman) algorithm proposed by Godard (1974) for fast adaptive equalization. The algorithm is quoted after Proakis (2000).

For convenience let us denote $P_n = R_n^{-1}$. Let us also assume that before adaptation at the nth moment we have the filter coefficients $\mathbf{c}_{n-1}$ and the inverse matrix P_{n-1} at our disposal. The algorithm steps are as follows:

- Initialization: $\mathbf{c}_0 = [0, \ldots, 0]^T$, $\mathbf{x}_0 = [0, \ldots, 0]^T$ and $R_n = \delta I$, where δ is a small positive constant.
  Do the following for $n \geq 1$
- Shift the contents of the filter tapped delay line by one position and accept the new input signal x_n
- Compute the filter output signal:

$$y_n = \mathbf{c}_{n-1}^T \mathbf{x}_n \tag{6.30}$$

- Compute the error at the filter output:

$$e_n = d_n - y_n \qquad (6.31)$$

- Compute the *Kalman gain* vector $\mathbf{k}_n = P_n \mathbf{x}_n$:

$$\mathbf{k}_n = \frac{P_{n-1} \mathbf{x}_n}{\lambda + \mathbf{x}_n^T P_n \mathbf{x}_n} \qquad (6.32)$$

- Update the inverse of the autocorrelation matrix:

$$P_n = \frac{1}{\lambda} \left[P_{n-1} - \mathbf{k}_n \mathbf{x}_n^T P_{n-1} \right] \qquad (6.33)$$

- Update the filter coefficients:

$$\mathbf{c}_n = \mathbf{c}_{n-1} + \mathbf{k}_n e_n \qquad (6.34)$$

Formulas (6.30)–(6.34) summarize the RLS Kalman algorithm for a real equalizer. The complex version of this algorithm can be found in Proakis (2000). Knowing that $\mathbf{k}_n = P_n \mathbf{x}_n$, we find that the coefficients' update is equivalent to the formula

$$\mathbf{c}_n = \mathbf{c}_{n-1} + R_n^{-1} \mathbf{x}_n e_n \qquad (6.35)$$

Comparing the equalizer update using the LMS algorithm (6.24) and the RLS algorithm (6.35) we see that the Kalman algorithm speeds up its convergence because of the inverse matrix $P_n = R_n^{-1}$ used in each iteration. In the LMS algorithm this matrix is replaced by a single scalar γ_n. Figure 6.6 presents the convergence rate for both the LMS and RLS algorithms used in the linear transversal equalizer. The step size of the LMS algorithm was constant and selected to ensure the same residual mean square error as that achieved by the RLS algorithm. The difference in the convergence rate is evident. However, we have to admit that for the channel model used in the simulations shown in Figure 6.6 the application of the step size according to formula (6.25) and switching it to a small fraction of the initial value after the appropriate number of iterations improves the convergence of the LMS equalizer considerably. On the other hand, the tracking abilities of the Kalman algorithm are much better than those of the LMS algorithm. However, the RLS Kalman algorithm is much more demanding computationally. Moreover, due to the roundoff noise it becomes numerically unstable in the long run, particularly if the forgetting factor λ is lower than 1. Solving the problem of excessive computational complexity and ensuring numerical stability have been the subject of intensive research. Cioffi and Kailath's paper (1984) is only one representative example of numerous publications in this area.

Besides the transversal filter, a lattice filter can also be applied in the adaptive equalizer using both the LMS (Satorius and Alexander 1979) and RLS (Satorius and Pack 1981) adaptation algorithms.

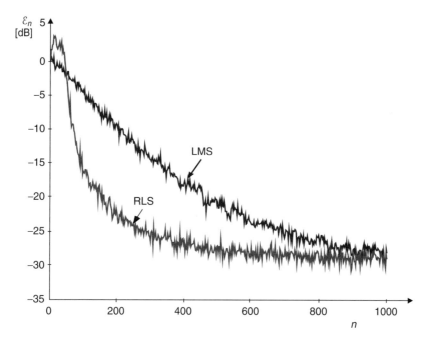

Figure 6.6 Convergence of the constant step size LMS and Kalman RLS equalizer of length $2N = 30$

6.5.4 Choice of Reference Signal

The choice of reference signal plays an important role in the equalizer adaptation process. In fact, the reference signal tests the unknown channel. Its spectral properties should be selected in such a way that the channel characteristics are fully reflected in the spectrum of the signal at the input of the equalizer. So far in our analysis we have assumed that the data symbols are uncorrelated and equiprobable, i.e.

$$E\left[d_n d_{n-k}^*\right] = \begin{cases} \sigma_d^2 & \text{for } k = 0 \\ 0 & \text{for } k \neq 0 \end{cases} \tag{6.36}$$

This means that the power spectrum of the test signal is flat and the channel characteristic is "sampled" by a constant power spectrum of the input signal. In practice this theoretical assumption is only approximately fulfilled. Typically, the data sequence is produced on the basis of the *Maximum-Length* sequence generator. The test generator is usually implemented by a scrambler contained in the transmitter that is based on a *Linear Feedback Shift Register* (LFSR). As a result, a *pseudonoise* (PN) binary sequence is generated. Typically, subsets of very long PN sequences are used as a training sequence.

Special attention has focused on very short test sequences that allow for fast, coarse setting of the equalizer coefficients. These sequences are periodic and are constructed in such a way that their deterministic autocorrelation function is zero apart from its zero argument.

6.5.5 Fast Linear Equalization using Periodic Test Signals

In certain applications extremely fast initial equalization is of major importance. A good example is the master modem in a computer network, which receives data blocks from many tributary modems communicating through different channels. Such communication can be effective if the block header is a small part of the whole transmitted data block. A part of the header is a training sequence necessary to acquire the equalizer settings. Let us neglect the influence of noise for a while. In many cases the signal-to-noise ratio (SNR) in the channel is so high that the ISI plays a dominant role in the signal distortion. Let the reference signal be periodic. In fact, no more than two periods of the test signal should be transmitted in order to acquire coarse equalizer settings. The period M of the test signal is at least as long as the highest expected length of the channel impulse response L. With periodic excitation the channel output (neglecting the influence of the additive noise) is also a periodic signal. This fact is reflected by the formula

$$
\begin{bmatrix} x_0 \\ x_1 \\ \vdots \\ x_{M-1} \end{bmatrix} = \begin{bmatrix} d_0 & d_1 & d_2 & \cdots & d_{M-1} \\ d_{M-1} & d_0 & d_1 & \cdots & d_{M-2} \\ \vdots & & \ddots & & \vdots \\ d_1 & d_2 & \cdots & d_{M-1} & d_0 \end{bmatrix} \cdot \begin{bmatrix} h_0 \\ h_1 \\ \vdots \\ h_{M-1} \end{bmatrix} \tag{6.37}
$$

If the length of the channel impulse response is shorter than the length of the test signal, we can assume that some of the last elements in the vector $\mathbf{h}^T = [h_0, h_1, \ldots, h_{M-1}]$ are equal to zero. Due to the periodic nature of the signal transmitted through the channel, a cyclic convolution of the sequence $\mathbf{h}$ and the data sequence $\mathbf{d} = [d_0, d_1, \ldots, d_{M-1}]$ is realized. In the frequency domain this operation is equivalent to the multiplication of two *Discrete Fourier Transform* (DFT) spectra, i.e.

$$
X(k\Delta f) = D(k\Delta f) \cdot H(k\Delta f), \quad k = 0, 1, \ldots, M - 1 \tag{6.38}
$$

where $\Delta f T = 1/M$ and

$$
X(k\Delta f) = \frac{1}{M} \sum_{i=0}^{M-1} x(iT) \exp(-j2\pi k\Delta f iT) \tag{6.39}
$$

Dependencies similar to (6.39) are held for the data and channel impulse response sequences. Knowing the spectrum of the data sequence, one can easily calculate the spectrum of the channel and, after reversing it, the characteristics of the ZF equalizer can be achieved, i.e.

$$
C(k\Delta f) = \frac{1}{H(k\Delta f)} = \frac{D(k\Delta f)}{X(k\Delta f)}, \quad k = 0, 1, \ldots, M - 1 \tag{6.40}
$$

On the basis of the equalizer characteristics $\mathbf{C}^T = [C(0), C(\Delta f), \ldots, C((M-1)\Delta f)]$ the equalizer coefficients $\mathbf{c}^T = (c_0, c_1, \ldots, c_{M-1})$ can be calculated using the inverse DFT. If the length of the training sequence and of the equalizer is a power of 2, then all the DFT and IDFT calculations can be effectively performed by the FFT/IFFT algorithms. More detailed considerations on fast start-up equalization using the periodic training sequence can be found in Chevillat *et al.* (1987).

6.5.6 Symbol-Spaced versus Fractionally Spaced Equalizers

So far we have considered equalizers that accepted one sample per symbol period at their input. In fact the spectrum of the transmitted signal, although usually carefully shaped, exceeds half of the signaling frequency by 10–50%. Thus, the Nyquist theorem is not fulfilled and, as a result of sampling at the symbol rate, the input signal spectra overlap. In consequence, the symbol-spaced equalizer is able to correct the overlapped spectrum only. In some disadvantageous cases the overlapping spectra can result in deep nulls in the sampled channel characteristic, which is the subject of equalization. In these spectral intervals the noise will be substantially amplified by the equalizer, which results in deterioration of the system performance.

Derivation of the optimum MSE receiver in the class of linear receivers results in the receiver structure consisting of a filter matched to the impulse observed at the receiver input and an infinite T-spaced transversal filter (see Gitlin *et al.* 1992 for details). This derivation also shows that the characteristics $W_0(f)$ of the optimum MSE linear receiver are given by the formula

$$W_0(f) = \frac{\sigma_d^2}{\sigma_v^2} H^*(f) \left[\sum_{i=-\infty}^{\infty} c_i \exp\left(-j2\pi f i T\right) \right] \exp\left(-j2\pi f t_0\right) \qquad (6.41)$$

where σ_d^2 is the data symbol mean power and σ_v^2 is the noise power. Lack of a matched filter preceding the transversal filter results in the suboptimality of the receiver and in performance deterioration. In practice, a sufficiently long but finite transversal filter is applied.

The question of whether an optimum receiver can be implemented more efficiently was answered by Macchi and Guidoux (1975) as well as by Qureshi and Forney (1977).

As we have mentioned, typically the spectrum of the received input signal is limited to the frequency $f_{max} = \frac{1}{2T}(1 + \alpha)$, where $\alpha \leq 1$ (cf. a typical square root raised cosine pulse shaping filter characteristics in the transmitter and similar characteristics of the receive filter). Let us assume that the noise is also limited to the same bandwidth because of the band-limiting filter applied in the receiver front-end. Thus, the bandwidth of the optimal receiver is also limited to the same frequency f_{max}. Because the input signal is spectrally limited to f_{max}, the optimum linear receiver can be implemented by the transversal filter working at the input sampling frequency equal at least to $2f_{max}$. Let the sampling period $T' = \frac{KT}{M}$ be selected to fulfill this condition, i.e. $\frac{1}{2T'} \geq f_{max}$, and K and M are integers of possibly small values. As a result, the following equation holds

$$H(f) \cdot W_0(f) = H(f) \cdot C_{opt}(f) \qquad (6.42)$$

where

$$C_{opt}(f) = \sum_i W_0\left(f - i\frac{1}{T'}\right) \qquad (6.43)$$

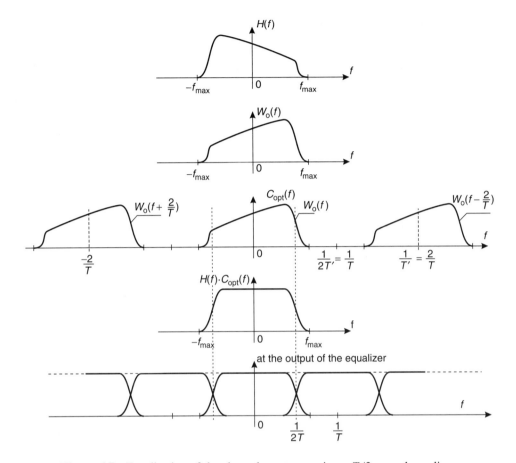

Figure 6.7　Equalization of the channel spectrum using a $T/2$-spaced equalizer

We must stress that although the input sampling frequency is $\frac{1}{T'}$, the data symbols are detected every T seconds, so the output of the equalizer is processed at the rate of $\frac{1}{T}$. It is important to note that the channel characteristics are first equalized by the T'-spaced filter and then the output spectrum components overlap due to sampling the output at the symbol rate. Figure 6.7 illustrates these processes for $K = 1$ and $M = 2$; specifically, the equalizer is $T/2$-spaced. One can also show that the performance of the fractionally spaced equalizer is independent of the sampling phase (Ungerboeck 1976).

Because the input signal spectrum is practically limited to $|f_{max}|$, the equalizer can synthesize any characteristics in the frequency range $(-\frac{1}{2T'}, -f_{max}) \cup (f_{max}, \frac{1}{2T'})$ without any consequences for the system performance. Therefore, the optimum fractionally spaced equalizer can have many sets of optimum coefficients. This phenomenon is disadvantageous from the implementation point of view because the values of the coefficients can slowly drift to unacceptable values. To stabilize the operation of the LMS gradient algorithm, a *tap leakage algorithm* was introduced (Gitlin *et al.* 1982).

6.6 Decision Feedback Equalizer

The decision feedback equalizer (DFE) is the simplest nonlinear equalizer with a symbol-by-symbol detector. It was first described by Austin (1967). The in-depth treatment of decision feedback equalization can be found in Belfiore and Park (1979). Austin noticed that intersymbol interference arising from past symbols can be cancelled by synthesizing it using already detected data symbols and subtracting the received value from the sample entering the decision device. Figure 6.8 presents the basic scheme of the decision feedback equalizer.

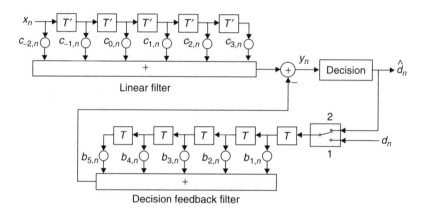

Figure 6.8 Structure of the decision feedback equalizer

The equalizer input samples are fed to the linear (usually fractionally spaced) adaptive filter, which performs matched filtering and shapes the ISI on its output in such a way that the symbol-spaced samples given to the decision device contain the ISI arising from the past symbols only. The ISI resulting from the joint channel and linear filter impulse response is synthesized in the transversal decision feedback filter. The structure of the DFE is very similar to the infinite impulse response filter; however, the decision device is placed inside the filter loop, causing the whole structure to be nonlinear. Generally, the operation of the decision feedback equalizer is described by the equation

$$y_n = \sum_{k=-N_1}^{N_2} c_{k,n} x(nT - kT') - \sum_{j=1}^{N_3} b_{j,n} \widehat{d}_{n-j} \tag{6.44}$$

where $c_{k,n}$ are the tap coefficients of the linear filter, $b_{j,n}$ are the tap coefficients of the decision feedback filter and $\widehat{d}_n$ is a data symbol estimate produced by the decision device. In training mode the data estimates are replaced by the training data symbols.

The decision feedback equalizer is applied in digital systems operating on channels with deep nulls (Monsen 1971). Such channels cannot be effectively equalized by the linear equalizers attempting to synthesize the reverse channel characteristics. Instead, the DFE cancels a part of the ISI without inverting the channel and, as a result, the noise in the frequency regions in which nulls in channel characteristics occur is not amplified.

Although the DFE structure is very simple and improves the system performance in comparison to that achieved for the linear equalizer, it has some drawbacks as well. First, part of the signal energy is not used in the decision process because of its cancellation by the decision feedback filter. Second, because of the decision feedback, errors made in the decision device take part in the synthesis of the ISI as they propagate along the decision feedback filter delay line. Thus, the errors contained in the tapped delay line increase the probability of occurrence of further errors. The phenomenon of error propagation can be observed if the signal-to-noise ratio is not sufficiently high. This effect is discussed in Barry *et al.* (2003).

The DFE tap coefficients can be adjusted according to the ZF or MSE criterion. As for the linear equalizer, the LMS and RLS adaptation algorithms can be used in the DFE. The DFE can be based on transversal or lattice filter structures (Ling and Proakis 1985). Let us concentrate on the LMS algorithm only. We can combine the contents of the tapped delay lines of the linear and decision feedback filters as well as the filter coefficients into single vectors, i.e.

$$\mathbf{z}_n = \begin{bmatrix} \mathbf{x}_n \\ \cdots \\ \mathbf{d}_n \end{bmatrix} \quad \mathbf{w}_n = \begin{bmatrix} \mathbf{c}_n \\ \cdots \\ -\mathbf{b}_n \end{bmatrix} \tag{6.45}$$

where $\mathbf{x}_n = [x_{n+N_1}, \ldots, x_{n-N_2}]^T$, $\mathbf{d}_n = [d_{n-1}, \ldots, d_{n-N_3}]^T$, $\mathbf{c}_n = [c_{-N_1,n}, \ldots, c_{N_2,n}]^T$ and $\mathbf{b}_n = [b_{1,n}, \ldots, b_{N_3,n}]^T$. Then equation (6.44) can be rewritten in the form

$$y_n = \mathbf{z}_n^T \mathbf{w}_n \tag{6.46}$$

and the LMS gradient algorithm can be described by the recursive expression

$$\mathbf{w}_{n+1} = \mathbf{w}_n - \beta_n e_n \mathbf{z}_n^* \tag{6.47}$$

where $e_n = y_n - d_n$. Knowing (6.45), we can break equation (6.47) into two separate LMS adjustment formulas for the feedforward and feedback filters

$$\mathbf{c}_{n+1} = \mathbf{c}_n - \beta_n e_n \mathbf{x}_n^* \tag{6.48}$$

$$\mathbf{b}_{n+1} = \mathbf{b}_n + \gamma_n e_n \mathbf{d}_n^* \tag{6.49}$$

where we have applied different values of the algorithm step size for both filters.

Besides the regular DFE structure shown in Figure 6.8 there exists the so-called *predictive DFE* (Belfiore and Park 1979; Proakis 2000), which, although featuring slightly lower performance, has some advantages in certain applications. Figure 6.9 presents the block diagram of this structure. The feedforward filter works as a regular linear equalizer according to the ZF or MSE criterion. Its adaptation algorithm is driven by the error signal between the filter output and the data decision (or training data symbol). As we remember, the linear equalizer more or less inverts the channel characteristics, which results in noise amplification. The noise contained in the feedforward filter output samples is correlated due to the filter characteristics. Therefore, its influence can be further minimized by applying the linear predictor. Assuming that the decision device makes correct decisions,

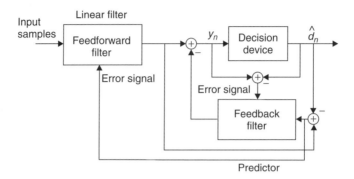

Figure 6.9   Predictive DFE

the noise samples contained in the feedforward filter output are the error samples used in the adaptation of this filter. The linear combination of the previous noise samples allows to predict the new sample, which is subsequently subtracted from the feedforward filter output. In this way the effective SNR is increased. The result of subtraction constitutes the basis for decision-making.

Let us note (see Figure 6.9) that the feedforward filter and the predictor are adjusted separately, so the performance of the predictive DFE is worse than the performance of the conventional DFE for which the taps adjustments are realized on the basis of the final output error. It has been shown that the predictive DFE is useful in realization of the joint trellis code decoder and channel equalizer (Chevillat and Elefteriou 1989).

6.7   Equalizers using MAP Symbol-by-Symbol Detection

The decision feedback equalizer is a particularly simple version of a nonlinear receiver in which the decision device is some kind of an M-level quantizer, where M is the number of data symbols. Much more sophisticated detectors have been developed that minimize the symbol error probability. This goal is achieved if the *Maximum a Posteriori Probability* (MAP) criterion is applied. Let us consider the receiver structure shown in Figure 6.10. The linear filter preceding the detection algorithm is a *Whitened Matched Filter* (WMF). Its function is very similar to the function of the linear filter applied in the decision feedback equalizer. It shapes the joint channel and linear filter impulse response to receive ISI arising from the past data symbols only. At the same time the noise samples at the output of the WMF are white. We say that the signal at the output of the WMF constitutes a *sufficient statistic* for detection, which roughly means that the

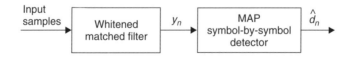

Figure 6.10   Basic scheme of the MAP symbol-by-symbol equalizer

part of the received signal that has been removed by the WMF is irrelevant for detection. Assuming that the number of interfering symbols is finite we can write the following equation describing the sample y_n at the detector input

$$y_n = \sum_{i=0}^{N} b_i d_{n-i} + v_n \tag{6.50}$$

where v_n is a white Gaussian noise sample. Let us note that the information on the data symbol d_n is "hidden" in the samples $y_n, y_{n+1}, \ldots, y_{n+N}$. Generally, according to the MAP criterion the detector finds that $\widehat{d}_n$ among all possible M data symbols for which the following *a posteriori* probability is maximum

$$\Pr\{d_n | \mathbf{y}_{n+N}\} \tag{6.51}$$

where $\mathbf{y}_{n+N} = [y_{n+N}, y_{n+N-1}, \ldots, y_1]$ is the vector of the observed input samples. From Bayes' theorem we know that for expression (6.51) the following equality holds

$$\Pr\{d_n = m | \mathbf{y}_{n+N}\} = \frac{p(\mathbf{y}_{n+N} | d_n = m) \Pr\{d_n = m\}}{p(\mathbf{y}_{n+N})} \tag{6.52}$$

Because $p(\mathbf{y}_{n+N})$ is common for all possible probabilities (6.52), it has no meaning in the search for the data symbol featuring the MAP probability. Thus, the task of the MAP detector can be formulated in the following manner

$$\widehat{d}_n = \arg\left\{\max_{d_n} p(\mathbf{y}_{n+N} | d_n) \Pr\{d_n\}\right\} \tag{6.53}$$

Finding the data estimate (6.53) is usually computationally complex. Several algorithms have been proposed to realize (6.53). Abend and Fritchman (1970) as well as Chang and Hancock (1966) algorithms [the latter being analogous to the well-known BCJR algorithm (Bahl *et al.* 1974) applied in convolutional code decoding and shown in Chapter 2] are good examples of these methods. We have to stress that all of them require knowledge of the impulse response $\{b_i\}$ ($i = 1, \ldots, N$) to calculate values of the appropriate conditional probability density functions. This problem will also appear in the MLSE receiver discussed in the next section.

6.8   Maximum Likelihood Equalizers

Instead of minimizing the data symbol error probability, we could select minimization of the probability of error of the whole data sequence as the optimization goal of the receiver. Thus, the MAP criterion yields the form

$$\max_{\mathbf{d}_n} P(\mathbf{d}_n | \mathbf{y}_n) \tag{6.54}$$

If the data sequences are equiprobable, our criterion is equivalent to the selection of such a data sequence that maximizes the conditional probability density function $p(\mathbf{x}_n|\mathbf{d}_n)$. Namely, we have

$$\widehat{\mathbf{d}}_n = \arg\left\{\max_{\mathbf{d}_n} P(\mathbf{d}_n|\mathbf{y}_n)\right\} = \arg\left\{\max_{\mathbf{d}_n} \frac{p(\mathbf{y}_n|\mathbf{d}_n)P(\mathbf{d}_n)}{p(\mathbf{y}_n)}\right\} = \arg\left\{\max_{\mathbf{d}_n} p(\mathbf{y}_n|\mathbf{d}_n)\right\} \quad (6.55)$$

where, as before, $\mathbf{y}_n = [y_1, \ldots, y_n]^T$, $\mathbf{d}_n = [d_1, \ldots, d_n]^T$. Because noise at the WMF output is white and Gaussian, its samples are statistically independent and the conditional probability density function can be expressed by the formula

$$p(\mathbf{y}_n|\mathbf{d}_n) = \prod_{i=1}^{n} p(y_i|\mathbf{d}_i) = \frac{1}{(2\pi\sigma^2)^{n/2}} \prod_{i=1}^{n} \exp\left(-\frac{\left|y_i - \sum_{k=0}^{N} b_k d_{i-k}\right|^2}{2\sigma^2}\right) \quad (6.56)$$

Calculating the natural logarithm of both sides of (6.56) we obtain

$$\widehat{\mathbf{d}}_n = \arg\left\{\max_{\mathbf{d}_n} \ln p(\mathbf{y}_n|\mathbf{d}_n)\right\} = \arg\left\{\min_{\mathbf{d}_n} \sum_{i=1}^{n} \left|y_i - \sum_{k=0}^{N} b_k d_{i-k}\right|^2\right\} \quad (6.57)$$

From all possible equiprobable data sequences $\mathbf{d}_n$ this sequence $\widehat{\mathbf{d}}_n$ is selected for which the sum

$$S_n = \sum_{i=1}^{n} \left|y_i - \sum_{k=0}^{N} b_k d_{i-k}\right|^2 \quad (6.58)$$

is minimum. It was found by Forney (1972) that the effective method of searching for such a sequence is the Viterbi algorithm. Let us note that in order to select the data sequence the samples of the impulse response $\{b_k\}$ ($k = 0, \ldots, N$) have to be estimated. They are usually derived on the basis of the channel impulse response $\{h_k\}$ ($k = -N_1, \ldots, N_2$). The scheme of such a receiver is shown in Figure 6.11. The heart of the receiver is the Viterbi detector fed with the impulse response samples $\{b_k\}$ calculated on the basis of the channel impulse response samples $\{h_k\}$ estimated in the *channel estimator*. The channel estimator is usually an adaptive filter using the LMS or RLS algorithm for deriving the impulse response samples. From the system theory point of view it performs system identification. The channel estimator input signal is the data reference signal or the final or preliminary decision produced by the Viterbi detector. The channel output signal acts as a reference signal for the channel estimator. Usually, the reference signal has to be appropriately delayed in order to accommodate the decision delay introduced by the Viterbi detector. In transmission over fast time-varying channels the FIR channel estimator is sometimes supported with a predictor that helps to decrease the delay in channel estimation caused by the decision delay introduced by the Viterbi detector.

For example, let us consider the channel estimator using the LMS algorithm and driven by ideal data symbols. Let us neglect the delay with which the data symbols are fed to

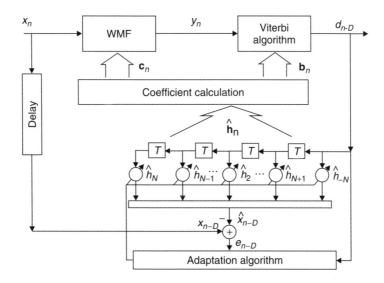

Figure 6.11 Basic scheme of the MLSE receiver with the WMF and Viterbi algorithm

the estimator. Assume that the data symbols are uncorrelated. Then, applying the mean square error as the criterion for the estimator, we have

$$\mathcal{E}_n = E\left[|e_n|^2\right] = E\left[\left|x_n - \sum_{j=-N}^{N} \widehat{h}_{j,n} d_{n-j}\right|^2\right] \tag{6.59}$$

where x_n is the channel output sample [see (6.3)] and $\widehat{h}_{j,n}$ $(j = -N, \dots, N)$ are the estimates of the channel impulse response at the nth moment. The calculation of the gradient of error $\mathcal{E}_n$ with respect to the channel impulse response estimate $\widehat{h}_j$ gives

$$\frac{\partial \mathcal{E}_n}{\partial \widehat{h}_{j,n}} = -2E\left[e_n d_{n-j}^*\right] \tag{6.60}$$

Therefore the stochastic gradient algorithm for the adjustment of channel impulse response estimates is

$$\widehat{h}_{j,n+1} = \widehat{h}_{j,n} + \alpha_n e_n d_{n-j}^*, \quad j = -N, \dots, N \tag{6.61}$$

where α_n is an appropriately selected step size. It can be shown that the initial step size should be $\alpha_0 = 1/((2N+1)E[|d_n|^2])$.

Another solution for deriving the channel impulse response is to use a zero-autocorrelation periodic training sequence. A fast channel estimator using such a sequence is applied, for example, in the GSM receiver. Part of the known sequence placed in the middle of the data burst, called *midamble*, is a zero-autocorrelation periodic training sequence. In this case the channel impulse response samples are estimated on the basis of the following formula

$$\widehat{h}_i = \sum_{j=-N}^{N} x_j d_{i-j}^* \qquad (6.62)$$

Thus, the received signal, which is the response of the channel to the periodic training signal, is cross-correlated with the complex conjugate of the training sequence. On the basis of the estimated impulse response samples $\widehat{h}_i$ the receiver calculates the WMF coefficients and the weights $\{b_k\}$ used by the Viterbi detector.

Closer investigation of formula (6.57) allows to conclude that in order to minimize the cost function and find the optimum data sequence, M^N operations (multiply and add, compare, etc.) have to be performed in each timing instant. M is the size of the data alphabet. If modulation is binary ($M = 2$) and the length of ISI is moderate, the detection algorithm is manageable. This is the case of the GSM receiver. However, if M is larger and/or ISI corrupts a larger number of modulation periods, the number of calculations becomes excessive and suboptimal solutions have to be applied. Papers by Eyuboglu and Qureshi (1988), Duel-Hallen (1992) and Wesołowski (1987) show examples of suboptimum MLSE receivers.

An alternative equivalent structure of the MLSE equalizer was proposed by Ungerboeck (1974). Its derivation following Ungerboeck's considerations can also be found in Proakis (2000).

As in (6.1), let the received signal have the form

$$x(t) = \sum_{i=0}^{n} d_i h(t - iT) + v(t) \qquad (6.63)$$

Let us note that this time we consider the received signal starting from the data symbol indexed with 0 and finishing at the current timing instant n. Let us represent the above signal in the form of an expansion, using a set of orthonormal functions. Then (6.63) can be written in the form

$$x(t) = \lim_{K \to \infty} \sum_{k=1}^{K} x_k f_k(t)$$

where $\{f_k(t)\}$ is a complete set of orthonormal functions and K is the number of functions used in the approximation of the continuous signal $x(t)$. The weights x_k are in fact linear combinations of data symbols d_i ($i = 0, \ldots, n$) and expansion coefficients h_{ki} ($i = 0, \ldots, n, k = 1, 2, \ldots, K$), i.e.

$$x_k = \sum_{i=0}^{n} d_i h_{ki} + v_k, \quad k = 1, 2, \ldots \qquad (6.64)$$

In turn, the expansion coefficients h_{ki} and v_k result from the formulas

$$h_{ki} = \int h(t - iT) f_k(t) dt \qquad (6.65)$$

and

$$v_k = \int v(t) f_k(t) dt \qquad (6.66)$$

Let us write the expansion coefficients and the data sequence in vector form as

$$\mathbf{x}_1^K = [x_1, x_2, \ldots, x_K] \qquad (6.67)$$

$$\mathbf{d}_0^n = [d_0, d_1, \ldots, d_n] \qquad (6.68)$$

This time, we explicitly show the size of appropriate vectors. The MLSE receiver selects the data vector $\widehat{\mathbf{d}}_1^n$ for which the conditional probability density function $p(\mathbf{x}_1^K | \mathbf{d}_0^n)$ is maximized. Let us note that coefficients $\mathbf{x}_1^K$ represent the received signal $x(t)$. The maximized probability density function has the form of a multidimensional Gaussian pdf, i.e.

$$p(\mathbf{x}_1^K | \mathbf{d}_0^n) = \left(\frac{1}{2\pi\sigma^2} \right)^K \exp\left(-\frac{1}{\sigma^2} \sum_{k=1}^K \left| x_k - \sum_{i=0}^n d_i h_{ki} \right|^2 \right) \qquad (6.69)$$

Maximization of the logarithm of $p(\mathbf{x}_1^K | \mathbf{d}_0^n)$ is much more convenient than the maximization of $p(\mathbf{x}_1^K | \mathbf{d}_0^n)$ itself. The logarithm of the conditional probability density function takes the form

$$\ln p(\mathbf{x}_1^K | \mathbf{d}_0^n) = \text{const} - \frac{1}{2\sigma^2} \sum_{k=1}^K \left| x_k - \sum_{i=0}^n d_i h_{ki} \right|^2 \qquad (6.70)$$

where const does not depend on the choice of data sequence and can be omitted in further considerations. As the number of applied orthonormal functions K tends to infinity, the finite sum evolves into an integral and discrete coefficients change into continuous functions. As a result, we have

$$
\begin{aligned}
\ln p(\mathbf{x}_1^K | \mathbf{d}_0^n) \sim & -\int_{-\infty}^{\infty} \left| x(t) - \sum_{i=0}^n d_i h(t - iT) \right|^2 dt \\
= & -\int_{-\infty}^{\infty} |x(t)|^2 dt + 2\,\mathrm{Re}\left\{ \sum_{i=0}^n \left[d_i^* \int_{-\infty}^{\infty} x(t) h^*(t - iT) dt \right] \right\} \\
& -\sum_{i=0}^n \sum_{j=0}^n d_i^* d_j \int_{-\infty}^{\infty} h(t - jT) h^*(t - iT) dt \qquad (6.71)
\end{aligned}
$$

The integral $-\int_{-\infty}^{\infty} |x(t)|^2 \, dt$ does not have any influence on the choice of the data sequence $\mathbf{d}_0^n$, so in order to maximize $\ln\ p(\mathbf{x}_1^K | \mathbf{d}_0^n)$ we search for such $\mathbf{d}_0^n$ for which the sum of the second and third component of (6.71) is maximized. Let us denote the samples of the autocorrelation function of the channel impulse response as

$$r_n = r(nT) = \int_{-\infty}^{\infty} h^*(t)h(t+nT)dt \qquad (6.72)$$

The variable y_n is then the sample of the filter matched to the channel impulse response taken at modulation period intervals when the channel output signal $x(t)$ is given to its input, i.e.

$$y_n = y(nT) = \int_{-\infty}^{\infty} x(t)h^*(t-nT)dt = \sum_{i=0}^{n} d_i r_{n-i} + v_n \qquad (6.73)$$

As a result, maximization of the logarithm of the conditional probability density function $p(\mathbf{x}_1^M | \mathbf{d}_1^n)$ is equivalent to finding the sequence $\hat{\mathbf{d}}_0^n$ for which the following cost function $C_n(\mathbf{d}_0^n)$ is maximized

$$C_n(\mathbf{d}_0^n) = 2\,\mathrm{Re}\left[\sum_{i=0}^{n} d_i^* y_i\right] - \sum_{i=0}^{n}\sum_{j=0}^{n} d_i^* d_j r_{i-j} \qquad (6.74)$$

The maximized cost function can be calculated recursively using the expression

$$C_n(\mathbf{d}_0^n) = C_{n-1}(\mathbf{d}_1^{n-1}) + \mathrm{Re}\left[d_n^*\left(2y_n - r_0 d_n - 2\sum_{m=1}^{L} r_m d_{n-m}\right)\right] \qquad (6.75)$$

where y_n is the sample at the matched filter output in the nth timing instant and r_m ($m = 0, \ldots, L$) are the samples of the autocorrelation function of the channel impulse response. In (6.75) we have assumed that at most L samples of the channel autocorelation function have significant values.

As in the case of the regular MLSE receiver, Ungerboeck's MLSE receiver (Figure 6.12) can apply the Viterbi algorithm to recursively calculate the cost function for each trellis state. This time the number of states is equal to M^L, where M is the number of data signal constellation points.

The performance of the Ungerboeck receiver and the Forney receiver is the same. The advantage of Ungerboeck's receiver is that there is no need to calculate the whitened matched filter impulse response. Instead, the matched filter is applied and the path metrics are appropriately modified.

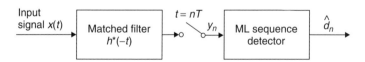

Figure 6.12   Block diagram of the Ungerboeck version of the MLSE receiver

6.9 Examples of Suboptimum Sequential Receivers

Sequential receivers applying the Viterbi algorithm achieve high performance, which is measured by a low probability of erroneous decisions. They are among the best receivers considered so far in this chapter. However, the performance quality of sequential receivers is usually bought by their high computational requirements resulting from the number M^L of states of the the trellis diagram. As we see, the number of states grows exponentially with the number of interfering symbols L. However, the base of the state number M^L is the size of digital modulation M. Therefore the receiver that is based on the pure Viterbi algorithm is impractical if 16- or 64-QAM modulations are applied. As we have said earlier, in such cases simplifications of sequential receivers are needed that will substantially decrease the computational requirements without unduly decreasing the performance. Two such solutions will be described below.

The first one is called the *M algorithm*, but it is worth noting that this symbol M has a different meaning than the modulation size. The algorithm was presented by Anderson and Mohan (1984). Its functioning is very similar to the Viterbi algorithm and it is based on the observation that in the Viterbi algorithm in a given time instant only a limited number of states are realistic candidates to be the source of paths that feature small or even the lowest metric. The performance quality should only be insignificantly decreased if in each algorithm step only a small subset of states, for which the cost of reaching them is among the smallest, is extended. Let us denote the number of such states as M_S. The M algorithm works as follows.

Let us assume that in a given time instant we know the subset of M_S states featuring the lowest metrics of their survivors. In each modulation period the algorithm extends the paths reaching each of these M_S states to those states that are available in the next timing instant. The number of possible paths originating from each of these M_S states is equal to the modulation size M. For all achievable states the algorithm calculates the metrics of the path reaching them. The states are subsequently sorted in decreasing order of the costs of their surviving paths and among at most $M \times M_S$ such states only M_S states are retained for processing in the next step.

Let us denote the cost of reaching the kth state in the $(n-1)$st moment as S_{n-1}^k. Let the data sequence associated with the kth state survivor in the $(n-1)$st moment be $\mathbf{d}_{n-1}^{(k)} = (d_{n-1}^{(k)}, d_{n-2}^{(k)}, \ldots, d_{n-L}^{(k)})$. Let us assume that the algorithm processes the samples from the output of the whitened matched filter. As a result, intersymbol interference on its output is only caused by the past data symbols. Then, the hypothetical intersymbol interference resulting from the data sequence associated with the kth state can be presented as (cf. Figure 6.11)

$$ISI_{n-1}^{(k)} = \sum_{j=1}^{L} b_j d_{n-j}^{(k)} \tag{6.76}$$

For each state considered in the $(n-1)$st moment the cost of reaching every possible state in the nth moment is calculated according to the formula

$$S_n^j = |y_n - b_0 d_n^i - ISI_{n-1}^{(k)}|^2 + S_{n-1}^{(k)}$$
$$\text{for } i = 1, 2, \ldots, M, \ j = 1, 2, \ldots, M_S \times M \tag{6.77}$$

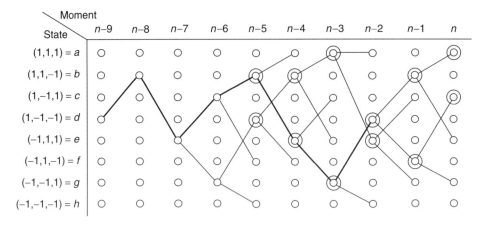

Figure 6.13 Illustration of the M algorithm

where j is the state index determined by the data vector $[d_n^i, d_{n-1}^{(k)}, d_{n-2}^{(k)}, \ldots, d_{n-(L-1)}^{(k)}]$. As we said earlier, the algorithm sorts the survivor costs in decreasing order and only M_S lowest costs and associated states are taken into account in the next moment. Figure 6.13 illustrates the operation of the M algorithm for a binary modulation ($M = 2$), when intersymbol interference results from $L = 3$ past data symbols and the algorithm retains $M_S = 2$ states for processing in the next step. Let us note that among $M^L = 8$ possible states the M algorithm considers only two best states and it extends their potential list to $M_S \times M = 4$ states. In the example shown in Figure 6.13 we assume that until the $(n - 7)$th moment the data sequence is known to the receiver (the data sequence can be a known preamble or a test sequence). Starting from the $(n - 7)$th moment the receiver uses the sequential algorithm. In the $(n - 7)$th moment the algorithm investigates the path from state e to two possible states, i.e. states c and g. Starting from the $(n - 6)$th moment the algorithm checks the possible paths from both states to two states in the next moment. Out of the resulting four states the best two states are retained. In Figure 6.13 they are denoted by circles. Let us note that all survivors of processed states in the nth moment originate from the same route. In Figure 6.13 this is marked in bold. The appearance of the common route indicates that the algorithm has already made the final decision upon the past symbols [up to the $(n - 2)$nd moment, as shown in Figure 6.13].

It has been noticed that the M algorithm operates well if it is preceded by the whitened matched filter. In consequence, calculation of the WMF coefficients is required. The next drawback is the necessity of sorting the state costs and creation of the state list. If the number of states extended in each step of the algorithm is large, such an operation can be cumbersome.

The second suboptimum sequential algorithm is DDFSE – *Decision-Delayed Feedback Sequence Estimation* – proposed in 1989 (Duel-Hallen and Heegard 1989). Assume again that the whitened matched filter precedes the functional block implementing the sequential algorithm. In the DDFSE receiver the effective decrease of the number of states of the original Viterbi algorithm is obtained through subtraction of the partial hypothetical intersymbol interference associated with the shortest path to a given state from the input sample y_n. Let us consider the calculation of the path metric λ_n^{kj} from the kth state in the $(n-1)$st moment to the jth state in the nth moment. In the DDFSE receiver this metric is calculated from the formula

$$\lambda_n^{kj} = \left| y_n - \sum_{i=0}^{L'} b_i d_{n-i}^{(kj)} - \sum_{i=L'+1}^{L} b_i d_{n-i}^{(k)} \right|^2 \tag{6.78}$$

where $d_{n-i}^{(kj)}$ $(i = 0, 1, \ldots, L')$ are data symbols associated with the path from the kth to the jth state in the nth moment, whereas $d_{n-i}^{(k)}$ $(i = L'+1, \ldots, L)$ is the data sequence uniquely determined on the trellis diagram by the shortest path to the kth state in the $(n-1)$st moment. As we can see, owing to this operation the number of algorithm states decreases from M^L to $M^{L'}$ $(L' < L)$. The characteristic feature of the DDFSE algorithm is the individual feedback applied in each trellis state. The choice of the number of states to be considered is made by selection of the number L' of interfering samples acquired from the output of the whitened matched filter and seen on the input of the DDFSE block. This number determines the compromise between the computational requirements and the performance quality of the suboptimal sequential detector.

6.10   Case Study: GSM Receiver

In order to illustrate the operation of a GSM receiver we have to describe the basic scheme of GSM transmission first. In the GSM system data are transmitted in the form of bursts. There are several types of bursts, resulting from different operations performed during the setting of connection, signaling, synchronization, paging request, etc. We will consider a typical situation in which the link has already been established and traffic data are transmitted.

The GSM system is based on the *Time Division Multiple Access* (TDMA) principle, which will be explained in detail in Chapter 9. In the GSM system time is divided into frames that last for 4.6125 ms each. A single frame is further divided into eight slots. Each slot has a duration of 0.577 ms. A slot in each frame is assigned to a particular link. Thus, in a typical connection the mobile terminal transmits and receives data in 1/8 of a frame. The basic GSM system applies a binary modulation called a GMSK (see Chapter 4). The modulation period is equal to 3.69 µs, so the modulation rate is 270.833 kbit/s and a single slot has a duration of 156.25 bits. Actually, only 148 bits are transmitted and the remaining time is a guard period. The guard time is needed for switching the transmit amplifier on and off and for protection against timing inaccuracies. Figure 6.14 shows a normal burst used for data transmisison after the link has been established. As already mentioned, the burst consists of 148 bits. Three bits at the beginning and at the end of the burst are the so-called zero tail bits. Two 57-bit blocks on both ends of the burst transmit the user data. Two separate flag bits inform the receiver what type of data is currently transported. Finally, a 26-bit training sequence called *midamble* is transmitted in the middle of the burst. The ETSI/GSM standard EN 300 908 (ETSI 1999) lists eight possible training sequences (see Table 6.1). A particular training sequence is assigned to the link during the process of its establishment.

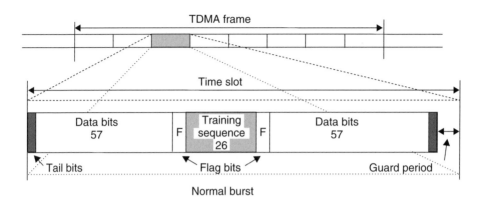

Figure 6.14 Structure of the normal GSM burst

Table 6.1 Training sequences for normal GSM burst

Training sequence code	Training sequence bits
0	0,0,1,0,0,1,0,1,1,1,0,0,0,0,1,0,0,0,1,0,0,1,0,1,1,1
1	0,0,1,0,1,1,0,1,1,1,0,1,1,1,1,0,0,0,1,0,1,1,0,1,1,1
2	0,1,0,0,0,0,1,1,1,0,1,1,1,0,1,0,0,1,0,0,0,0,1,1,1,0
3	0,1,0,0,0,1,1,1,1,0,1,1,0,1,0,0,0,1,0,0,0,1,1,1,1,0
4	0,0,0,1,1,0,1,0,1,1,1,0,0,1,0,0,0,0,0,0,1,1,0,1,0,1,1
5	0,1,0,0,1,1,1,0,1,0,1,1,0,0,0,0,0,1,0,0,1,1,1,0,1,0
6	1,0,1,0,0,1,1,1,1,1,0,1,1,0,0,0,1,0,1,0,0,1,1,1,1,1
7	1,1,1,0,1,1,1,1,0,0,0,1,0,0,1,0,1,1,1,0,1,1,1,1,0,0

Training sequences have a special form. Each of them consists of a 16-bit word plus five cyclicly repeated bits at both ends for protection purposes. The aim of the training sequence is to allow the receiver to position the received burst on its own time axis, i.e. to synchronize its own operation with the received signal and to estimate the channel impulse response that is needed for appropriate operation of the data sequence estimator. Assuming that the received samples can be accumulated in the memory, placement of the training sequence is arbitrary – it does not need to be located at the start of the burst. If placed in the middle of the burst, it minimizes the effect of channel time variability within the slot. If the channel is time varying, its change is substantially smaller when observed from the middle of the burst towards both ends compared with a potential change from the beginning of the burst to its end.

In the transmitter, 148 bits of the normal burst are fed to the GMSK modulator. According to ETSI Standard EN 300 959 (ETSI 2001a), functioning of the modulator is described by the following formulas. First, data bits d_i are subject to some kind of differential encoding in the form

$$a_i = d_i \oplus d_{i-1} \tag{6.79}$$

where $\oplus$ denotes a modulo-2 sum. Then, the resulting bits are converted into bipolar form, using the expression

$$\alpha_i = 1 - 2a_i \tag{6.80}$$

Data symbols are shaped by an appropriate filter characterized by the impulse response $g(t)$ given by formula (5.102) with parameters $BT = 0.3$, where B is a 3-dB bandwidth and T is the modulation period. Finally, the GMSK signal is described by the formula

$$s(t) = \sqrt{\frac{2E_c}{T}} \cos[2\pi f_c t + \varphi(t) + \varphi_0] \tag{6.81}$$

where the phase function containing data symbols is given by the expression

$$\varphi(t) = \sum_i \pi h \alpha_i \int_{-\infty}^{t-iT} g(\tau) d\tau \tag{6.82}$$

As we explained in Chapter 4, signal $s(t)$ given by (6.81) and (6.82) with $BT = 0.3$ and $h = 0.5$ is well approximated by a linear modulation with the appropriately selected pulse-shaping filter and a numerically calculated impulse response $p(t)$ that has a similar shape to the Gaussian one. Unfortunately this pulse spans up to five modulation periods, so the modulator itself introduces intersymbol interference.

The burst in the form of a GMSK-modulated signal (6.81) is transmitted over the multipath channel, received in distorted form by the receive antenna (Figure 6.15) and down-converted to the baseband. The burst samples of the in-phase and quadrature baseband components are stored in RAM. The whole transmission chain after conversion to the baseband can be approximately described by the following equation

$$x(t) = \sum_{k=0}^{N} c_k h(t - kT) + v(t) \tag{6.83}$$

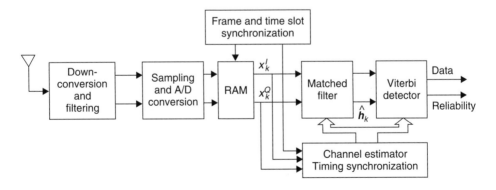

Figure 6.15   General scheme of the GSM receiver

where $h(t)$ is a convolution of the pulse-shaping filter $p(t)$, the baseband equivalent channel and the receive filter, and $v(t)$ is the additive Gaussian noise. The data symbols are given by the formula

$$c_k = c_{k-1} \exp(j\alpha_k \pi/2) = c_0 \exp\left(j\frac{\pi}{2}\sum_{i=0}^{k}\alpha_i\right) \tag{6.84}$$

After the whole burst is accumulated, its contents can be processed off-line. First, the middle of the burst is found and the joint channel, transmit and receive filter impulse response is estimated on the basis of the known midamble.

Estimation of the joint channel impulse response $h(t)$ is an important step in the receiver operation. Let us consider it when taking the midamble sequence No. 0 as an example. As we know from (6.79) the differential encoding is performed in the transmitter. Let us arbitrarily assume that $d_0 = 0$, i.e. the value of the last bit preceding the midamble in the burst. Let us note that it could be equal to "1" as well. In Tables 6.2 and 6.3 we show the original midamble d_k ($k = 1, \ldots, 26$), the midamble a_k ($k = 1, \ldots, 26$) after differential encoding and related symbols c_k ($k = 1, \ldots, 26$), assuming that $c_0 = 1$.

The specific 16-symbol word is contained between positions 6 and 21. The first five symbols are copied from positions 17–21, whereas the last five symbols are copied from positions 6–10. It is assumed that the first five samples contain echoes from the data symbols preceding the midamble, so they are not taken into account in channel estimation. It is also assumed that the channel impulse response spans at most five modulation periods, so we estimate six channel impulse response samples and we assume $N = 5$ in (6.83). After sampling the baseband equivalent channel output at the moments $t = nT$, and if we neglect noise, we obtain formula (6.83) in the form

$$x_n = x(nT) = \sum_{k=0}^{N} c_k h_{n-k} = \sum_{k=0}^{N} h_k c_{n-k} \tag{6.85}$$

Table 6.2   Several forms of GSM training sequence No. 0 – Part 1

k	1	2	3	4	5	6	7	8	9	10	11	12	13
d_k	0	0	1	0	0	1	0	1	1	1	0	0	0
a_k	0	0	1	1	0	1	1	1	0	0	1	0	0
c_k	j	-1	j	1	j	1	$-j$	-1	$-j$	1	$-j$	1	j

Table 6.3   Several forms of GSM training sequence No. 0 – Part 2

k	14	15	16	17	18	19	20	21	22	23	24	25	26
d_k	0	1	0	0	0	1	0	0	1	0	1	1	1
a_k	0	1	1	0	0	1	1	0	1	1	1	0	0
c_k	-1	j	1	j	-1	j	1	j	1	$-j$	-1	$-j$	1

If the sequence c_k has the following property

$$\sum_{m=0}^{N} c_m^* c_{m+k} = \begin{cases} 0 & \text{for } k \neq 0 \\ (N+1)|c|^2 & \text{for } k = 0 \end{cases} \tag{6.86}$$

then on the basis of the received samples x_n we can derive the channel impulse response sample, using the formula

$$y_i = \sum_{m=0}^{N} c_m^* x_{i+m} \tag{6.87}$$

Substituting (6.85) into (6.87) we obtain

$$y_i = \sum_{m=0}^{N} c_m^* \sum_{k=0}^{N} h_k c_{i+m-k} = \sum_{k=0}^{N} h_k \sum_{m=0}^{N} c_m^* c_{i+m-k}$$

$$= \begin{cases} 0 & \text{for } k \neq i \\ (N+1)|c|^2 h_i & \text{for } k = i \end{cases} \tag{6.88}$$

If we wish to estimate the channel impulse response using (6.87), we have to apply a training sequence fulfilling the zero-autocorrelation property expressed by (6.86). Let us note that if we extract data symbols c_k from positions 11–16 shown in Tables 6.2 and 6.3, we obtain such a sequence. Equation (6.87), which allows us to estimate the channel impulse response, can be easily implemented as a FIR filter in which the tap coefficients are complex conjugates of the elements of the training sequence. In fact, such a filter implements the filter matched to the training sequence and its scheme is shown in Figure 6.16. In this way a sliding window over the sample sequence from the matched filter output can be created. The moment in which the signal energy within the window is maximum indicates synchronization with the burst midamble and this point determines the middle of the burst. In this way slot (burst) synchronization is performed.

Knowing the channel impulse response estimates h_i $(i = 0, \ldots, N)$ and the midamble sequence, the Viterbi detector can start its operation. It is preceded by the matched filter, so the optimum solution is to apply the Viterbi detector in the version proposed by

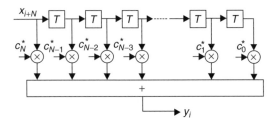

Figure 6.16   The filter matched to the GSM midamble used for slot synchronization

Ungerboeck that has been shown earlier in this chapter. Let us note that searching for the best route on the trellis is performed in both ways towards the ends of the burst, i.e. in the regular and reversed order. Despite $N = 5$ being assumed in the above considerations, five samples of the channel impulse response are usually taken into account so the equivalent channel model has four memory cells. In consequence, the Viterbi detector operates on the trellis that has 16 states, starting from the known state determined by the current training sequence c_k in the middle of the burst. In a typical GSM receiver the deinterleaver and convolutional code decoder follow the Viterbi detector, so it is very desirable that the Viterbi detector produces not only data symbols but also their reliability measures, i.e. it generates soft outputs. Consequently, the convolutional code decoder can be implemented in the soft-input version and its performance can be increased compared with the case when only hard-output bits are generated by the Viterbi detector.

6.11   Equalizers for Trellis-Coded Modulations

Trellis-coded modulation (TCM) has already been considered in Chapter 4. However, at that time we considered TCM in a nondistorting channel with additive white Gaussian noise as the only impairment. As we remember, detection of the trellis-coded data stream requires a sequential algorithm, such as the Viterbi algorithm.

Using TCM signals on the ISI channels requires both adaptive equalization and TCM decoding. The TCM detection process of the whole symbol sequences creates problems in selection of the equalizer structure and in adjustment of the equalizer coefficients. The standard solution is to apply a linear equalizer minimizing ISI, followed by the TCM Viterbi decoder. The equalizer coefficient updates can be done using either unreliable tentative decisions or the reliable but delayed decisions from the TCM Viterbi decoder (Long *et al.* 1989). In the case of the LMS algorithm applied in the equalizer, the consequence of using the delayed error signal (see Figure 6.17) is the need to decrease the step size (Long *et al.* 1989).

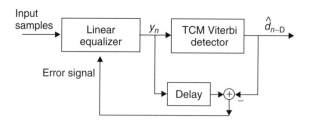

Figure 6.17   Linear equalizer with TCM decoder

On some channels, in particular those featuring a long tail in the channel impulse response or possessing deep nulls in their characteristics, applying a decision feeedback equalizer is more advantageous. Using joint DFE and trellis coding requires some special solutions due to the fact that in its feedback filter the DFE uses symbol-by-symbol decisions with a single delay. One solution is to apply an interleaver between the TCM encoder and the modulator at the transmitter, and the predictive DFE with the deinterleaver between the linear part of the equalizer and the decision feedback part incorporating the

TCM Viterbi decoder and predictor (Proakis 2000). Another solution applicable in systems with a feedback channel and operating on transmission channels that are stationary or slowly varying in time is to share the DFE equalization between transmitter and receiver. In this case the concept of *Tomlinson precoding* applied jointly with the TCM coding is very useful (Aman *et al.* 1991).

The optimum receiver for TCM signals corrupted by ISI was shown in Chevillat and Elefteriou (1989). Its structure is basically the same as that shown in Figure 6.11; however, now the Viterbi detector operates on the supertrellis resulting from concatenation of ISI and TCM code trellises. Because both the number of supertrellis states and the resulting computational complexity are very high, suboptimum solutions have to be applied. The most efficient relies on incorporating ISI into the decision feedback for each supertrellis state, using the data sequences that constitute the "oldest" part of the maximum likelihood data sequence associated with each state (the so-called *survivor*). In fact, this idea is already known from the decision-delayed feedback sequence estimation (Duel-Hallen 1992) used for uncoded data.

6.12 Turbo Equalization

In modern digital communication systems data transmission is often supplemented with FEC error correction coding, which ensures the required overall system performance. A simplified configuration of a transmitter of such a system is shown in Figure 6.18. The binary data stream a_k $(k = 1, \ldots, N)$ is encoded using the convolutional code encoder that produces binary coded symbols $x_{k,l}$ $(k = 1, \ldots, N; l = 1, \ldots, n)$, where $1/n$ is the coding rate. This stream is then subject to interleaving. Data reordering performed by the interleaver is denoted by a new index i on the interleaver output, so the interleaved data stream is denoted as x_i $(i = 1, \ldots, nN)$. If a BPSK modulator is applied in the system, as shown in Figure 6.18, each output BPSK symbol represents a single bit from its input. The BPSK output symbols are $d_i = 2x_i - 1$ $(i = 1, \ldots, nN)$. In vector notation we denote respective data streams as **a**, **x**, and **d**. The resulting BPSK data stream is transmitted through the channel. The channel block shown in Figure 6.18 represents the cascade of the transmit filter, the transmission channel and the whitened matched filter. The received signal is sampled with the frequency equal to the modulation rate $1/T$. The signal seen at the input of the equalizer is described by (6.50), repeated here for our convenience in a slightly changed version

$$y_i = \sum_{j=0}^{L} b_j d_{i-j} + v_i \tag{6.89}$$

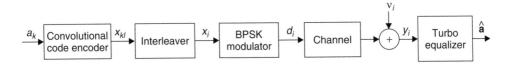

Figure 6.18 Simplified scheme of the transmission system with coding and turbo equalization

where v_i is, as before, the white Gaussian noise sample of variance σ_v^2. Let us denote the whole block of received samples y_i $(i = 1, \ldots, nN)$ as **y**. The goal of the receiver is to find the input block vector **a** on the basis of the received block **y**.

In the conventional approach, FEC decoding and equalization of the transmission channel are treated separately. This disjoint strategy results in performance loss compared with the optimal receiver, which would treat equalization and decoding jointly. The optimal receiver is an MLSE or symbol-by-symbol MAP detector that operates on the super-trellis resulting from joint treatment of the channel and FEC code. Let us note that the channel, whose operation is described by equation (6.89), can be considered as a convolutional code with analog coefficients and coding rate $R = 1$. Thus, in fact we can interpret the transmission system shown in Figure 6.18 as a concatenated coding system with the outer code in the form of a convolutional code and the inner code in the form of a finite impulse response channel.

The optimal MLSE or symbol-by-symbol MAP detector would have excessive complexity because of the large size of the super-trellis characterizing the whole system. Instead, a suboptimal solution similar to the turbo decoder can be applied. The technique, which uses the turbo principle, is called *turbo equalization*. A general scheme of the receiver based on this technique is shown in Figure 6.19. It is very similar to the turbo decoder described in Chapter 2.

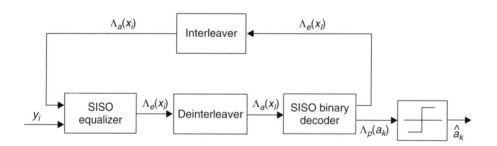

Figure 6.19   General scheme of the turbo equalizer

The turbo equalizer consists of the SISO (*Soft-Input Soft-Output*) equalizer, the deinterleaver, the SISO binary decoder and the interleaver in the feedback loop connected to the SISO equalizer input. The SISO equalizer acquires the received sequence y_i of length nN and the sequence of *a priori* log-likelihood ratios $\Lambda_a(x_i)$ of the code symbols x_i of the same length. Namely, we have

$$\Lambda_a(x_i) = \ln \frac{\Pr\{x_i = 1\}}{\Pr\{x_i = -1\}} = \ln \frac{\Pr\{d_i = 1\}}{\Pr\{d_i = -1\}} \tag{6.90}$$

The log-likelihood ratio $\Lambda_a(x_i)$ is obtained by interleaving the extrinsic information sequence calculated by the SISO decoder in the previous iteration of the turbo equalizer. We assume that in the first iteration the data symbols d_i, or equivalently x_i, are equiprobable, so at the beginning $\Lambda_a(x_i) = 0$ $(i = 1, \ldots, nN)$.

The SISO equalizer can operate according to different criteria. The optimal one is MAP symbol-by-symbol detection, whereas the suboptimal one is soft-output MLSE detection. Assuming application of the MAP criterion, which operates according to the formula

$$\widehat{d}_i = \arg\max_{\{d_i\}} P(d_i | \mathbf{y}) \tag{6.91}$$

in each iteration the SISO equalizer calculates the *a posteriori* log-likelihood ratios $\Lambda_p(x_i)$ of the data symbols d_i $(i = 1, \ldots, nN)$ on the basis of the received data block $\mathbf{y}$, i.e.

$$\Lambda_p(d_i) = \ln \frac{\Pr\{d_i = 1 | \mathbf{y}\}}{\Pr\{d_i = -1 | \mathbf{y}\}} = \ln \frac{\Pr\{x_i = 1 | \mathbf{y}\}}{\Pr\{x_i = 0 | \mathbf{y}\}} = \Lambda_p(x_i) \tag{6.92}$$

As in the case of turbo decoding, one can show that $\Lambda_p(x_i)$ can be split into two components: the *a priori* information related to the data symbol d_i, or equivalently to the code symbol x_i, and the extrinsic information about the code symbol x_i, i.e.

$$\Lambda_p(x_i) = \Lambda_a(x_i) + \Lambda_e(x_i) \tag{6.93}$$

As in the case of turbo decoding, only the extrinsic information $\Lambda_e(x_i) = \Lambda_p(x_i) - \Lambda_a(x_i)$, which is changed to $\Lambda_e(x_l)$ after deinterleaving, is supplied as the input sequence to the SISO decoder, which treats it as the *a priori* information $\Lambda_a(x_l)$. On the basis of the sequence $\Lambda_a(x_l)$ $(l = 1, \ldots, nN)$, the SISO decoder calculates the sequence of *a posteriori* log-likelihood ratios, which again can be treated as the sum of *a priori* information $\Lambda_a(x_l)$ and extrinsic information $\Lambda_e(x_l)$ about code symbol x_l, i.e.

$$\Lambda_p(x_l) = \ln \frac{\Pr\{a_k = 1 | \mathbf{y}\}}{\Pr\{a_k = 0 | \mathbf{y}\}} = \Lambda_a(x_l) + \Lambda_e(x_l) \tag{6.94}$$

The extrinsic information $\Lambda_e(x_l)$ is, after interleaving, supplied back to the SISO equalizer as the *a priori* information $\Lambda_a(x_i)$. In this way the feedback loop of the turbo equalizer has been closed. The detection and decoding operations are performed iteratively an appropriate number of times. In the last iteration, instead of $\Lambda_p(x_l)$ the SISO decoder generates a sequence of the *a posteriori* log-likelihood ratios

$$\Lambda_p(a_k) = \ln \frac{\Pr\{a_k = 1 | \mathbf{y}\}}{\Pr\{a_k = 0 | \mathbf{y}\}} \tag{6.95}$$

The final decision concerning the data block a_k $(k = 1, \ldots, N)$ results from the rule

$$\widehat{a}_k = \begin{cases} 1 & \text{if } \Lambda_p(a_k) \geq 0 \\ 0 & \text{if } \Lambda_p(a_k) < 0 \end{cases} \tag{6.96}$$

The way in which different log-likelihood ratios applied in the turbo equalizer are calculated depends on the applied detection and decoding algorithms. The performed calculations are similar to those shown for the turbo decoder in Chapter 2, so we will not repeat them here.

6.13   Blind Adaptive Equalization

As we have already mentioned, in some cases sending a known data sequence to train the equalizer can result in wasting of a considerable part of transmission time. One of the cases where adaptive equalization without a training sequence is applied is the transmission of a *Digital Video Broadcasting* (DVB) data stream in a DVB cable distribution system. A DVB cable receiver, after being switched on, has to compensate for intersymbol interference on the basis of the received signal and general knowledge of the transmitted signal properties.

Blind equalization algorithms can be divided into three groups:

- the Bussgang (Bellini 1988) algorithms, which apply the gradient-type procedure with nonlinear processing of the filter output signal in order to obtain a reference signal conforming to the selected criterion;
- second- and higher order spectra algorithms, which apply higher order statistics of the input signals in order to recover the channel impulse response and subsequently to calculate the equalizer coefficients;
- probabilistic algorithms, which realize the ML or MAP sequence estimation or suboptimum methods.

The algorithms belonging to the first category are easiest to implement and will be described below. A short overview of the remaining categories can be found in Wesołowski (2003).

The theory of blind equalization presented in Benveniste *et al.* (1980) shows that in order to adjust the linear equalizer properly one should drive its coefficients in such a way that the instantaneous probability distribution of the equalizer output y_n converges to the data input signal probability distribution $p_D(y)$. However, one important condition has to be fulfilled: the probability density function of the input signal d_n must be different from the Gaussian one. It has been found that the ISI introduced by the channel distorts the shape of the input probability density function unless it is Gaussian.

The main difficulty in designing the equalizer's adaptation algorithm is finding a criterion that, when minimized with respect to the equalizer's coefficients, results in (almost) perfect channel equalization. One approach is to calculate the error

$$e_n = y_n - g(y_n) \tag{6.97}$$

which is to be minimized in the mean square error sense, where $g(y_n)$ is an "artificially" generated "reference signal" and $g(.)$ is the memoryless nonlinearity. Thus, the general criterion that is the subject of minimization with respect to the coefficient vector $\mathbf{c}_n$ is

$$\mathcal{E}_n = E\left[|e_n|^2\right] = E\left[|y_n - g(y_n)|^2\right] \tag{6.98}$$

A typical approach to finding the minimum of $\mathcal{E}_n$ is to change the equalizer's coefficients in a direction opposite to that indicated by the current gradient of $\mathcal{E}_n$, calculated with respect to $\mathbf{c}_n$. If we assume that all the signals and filters are complex, we get the following "reference" and error signals

$$\tilde{y}_n = g[\text{Re}(y_n)] + jg[\text{Im}(y_n)] \qquad \tilde{e}_n = y_n - \tilde{y}_n \tag{6.99}$$

Calculation of the gradient of $\mathcal{E}_n$ leads to the result

$$\text{grad}\mathcal{E}_n = 2E\left[\langle \text{Re}(\widetilde{e}_n)\{1 - g'[\text{Re}(y_n)]\} + j\,\text{Im}(\widetilde{e}_n)\{1 - g'[\text{Im}(y_n)]\}\rangle \mathbf{x}_n^*\right] \qquad (6.100)$$

In practice the derivative $g'(.)$ is equal to zero except for a few discrete values of its argument. Thus, the stochastic version of the gradient algorithm achieves the well-known form

$$\mathbf{c}_{n+1} = \mathbf{c}_n - \alpha \widetilde{e}_n \mathbf{x}_n^* \qquad (6.101)$$

where this time the error signal $\widetilde{e}_n$ is described by (6.99). Unfortunately, the optimal non-linear function $g(.)$ is difficult to calculate. Bellini (1988) investigated this function with several simplifying assumptions. Generally, function $g(.)$ should vary during the equalization process. Most of the gradient-based adaptation algorithms are in fact examples of the Bussgang technique, although they were found independently of it. Below we list the most important versions of the gradient algorithms, quoting the error signals that are characteristic for them:

- Sato algorithm: $\widetilde{e}_n = e_n^S = y_n - A_S \text{csgn}(y_n)$, where $\text{csgn}(y_n) = \text{sgn}[\text{Re}(y_n)] + j\,\text{sgn}[\text{Im}(y_n)]$ and A_S is the weighting center of the in-phase and quadrature data signal components;
- Benveniste–Goursat algorithm: $\widetilde{e}_n = e_n^B = k_1 e_n + k_2 |e_n| e_n^S$, $e_n = y_n - \text{dec}(y_n)$, where k_1 and k_2 are properly selected weighting coefficients;
- Stop-and-Go algorithm: $\widetilde{e}_n = e_n^{SG} = f_n^R[\,Re(e_n)] + jf_n^I[[\,Im(e_n)]$, where $e_n = y_n - \text{dec}(y_n)$ and the weighting factors f_n^R and f_n^I turn on and off the in-phase and quadrature components of the decision error, depending on the probability of the event that these components indicate the appropriate direction of the coefficients' adjustment;
- *Constant Modulus* (CM) algorithm: $\widetilde{e}_n = e_n^G = (|y_n|^2 - R_2)\,y_n$, where R_2 is a properly selected data constellation radius.

The CM algorithm, which is explicitly described by the equation

$$c_{i,n+1} = c_{i,n} - \alpha\left(|y_n|^2 - R\right)y_n x_{n-i}^*, i = -N, -(N-1), \ldots, N \qquad (6.102)$$

is the most popular among the four described above but loses information about the phase of the received signal. Therefore, it has to be supported by the phase-locked loop in order to compensate for the phase ambiguity.

Example 6.13.1 *Let us consider application of the CM algorithm in the transmission of DVB signals over a cable used before for transmission and distribution of analog TV signals. Our example is similar to that reported in Karam et al. (1996). Let the single TV 8-MHz channel have the characteristics shown in Figure 6.20, which results from the echoes occurring in the cable and from the characteristics of the transmit and receive filters. Let the transmit and receive filters have the square root raised cosine characteristics with a roll-off factor of $\alpha = 0.15$, as it is set by the standard. The DVB-C signal is transmitted using 64-QAM. Figure 6.21a presents the signal received on the equalizer input*

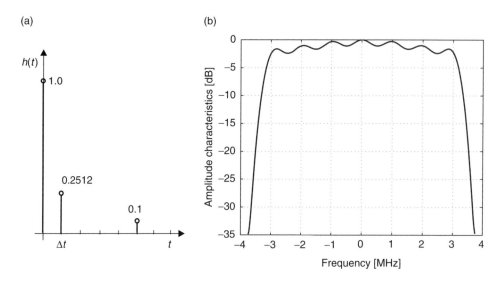

Figure 6.20 Characteristics of the cascade connection of the transmit and receive square root raised cosine filters and the exemplary cable channel used in DVB-C digital transmission: (a) impulse response, (b) amplitude characteristics (Karam *et al.* 1996)

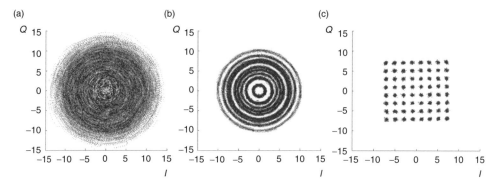

Figure 6.21 Illustration of functioning of the linear equalizer with CM algorithm compensating for the channel from Figure 6.20 after achieving tap settings close to the optimum: (a) signal constellation on the equalizer input, (b) signal constellation on the filter output, (c) signal constellation after carrier phase adaptation

in the in-phase and quadrature plane. As we see, the signal is distorted by the channel introducing ISI, noise and frequency offset to such an extent that correct decisions upon received constellation points are not possible. Figure 6.21b presents the results of the channel equalization performed by the linear equalizer operating according to the CMA citerion, whereas Figure 6.21c shows the signal constellation on the output of the carrier phase correction block placed on the output of the equalizer filter. Such constellation points enable reliable data decisions to be made.

6.14   Equalizers for MIMO Systems

So far we have considered several configurations of digital receivers operating in single-input single-output (SISO) systems. In Chapter 1 we showed that the system capacity can be substantially increased with respect to SISO systems if the transmitter emits signals over N_T antennas and the receiver uses N_R antennas to acquire all the signals generated by the transmitter. Such a system is called a MIMO (*Multiple-Input Multiple-Output*) system. Let us recall that all the signals are emitted simultaneously by N_T antennas in the same band. Because there are N_R receive antennas, $N_T \times N_R$ channels are established in such a system (see Figure 6.22). The MIMO scheme can be supplemented by channel coding if necessary; however, in the following we consider a pure MIMO system without coding.

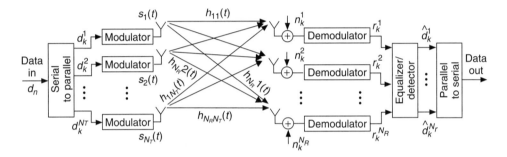

Figure 6.22   General configuration of the MIMO system

M-ary data symbols, which are generated serially, are initially converted into the data vector $\mathbf{d}_k = [d_k^1, d_k^2, \ldots, d_k^{N_T}]$ of length N_T, where index k denotes the current timing instant. Each vector entry d_k^i $(i = 1, \ldots, N_T)$ is subsequently emitted by a separate transmit antenna. The signals generated by all N_T antennas are jointly received by N_R receive antennas. Let us denote the channel impulse response between the ith transmit antenna and jth receive antenna, in which the transmit pulse-shaping filter $g(t)$ is included, as $h_{ji}(t)$. The baseband equivalent signal generated by the ith transmit antenna is

$$s_i(t) = \sum_{k=-\infty}^{\infty} d_k^i g(t - kT), \quad i = 1, \ldots, N_T \quad (6.103)$$

whereas the signal observed on the jth receive antenna is

$$x_k^j = x^j(kT) = \sum_{i=1}^{N_T} \sum_{m=0}^{L_{ji}-1} h_{ji}^{(m)} d_{k-m}^i + n_k^j = r_k^j + n_k^j \quad (6.104)$$

where $h_{ji}^{(m)} = h_{ji}(mT)$ $(m = 0, \ldots, L_{ji} - 1)$ are samples of the impulse response of the channel existing between the ith input and jth output antennas. Variable n_k^j is a sample of white Gaussian noise observed in the kth timing instant on the jth receive antenna and r_k^j is an unobservable "noiseless" channel output sample. Let us assume that channel coefficients $h_{ji}^{(m)}$ are complex-valued, mutually statistically independent Gaussian random variables with zero mean and variance σ_{jim}^2.

6.14.1 MIMO MLSE Equalizer

A MIMO MLSE equalizer is a generalization of the MLSE equalizer considered previously for SISO systems. The MLSE receiver treats the equalized channel as a finite state machine. The channel states are determined by the contents of the tapped delay lines of the filters that model each composite channel. The filter weighting coefficients are given by vectors $\mathbf{h}_{ji} = [h_{ji}^{(0)}, h_{ji}^{(1)}, \ldots, h_{ji}^{(L_{ji}-1)}]^T$ ($i = 1, \ldots, N_T$, $j = 1, \ldots, N_R$). Let us assume for simplicity that $L_{ji} = L$ for $i = 1, \ldots, N_T$ and $j = 1, \ldots, N_R$. As in each timing instant the data vector $\mathbf{d}_k = [d_k^1, d_k^2, \ldots, d_k^{N_T}]$ enters all the composite channels, and since their tapped delay line length is equal to L, the MIMO channel state in the kth timing instant is determined by the vector

$$S_k = [\mathbf{d}_{k-1}, \mathbf{d}_{k-2}, \ldots, \mathbf{d}_{k-L+1}] \tag{6.105}$$

We conclude from (6.105) that the number of channel states is $M^{N_T(L-1)}$ and it exponentially depends on the number of transmit antennas N_T and the channel impulse response length L. The channel states and transitions between them can be represented by the trellis diagram with M^{N_T} paths going out of each state. Thus, as in SISO transmission, finding the maximum likelihood sequence of data vectors $\mathbf{d}_k$ ($k = 1, \ldots, K$) is equivalent to searching for the maximum likelihood sequence of states S_k ($k = 1, \ldots, K$) on the trellis diagram. Formally, the MIMO MLSE receiver finds the sequence of data vectors $\mathbf{d}_k$ ($k = 1, \ldots, K$) that maximizes the probability density function

$$p(\mathbf{x}_1^K | \mathbf{d}_1^K) = \prod_{k=1}^{K} p(\mathbf{x}_k | \mathbf{d}_k, \mathbf{d}_{k-1}, \mathbf{d}_{k-2}, \ldots, \mathbf{d}_{k-L+1}) \tag{6.106}$$

where $\mathbf{x}_1^K = [\mathbf{x}_1, \mathbf{x}_2, \ldots, \mathbf{x}_K]$ and $\mathbf{x}_k = [x_k^1, x_k^2, \ldots, x_k^{N_R}]$ ($k = 1, \ldots, K$). We replaced the joint conditional probability density function $p(\mathbf{x}_1^K | \mathbf{d}_1^K)$ by the products of the terms $p(\mathbf{x}_k | \mathbf{d}_k, \mathbf{d}_{k-1}, \mathbf{d}_{k-2}, \ldots, \mathbf{d}_{k-L+1})$ owing to the assumption that additive noise samples are Gaussian and white, so they are statistically independent. As a result we can write

$$p(\mathbf{x}_1^K | \mathbf{d}_1^K) = \left(\frac{1}{2\pi\sigma^2}\right)^{N_R K} \exp\left(-\frac{1}{2\sigma^2} \sum_{k=1}^{K} \sum_{j=1}^{N_R} \left|x_k^j - r_k^j\right|^2\right) \tag{6.107}$$

Maximization of (6.107) is equivalent to minimization of the cost metric

$$C_K = \sum_{k=1}^{K} \sum_{j=1}^{N_R} \left|x_k^j - r_k^j\right|^2 \tag{6.108}$$

Cost metric (6.108) can be calculated recursively as

$$C_k = C_{k-1} + \sum_{j=1}^{N_R} \left| x_k^j - r_k^j \right|^2 = C_{k-1} + \lambda_k(\mathbf{d}_k) \tag{6.109}$$

where

$$\lambda_k(\mathbf{d}_k) = \sum_{j=1}^{N_R} \left| x_k^j - \sum_{i=1}^{N_T} \sum_{m=0}^{L_{ji}-1} h_{ji}^{(m)} d_{k-m}^i \right|^2 \tag{6.110}$$

We can efficiently search for the sequence of trellis states, or equivalently the sequence of data symbols $\mathbf{d}_k$ ($k = 1, \ldots, K$), for which cost metric (6.108) is minimized if we use the Viterbi algorithm. What we need to determine is the shortest route to each trellis state. The only difference between the Viterbi algorithm applied in the MIMO MLSE receiver and that used in the SISO MLSE receiver is the calculation of the path metric according to (6.110). For that reason there is no need to present the whole Viterbi algorithm again. However, we have to stress that the Viterbi algorithm becomes impractical due to the excessive number of trellis states $M^{N_T(L-1)}$ even for short ISI channels when an M-ary modulation of data symbols is applied and the number of transmit antennas N_T is moderate, e.g. 2–4, as in MIMO systems. Thus, suboptimal receiver structures are necessary, or reception has to be performed according to another criterion.

6.14.2 Linear MIMO Receiver

As the MLSE receiver is computationally intensive, it can be replaced by simpler structures. The first one is the linear MIMO equalizer and its scheme is shown in Figure 6.23. The signals received from N_R antennas are down-converted and sampled, resulting at the kth moment in the sample vector $\mathbf{x}_k = [x_k^1, x_k^2, \ldots, x_k^{N_R}]^T$. The vector entries constitute the input signals for $N_T \times N_R$ FIR filters with coefficient vector $\mathbf{a}_{ji} = [a_{ji}^{(-N/2)}, a_{ji}^{(-N/2+1)}, \ldots, a_{ji}^{(0)}, \ldots, a_{ji}^{(N/2)}]$, where the length of FIR filters is equal to $N + 1$. Strictly speaking, sample x_k^1 is the input for FIR filters $\mathbf{a}_{1i}(k)$ ($i = 1, \ldots, N_T$), sample x_k^2 is the input for FIR filters $\mathbf{a}_{2i}(k)$ ($i = 1, \ldots, N_T$), etc. On the other hand the estimate $\widetilde{d}_k^1$ of the transmitted data symbol d_k^1 is the sum of the outputs from FIR filters $\mathbf{a}_{j1}(k)$ ($j = 1, \ldots, N_R$). The operation of the linear MIMO equalizer can be described in mathematical form by

$$\widetilde{d}_k^i = \sum_{j=1}^{N_R} \left[\sum_{m=-N/2}^{N/2} a_{ji}^{(m)} x_{k-m}^j \right], \quad i = 1, \ldots, N_T \tag{6.111}$$

or in matrix form as

$$\widetilde{\mathbf{d}}_k = \sum_{m=-N/2}^{N/2} A_m \mathbf{x}_{k-m} \tag{6.112}$$

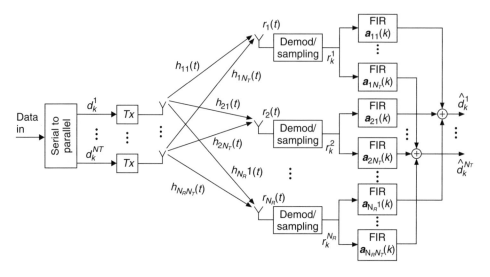

Figure 6.23   MIMO system with a linear equalizer

where matrix A_m is described by the expression

$$A_m = \begin{bmatrix} a_{11}^{(m)} & a_{21}^{(m)} & \cdots & a_{N_R1}^{(m)} \\ a_{12}^{(m)} & a_{22}^{(m)} & \cdots & a_{N_R2}^{(m)} \\ \vdots & \vdots & \ddots & \vdots \\ a_{1N_T}^{(m)} & a_{2N_T}^{(m)} & \cdots & a_{N_RN_T}^{(m)} \end{bmatrix}, m = -N/2, \ldots, 0, \ldots, N/2 \qquad (6.113)$$

Equation (6.112) describes the operation of the generalized FIR filter, for which matrices A_m are the filter coefficients, the received sample vectors $\mathbf{x}_{k-m}$ are the tapped delay line contents, and the vector of equalizer data estimates $\tilde{\mathbf{d}}_k$ is the output the equalizer. Thus, finding the best coefficients is in fact equivalent to finding such a matrix set A_m ($m = -N/2, \ldots, N/2$) for which the selected criterion is fulfilled. Typical optimization criteria are Zero Forcing (ZF) or Minimum Mean Square Error (MMSE).

Let us consider the MMSE criterion. Inspection of Figure 6.23 and formula (6.111) indicates that separate sets of composite FIR filters participate in the minimization of each mean square error $E[|\tilde{d}_k^i - d_k^i|^2]$ ($i = 1, \ldots, N_T$), so minimization of the mean square error for each data symbol separately leads to the joint minimum mean square error. For the ith output data symbol the mean square error, which has to be minimized, is described by the formula

$$\mathcal{E}_k^i = E\left[\left| \sum_{j=1}^{N_R} \mathbf{a}_{ji}^T \mathbf{x}_k^j - d_k^i \right|^2\right] \qquad (6.114)$$

where expectation is calculated jointly with respect to data symbols, noise samples and composite channel coefficients. Denoting the error between the ith transmitted data symbol and its equalizer estimate as

$$e_k^i = \sum_{j=1}^{N_R} \mathbf{a}_{ji}^T \mathbf{x}_k^j - d_k^i \qquad (6.115)$$

we can express the mean square error for the ith data symbol at the kth moment as $\mathcal{E}_k^i = E\left[\left|e_k^i\right|^2\right]$ and we can calculate its gradient with respect to the coefficient vector $\mathbf{a}_{ni}$ $(n = 1, \ldots, N_R)$

$$\frac{\partial \mathcal{E}_k^i}{\partial \mathbf{a}_{ni}} = E\left[\frac{\partial e_k^i}{\partial \mathbf{a}_{ni}}\left(e_k^i\right)^* + \frac{\partial \left(e_k^i\right)^*}{\partial \mathbf{a}_{ni}} e_k^i\right] \qquad (6.116)$$

If we calculate the gradient (6.116) and set it to zero we obtain the equation

$$E\left[\left(\mathbf{x}_k^n\right)^* \sum_{j=1}^{N_R}\left(\mathbf{x}_k^j\right)^T \mathbf{a}_{ji}\right] = E\left[d_k^i\left(\mathbf{x}_k^n\right)^*\right], n = 1, \ldots, N_R \qquad (6.117)$$

So far, we have assumed that the composite channel coefficients are statistically independent, zero-mean and Gaussian. Based on these assumptions, the left-hand side of (6.117) reduces to a single matrix, so the equation takes the form

$$E\left[\left(\mathbf{x}_k^n\right)^*\left(\mathbf{x}_k^n\right)^T\right]\mathbf{a}_{ni} = E\left[d_k^i\left(\mathbf{x}_k^n\right)^*\right], \quad n = 1, \ldots, N_R \qquad (6.118)$$

and the optimum coefficient vector can be calculated from the equation

$$\mathbf{a}_{ni,opt} = E\left[\left(\mathbf{x}_k^n\right)^*\left(\mathbf{x}_k^n\right)^T\right]^{-1} \cdot E\left[d_k^i\left(\mathbf{x}_k^n\right)^*\right], \quad n = 1, \ldots, N_R \qquad (6.119)$$

As we can see, the form of this equation is analogous to that derived for the SISO linear equalizer. Equation (6.119) allows us to calculate the optimum MMSE coefficients for the case when the channel coefficients are Gaussian, zero-mean and statistically independent. If this assumption is not fulfilled the composite equalizer filter coefficients depend not only on the contents of its own tapped delay line $\mathbf{x}_k^n$, but also on all other sample vectors $\mathbf{x}_k^j$. This is caused by possible correlation of impulse responses of the composite MIMO channels. In order to find a solution for this case, we define the joint vectors of input samples and equalizer coefficients in the form

$$\mathbf{x}_k = \left[\left(\mathbf{x}_k^1\right)^T, \left(\mathbf{x}_k^2\right)^T, \ldots, \left(\mathbf{x}_k^{N_R}\right)^T\right]^T$$

$$\mathbf{a}_i = \left[\mathbf{a}_{1i}^T, \mathbf{a}_{2i}^T, \ldots, \mathbf{a}_{N_R i}^T\right] \qquad (6.120)$$

and calculate the optimum joint vector of the equalizer coefficients $\mathbf{a}_{i,opt}$. Similar calculations as sketched above lead us to the following solution

$$\mathbf{a}_{i,opt} = E\left[\mathbf{x}_k^* \mathbf{x}_k^T\right]^{-1} \cdot E\left[d_k^i \mathbf{x}_k^*\right] \qquad (6.121)$$

in which cross-correlations of signal samples received from different receive antennas appear in matrix $E\left[\mathbf{x}_k^* \mathbf{x}_k^T\right]$.

Instead of calculating the equalizer coefficients directly from (6.119) or (6.121) we can apply recursive algorithms based on the LMS or LS algorithms. They are similar to SISO equalizers, so we do not consider them in this chapter.

6.14.3 Decision Feedback MIMO Equalizer

As in SISO data transmission over channels corrupted by intersymbol interference, a *Decision Feedback Equalizer* (DFE) can be considered in MIMO systems (see Figure 6.24). In this case the MIMO linear equalizer is supplemented with the set of decision feedback filters synthesizing the ISI caused by past data symbols. Therefore, N_T FIR feedback filters are used for each data symbol estimate $\widehat{d}_k^i$ ($i = 1, \ldots, N_T$). Let us denote the data vector of past decisions made at the ith equalizer output in the kth timing instant as

$$\widehat{\mathbf{d}}_k^i = \left[\widehat{d}_{k-1}^i, \widehat{d}_{k-2}^i, \ldots \widehat{d}_{k-F}^i\right]^T \tag{6.122}$$

where F is the number of data symbols taking part in the synthesis of ISI. The coefficient vector of the decision filter synthesizing ISI caused by past data symbols from the lth data output to the ith one is

$$\mathbf{b}_{li} = \left[b_{li}^{(1)}, b_{li}^{(2)}, \ldots, b_{li}^{(F)}\right]^T \tag{6.123}$$

so for the ith output data symbol estimate we can write the equation that describes the operation of the DFE equalizer as

$$\widetilde{d}_k^i = \sum_{j=1}^{N_R}\left[\sum_{m=-N/2}^{N/2} a_{ji}^{(m)} x_{k-m}^j\right] - \sum_{i=1}^{N_T}\left[\sum_{m=1}^{F} b_{ji}^{(m)} \widehat{d}_{k-m}^i\right], \ i = 1, \ldots, N_T \tag{6.124}$$

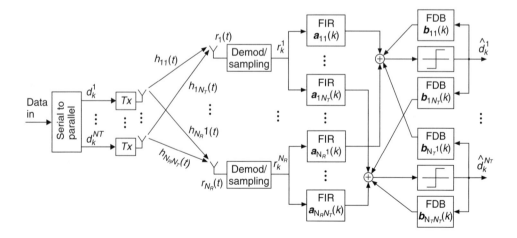

Figure 6.24 Structure of MIMO DFE equalizer

or in vector form

$$\tilde{d}_k^i = \sum_{j=1}^{N_R} \mathbf{a}_{ji}^T \mathbf{x}_k^j - \sum_{l=1}^{N_T} \mathbf{b}_{li}^T \widehat{\mathbf{d}}_k^i, \, i = 1, \ldots, N_T \qquad (6.125)$$

For each timing instant k, decisions are made for all N_T received data symbols $\widehat{d}_k^i = \mathrm{dec}(\tilde{d}_k^i)$ $(i = 1, \ldots, N_T)$. Again, the MMSE or ZF criterion can be applied in setting the linear and decision feedback filter coefficients, as in (6.119) and (6.121).

6.14.4 Equalization in the MIMO-OFDM System

As we see, the linear or decision feedback equalizers of MIMO channels in single-carrier systems corrupted by ISI are very complicated and require many matrix calculations. Thus, the implementation of these equalizers is highly complex. Fortunately, there is an alternative to this solution. As we have alredy mentioned in Chapter 4, multicarrier transmission can be considered as the way to avoid the influence of intersymbol interference. This is a typical solution in modern wireless systems, which, if a single carrier transmission were applied, would suffer from multipath propagation creating the ISI. Figure 6.25 presents a scheme of the multicarrier system with N_T transmit and N_R receive antennas.

In the transmitter a binary data stream is represented in the form of parallel blocks that are subsequently mapped onto blocks of data symbol constellation points. If there are N_T transmit antennas and if the OFDM system applies the N-point IFFT/FFT, there are N_T N-element signal constellation blocks. Let us denote the ith block $(i = 1, \ldots, N_T)$ as

$$\mathbf{D}_{k,i} = [D_{ki}^1, \ldots, D_{ki}^N]^T, i = 1, \ldots, N_T \qquad (6.126)$$

where index k denotes the current time instant. Owing to the application of an appropriately long cyclic prefix, the channel characteristics for each subcarrier appears to be flat

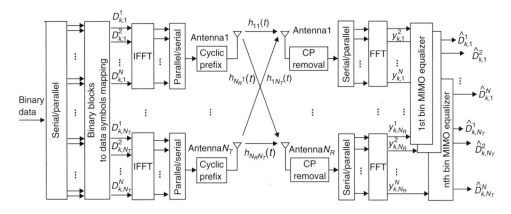

Figure 6.25 MIMO-OFDM system with the equalizer

and the complex samples at the output of the FFT demodulator can be represented by the equation

$$Y_{k,j}^n = \sum_{i=1}^{N_T} H_{ji}^n D_{ki}^n + N_{k,j}^n, \quad n = 1, \ldots, N \tag{6.127}$$

where H_{ji}^n is the gain of the channel between the ith transmit antenna and jth receive antenna observed on the nth OFDM subcarrier and $N_{k,j}^n$ is the additive noise sample at the output of the nth FFT bin of the jth antenna at the kth moment. As the subcarriers remain mutually orthogonal, we can consider each output bin separately. As we consider MIMO transmission, we jointly process FFT output samples from a given nth bin received from all N_R OFDM demodulators. The goal of this operation is to recover all N_T data symbols placed in the nth bin of the IFFT block in the transmitter. If we apply the following vector notation

$$\mathbf{Y}_k^n = [Y_{k,1}^n, \ldots, Y_{k,N_R}^n]^T$$

$$\mathbf{D}_k^n = [D_{k,1}^n, \ldots, D_{k,N_T}^n]^T$$

$$\mathbf{N}_k^n = [N_{k,1}^n, \ldots, N_{k,N_R}^n]^T \tag{6.128}$$

$$\mathbf{H}^n = \begin{bmatrix} H_{11}^n & \cdots & H_{1N_T}^n \\ \vdots & \ddots & \vdots \\ H_{N_R 1}^n & \cdots & H_{N_R N_T}^n \end{bmatrix}$$

we are able to express the outputs of the nth bin of the FFT demodulator as

$$\mathbf{Y}_k^n = \mathbf{H}^n \mathbf{D}_k^n + \mathbf{N}_k^n, \quad n = 1, \ldots, N \tag{6.129}$$

Our task is to find the data block $\mathbf{D}_k^n$ on the basis of the sample vector $\mathbf{Y}_k^n$ that is observed at the outputs of the nth bin of N_R FFT OFDM demodulators. Let us note that in general matrix $\mathbf{H}^n$ is rectangular, because usually $N_R \geq N_T$. One can prove that if the MMSE criterion is applied, the estimates $\widetilde{\mathbf{D}}_k^n$ of the data vectors $\mathbf{D}_k^n$ can be obtained from the formula

$$\widetilde{\mathbf{D}}_k^n = \left[\left(\mathbf{H}^n\right)^\dagger \mathbf{H}^n \right]^{-1} \cdot \left(\mathbf{H}^n\right)^\dagger \mathbf{Y}_k^n, \quad n = 1, \ldots, N \tag{6.130}$$

where $(.)^\dagger$ denotes Hermitian transposition (i.e. matrix transposition with complex conjugation of its elements). Such a matrix operation has to be made for all FFT subcarriers used for data transmission. On the other hand, if the MLSE criterion is applied, we search for the data vector $\widehat{\mathbf{D}}_k^n$ for which the following expression holds

$$\widehat{\mathbf{D}}_k^n = \arg\min_{\mathbf{D}_k^n} \left\| \mathbf{Y}_k^n - \mathbf{H}^n \mathbf{D}_k^n \right\|^2 \tag{6.131}$$

where $\|.\|$ denotes the vector norm. We also note that estimation of matrices $\mathbf{H}^n$ ($n = 1, \ldots, N$) has a crucial meaning for the detection quality if any of the above-mentioned criteria is applied.

Generally, although the MIMO-OFDM system is computationally more efficient than a single-carrier MIMO system, its complexity is still very high. However, it is already implementable in currently available digital technology.

6.15 Conclusions

In this chapter we have concentrated on the problem of equalization in point-to-point transmission. In the early years adaptive equalizers were mostly applied in voiceband acoustic modems; however, fast development of wireless communications substantially extended applications of equalizers. In our chapter we have considered the GSM receiver as an example of the system in which channel equalization in the form of sequential detection is performed. Introduction of the turbo principle and MIMO systems opened up new problems for equalization in such systems. We have simply sketched these problems only, leaving the details for self-study by motivated readers.

Problems

Problem 6.1 *Consider the simplified model of the baseband transmission system shown in Figure 6.26. The signal source generates bipolar, equiprobable, statistically independent data symbols at the rate R. The baseband transmitter applies a pulse-shaping filter and the receiver uses a receive filter, both having the square root raised cosine characteristics with a roll-off factor of $\alpha = 0.25$. Let the amplitude characteristics $|H_B(f)|$ be linear in the dB scale, and let $|H_B(f)|_{dB} = 0$ for $f = 0$ and $|H_B(f)|_{dB} = -a$ for $|f| = 1/T$, as shown in Figure 6.26. Let the delay characteristics have a parabolic shape, i.e. $\tau(f) = \beta \cdot f^2$. The level of the AWGN is reflected in the given SNR. Write a computer program (e.g. using Matlab or a similar package) that simulates the operation of the above system. Apply the oversampling factor $N_s = 16$ to all the signals (i.e. $N_s = 16$ samples per signalling period $T = 1/R$).*

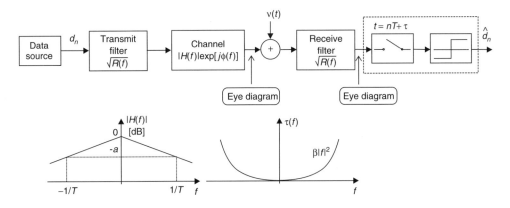

Figure 6.26 Simple model of the baseband digital communication system

1. *Calculate and subsequently plot the impulse response of the cascade connection of the pulse-shaping filter, the channel and the receive filter as a function of the filter and channel parameters. To simplify the calculations, truncate the joint channel impulse response to* 10 *signalling periods* T.

2. *Simulate the transmission of* 10 000 *data symbols and accumulate the waveform at the receive filter output. Plot the eye diagrams for the signals at the channel output and the receive filter output. In Matlab you can use the function called* eyediagram. *In your simulation apply the following sets of system and channel parameters:*

   (a) $R = 2400\,symb/s$, $a = 0\,dB$, $\beta = 0$, $SNR = 20$, 15, 10 *and* 5 dB. *For this set of parameters we simulate the system with the AWGN channel, as the transmisson channel introduces the additive noise only.*

   (b) $R = 2400\,symb/s$, $a = 0$, 5 *and* 10 dB, $\beta = 0$, $SNR = 15\,dB$. *For this set of parameters we check the influence of amplitude distortions on the system performance.*

   (c) $R = 2400\,symb/s$, $a = 0\,dB$, $SNR = 15\,dB$. *The values of* β *are such that the group delay* τ *is equal to* 0.5, 1, 1.5 *and* 2 ms *at the frequency* $R/2\,Hz$. *For this set of parameters we check the influence of the delay characteristics on the system performance.*

   (d) *Select your own set of parameters to show how amplitude and delay distortions as well as additive noise change the eye diagram. Find the set of parameters for which the eye is practically closed.*

3. *For each of* $N_s = 16$ *possible sampling phases, calculate the measure of intersymbol interference given by formula (6.10). Find the best sampling phase resulting in the lowest intersymbol interference according to the applied measure.*

4. *Add a block of down-sampling at a selected time phase, resulting in one sample per signaling period* T. *The block is placed at the output of the receive filter. Also apply the decision device block following the down-sampling block, as shown in Figure 6.26. Next extend your simulation program to include the functionality of counting the decision errors and the estimation of the bit error rate (BER). For a few selected sets of transmission parameters estimate the BER curve as a function of SNR, plot the results and draw conclusions. Apply the sampling phase to the receive filter output signal for which the ISI is the lowest.*

Problem 6.2   *Consider the minimum MSE linear equalizer. As we know, the MSE error at the output of its FIR filter is given by formula (6.18) whereas the optimal equalizer coefficient vector is described by expression (6.20).*

1. *Derive the formula for the minimum MSE error that is achieved for the optimal equalizer settings.*

2. *Recall that the autocorrelation matrix A and the cross-correlation vector $\mathbf{b}$ are described by the expressions $A = E[\mathbf{x}_n \mathbf{x}_n^T]$ and $\mathbf{b} = E[d_n \mathbf{x}_n]$, respectively. Assuming that the data symbols d_n are equiprobable and uncorrelated, knowing that the equalizer input samples x_n are given by formula (6.2) and assuming the uncorrelated noise samples in (6.2), calculate the expression for the entries of matrix A and vector $\mathbf{b}$ depending on the joint channel, transmit and receive filters' impulse response $\{h_i\}$ and the noise variance σ^2.*

3. *Based on the expression derived in the previous point, write a script of the function that calculates matrix A and vector **b** for the system and equalizer parameters: equalizer length $2N + 1$, data symbol power $\sigma_d^2 = E[d_n^2]$, vector of the impulse response $\mathbf{h} = [h_0, h_1, \ldots, h_L]$ and noise variance σ^2. The function subsequently finds the optimal coefficients according to the MMSE criterion.*

4. *Using the function derived in the previous point, calculate the minimum MSE error at the equalizer output for the system transmitting the uncorrelated and equiprobable data symbols d_n of the mean power σ_d^2 over the channel with impulse response $\mathbf{h} = [0.407, 0.815, 0.407]$ and additive uncorrelated and zero-mean Gaussian noise with variance σ_v^2 resulting in SNR $= 20\,dB$. Assume that the lengths of the equalizer are $2N + 1 = 5, 7, 9, 11, 13$ and 15. Plot the resulting minimum MSE error as a function of the equalizer length.*

5. *Apply the program to the system model described in Problem 6.1 in which a given set of parameters has been selected. For that purpose calculate the joint channel impulse response of the transmit filter, the channel and the receive filter for the applied sampling phase. Calculate the minimum MSE for each of $N_s = 16$ possible sampling phases and for the given equalizer length $2N + 1$. For the best sampling phase find the sufficient length of the equalizer and calculate the minimum mean square error achievable by it. The equalizer length can be considered sufficient if further extension does not bring about any meaningful decrease in the MSE.*

Problem 6.3 *Consider again the data transmission system model derived in Problem 6.1. Let the SNR be equal to 20 dB and let the system parameters be $R = 2400\,symb/s$, $a = 10\,dB$, $\beta = 6.9 \times 10^{-9}$. Extend the chain of the simulated blocks by an adaptive equalizer of length $2N + 1 = 21$ (see Figure 6.27). Apply the appropriately delayed data stream from the transmitter as the reference data sequence in the equalizer. Find the correct delay of the data stream in order to treat the signal sample at the central point of the tapped delay line of the equalizer as the main one. In two simulation experiments simulate the convergence process of the equalizer. Apply the ZF algorithm given by (6.15) in the first simulation experiment and the LMS algorithm given by (6.23) in the second one. Select the appropriate values of the step sizes α and γ, respectively [recall formula (6.25) describing the initial value of the step size for the LMS equalizer]. Assume the initial equalizer settings to be equal to the vector $\mathbf{c}_0 = [c_{-N,0} = 0, \ldots, 0, c_{0,0} = 1, 0, \ldots c_{N,0} = 0]$. In order to obtain statistically valuable results, perform simulation runs 50 times and calculate the mean value of the square error for each iteration number.*

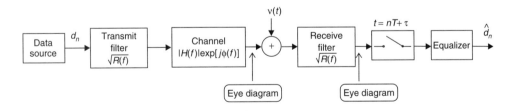

Figure 6.27   Scheme of the baseband transmission system with the equalizer

1. *How many reference data symbols have to be sent for application of each type of adaptation algorithm in order to obtain equalizer settings close to the optimal ones?*
2. *Experiment with the value of the step size applied in the LMS algorithm. Let the constant step size be applied in the convergence process, equal to γ_0, $0.8\gamma_0$, $0.5\gamma_0$ and $0.3\gamma_0$. Compare the rate of convergence and the steady state MSE error for all applied values of the step size.*

Problem 6.4 *Consider the MSE decision feedback equalizer with N_L linear filter taps and N_F decision feedback taps. We have shown in (6.46) that the equalizer output signal sample is the scalar product of the coefficient vectors $\mathbf{w}_n$ and the signal samples $\mathbf{z}_n$ contained in the tapped delay lines of both filters. Let the reference data symbol be related to the signal sample contained in the middle of the tapped delay line of the linear filter. It can be easily shown that the optimum coefficient vector $\mathbf{w}_{opt} = A^{-1}\mathbf{b}$, where $A = E[\mathbf{z}_n\mathbf{z}_n^T]$ and $\mathbf{b} = E[d_n\mathbf{z}_n]$.*

1. *Calculate the entries of matrix A and vector $\mathbf{b}$ as a function of the channel impulse response samples $\{h_i\}$, the additive Gaussian noise variance σ_v^2 and the data symbol mean power $\sigma_d^2 = E[d_n^2]$.*
2. *Derive the function that calculates the minimum MSE error for the given linear and decision feedback filter lengths, the channel impulse response vector, the noise variance and the data symbol mean power. Apply this function to calculate the minimum MSE error for the same system parameters as in Problem 6.2, Point 5, for all possible sampling phases. Assume the lengths of the linear and decision feedback filters to be $N_L = 10$ and $N_F = 11$. Compare the results with those achieved in Problem 6.2 for the linear MSE equalizer.*
3. *Repeat the calculations performed in the previous point for different lengths of the linear and decision feedback filters N_L and N_F, keeping the number of all coefficients $N_L + N_F = 21$.*

Problem 6.5 *Replace the linear equalizer applied in Problem 6.3 by the decision feedback equalizer with $N_L = 10$ linear filter taps and $N_F = 11$ decision filter taps. Treat the sample in the middle of the tapped delay line of the linear filter as the main one. Adjust the delay of the reference data symbols appropriately. Apply the LMS algorithm described by (6.48) and (6.49) to both linear and feedback filters. Select the appropriate step sizes and show the abilty of the equalizer to converge. Set the system parameters as in the previous problem. Repeat the simulation experiments for different lengths of the linear and decision feedback filters, e.g. $N_L = 7$ and $N_F = 14$ or $N_L = 15$ and $N_F = 6$. Does the choice of the filter lengths have any influence on the equalizer performance?*

Problem 6.6 *Consider the simple tapped delay line model of a transmission system. Assume that the cascade of the transmit filter, channel and receive filter has the impulse response in the form $\mathbf{h} = [h_0, h_1, h_2]$, where $h_0 = 1$, $h_1 = 0.5$ and $h_2 = -0.2$. The transmitted data symbols are binary, bipolar and equiprobable. The channel output signal samples are distorted by additive Gaussian noise v_n. Denote the samples observable on the joint channel output as x_n.*

1. *Find the trellis diagram of this model and calculate all possible unobservable channel outputs $r_n = x_n - v_n$.*

2. *Perform the operation of the Viterbi detector and find the ML data sequence if the following samples are observed on the detector input (0.7, -0.3, 1.1, 0.6, 0.3, -0.1, 0.2, 1.6, 0.4, 1.8)*

Problem 6.7 *Use the program derived in Problem 6.1 to demonstrate the operation of the channel estimator. Configure the simulated system as shown in Figure 6.28. Set the simulated system parameters as in Problem 6.3. Apply the LMS algorithm given by formula (6.61) in the adaptation process of the channel estimator shown in Figure 6.11.*

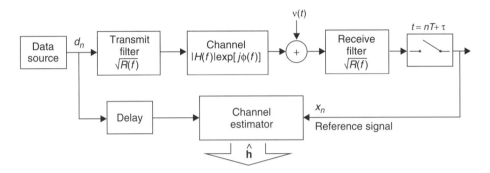

Figure 6.28   Block diagram of a baseband transmission system with the channel estimator

1. *Demonstrate the convergence of the LMS algorithm with properly selected step size α applied in the estimator of length $N_e = 12$ by performing 50 independent simulation runs and averaging the received squared error values along the time axis. What is the best step size value and the steady state MSE error?*
2. *Repeat the experiments for $N_e = 10$ and $N_e = 14$.*

7

Spread Spectrum Systems

7.1 Introduction

Spread spectrum systems were invented in the late 1940s. Originally they were applied in military communications. Today the concept of spread spectrum systems is used in many contemporary communication systems. The most important ones are the third generation cellular systems UMTS (*Universal Mobile Telecommunications System*; Holma and Toskala 2004) and cdma2000 (Garg 2000), GPS (*Global Positioning System*), personal satellite communication system GLOBALSTAR (Lagarde *et al.* 1995) and the second generation cellular telephony system based on the IS-95 standard (Garg 2000).

The basic concept of the spread spectrum system can be derived from the formula that describes the capacity of a continuous channel of bandwidth B Hz and with the signal-to-noise ratio $\gamma = P/(N_0 B)$

$$C = B \log (1 + \gamma) \tag{7.1}$$

Changing the logarithm base from 2 to e, we get

$$\frac{C}{B} = 1.44 \ln (1 + \gamma) \tag{7.2}$$

Let us note that if the SNR is small, e.g. $\gamma < 0.1$, we can represent logarithmic function (7.2) in the form of a Taylor expansion with respect to γ. We obtain

$$\frac{C}{B} = 1.44 \left(\gamma - \frac{1}{2}\gamma^2 + \frac{1}{3}\gamma^3 - \frac{1}{4}\gamma^4 + \ldots \right) \tag{7.3}$$

Neglecting higher order components in (7.3), for small values of γ we get an approximate expression for the system bandwidth as a function of the channel capacity and the channel SNR

$$B \approx \frac{C}{1.44\gamma} \tag{7.4}$$

Introduction to Digital Communication Systems   Krzysztof Wesołowski
© 2009 John Wiley & Sons, Ltd

Therefore, if the SNR is low, i.e. the signal level is significantly lower than the noise level, then the required channel capacity can be achieved by sufficient widening of the signal bandwidth. For example, in order to ensure a channel throughput of 10 kbit/s at the SNR of $\gamma = 0.01$ (i.e. -20 dB), a 690 kHz bandwidth is needed. This finding indicates how to transmit data at a given rate in a completely different way from what we did earlier. So far we have attempted to use as narrow band as possible when simultaneously ensuring a possibly high value of the SNR. However, we can proceed in the opposite way: We can substantially widen the applied signal band, significantly decreasing the SNR. If the signal is transmitted in a wide band at a low SNR it is hardly detectable and can be applied even if other systems use parts of the same spectrum range. These features are particularly valuable in military applications. As a result of the assumed low SNR the following problem arises: how to spread the spectrum of the signal carrying the user's messages and how to enable reliable reception of this signal in the presence of several disturbances introduced by the transmission channel and its other users. We will sketch the answers to these questions below.

There are several kinds of spread spectrum systems. The most important among them are:

- *Direct Sequence Spread Spectrum* (DS-SS) system, in which spectral spreading is performed by direct modulation of the data signal by a pseudorandom sequence;
- *Frequency-Hopping Spread Spectrum* (FH-SS) system, in which spectral spreading is performed by pseudorandom hopping of the carrier frequency of the data-modulated signal;
- *Time-Hopping Spread Spectrum* (TH-SS) system, in which data-carrying pulses are located in a pseudorandom manner within the defined time slots.

The first type of spread spectrum system is widely used in the radio systems listed at the beginning of this chapter. The second one dominates in military applications, although it is also the basis of one of the standards of radio access to local area networks (IEEE 802.11: IEEE 2007; Bing 2000). The popular Bluetooth standard (Haartsen 1998) of short-range radio links is also based on this type of spread spectrum system. The third type of spreading has found application in ultra-wideband communication systems.

Below we will shortly describe the DS-SS, FH-SS and TH-SS systems. A distinctive feature of all such systems is the application of a source of pseudorandom signals that makes spectrum spreading in a given band possible. For that reason we will start our description from the generation of spreading sequences.

7.2 Pseudorandom Sequence Generation

Generation of pseudorandom sequences with given correlation properties was considered by scientists and engineers a long time ago (Golomb 1967). This problem is closely related to the polynomial theory described briefly in Chapter 2. Generation of a fully random sequence in a repetitive manner is technically not realizable. Repetitiveness is a key property of the sequence, because, as we will soon learn, both transmitter and receiver apply the same sequence, which is used for spectral spreading in the transmitter and spectral de-spreading in the receiver. Thus, instead of random sequences, pseudo-random ones are applied. The latter are in fact deterministic symbol sequences. Except for a few special cases, these sequences have a very long period compared with a single sequence element. Although they are fully deterministic, they appear random to an external observer. It is desirable that the spectrum of such a sequence is flat. In the ideal case it could be white, which means that the autocorrelation function of such a sequence should have the form of an ideal pulse for a zero argument, with zero values for any nonzero argument.

The above-mentioned features of an ideal pseudorandom sequence are essential in the case of its application by a single transmitter and receiver pair. However, the same concept of spectrum spreading is also applied when the same spectrum is used by many users at the same time. Thus, many transmitter – receiver pairs simultaneously operate in the same spectrum and zero cross-correlation between sequences used by different transmitter – receiver pairs is of key importance. Summarizing, the pseudorandom sequences are useful if:

- there exists a sufficiently large set of pseudorandom sequences that can be applied simultaneously;
- each sequence belonging to this set has good autocorrelation properties, i.e. its auto-correlation function is an impulse for a zero argument and is equal to zero or at least very close to zero otherwise;
- any two different sequences from the set are mutually uncorrelated.

Naturally, typical pseudorandom sequences do not feature ideal properties. Nevertheless, they are often applied in practical systems. We will review several kinds of pseudorandom sequences now.

7.2.1 Maximum Length Sequences

Maximum length sequences, often called m-sequences, are codewords of the maximum length code described in Chapter 2. As we remember, codewords of length $n = 2^m - 1$ of that code are generated by the encoder based on the parity check polynomial $h(x)$, which is a primitive polynomial and the divisor of the polynomial $x^n - 1$ and cannot be factorized into polynomials of lower degrees. The structure of such an encoder is called a *Linear Feedback Shift Register* (LFSR). An example of such a sequence generator is shown in Figure 7.1.

For rising degree m the number of different primitive polynomials useful in generating the maximum length sequences quickly increases. The m-sequences have the following properties.

Property 7.2.1 *The balance property. The number of binary "1"s occurring in a full period of an m-sequence of length $2^m - 1$ is equal to 2^{m-1}, whereas the number of zeros is equal to $2^{m-1} - 1$.*

This property is a direct consequence of the fact that m-sequences are codewords of a maximum length code. The Hamming weights of all nonzero codewords of such codes are exactly equal to 2^{m-1}.

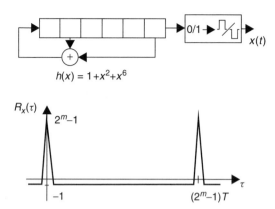

Figure 7.1 Example of the LFSR generating a pseudorandom sequence (a) and the autocorrelation function of this sequence (b)

Property 7.2.2 *The run property. There exist 2^{m-1} different runs consisting of only "1"s or zeros, half of which have length equal to 1, $1/2^2$ have length equal to 2, and generally $1/2^k$ have length k. There is a single run of length $m-1$ containing only zeros and a single run of the same length consisting of only "1"s.*

Property 7.2.3 *The correlation property. The autocorrelation function of the m-sequence, whose symbols are represented by bipolar symbols ± 1, takes the value $R(n) = -1$ for all arguments n different from zero and equals $R(0) = 2^m - 1$ for the zero argument.*

The last property indicates that the autocorrelation function is not exactly equal to zero for a nonzero argument, although its form approximates well to an ideal pulse if the length of the m-sequence is large. Recall that as the m-sequence $(c_0, c_1, \ldots, c_{2^m-2})$ is a periodical deterministic sequence with the period $P = 2^m - 1$, its autocorrelation function $R(n)$ can be calculated from the formula

$$R(n) = \sum_{i=0}^{P-1} c_i c_{(i+n) \bmod P} \tag{7.5}$$

Pseudorandom sequences applied in spread spectrum systems often have such a long period that only fragments of them are used. Then the autocorrelation function calculated over these fragments is crucial for the whole system properties. It turns out that for nonzero arguments the autocorrelation function can take much higher values than those calculated on the basis of the whole sequence period. Cross-correlation of two equal length fragments of different m-sequences takes values substantially different from zero. For that reason Gold (1966) proposed a new type of pseudorandom sequence for which the above-mentioned disadvantages appear to a much more limited extent.

7.2.2 Gold Sequences

Gold discovered that in a set of m-sequences of length $2^m - 1$ there exist such sequence pairs x, y that have three-level values of the cross-correlation function $R_{xy}(n)$. These values belong to the set $\{-1, -t(m), t(m) - 2\}$, where

$$t(m) = \begin{cases} 2^{(m+1)/2} + 1 & \text{for odd } m \\ 2^{(m+2)/2} + 1 & \text{for even } m \end{cases} \tag{7.6}$$

The pairs x, y featuring this property are called *preferred sequences*, and possible values of their cross-correlation function are substantially lower than in the case of regular m-sequences of the same length. Gold sequences are created on the basis of preferred sequences by summing modulo-2 the first sequence x with a cyclically shifted version of the second sequence y. The sequences created by this operation have length equal to $2^m - 1$. In this way we achieve a family of $2^m + 1$ Gold sequences, because there are $2^m - 1$ cyclic shifts of one of the component sequences and single preferred sequences x and y belong to the family as well. Figure 7.2 presents an example of the scheme for the generation of Gold sequences of length $2^6 - 1$. Each Gold sequence is created by summing modulo-2 two m-sequences generated in the LFSRs, determined by polynomials $h_1(x) = 1 + x + x^6$ and $h_2(x) = 1 + x + x^2 + x^5 + x^6$, respectively.

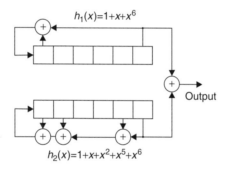

Figure 7.2 Generator of Gold sequences of length $2^6 - 1$

Besides Gold sequences there are other pseudorandom sequences that will not be described here. Among them, the sequences discovered by Kasami *et al.* (1968) are well known. Interested readers are asked to study Wesołowski (2002) and the bibliography therein.

Pseudorandom sequences are not the only ones used for spreading the spectrum of information-bearing signals. In practical wireless systems, deterministic sequences constituting a set of mutually orthogonal sequences are also applied for that purpose. Walsh-Hadamard sequences (Garg 2000; Wesołowski 2002), for example, are applied in cellular telephony for assigning a channel in a given cell.

7.2.3 Barker Sequences

Besides long period pseudorandom sequences, short sequences of good autocorrelation properties are used in some systems for spreading or synchronization purposes. *Barker sequences* are a good example of such systems. These sequences are strictly deterministic bipolar sequences that have an almost ideal form of the so-called aperiodic autocorrelation function defined by the formula

$$
R_{ap}(n) = \begin{cases}
\displaystyle\sum_{i=1}^{P-n} c_{i+n}c_i & \text{for } 0 \le n \le P-1 \\[4mm]
\displaystyle\sum_{i=1}^{P+n} c_i c_{i-n} & \text{for } -P+1 \le n \le 0 \\[4mm]
0 & \text{for } |n| \ge P
\end{cases}
\tag{7.7}
$$

Let us note that the aperiodic autocorrelation function is in fact a regular autocorrelation function of a deterministic sequence supplemented by a sequence of zeros.

Barker sequences exist for strictly limited lengths that are equal to 2, 3, 4, 5, 7, 11 and 13. They do not exist for other lengths. The sequences are listed below.

$$
\begin{aligned}
\mathbf{c}_2 &= (+1, -1) \\
\mathbf{c}_3 &= (+1, +1, -1) \\
\mathbf{c}_4 &= (+1, +1, -1, +1) \\
\mathbf{c}_5 &= (+1, +1, +1, -1, +1) \\
\mathbf{c}_7 &= (+1, +1, +1, -1, -1, +1, -1) \\
\mathbf{c}_{11} &= (+1, +1, +1, -1, -1, -1, +1, -1, -1, +1, -1) \\
\mathbf{c}_{13} &= (+1, +1, +1, +1, +1, -1, -1, +1, +1, -1, +1, -1, +1)
\end{aligned}
\tag{7.8}
$$

The sequences reordered in the backward direction are also Barker sequences. The aperiodic autocorrelation function of Barker sequences is

$$
R_{ap}(n) = \begin{cases}
1 & \text{for } n = 0 \\
0, 1/P \text{ or } -1/P & \text{for } 1 \le |n| \le P-1
\end{cases}
\tag{7.9}
$$

where P is the period of a sequence.

7.3 Direct Sequence Spread Spectrum Systems

Figure 7.3 presents a basic block diagram of the direct sequence spread spectrum (DS-SS) system applying a pseudorandom sequence. Assume that the data symbols are bipolar, i.e. $a_n = \pm 1$. Index n is related to the current data symbol. Data symbols appear on the DS-SS transmitter input every T_b seconds. They can be treated as a signal modulating the sequence c_i, which is achieved from the output of a pseudorandom generator whose pulses, so-called *chips*, are generated every T_c seconds. Period T_b is usually an integer multiple of period T_c, i.e., $T_b = KT_c$. As a result, the data signal spectrum is widened K times. The obtained bipolar signal constitutes a modulating signal for a PSK modulator.

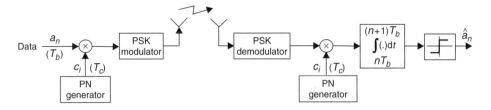

Figure 7.3 Basic scheme of a direct sequence spread spectrum system

In the simplest version a BPSK is applied, although higher level modulations such as QPSK can be used as well. Let the transmission channel be characterized by constant attenuation in the whole range of the signal spectrum and let the additive white Gaussian noise with power density spectrum equal to $N_0/2$ be the only disturbance. In the receiver a synchronous demodulator converts the received signal to the baseband. Let us note that the considered system can be treated as a bipolar PAM system in which a data symbol a_n modulates the amplitude of the baseband pulse that has the form of a pseudorandom bipolar sequence c_i. Figure 7.4 presents the process of modulation of the pseudorandom sequence by the data sequence. The data sequence $\{a_n\}$ is shown as a continuous function $a(t)$, whereas $c(t)$ is a waveform representation of the chip sequence $\{c_i\}$. In accordance with the theory of the optimum synchronous receiver, for equiprobable bipolar data symbols one should apply the receiver that correlates the received signal with the baseband pulse, i.e. with the pseudorandom sequence c_i. Obviously, if the receiver is to function correctly, it must know the shape of the baseband pulse. In our case the receiver has to know the exact pseudorandom sequence and this sequence has to be synchronized with the signal it receives. These considerations result in the receiver structure shown in Figure 7.3.

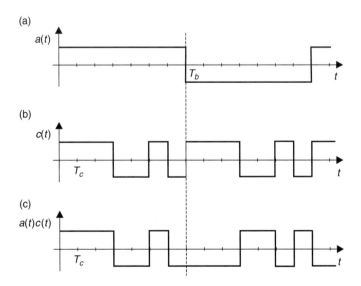

Figure 7.4 Creating the spread spectrum signal using the Barker sequence c_5: (a) data sequence, (b) spreading sequence, (c) resulting signal

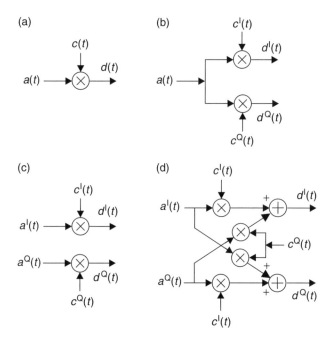

Figure 7.5 Several spreading schemes: (a) spreading a single data sequence by a single spreading sequence, (b) spreading a single data sequence by two spreading sequences, (c) independent spreading of two data sequences by a pair of spreading sequences, (d) complex spreading

As we said earlier, the spreading scheme shown in Figure 7.3, in which a single spreading sequence is applied and data are in the form of a single sequence, is particularly simple. Obviously, more advanced schemes are applied in real systems. We can encounter the systems in which two multilevel data sequences are the subject of spreading at the same time. Several other spreading configurations are also possible. Figure 7.5 presents typical spreading schemes. So far we have considered the scheme from Figure 7.5a. In the case of transmission of a single data sequence and QPSK modulation we can select the spreading scheme shown in Figure 7.5b. If two data sequences $a^I(t)$ and $a^Q(t)$ are the subject of spreading, we can apply two independent spreading sequences $c^I(t)$ and $c^Q(t)$ (Figure 7.5c) or the so-called complex spreading (Figure 7.5d) for which the following operation is performed

$$d^I(t) + jd^Q(t) = [a^I(t) + ja^Q(t)][c^I(t) + jc^Q(t)] \tag{7.10}$$

In the last two cases the resulting spread symbols $d^I(t)$ and $d^Q(t)$ are subsequently treated as a pair of modulating signals in a QPSK or OQPSK modulator.

Let us come back to the analysis of the simplest system (see Figure 7.3) with direct sequence spreading and let us compare the error probability in two cases. We assume that data symbols are equiprobable and the channel noise is Gaussian and white. In the first case a traditional bipolar transmission system is applied in which data symbols a_n directly modulate the carrier without application of spreading, i.e. the BPSK signal is applied that has the form

$$x(t) = a_n u(t) = a_n \sqrt{\frac{2E_b}{T_b}} \cos(2\pi f_c t + \theta) \quad \text{for} \quad nT_b \leq t < (n+1)T_b \tag{7.11}$$

In the optimal synchronous receiver, the received signal given by the formula $r(t) = x(t) + v(t)$ is correlated with the pulse $u(t)$, whose energy in the signaling period T_b is equal to E_b. As we know, the probability of error on the output of such a receiver is

$$P(\mathcal{E}) = \frac{1}{2}\text{erfc}\left(\sqrt{\frac{E_b}{N_0}}\right) \tag{7.12}$$

In the second case direct sequence spreading using the pseudorandom sequence c_i is applied. Let the spreading signal in the nth modulation period, in which data symbol a_n is transmitted, have the form

$$c(t) = \sum_{i=0}^{K-1} c_i p(t - iT_c - nT_b) \quad \text{for} \quad nT_b \leq t < (n+1)T_b \tag{7.13}$$

where $c_i = \pm 1$, and $p(t)$ describes the shape of a single chip. Assume that $p(t)$ is a gate function of unit amplitude ($|c(t)| = 1$) and duration T_c. Thus, the pulse $u(t)$ whose amplitude is modulated by data symbol a_n is described by the formula

$$u(t) = c(t)\sqrt{\frac{2E_c}{T_c}} \cos(2\pi f_c t + \theta) \tag{7.14}$$

where $E_c = E_b/K$. As a result, the energy of the signal transmitted within the duration of a single data symbol a_n is the same in both cases, because

$$\int_0^{T_b} u^2(t)\mathrm{d}t = \frac{2E_c}{T_c}|c(t)|^2 \int_0^{T_b} \cos^2(2\pi f_c t + \theta)\mathrm{d}t = \frac{2E_c}{T_c}\frac{T_b}{2} = KE_c = E_b \tag{7.15}$$

Thus, if in the system with spreading sequence $c(t)$ the optimal receiver correlating the received signal with the reference signal (7.14) is applied, the probability of an erroneous decision on the data symbol a_n is again given by formula (7.12). Therefore we can conclude that spreading of the signal over a much wider spectrum theoretically does not bring any advantage. So, what is the aim of using spread spectrum systems?

We can formulate at least four answers to this question:

1. If the mean power of the signal is constant, by performing K-fold spreading of the signal spectrum we decrease the power density spectrum of the signal K times. If K

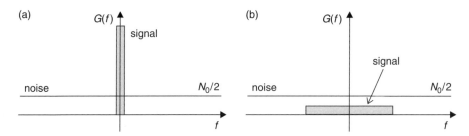

Figure 7.6 Illustration of the influence of spreading on spectral properties of the signal in the presence of additive white Gaussian noise: (a) power density spectrum of the narrow-band signal and noise, (b) power density spectrum of the spread spectrum signal and noise when preserving the signal power

is sufficiently high, the power density spectrum of the transmitted signal can fall down below the level of the noise power density spectrum. Thus signal transmission can be hidden from unauthorized users, which can potentially have a military application. This property is illustrated in Figure 7.6.

2. Spreading of the data-carrying signal in the transmitter and de-spreading of the received signal in the receiver are dual operations that compensate for each other; however, the disturbance appearing within the spectrum of the transmitted spread spectrum signal is the subject of de-spreading only. If the disturbance and the de-spreading signal are mutually uncorrelated, then the result of correlation performed by the de-spreader is close or equal to zero. Thus, spreading of the transmitted signal improves its robustness against uncorrelated disturbance. This process is illustrated in Figure 7.7.

3. It follows from the above conclusion that different links may use the same spectrum simultaneously if the signals applied in one link are orthogonal to any signals used in any other links. This observation is the basis for *Code Division Multiple Access* (CDMA) applied in different wireless communication systems, e.g. in UMTS and cdma2000.

4. Radio channels on which spread spectrum systems are applied can rarely be represented by an AWGN channel model. As we know from our studies on channel properties, a typical radio channel features multipath propagation. If propagation delays between particular channel paths differ by more than the chip period T_c, then, assuming that the applied spreading sequence has an ideal autocorrelation function, the signals arriving at the receiver along particular paths are mutually uncorrelated. As a result, each path signal can be individually extracted using correlators, one of which has been shown in Figure 7.3. In order to take full advantage of each path signal, these signals should be appropriately combined in order to minimize the probability of an erronous decision upon transmitted data symbols. Such operations are performed by the so-called RAKE receiver described in the next section.

Let us come back to the second answer to the question concerning the meaning of the spread spectrum systems. Let us analyze the spectral properties of the signals in the most important locations of the considered DS-SS system (Figure 7.7). The following notation has been applied: $S_a(f)$ – data signal spectrum, $S_c(f)$ – spreading signal spectrum, $S_{\mathrm{mod}}(f)$ – spectrum at the output of the BPSK modulator, $S_{\mathrm{dem}}(f)$ – spectrum

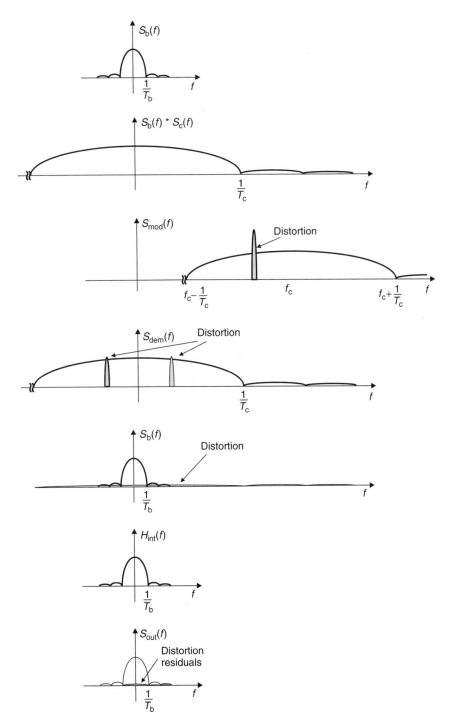

Figure 7.7 Spectra in particular locations of the DS-SS system operating in the presence of narrow-band disturbance (Wesołowski 2002)

at the output of the BPSK demodulator, $H_{\text{int}}(f)$ – transfer function of the integrator, and $S_{\text{out}}(f)$ – signal spectrum at the system output. As we can see, in the receiver the narrow-band disturbance located within the band of the spread signal is the subject of correlation consisting of multiplication and integration. Multiplication results in spreading of the disturbance over the whole spectrum of the spread signal. In turn, integration results in cutting off a significant part of the spectrum of the spread disturbance. Owing to both operations only a small part of the disturbance power located in the band of the data signal is contained in the spectrum of the output signal.

7.4 RAKE Receiver

In the case of spread spectrum transmission over an ideal AWGN channel the optimal receiver is a single correlator that correlates the received signal with the spreading sequence. However, this is not an optimal solution if the transmission channel features multipath propagation and if its channel impulse response is given in a short time interval by formula (5.48). The optimal receiver for such a channel, called RAKE, was first described by Price and Green (1958). Its description can be found in many handbooks on digital communications, including Proakis (2000) and Wesołowski (2002), but for completeness of the spread spectrum system description we repeat its derivation below.

Let the signal on the demodulator output be described in the baseband by the equation

$$r(t) = h(\tau; t) * x(t) + v(t) \tag{7.16}$$

We assume that all the functions of time in (7.16) are complex. Owing to this assumption this description refers to the systems in which BPSK as well as two-dimensional modulations such as QPSK are applied. Function $h(\tau; t)$ is the baseband equivalent channel impulse response. The channel contains not only the multipath physical transmission channel but also the transmit and receive filters. Function $x(t)$ describes the baseband equivalent transmitted signal, whereas $v(t)$ represents the additive noise. As in the multipath channel description presented in Chapter 4, the variable τ reflects the time running from the moment of channel excitation, whereas the variable t describes slow changes of the channel impulse response in time. For simplicity let us consider the bipolar transmission in the form

$$x(t) = a_i u(t) \quad \text{for } 0 \le t < T_b \tag{7.17}$$

where T_b is, as in our previous considerations, the duration of the information symbol, whereas $u(t)$ is a signal that is the subject of modulation by a sequence of information symbols according to one of the rules shown in Figure 7.5. One of the possible forms of signal $u(t)$ is then a bipolar spreading sequence. Our derivation is based on that shown in Proakis (2000).

Let us assume that the bandwidth of the signal $u(t)$ is limited to $W/2$ Hz. Then, according to the sampling theorem, this signal can be characterized by the following expression

$$u(t) = \sum_{n=-\infty}^{\infty} u\left(\frac{n}{W}\right) \frac{\sin\left[\pi W(t - n/W)\right]}{\pi W(t - n/W)} \tag{7.18}$$

As a result, its density spectrum is described by the formula

$$
U(f) = \begin{cases} \dfrac{1}{W} \displaystyle\sum_{n=-\infty}^{\infty} u\left(\dfrac{n}{W}\right) \exp\left(-j2\pi fn/W\right) & \text{for } |f| \le W/2 \\ 0 & \text{otherwise} \end{cases} \tag{7.19}
$$

When the signal $u(t)$ is given to the channel input, the signal received on the output of the time-varying channel characterized by the transfer function $H(f;t)$ is given by the inverse Fourier transform

$$
r(t) = \int_{-\infty}^{\infty} H(f;t)U(f) \exp(j2\pi ft)\,df \tag{7.20}
$$

If we apply (7.19) in expression (7.20), we get

$$
r(t) = \frac{1}{W} \sum_{n=-\infty}^{\infty} u\left(\frac{n}{W}\right) \int_{-\infty}^{\infty} H(f;t) \exp[-j2\pi f(t-n/W)]df
$$

$$
= \frac{1}{W} \sum_{n=-\infty}^{\infty} u\left(\frac{n}{W}\right) h\left(t - \frac{n}{W};t\right) \tag{7.21}
$$

where $h(\tau;t)$ is the channel impulse response of the time-varying channel characterized by the transfer function $H(f;t)$. If we interchange the variables in (7.21), we obtain

$$
r(t) = \frac{1}{W} \sum_{n=-\infty}^{\infty} u\left(t - \frac{n}{W}\right) h\left(\frac{n}{W};t\right) \tag{7.22}
$$

Defining

$$
h_n(t) = \frac{1}{W} h\left(\frac{n}{W};t\right) \tag{7.23}
$$

we obtain

$$
r(t) = \sum_{n=-\infty}^{\infty} h_n(t) u\left(t - \frac{n}{W}\right) \tag{7.24}
$$

It can be easily seen from (7.24) that by applying notation (7.23) the channel impulse response is, as in (5.48), described by the formula

$$
h(\tau;t) = \sum_{n=-\infty}^{\infty} h_n(t)\delta\left(t - \frac{n}{W}\right) \tag{7.25}
$$

Let the effective duration of the channel impulse response be T_m. Then summation of all components in formula (7.25) is limited to $L = \lfloor T_m W \rfloor + 1$ terms with the weights

$h_n(t)$ $(n = 0, \ldots, L - 1)$ and the channel impulse response achieves the form

$$h(\tau; t) = \sum_{n=0}^{L-1} h_n(t)\delta\left(t - \frac{n}{W}\right) \tag{7.26}$$

Now let us come back to the analysis of the signal on the channel output if signal (7.17) is transmitted. Let the duration of the pulse $u(t)$ be much longer than the duration of the channel impulse response. In practice this means that channel responses to subsequent pulses $u(t)$ almost do not overlap. The signal received on the channel output is then approximately described by the formula

$$r(t) = \sum_{n=0}^{L-1} a_i h_n(t) u\left(t - \frac{n}{W}\right) + v(t) \quad \text{for } iT_b \le t < (i + 1)T_b \tag{7.27}$$

Assuming that the coefficients of the channel impulse response are known, the optimum receiver is a correlator in which the reference signal has the form

$$q(t) = \sum_{n=0}^{L-1} h_n(t) u\left(t - \frac{n}{W}\right) \tag{7.28}$$

The operation performed by the correlator is described by the formula

$$\tilde{a}_i = \int_0^{T_b} r(t)q^*(t)\mathrm{d}t = \sum_{n=0}^{L-1} \int_0^{T_b} h_n^*(t)r(t)u^*\left(t - \frac{n}{W}\right) \mathrm{d}t \tag{7.29}$$

The change in channel impulse response within a single data symbol period T_b is usually negligibly small, so the correlator output signal can be expressed by the formula

$$\tilde{a}_i = \sum_{n=0}^{L-1} h_n^*(t) \int_0^{T_b} r(t)u^*\left(t - \frac{n}{W}\right) \mathrm{d}t, \quad iT_b \le t < (i + 1)T_b \tag{7.30}$$

Expression (7.30) implies the scheme of the receiver shown in Figure 7.8. The complex conjugate version of the pulse $u(t)$ propagates along the tapped delay line. The time span between neighboring delay line taps is equal to $1/W$ whereas the tap weighting coefficients are equal to the complex conjugates of the channel impulse response coefficients $h_n^*(t)$ $(n = 0, \ldots, L - 1)$. All tap signals are, after their appropriate weighting, correlated with the received signal $r(t)$. The correlator outputs are summed and, based on the resulting value, a decision upon the received data symbol is made.

For DS-SS systems the pulse $u(t)$ is simply the spreading signal and sampling with frequency W can be approximated by sampling with a frequency equal to the chip rate $1/T_c$. From the technical implementation point of view the functioning of the optimum receiver shown in Figure 7.8 is equivalent to supplying the received signal to the input of the delay line and correlating the delay line tap signals with the complex conjugated form

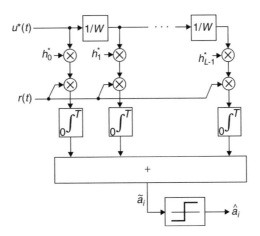

Figure 7.8 Optimum receiver for multipath channel under the assumption that the duration of the channel impulse response is significantly shorter than the information symbol duration

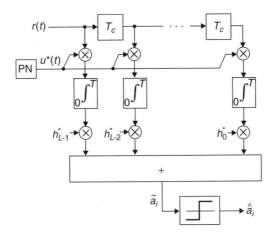

Figure 7.9 Basic scheme of the RAKE receiver

of the spreading sequence $u^*(t)$. The weighting coefficients assigned to the appropriate taps appear along the tapped delay line in reverse order. This version of the receiver is shown in Figure 7.9 and is known as the *RAKE receiver*. The correlators applied in appropriate taps are called *RAKE fingers*. Each finger "collects" the signal from a particular path of the multipath channel. Subsequently, these signals are optimally combined, which is performed by weighting them by the conjugated channel coefficients h_i^* $(i = 0, \ldots, L - 1)$.

Let us note that the RAKE structure is equivalent to the filter matched to the signal that is a convolution of the channel impulse response and the spreading sequence $u(t)$. In practice, when the chip rate is very high, the number of RAKE fingers is often limited to 3–4. The receiver automatically selects three to four strongest channel paths and places the fingers in appropriate locations of the delay line in order to extract these path signals.

This automatic process is realized by applying an additional correlator that sequentially scans the possible delay line taps in order to measure the level of the signal received on a given path and to find the strongest paths.

Calculation of the error probability on the RAKE receiver output is a relatively complex task. This probability depends on the statistical properties of the channel, the time delay distribution of particular channel propagation paths and the time variability of the channel weighting coefficients $h_n(t)$ $(n = 0, 1, \ldots, L-1)$. We leave the derivation of this error probability to particularly motivated readers. For this purpose the author advises the readers to study appropriate sections in Proakis (2000) or Stüber (2001).

We have to stress that the RAKE receiver is optimal in a single link transmission when the duration T_m of the channel impulse response is much shorter than the data symbol interval T_b. The reception quality of the RAKE receiver degrades if the same band is occupied by more users, i.e. there are more transmission links, as it occurs in CDMA. In reality, the spreading signals of other users that are applied in the same band are not fully mutually orthogonal due to the inherent signal and channel properties. As a result, these signals infiltrate into other receivers where they are treated as a disturbance. This negative effect may be counteracted by applying *joint detection*. Basic information on that topic can be found in Wesołowski (2002), and full treatment of this subject is contained in Verdú (1998) and Castoldi (2002).

7.5   Frequency-Hopping Spread Spectrum Systems

The DS-SS systems considered so far have high requirements with respect to sequence synchronization and their receivers are synchronous. In some applications receiver synchronization with an accuracy of a fraction of a chip is difficult to ensure or can be expensive to implement, therefore other spread spectrum systems have been proposed in which the signal carrier hops in a pseudorandom manner. Such systems are called *Frequency-Hopping Spread Spectrum* (FH-SS) systems. The general scheme of the basic FH-SS system is shown in Figure 7.10.

Data, possibly protected by an error correction code, are given to the input of the M-FSK modulator. In Chapter 4 we considered FSK modulation for which $M = 2$. In FH-SS systems it is possible to apply a higher modulation level, for which $M = 2^k$. A pseudorandom

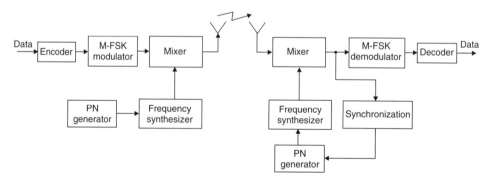

Figure 7.10   General scheme of the system with spectrum spreading using pseudorandom hopping of carrier frequency

generator produces a binary sequence, which is segmented into binary blocks and controls the choice of the frequency generated by the carrier frequency synthesizer. Owing to the applied frequency mixer, the M-FSK signal is placed on the carrier currently generated by the frequency synthesizer. If the number of possible carrier frequencies is equal to 2^L, then the length of the blocks onto which the generated pseudorandom sequence is divided is exactly equal to L. In this way the frequency synthesizer implements the process of frequency hopping. The receiver performs dual operations to those performed in the transmitter. After synchronization of the pseudorandom generator in the receiver with respect to the received signal, the L-bit binary block is given to the synthesizer input. As a result, the synthesizer generates an appropriate carrier frequency. This in turn allows the received signal to be converted to the band in which the M-FSK-modulated signal is detected. Finally, k-bit data blocks appear on the detector output.

Let us consider the mathematical description of an FH-SS signal. An M-FSK-modulated signal can be described by the formula

$$x_{MFSK}(t) = \text{Re}\left\{ A \exp\left[j2\pi \left(f_c + \Delta f(t) \right) t \right] \right\} \tag{7.31}$$

in which the frequency keying is denoted by the expression

$$\Delta f(t) = \sum_{n=-\infty}^{\infty} a_n p_d(t - nT_d) \tag{7.32}$$

and data symbols a_n are selected from the set $\{-(M-1), \ldots, -1, +1, \ldots, (M-1)\}$ every T_d seconds. The spreading signal given to the input of the frequency mixer is described by the formula

$$c_{FH}(t) = \sum_{i=-\infty}^{\infty} p_h(t - iT_h) \exp\left[j(2\pi f_i t + \varphi_i) \right] \tag{7.33}$$

where both $p_d(t)$ and $p_h(t)$ functions describe the frequency pulses as functions of time and mostly have the form of a gate function of duration T_d and T_h, respectively. Finally, an FH-SS signal is given by the expression

$$x_{FH}(t) = c_{FH}(t)x_{MFSK}(t) \tag{7.34}$$

The phase φ_i that appears in formula (7.33) results from the fact that the frequency synthesizer of the dense frequency grid is usually not able to preserve phase continuity.

Depending on the relation that exists between the time periods T_d and T_h, the FH-SS systems can be divided into *slow* (SFH – *Slow Frequency Hopping*) and *fast* (FFH – *Fast Frequency Hopping*) frequency-hopping systems. The key difference between these systems is illustrated in Figure 7.11. We have assumed that the M-FSK modulation is binary, i.e. $M = 2$. In the case of fast frequency hopping a few carrier frequency hops (four in Figure 7.11) occur within a single data symbol. In turn, in the case of slow frequency hopping a single frequency hop is performed after a few data symbols. Therefore, in the first case $T_d = NT_h$, whereas in the second case $T_h = KT_d$.

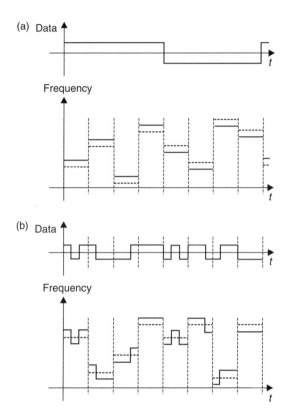

Figure 7.11 Illustration of the FH-SS system operation with fast (a) and slow (b) carrier frequency hops (current carrier frequency is denoted by a dashed line)

The data symbol period T_d and the frequency-hopping interval T_h are strictly associated with *chip frequency* f_{chip}. It is the greater value out of two values $1/T_d$ and $1/T_h$, so it can be meant as the highest timing frequency applied in a given FH-SS system. The frequency f_{chip} is also associated with the frequency deviation applied in the M-FSK modulation and the frequency separation between carriers selected during the hopping process.

In the SFH systems the interval between the applied nominal frequencies is the same as between nominal frequencies in the M-FSK modulation. The applied tones are equally distributed on the frequency axis. In turn, in order to ensure orthogonality between tones applied in M-FSK, which is a desired feature due to the applied noncoherent detection, the distance between tones should be a multiple of the chip frequency $f_{chip} = 1/T_d$. If this distance were exactly equal to f_{chip}, then for N_c carrier frequencies the bandwidth used by the system would be equal to $N_c M f_{chip}$. It is indeed exactly equal to this value if the used band is divided into disjoint subbands, which are used by M-FSK-modulated signals with a momentary carrier frequency located exactly in the middle of a subband. Then the distance between possible neighboring carrier frequencies is $M f_{chip}$. As FH-SS systems are often applied for military purposes, they should be robust against intentional disturbance (*jamming*). Robustness can be improved compared with the previous choice of carrier frequencies if the distance between carriers is equal to f_{chip}, i.e. if it is the same as between the nominal frequencies of M-FSK-modulated signals. Neighboring bands of M-FSK signals will substantially overlap and the jammer will have a more difficult task facing a much higher number of carrier frequencies that can be potentially selected.

For FFH systems with fast frequency hopping the chip frequency is equal to the hopping frequency, $f_{chip} = 1/T_h$. Thus, the distance between the applied tones in M-FSK signals is equal to $f_{chip} = 1/T_h$ and the whole band used by the FFH system is divided into subbands with carrier frequencies placed in the middle of them and selectable by the pseudorandom generator. Since a few frequency hops occur for each data symbol, the decision upon a given data symbol has to be based on a sequence of tones that are generated by the modulator within the duration T_d of the whole data symbol.

Although FH-SS systems are mainly used in military communication systems, they have also found civil applications. The most popular one is Bluetooth (Haartsen 1998). The idea of FH-SS is also a basic rule for operation of one of the alternative versions of IEEE 802.11 (IEEE 2007) wireless access to LANs. Both systems are classified as HF systems with slow hopping and both systems take advantage of nonlicensed bands. These bands are not reserved for any particular wireless system so several systems can operate in them simultaneously. Robustness against disturbances becomes a key feature of such FH-SS systems, additionally supported by the application of channel coding.

FH-SS systems, as with DS-SS systems, can operate in a multiple access mode. In a given band more than a single link between users can be established without mutual disturbances if mutually orthogonal signals are used in different links. This condition is fulfilled if the selection rules of M-FSK nominal and carrier frequencies presented above are preserved and if the users apply synchronized pseudorandom generators in such a way that a simultaneous generation of two identical tones by different transmitters does not occur. In the case of lack of synchronism among users, collisions are possible and channel coding becomes a remedy against them.

7.6 Time-Hopping Spread Spectrum System with Pseudorandom Pulse Position Selection

The time-hopping spread spectrum (TH-SS) system is probably the least popular of the three main types. Its general scheme is shown in Figure 7.12. User data, which can be

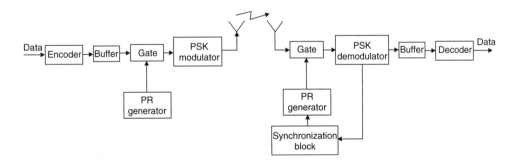

Figure 7.12   General scheme of a TH-SS system

the subject of channel coding at the coding rate R_c, are accumulated in block form in the buffer. Time is divided into frames of length T seconds. The frames in turn are divided into time slots. Let the number of time slots in a single frame be M_T. A pseudorandom generator selects one of M_T time slots, in which a PSK modulator generates a signal representing the binary sequence stored in the buffer. If the information data rate is equal to R bit/s, then application of the channel coding at the coding rate of R_c results in the data stream of the rate being equal to R/R_c bit/s. Because a single time slot out of M_T possible slots in a frame can be applied in user data transmission, the transmission rate is $M_T R/R_c$. This results in spectrum spreading of around $M_T R/R_c$ Hz, compared with the bandwidth of the order of R/R_c, if pseudorandom selection of the pulse position is not used.

Due to synchronization difficulties and the bursty nature of transmission, TH-SS systems are not so popular as DS-SS and FH-SS systems. However, we have to admit that a similar rule to that used in TH-SS systems has been applied in one of the alternative solutions of *Ultra-WideBand* (UWB) communication systems.

The main idea of UWB systems is to apply signals with an extremely wide bandwidth. They are used to transmit data over short distances. In a UWB system, spreading of the energy of a transmitted signal is so wide that very low spectral density allows the system to share the spectrum simultaneously with other systems using some parts of the same spectrum. One of the UWB transmission methods is application of OFDM-like transmission, however a powerful alternative is to use the TH-SS principle.

* * *

Spread spectrum systems are an important category of digital communication systems, particularly in their radio segment. Owing to multiple use of the same band, which is possible by applying mutually orthogonal spreading sequences, DS-SS systems have been particularly useful in mobile cellular systems. They ensure higher capacity (i.e. number of users per unit area and spectrum unit) than other cellular systems based on other multiple access methods such as Time Division Multiple Access (TDMA) or Frequency Division Multiple Access (FDMA). The main standards of third generation cellular telephony (UMTS, cdma2000) apply Code Division Multiple Access (CDMA) using DS-SS transmission. However, such systems require a sufficiently wide spectrum, which is possible when the data transmission is of the order of a few Mbit/s but if we wish to consider data transmission at rates of several hundred Mbit/s then signal spreading would lead to a spectral occupancy that is too high. The electromagnetic spectrum is a scarce resource and requires particularly careful usage, therefore it is better to use spectrally efficient transmission methods for higher transmission rates. The main candidates for such applications are solutions based on OFDM transmission.

Problems

Problem 7.1 *In DS-SS systems m-sequences are often used as spreading sequences. For simplicity of calculations, let us consider an m-sequence based on the linear feedback shift register (LSFR) defined by the polynomial $x^4 + x^3 + 1$. Draw the scheme of this LFSR. What is the period of the generated sequences? How many different sequences can be generated starting from the time instant $n = 0$? Generate these sequences and compare* *them. Let us treat the sequence received at the highest LFSR position (described by the power x^4), when a single "1" is contained in the shift register at this position at the moment $n = 0$, as the reference sequence. How can we easily generate the m-sequences originating from the same LFSR and shifted in time with respect to the reference sequence? Recall that m-sequences are in fact the codewords of the maximum length code (cf. Chapter 2). Design an appropriate logical circuit generating the sequence shifted in time by an appropriate number of clock cycles with respect to the reference sequence.*

Problem 7.2 *Write a simple program (e.g. applying Matlab) that generates an m-sequence based on the polynomial $x^9 + x^4 + 1$. Let a single "1" be contained in the LFSR at its highest position at the moment $n = 0$ and let the reference LFSR output be the output of the memory cell at this position. Assuming representation of the generated zeros and "1"s in the form of bipolar rectangular pulses of length T_c, calculate the power density spectrum of the generated sequence and plot it. For computational purposes assume that the oversampling rate is $N_s = 8$. Recall that the m-sequence is in fact a periodic sequence with the period equal to $2^N - 1$ ($N = 9$ in our case). For comparison derive the formula for the power spectral density analytically for any m-sequence length $m = 2^N - 1$.*

Problem 7.3 *Consider the m-sequence from Problem 7.2. Let us generate the m-sequence as a modulo-2 sum of the 9th, 7th and 5th positions of the LFSR.*

1. *Calculate the cross-correlation of this sequence with the reference sequence generated at the highest LFSR position. Perform this calculation over the whole m-sequence period.*
2. *Repeat the same calculations for a shorter period of time, i.e. calculate the cross-correlation of both sequences in bipolar form (assume the oversampling rate $N_s = 1$) performed over the sliding window of length $W = 2^8$, 2^7 and 2^6 symbols. Find the maximal values of cross-correlation for all window lengths and compare them with the result of cross-correlaton performed over a full sequence period. Is almost perfect orthogonality of both sequences preserved when only partial sequences are considered?*

Problem 7.4 *Consider the Barker sequence of length 11, as shown in formula (7.8). Calculate its aperiodic autocorrelation function. Next calculate the autocorrelation function of the periodic version of the same Barker sequence. Compare both cases and draw conclusions.*

Problem 7.5 *Consider the family of Gold sequences of length $N = 511$ that are generated as an XOR sum of the output sequences of two LFSRs determined by the polynomials $h_1(x) = x^9 + x^4 + 1$ and $h_2(x) = x^9 + x^6 + x^4 + x^3 + 1$. For the experiments select two sequences belonging to this family. Let both LFSRs start at the moment $n = 0$ from a single one in the highest position. The first sequence is simply the XOR sum of both output sequences of the component LFSRs whereas the second one is the sequence that is cyclically shifted by three symbols with respect to the first one. Calculate the cross-correlation value over the whole sequence length. Next calculate the maximum of the cross-correlation values over the sliding window of length $W = 2^8$, 2^7 and 2^6. Compare the results with appropriate cross-correlation values found in Problem 7.3.*

Problem 7.6 *Consider the DS-SS system with bipolar signaling and the bipolar spreading sequence based on the m-sequence generated by the LFSR given by the polynomial $g(x) = x^9 + x^4 + 1$. For data signal spreading the whole period of the spreading sequence is used. The signal is transmitted over a two-tap channel with the tap coefficients $h_0 = 0.8$ and $h_1 = 0.6$. The additive Gaussian noise of the variance $\sigma^2 = 0.1$ added at the channel output models the system noise. Consider two types of receiver:*

1. a simple correlator with a bipolar reference signal tuned to the first signal path;
2. a two-tap RAKE receiver.

Draw the block diagrams of both receivers. Calculate the SNR at the input of the decision device applied in both receivers, assuming that the length of the channel impulse response is negligible compared with the length of the spreading sequence.

8

Synchronization in Digital Communication Systems

8.1 Introduction

In previous chapters we considered several questions related to digital communication systems. Some of them involved baseband transmission, whereas others were associated with digital modulations of a sinusoidal carrier and methods of demodulation of such signals. In our considerations, including those associated with spread spectrum systems, we assumed ideal synchronization of the receiver with the incoming signal, which is surely an idealization. In order for the system to work properly, synchronization is necessary on several levels of system functioning. Providing synchronization can be a difficult task. Synchronization is essential for starting and continuing transmission, and its quality often decides the actual level of bit error probability in the considered system.

Whole books are devoted to synchronization (Lindsey 1972; Meyr and Ascheid 1990; Mengali and D'Andrea 1997) but it is also the subject of large chapters in each significant academic handbook devoted to digital communication systems (Proakis 2000; Barry *et al.* 2003; Kurzweil 2000; Benvenuto and Cherubini 2002). In this chapter we will only sketch the most important synchronization topics. More interested readers are asked to study one of the above-mentioned handbooks or the tutorial (Luise *et al.* 2003) devoted to the role of synchronization and its structures in digital communication systems.

Let us consider digital transmission in which a sinusoidal carrier is used. The transmitted signal approaches the receiver in the form disturbed by noise. Moreover, it is often a subject of intersymbol interference resulting from amplitude and phase distortions introduced by the transmission channel. The Doppler effect, which shifts the signal spectrum along the frequency axis, can also occur. If a synchronous receiver is applied, the following types of sychronization become necessary:

- *carrier synchronization*, whose aim is to recover the carrier frequency of the received signal (*frequency synchronization*) and the carrier phase (*phase synchronization*);
- *timing synchronization*, whose aim is to recover the optimal sampling moments for data symbol detection or the starting moments of data symbol periods, both of which allow the received distorted data symbols to be analyzed;

Introduction to Digital Communication Systems   Krzysztof Wesołowski
© 2009 John Wiley & Sons, Ltd

- *frame synchronization*, whose aim is to synchronize the operation of the receiver with the hierarchical structure of the received information-carrying signal.

In some digital communication systems implemented mostly in digital signal processing technology, in which data transmission has a continuous character, a fourth type of synchronization is also needed. This is the so-called *sampling frequency synchonization*. A free-running oscillator, which would be the source of sampling clock, would cause a loss of quality if data were transmitted in long sequences. Therefore, it has to be synchronized on the basis of the data measurements extracted from the received signal.

Each of the above-mentioned types of synchronization can be implemented in several ways. The functioning of many communication systems is organized in such a way that it facilitates acquiring synchronization followed by its tracking. For example, ITU-T recommendations related to voiceband telephone modems standardize sequences of specially selected data signals that precede the user data transmission. These sequences enable gradual carrier and timing synchronization. Other examples of communication standards in which synchronization is an important issue are IEEE 802.11a/g (IEEE 1999, 2003) and ETSI HIPERLAN/2 (ETSI 2001b) which describe the functioning of wireless LAN modems with OFDM modulation. Transmission is organized in frames that start with a specially selected OFDM symbol sequence, called preamble, which allows the receiver to synchronize the carrier and timing of OFDM symbols. Additionally, during user data transmission some OFDM subcarriers are exclusively applied for transmission of the data symbols known to the receiver (the so-called *pilots*), which are used to maintain frequency and timing synchronization and enable estimation of the channel transfer function. The next example of the synchronization approach can be found in GSM cellular telephony. In this system transmission is organized according to the hybrid time division/frequency division multiple access method (cf. Chapter 9). Each GSM base station has a selected carrier on which in certain time slots it periodically sends a sinusoidal signal, which plays the role of a pilot. Its detection by a newly turned on mobile station allows it to synchronize its carrier and to determine the frame starting moment. In the detected time slot of subsequent frames the mobile station reads information that enables it to acquire system synchronization, i.e. frame synchronization on different levels of the complex GSM timing hierarchy.

Figure 8.1 presents an example of a synchronous receiver (Luise *et al.* 2003) with carrier and timing synchronization blocks. This is one of several possible configurations typical for digital communication systems. For the systems in which intersymbol interference occurs, the matched filter block is supplemented with or replaced by the channel equalizer.

As we conclude from the above discussion on several approaches to synchronization issues, acquisition and tracking of synchronization can be supported by application of special preambles or pilot signals. Sometimes synchronization has to be achieved exclusively on the basis of a continuously received information-bearing signal. Methods that extract synchronization signals from the received signal exclusively on the basis of the knowledge of statistical signal properties are called *blind synchronization* methods.

Most carrier and timing synchronization methods apply some form of *Phase-Locked Loop* (PLL). We will describe the basics of its functioning in the next section and in our analysis we will follow the book by Barry *et al.* (2003).

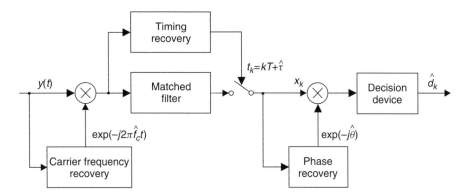

Figure 8.1 Example of configuration of carrier and timing synchronization blocks in a synchronous receiver (Luise *et al.* 2003)

8.2   Phase-locked loop for continuous signals

The phase-locked loop is a key element of various synchronization circuits. The basic PLL block diagram for continuous signals is shown in Figure 8.2. The PLL input signal is given to the *phase detector*, which compares the phase of the input signal with the phase of the reference signal generated by the *Voltage-Controlled Oscillator* (VCO). The phase detector output signal, which is a measure of phase difference between both signals, is filtered in the *PLL filter*. The filter output signal controls the VCO. The place from which the PLL signal is treated as an output signal depends on a particular PLL application. It can be the output of the VCO, as shown in Figure 8.2, or the PLL filter output.

Let us assume that the signal on the PLL input has the form

$$y(t) = A_{in} \cos \left[2\pi f_c t + \theta(t) \right] \tag{8.1}$$

where A_{in} is the amplitude of the cosinusoidal input signal and f_c is its frequency. The signal that approaches the receiver is usually modulated, distorted by the channel and disturbed by noise. However, we will temporarily apply the idealized signal (8.1) in our analysis. Let the VCO output signal be given by the formula

$$v(t) = A_{VCO} \cos \left[2\pi f_c t + \varphi(t) \right] \tag{8.2}$$

If the phase function $\varphi(t)$ were constant, the frequency of the VCO output signal would be equal to f_c. This frequency is often called the *free-running frequency* of the VCO.

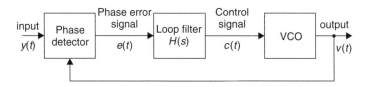

Figure 8.2   Block diagram of a phase-locked loop

Let us note that all possible frequency variations can be contained in the function $\varphi(t)$, because the instantaneous frequency of the VCO output signal is

$$f(t) = \frac{1}{2\pi} \frac{d[2\pi f_c t + \varphi(t)]}{dt} = f_c + \frac{1}{2\pi} \frac{d\varphi(t)}{dt} \tag{8.3}$$

Let us temporarily assume that the phase detector is ideal, i.e. its output signal is described by the formula

$$e(t) = W[\theta(t) - \varphi(t)] \tag{8.4}$$

where $W(.)$ denotes the input–output characteristics of the detector shown in Figure 8.3. The function from Figure 8.3 is called a *sawtooth characteristic* and reflects phase difference ambiguity equal to 2π. The slope of the detector characteristic shown in Figure 8.3 is equal to unity. If it were different from unity, then for the purpose of PLL analysis it can be incorporated into the gain of the loop filter. The time function $e(t)$, which is a measure of the phase difference between the input signal and the signal generated by the VCO, is filtered by the PLL filter of transfer function[1] $H(s)$, resulting in the output signal $c(t)$. This signal is subsequently applied in VCO control. In the case of an ideal VCO, its instantaneous frequency offset from the free-running frequency exactly reflects the control signal $c(t)$. Therefore, if we want the control signal to be proportional to the instantaneous frequency offset, we set

$$c(t) = \frac{d\varphi(t)}{dt} \tag{8.5}$$

Also in this case, if the slope of the VCO characteristics is different from unity, it can be incorporated into the PLL filter gain. If the VCO generates the signal in such a way that the following expression is valid

$$\varphi(t) = \theta(t) + \varphi_0 \tag{8.6}$$

we say that the PLL is *locked*, whereas if the PLL recovers the phase function of the input signal with the zero phase shift, we say that the PLL is *ideally locked*.

Let us determine the range of operation of the PLL. Let us assume that the frequency of the input signal differs from the free-running PLL frequency by a constant offset equal

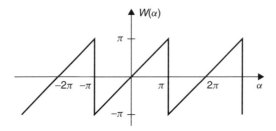

Figure 8.3   Characteristics of the ideal phase detector

---
[1] The Laplace transform is a useful tool in PLL analysis.

to f_0 Hz. This means that

$$\theta(t) = 2\pi f_0 t + \theta_0 \tag{8.7}$$

In order for the PLL to stay locked, the phase difference on the phase detector output cannot exceed the range of $\pm\pi$, which follows from the detector characteristics shown in Figure 8.3. The VCO has to generate the signal with the same frequency offset f_0. Thus we conclude that

$$\varphi(t) = 2\pi f_0 t + \varphi_0 \tag{8.8}$$

and the signal on the phase detector output resulting, according to formula (8.5), in the signal $c(t) = 2\pi f_0$ produced on the filter output has to fulfill the dependency

$$|e(t)| = |\theta(t) - \varphi(t)| = |\theta_0 - \varphi_0| \leq \pi \tag{8.9}$$

Therefore, in the PLL steady state, taking into account the PLL scheme and formulas (8.5) and (8.9), we obtain the following form of the condition for keeping the PLL locked

$$2\pi f_0 \leq H(0)\pi \tag{8.10}$$

where $H(0)$ is the filter gain for the DC signal component.

Let us analyze the dynamic properties of the PLL when it stays locked, i.e. when condition (8.9) is fulfilled. The loop operates in a linear range and mathematic tools appropriate for the description of linear systems can be applied. Let us derive the *phase transfer function* of the loop. Taking the Laplace transform of both sides of formula (8.5), we can write

$$s\Phi(s) = C(s) = E(s)H(s) \tag{8.11}$$

In turn, taking into account (8.4) and assuming a unity slope of the detector characteristics we have

$$E(s) = \Theta(s) - \Phi(s) \tag{8.12}$$

If we calculate the ratio $\Phi(s)/\Theta(s)$, we get the closed loop transfer function in the form

$$\frac{\Phi(s)}{\Theta(s)} = \frac{H(s)}{H(s) + s} \tag{8.13}$$

As we see, the PLL is a dynamic system whose properties highly depend on the applied loop filter. In the simplest case the loop filter is represented by a constant gain[2] denoted as K_{amp}. Then we have

$$\frac{\Phi(s)}{\Theta(s)} = \frac{K_{amp}}{K_{amp} + s} \tag{8.14}$$

---

[2] Let us recall that we have accumulated the slope coefficients of the phase detector and VCO characteristics, as well as the filter gain characteristics, into a single PLL gain coefficient.

This time $H(0) = K_{\mathrm{amp}}$, so from (8.10) we get the following inequality for the lock range of the PLL

$$|2\pi f_0| \le K_{\mathrm{amp}}\pi \tag{8.15}$$

As we see in (8.15), the lock range of the PLL is proportional to the filter gain, which in our analysis represents the real filter gain and the slope coefficients of the phase detector and VCO. The value of K_{amp} determines the bandwidth of the PLL, because if we substitute $s = j2\pi f$ in (8.14) we get the magnitude of the transfer function in the form

$$\left|\frac{\Phi(f)}{\Theta(f)}\right| = \frac{K_{\mathrm{amp}}}{\sqrt{K_{\mathrm{amp}}^2 + 4\pi^2 f^2}} \tag{8.16}$$

The magnitude of the transfer function in logarithmic scale is shown in Figure 8.4, in which we can observe how the gain coefficient K_{amp} influences the bandwidth of the PLL. Decreasing the PLL bandwidth is realized by lowering the value of K_{amp}. Unfortunately, the lock range decreases as well. In order to avoid this negative effect we can apply a higher order loop filter.

Let us now consider the response of the PLL in steady state to a given signal type. Let us take the frequency step function $2\pi f_0 u(t)$, where $u(t)$ is a unit step function. We are interested in the value of the phase error in the steady state. According to the Laplace

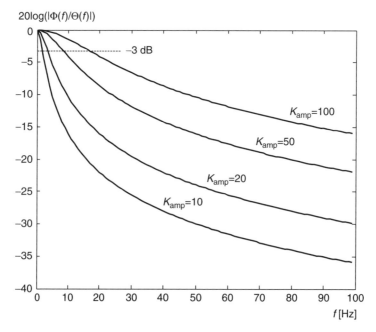

Figure 8.4 Magnitude of the phase transfer function for the PLL with the filter featuring the constant gain K_{amp}, for different values of K_{amp}

transform rules, this error can be calculated as

$$e_\infty = \lim_{t \to \infty} e(t) = \lim_{s \to 0} sE(s) \tag{8.17}$$

Deriving the Laplace transform of the error signal on the output of the phase detector as a function of the Laplace transform of the input signal phase $\Theta(s)$, we get

$$E(s) = \frac{s\Theta(s)}{H(s) + s} \tag{8.18}$$

Therefore

$$e_\infty = \lim_{s \to 0} sE(s) = \lim_{s \to 0} \frac{s^2\Theta(s)}{H(s) + s} \tag{8.19}$$

If a frequency shift by f_0 occurs in the input signal at the moment $t = 0$, the phase $\theta(t)$ changes as follows

$$\theta(t) = 2\pi f_0 u(t), \quad \text{i.e.} \quad \Theta(s) = \frac{2\pi f_0}{s^2} \tag{8.20}$$

Substituting (8.20) into (8.19) we obtain e_∞, in the general case as

$$e_\infty = \lim_{s \to 0} \frac{2\pi f_0}{H(s) + s} \tag{8.21}$$

If the filter transfer function $H(s) = K_{\text{amp}}$, the phase error in steady state is

$$e_\infty = \frac{2\pi f_0}{K_{\text{amp}}} \tag{8.22}$$

so it is nonzero. We conclude that the loop is able to compensate for the frequency shift appearing on its input, but the price for this is a nonzero phase error on the output of the phase detector.

Let us now consider a higher order filter. Let its transfer function be

$$H(s) = K_{\text{amp}} \frac{s + K_1}{s + K_2} \tag{8.23}$$

For $s = 0$ we obtain $H(0) = K_{\text{amp}} K_1 / K_2$. Therefore, for this filter type we have the following PLL lock range

$$|2\pi f_0| \leq \pi K_{\text{amp}} \frac{K_1}{K_2} \tag{8.24}$$

The closed loop transfer function is then given by the formula

$$\frac{\Phi(s)}{\Theta(s)} = \frac{K_{\text{amp}}s + K_{\text{amp}}K_1}{s^2 + (K_{\text{amp}} + K_2)s + K_{\text{amp}}K_1} \tag{8.25}$$

As known from control theory, a linear system is stable if the roots of its transfer function denominator are located in the left half-plane of the complex variable s. One can show that in order for the stability criterion to be satisfied, the following conditions must be fulfilled

$$K_2 > -K_{\text{amp}} \quad \text{and} \quad K_{\text{amp}} K_1 > 0 \tag{8.26}$$

If the stability is ensured, the dynamic PLL properties result from the choice of parameters K_1, K_2 and K_{amp}. As we hopefully remember from the course devoted to control theory, transfer function (8.25) can be presented in the following form

$$\frac{\Phi(s)}{\Theta(s)} = \frac{(2\zeta\omega_n - K_2)s + \omega_n^2}{s^2 + 2\zeta\omega_n s + \omega_n^2} \tag{8.27}$$

where $\omega_n = \sqrt{K_{\text{amp}} K_1}$ is the so-called *natural angular frequency* of the system and ζ is the *damping factor*. If parameters K_1, K_2 and K_{amp} are selected in such a way that if conditions (8.26) are simultaneously fulfilled we have $\zeta < 1$, then a gain higher than unity is observed around the angular frequency ω_n. This is easily seen as a ripple on the transfer function magnitude plot (see Figure 8.5). The lower the damping factor ζ, the higher the ripple. If $\zeta > 1$, this phenomenon does not occur. In practice the choice of the loop parameters for which $\zeta < 1$ is often disadvantageous. The bandwidth of phase synchronization is in practice determined by $\omega_n = \sqrt{K_{\text{amp}} K_1}$, whereas the lock range depends on all three parameters, as it is in formula (8.24). Therefore, by selecting the values of $K_{\text{amp}} K_1$ we determine the bandwidth, and by manipulating the parameter K_2 we determine the PLL lock range.

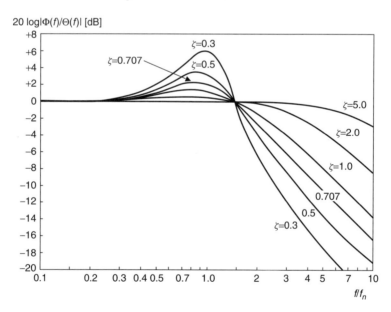

Figure 8.5 Magnitude of the transfer function of the second-order loop for several values of the damping factor ζ

If we take advantage of (8.21), for the given filter transfer function we can derive the phase error on the phase detector output if a frequency shift of f_0 Hz is applied to the loop input. Substituting the formula for the filter transfer function in expression (8.21) and determining the limit of the latter for s tending to zero, we get

$$e_\infty = \frac{K_2 2\pi f_0}{K_1 K_{amp}} \tag{8.28}$$

As we see, this time the error is also nonzero; however, it can be small if the parameter K_2 is also small. Let us note that the error is exactly equal to zero if $K_2 = 0$, i.e. the filter transfer function has the form

$$H(s) = K_{amp} \frac{s + K_1}{s} = K_{amp} \left(1 + \frac{K_1}{s} \right) \tag{8.29}$$

Such a filter is called a *proportional-integral filter*.

So far we have assumed linear characteristics of the phase detector shown in Figure 8.4, so our analysis was related to a linear system. This assumption is justified if the PLL is already locked. Typically, at the initial moment the PLL is not synchronized to the input signal and it is in the mode of synchronization search. The frequency range within which the PLL is able to synchronize with the input signal is called the *capture range*. The phenomena that occur during synchronization search are slip cycles. They result from modulo-2π discontinuities between two linear ranges of the phase detector characteristics and have a strongly nonlinear character, but they require much more advanced analysis so will not be covered in this short description of the PLL.

So far now we have assumed an ideal phase detector. A real phase detector can be implemented in many different ways, e.g. using a multiplier. If the signal on the PLL input and the signal generated by the VCO are logic signals, the phase detector can be implemented using the XOR gate.

Let us analyze the PLL phase detector based on the multiplier. If the signals given to its inputs have the form conforming to formulas (8.1) and (8.2), then the output of the multiplier resulting from the well-known trigonometric formula is

$$\frac{A_{in} A_{VCO}}{2} \left\{ \cos \left[\theta(t) - \varphi(t) \right] + \cos \left[4\pi f_c t + \theta(t) + \varphi(t) \right] \right\} \tag{8.30}$$

Assuming that the double-frequency component is eliminated by the loop filter, we can write the following expression for the phase error signal

$$e(t) = \frac{A_{in} A_{VCO}}{2} \cos \left[\theta(t) - \varphi(t) \right] \tag{8.31}$$

We conclude from (8.31) that the error signal is equal to zero if

$$\left[\theta(t) - \varphi(t) \right] \bmod 2\pi = \pi/2 \tag{8.32}$$

i.e. if the VCO output signal is shifted in phase by $\pi/2$ with respect to the received signal phase. However, if the VCO generates the signal

$$v(t) = A_{VCO} \cos\left[2\pi f_c t + \psi(t) - \frac{\pi}{2}\right] = A_{VCO} \sin\left[2\pi f_c t + \psi(t)\right] \qquad (8.33)$$

then the error signal on the output of the multiplying detector is

$$e(t) = \frac{A_{in} A_{VCO}}{2} \sin\left[\theta(t) - \psi(t)\right] \qquad (8.34)$$

We conclude from the above formula that the multiplying phase detector is a nonlinear block, because its error signal depends on the sine of the phase difference between the input and VCO signals, but it is not proportional to this phase difference. Fortunately, such a detector can be approximately treated as a linear circuit if the phase difference is small, because for a small angle α we have $\sin(\alpha) \approx \alpha$. Therefore

$$e(t) \approx \frac{A_{in} A_{VCO}}{2}\left[\theta(t) - \psi(t)\right] \qquad (8.35)$$

The above considerations allow us to conclude that it would be useful if the VCO were able to generate a pair of carrier signals: the regular one and the signal shifted by 90° with respect to the first one. One of them could be used for comparison in the phase detector, whereas the other could be the PLL output signal.

8.3   Phase-Locked Loop for Sampled Signals

With digital signal processing becoming a dominating technology in receiver implementation, synchronization algorithms are needed for sampled signals instead of continuous signals. For that reason it seems advantageous to consider the PLL that operates exclusively on the received signal samples (see Figure 8.6).

Let the sampling period be equal to T_S. The sequence of sinusoidal signal samples is then described by the formula

$$y_k = A_{in} \cos(2\pi f_c k T_S + \theta_k) \qquad (8.36)$$

Let the sequence of samples generated by the time-discrete VCO be determined by the formula

$$v_k = A_{VCO} \cos(2\pi f_c k T_S + \varphi_k) \qquad (8.37)$$

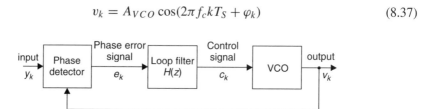

Figure 8.6   Block diagram of the time-discrete PLL

The error on the output of the time-discrete phase detector is

$$e_k = W(\theta_k - \varphi_k) \tag{8.38}$$

and the control signal c_k of the VCO is described by the following difference equation depending on the phase φ_k

$$c_k = \varphi_{k+1} - \varphi_k \tag{8.39}$$

Let us note that the above difference equation is analogous to equation (8.5), which contains a time derivative of a continuous phase signal. Let us calculate the $\mathcal{Z}$-transform of both sides of equation (8.39). We obtain

$$C(z) = z\Phi(z) - \Phi(z) \tag{8.40}$$

therefore, on the basis of Figure 8.6, we have

$$\Phi(z) = \frac{C(z)}{z-1} = \frac{H(z)E(z)}{z-1} \tag{8.41}$$

Assuming the linear form of the function $W(.)$ or alternatively assuming small differences of the angles θ_k and φ_k justifying a linear approximation of $W(.)$, we can write the following formula on the basis of equation (8.38)

$$e_k = \theta_k - \varphi_k \tag{8.42}$$

Therefore, we obtain

$$E(z) = \Theta(z) - \Phi(z) \tag{8.43}$$

Taking advantage of (8.41) we can derive the PLL phase transfer function in the form

$$\frac{\Phi(z)}{\Theta(z)} = \frac{H(z)}{H(z) + z - 1} \tag{8.44}$$

Let us now derive the PLL steady-state response for time-discrete signals. Let us check the error value on the output of the phase detector after an infinitely long time, i.e.

$$e_\infty = \lim_{k\to\infty} e_k \tag{8.45}$$

On the basis of the appropriate theorem related to the $\mathcal{Z}$-transform, assuming that $e_k = 0$ for $k < 0$, we obtain

$$e_\infty = \lim_{z\to 1}(z-1)E(z) = \lim_{z\to 1}\frac{(z-1)^2\Theta(z)}{H(z) + z - 1} \tag{8.46}$$

Let us investigate the steady-state PLL error if there is a phase shift by the angle θ in the $k = 0$ time instant, i.e.

$$\theta_k = \theta u_k \tag{8.47}$$

where u_k is a unit time-discrete step function. The $\mathcal{Z}$-transform of both sides of (8.47) is equal to

$$\Theta(z) = \frac{z\theta}{z - 1} \tag{8.48}$$

After substituting this expression in formula (8.46) we get

$$e_\infty = \lim_{z \to 1} \frac{(z - 1)z\theta}{H(z) + z - 1} \tag{8.49}$$

For the steady-state phase error to be zero, it is sufficient to have $H(1) \neq 0$. Let us also investigate the phase error if at the moment $k = 0$ the input frequency is shifted by f_0, which transforms to the phase function at the moment k in the following form

$$\theta_k = 2\pi f_0 k u_k \tag{8.50}$$

The $\mathcal{Z}$-transform of (8.50) is

$$\Theta(z) = \frac{2\pi f_0 z}{(z - 1)^2} \tag{8.51}$$

After substituting it in (8.46) we achieve the expression

$$e_\infty = \lim_{z \to 1} \frac{2\pi f_0 z}{H(z) + z - 1} \tag{8.52}$$

This error will be equal to zero if the loop transfer function has a pole for $z = 1$.

Let us stress once more that the performed analysis is valid for a linear system. The system can be considered linear if the loop is in the locked mode and the phase error is so small that even if a nonlinear phase detector is applied its functioning can be approximated by a linear operation.

As we show below, the PLL is a basic block used for the phase, frequency and timing recovery in receivers of digitally modulated signals.

8.4 Maximum Likelihood Carrier Phase Estimation

Let us recall that we derived the maximum likelihood rule when considering the decision rules in Chapter 2. We applied it in the decoding of the channel codes and in sequential detection of signals corrupted by intersymbol interference. However, the maximum likelihood rule can also be applied in the estimation of signal parameters such as carrier phase or signal timing.

First, let us consider a general model of the received signal for which the signal parameter θ is estimated. Let the signal $y(t)$ observed on the receiver input be given by the following general formula

$$y(t) = x(t, \theta) + v(t) \tag{8.53}$$

which is the sum of the transmitted signal $x(t, \theta)$, where θ is unknown, and of additive noise $v(t)$. Let us assume that $v(t)$ is a white Gaussian noise of power density equal to $N_0/2$. Let us also introduce the set of functions $\{\psi_i(t)\}$ that are orthonormal in the interval $(0, T_s)$. In other words, they have the following property

$$\int_0^{T_s} \psi_i(t)\psi_j(t)\mathrm{d}t = \begin{cases} 1 & \text{for} \quad i = j \\ 0 & \text{for} \quad i \neq j \end{cases} \tag{8.54}$$

The signal observed on the receiver input can be expressed in the form of an expansion using orthonormal functions (8.54) in the following way

$$y(t) = \sum_{k=1}^{\infty} x_k(\theta)\psi_k(t) + \sum_{k=1}^{\infty} v_k \psi_k(t)$$

$$= \sum_k [x_k(\theta) + v_k]\psi_k(t) = \sum_k y_k \psi_k(t) \tag{8.55}$$

where

$$x_k(\theta) = \int_0^{T_s} x(t, \theta)\psi_k(t)\mathrm{d}t \quad v_k = \int_0^{T_s} v(t)\psi_k(t)\mathrm{d}t \quad y_k = x_k(\theta) + v_k \tag{8.56}$$

The Gaussian additive noise is white, so its expansion coefficients are mutually uncorrelated, i.e.

$$E[v_k v_j] = \frac{N_0}{2}\delta_{k,j} \tag{8.57}$$

where $\delta_{k,j}$ is a Kronecker delta function. Let us also note that because $y_k = x_k(\theta) + n_k$, and the noise has a zero mean, we have $E[y_k|\theta] = x_k(\theta)$.

Now let us consider the random vector $\mathbf{y} = [y_1, y_2, \ldots]$ consisting of expansion coefficients in the set of orthonormal functions. Its joint conditional probability density function for a given value of parameter θ is given by the formula

$$p(\mathbf{y}|\theta) = \lim_{n \to \infty} \left(\frac{1}{\sigma\sqrt{2\pi}}\right)^n \exp\left\{-\frac{1}{2\sigma^2}\sum_{k=1}^{n}[y_k - x_k(\theta)]^2\right\} \tag{8.58}$$

whereas the probability density function of the vector $\mathbf{y}$ is described by the expression

$$p(\mathbf{y}) = \lim_{n \to \infty} \left(\frac{1}{\sigma \sqrt{2\pi}} \right)^n \exp\left[-\frac{1}{2\sigma^2} \sum_{k=1}^{n} (y_k)^2 \right] \tag{8.59}$$

In accordance with the maximum likelihood rule, we wish to find the value of θ that maximizes the conditional probability density function $p(\mathbf{y}|\theta)$ or, equivalently, maximizes the ratio $p(\mathbf{y}|\theta)/p(\mathbf{y})$ given by the formula

$$\frac{p(\mathbf{y}|\theta)}{p(\mathbf{y})} = \lim_{n \to \infty} \exp\left\langle \frac{1}{2\sigma^2} \sum_{k=1}^{n} \left\{ y_k^2 - \left[y_k - x_k(\theta) \right] \right\}^2 \right\rangle \tag{8.60}$$

Since the logarithmic function is monotonic, the result of the search for the optimum value of parameter θ will be the same if the maximized object is in the form of the logarithmic likelihood ratio, i.e.

$$L(\theta) = \ln \frac{p(\mathbf{y}|\theta)}{p(\mathbf{y})} = \lim_{n \to \infty} \left\langle \frac{1}{2\sigma^2} \sum_{k=1}^{n} \left\{ y_k^2 - \left[y_k - x_k(\theta) \right] \right\}^2 \right\rangle$$

$$= \lim_{n \to \infty} \left\{ \frac{1}{2\sigma^2} \sum_{k=1}^{n} \left[2y_k x_k(\theta) - x_k^2(\theta) \right] \right\} \tag{8.61}$$

Taking advantage of the Parseval theorem, we can show that

$$\int_0^{T_s} x^2(t, \theta) dt = \sum_{k=1}^{\infty} x_k^2(\theta) \quad \text{and} \quad \int_0^{T_s} x(t, \theta) y(t) dt = \sum_{k=1}^{\infty} x_k(\theta) y_k \tag{8.62}$$

therefore

$$L(\theta) = \frac{1}{2\sigma^2} \int_0^{T_s} \left[2x(t, \theta) y(t) - x^2(t, \theta) \right] dt \tag{8.63}$$

For the optimum value of parameter θ, for which the logarithmic likelihood ratio $L(\theta)$ is maximized, the following equation holds

$$\frac{\partial L(\theta)}{\partial \theta} = 0 \tag{8.64}$$

We conclude from (8.64) that θ can be calculated from the equation

$$\frac{1}{2\sigma^2} \int_0^{T_s} \left[2 \frac{\partial x(t, \theta)}{\partial \theta} y(t) - 2x(t, \theta) \frac{\partial x(t, \theta)}{\partial \theta} \right] dt$$

$$= \frac{1}{\sigma^2} \int_0^{T_s} \left[y(t) - x(t, \theta) \right] \frac{\partial x(t, \theta)}{\partial \theta} dt = 0 \tag{8.65}$$

Finally, the expression from which the estimate of θ can be calculated has the form

$$\int_0^{T_s} [y(t) - x(t, \theta)] \frac{\partial x(t, \theta)}{\partial \theta} dt = 0 \tag{8.66}$$

Similarly, on the basis of (8.61), the expression from which the optimum value of θ can be derived is given by the formula

$$\lim_{n \to \infty} \sum_{k=1}^{n} [y_k - x_k(\theta)] \frac{\partial x_k(\theta)}{\partial \theta} = 0 \tag{8.67}$$

The above general expressions can now be used to estimate parameters of particular signals. First, let us consider estimation of the carrier phase of the signal received in the presence of noise. Let the received signal have the form

$$y(t) = A \cos(2\pi f_c t + \theta) + v(t), \quad \text{where } f_c = \frac{1}{T_s} \tag{8.68}$$

In order to derive the maximum likelihood carrier phase estimate it is sufficient to apply formula (8.66) directly. We have

$$\frac{\partial x(t, \theta)}{\partial \theta} = -A \sin(2\pi f_c t + \theta) \tag{8.69}$$

and condition (8.66) achieves the form

$$-\int_0^{T_s} [y(t) - A \cos(2\pi f_c t + \widehat{\theta})] A \sin(2\pi f_c t + \widehat{\theta}) dt$$

$$= -A \int_0^{T_s} y(t) \sin(2\pi f_c t + \widehat{\theta}) dt + A^2 \int_0^{T_2} \cos(2\pi f_c t + \widehat{\theta}) \sin(2\pi f_c t + \widehat{\theta}) dt \tag{8.70}$$

$$= -A \int_0^{T_s} y(t) \sin(2\pi f_c t + \widehat{\theta}) dt + \frac{A^2}{2} \int_0^{T_2} \sin(4\pi f_c t + 2\widehat{\theta}) dt = 0$$

Since integration in the second integral of (8.70) is performed over an integer number of sinusoidal cycles, its result is equal to zero. Therefore the final form of the equation from which the maximum likelihood carrier phase estimate is derived is

$$\int_0^{T_s} y(t) \sin(2\pi f_c t + \widehat{\theta}) dt = 0 \tag{8.71}$$

Using the formula for the sine of a sum of two angles, we obtain

$$\cos\widehat{\theta}\int_0^{T_s} y(t)\sin 2\pi f_c t\, dt + \sin\widehat{\theta}\int_0^{T_s} y(t)\cos 2\pi f_c t\, dt = 0 \qquad (8.72)$$

From this formula we conclude that the phase estimate $\widehat{\theta}$ can be derived on the basis of the expression

$$\widehat{\theta} = -\arctan\frac{\displaystyle\int_0^{T_s} y(t)\sin 2\pi f_c t\, dt}{\displaystyle\int_0^{T_s} y(t)\cos 2\pi f_c t\, dt} \qquad (8.73)$$

Formula (8.73) implies the block diagram of the circuit calculating the carrier phase estimate $\widehat{\theta}$ and shown in Figure 8.7.

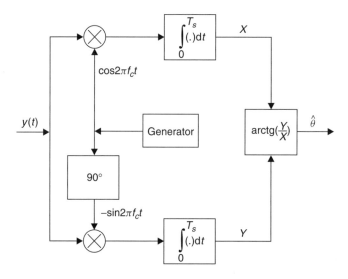

Figure 8.7   Block diagram of the circuit for maximum likelihood carrier phase estimation

As we see, the circuit calculating the carrier phase estimate consists of two correlators that use mutually orthogonal reference signals. The results of correlations are applied to calculate the arc tangent of their quotient.

Let us inspect once more formula (8.71), which suggests the application of the phase-locked loop in which the VCO generator produces the signal $\sin(2\pi f_c t + \widehat{\theta})$, and where the loop filter is an integrating circuit calculating the integral over the period T_s. As we see, the PLL can be applied to generate the reference signal whose phase is the maximum likelihood phase estimate of the unmodulated input signal that is disturbed by the additive noise, under the condition that the appropriate loop filter is applied (Figure 8.8). Typically, instead of the integrator, a carefully selected lowpass filter is applied, which approximates the operation of the integrator sufficiently well.

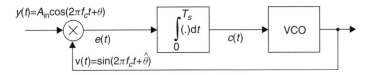

$y(t)=A_{in}\cos(2\pi f_c t+\theta)$

$e(t)$

$v(t)=\sin(2\pi f_c t+\hat{\theta})$

$\int_0^{T_s}(.)dt$

$c(t)$

VCO

Figure 8.8 Application of the PLL in derivation of the maximum likelihood phase estimate $\hat{\theta}$ of an unmodulated signal

8.5 Practical Carrier Phase Synchronization Solutions

As we know, the phase of the received signal is recovered on the basis of the signal containing a random component that is a transmitted data sequence. An additional random component is additive noise. As shown in Proakis (2000), in practice there are two approaches to finding the carrier phase in the presence of random data and noise. The first one relies on averaging the received signal, taking into account the statistical properties of data symbols. One can also perform some operations on the received signal, which aim to get rid of its dependence on the data symbols. The second approach relies on the assumption that sufficiently reliable estimates of the transmitted data symbols are already available in the receiver. These estimates are subsequently used in the carrier phase estimation and correction. Circuits belonging to the first group have no decision feedback, whereas the second group of solutions inherently use the feedback in the estimation and correction process. Below we consider some representative examples of both types of carrier phase estimators and correctors.

8.5.1 Carrier Phase Synchronization without Decision Feedback

So far we have considered carrier phase recovery for an unmodulated signal. Typically, the carrier synchronization block in the receiver has to estimate the phase on the basis of the digitally modulated signal with additive noise. Let us consider a simple example of synchronization when a cosinusoidal carrier is modulated by the PAM signal, i.e. when the modulated signal has the form

$$x(t, \theta) = A(t) \cos(2\pi f_c t + \theta) \qquad (8.74)$$

Let us note that this case comprises, among others, the BPSK modulation, for which $A(t) = \pm A$. Let us assume that $E[A(t)] = 0$. This means that we are not able to create a spectral line at frequency f_c or its multiple, which would allow us to estimate the carrier phase through averaging the received signal. However, such a spectral line can be generated by nonlinear processing of the received signal followed by application of the PLL, which generates the reference signal in an analogous way to that presented in the previous section. A nonlinear circuit that could be applied for this purpose is a memoryless block with a quadratic input/output characteristic. Figure 8.9 presents a typical scheme of carrier phase recovery with a quadratic circuit.

The signal (8.74) received in the presence of noise is squared and the result of this operation is

$$y(t) = x^2(t, \theta) + 2x(t, \theta)v(t) + v^2(t) \qquad (8.75)$$

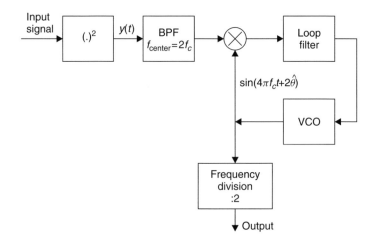

Figure 8.9 Block diagram of the carrier phase recovery system with the circuit that squares the received signal

The signal $x^2(t, \theta)$ is the desired component whereas the remaining two components can be considered as a disturbance. Assuming a lack of correlation between the modulated signal and noise and knowing that $E[v(t)] = 0$, we obtain

$$E[y(t)] = E[x^2(t, \theta)] + E[v^2(t)] \qquad (8.76)$$

In turn, taking into account (8.74) we have

$$E[x^2(t, \theta)] = \frac{E[A(t)^2]}{2} + \frac{E[A(t)^2]}{2} \cos(4\pi f_c t + 2\theta) \qquad (8.77)$$

The bandpass filter of center frequency $2f_c$ extracts the component concentrated around $2f_c$ from the signal $y^2(t)$. Its mean amplitude is equal to $E[A(t)^2]|H(2f_c)|/2$. The filter output signal is the input signal for the PLL, whose VCO generates a periodic signal of free-running frequency equal to $2f_c$. The PLL estimates the phase 2θ of the signal extracted by the bandpass filter. The loop bandwidth should be sufficiently narrow for the PLL to average the part of the noise that passes through this bandpass filter. Thus, the PLL approximately generates the signal $\sin(4\pi f_c t + 2\widehat{\theta})$. The frequency of this signal has to be divided by 2 in order for the signal to be useful in synchronous demodulation. However, the process of frequency division by 2 has a side effect, which is the phase ambiguity equal to 180°. Frequency division of the logical signal of a given phase $2\widehat{\theta}$ can be realized by a logical circuit based on a counter to 2. However, depending on the initial state of the counter flip-flop we can receive the counter output signal with the correct phase equal to $\widehat{\theta}$ or with the opposite phase equal to $\widehat{\theta} + \pi$. The latter can lead to decision errors. In order to counteract this effect, differential coding is usually applied. Alternatively, the user data sequence is preceded by a reference signal (preamble) that allows the receiver to estimate the carrier phase correctly.

The synchronization scheme shown above can be generalized. In general, the nonlinear ciruit raises the received signal to the Mth power, the bandpass filter has a center frequency

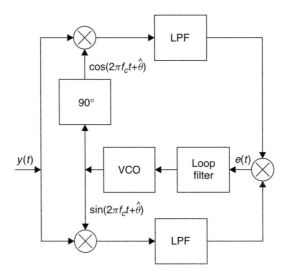

Figure 8.10 Costas loop used in carrier phase recovery

of Mf_c, the PLL operates around this frequency too, and on the output there is a frequency divisor by M. Owing to such a configuration, it is possible to perform carrier phase recovery of M-PSK signals.

Another example of a carrier phase recovery circuit for a PAM-modulated carrier signal is the so-called *Costas loop*, shown in Figure 8.10. Let us assume that the following signal is given to its input

$$y(t) = x(t, \theta) + v(t) = A(t) \cos(2\pi f_c t + \theta) + v(t)$$

$$= A(t) \cos(2\pi f_c t + \theta) + v^I(t) \cos(2\pi f_c t + \theta) + v^Q(t) \sin(2\pi f_c t + \theta) \quad (8.78)$$

This signal is multiplied in two parallel branches by $\cos(2\pi f_c t + \widehat{\theta})$ and $\sin(2\pi f_c t + \widehat{\theta})$, respectively. As a result, we obtain the following signals in the in-phase and quadrature branches

$$y^I(t) = [x(t, \theta) + v(t)] \cos(2\pi f_c t + \widehat{\theta})$$

$$= \frac{1}{2}[A(t) + v^I(t)] \cos(\widehat{\theta} - \theta) + \frac{1}{2}v^Q(t) \sin(\widehat{\theta} - \theta)$$

$$+ \text{components around frequency } 2f_c \quad (8.79)$$

$$y^Q(t) = [x(t, \theta) + v(t)] \sin(2\pi f_c t + \widehat{\theta})$$

$$= \frac{1}{2}[A(t) + v^I(t)] \sin(\widehat{\theta} - \theta) - \frac{1}{2}v^Q(t) \cos(\widehat{\theta} - \theta)$$

$$+ \text{components around frequency } 2f_c \quad (8.80)$$

Passband components of both signals around $2f_c$ are eliminated by the lowpass filters applied in each branch. The lowpass filter output signals are multiplied by each other, resulting in the error signal $e(t)$ that is subsequently filtered by the loop filter. The loop filter output signal controls the VCO. Simple calculations based on formulas (8.79) and (8.80) prove that the error signal $e(t)$ is given by the expression

$$e(t) = \frac{1}{8} \left\{ \left[A(t) + v'(t) \right]^2 - \left[v^{\mathcal{Q}}(t) \right]^2 \right\} \sin 2(\widehat{\theta} - \theta)$$
$$- \frac{1}{4} \left[A(t) + v'(t) \right] v^{\mathcal{Q}}(t) \cos 2(\widehat{\theta} - \theta) \tag{8.81}$$

The error signal comprises the desired component of form $\frac{1}{8} A^2(t) \sin 2(\widehat{\theta} - \theta)$ and the components that consist of the products of the data signal and noise or the products of noise signals. The loop filter should eliminate the latter by averaging. As stated in Proakis (2000), if the loop filters applied in the Costas loop and the squaring loop are identical, then both loops are equivalent. The optimal lowpass filter applied in both loop branches is a filter that is matched to the data pulse contained in the signal $x(t, \theta)$. If it is used, then both filter outputs can be sampled once per modulation period and the VCO can be controlled using these samples.

8.5.2 Carrier Phase Synchronization using Decision Feedback

Let us consider the practical case of digital transmission in which the passband digitally modulated signal reaching the receiver has the form

$$x(t) = \mathrm{Re} \left\{ \exp[j2\pi f_c t + \theta(t)] \sum_{i=-\infty}^{\infty} d_i h(t - iT) + v(t) \right\} \tag{8.82}$$

We assume that data symbols d_i, the pulse $h(t)$ and the noise $v(t)$ are complex-valued. Complex data symbols allow us to treat such modulations as BPSK, QPSK, QAM, etc. In digitally implemented receivers, the in-phase and quadrature passband components located around the carrier frequency f_c are achieved on the basis of the signal $x(t)$ owing to a pair of bandpass filters, for which the impulse response of the second filter is the Hilbert transform of the response of the first filter. Let us recall that if $g(t) = \mathcal{F}^{-1}\{G(f)\}$ is the impulse response of the first filter, the second filter has the impulse response given by the formula

$$\widetilde{g}(t) = \mathcal{F}^{-1}\{-j \, \mathrm{sgn}(f) G(f)\} \tag{8.83}$$

Operation in the frequency domain, which is described by the function $-j \, \mathrm{sgn}(f)$, is in fact the shifting of the signal phase in its whole spectral range by 90°. Therefore the signals on both filter outputs are in quadrature to each other, so they can be represented in the form of real and imaginary parts of the following complex function of time

$$x_P(t) = \exp[j2\pi f_c t + \theta(t)] \sum_{i=-\infty}^{\infty} d_i h_g(t - iT) + v_g(t) \tag{8.84}$$

where $h_g(t)$ is the impulse response of the complex joint channel and passband filters $G(f)$ and $\widetilde{G}(f)$. In turn, $v_g(t)$ is filtered noise.

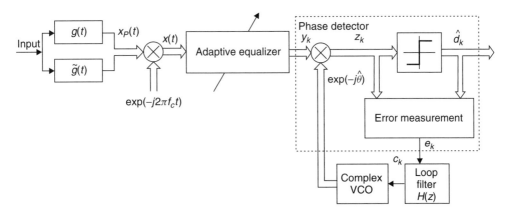

Figure 8.11 Block diagram of the receiver with the adaptive equalizer and the carrier phase correction on the equalizer output

The block diagram of a typical receiver in which both filters considered above are applied is shown in Figure 8.11. The passband signal $x_P(t)$ is converted to the baseband. This operation is performed by multiplying signal $x_P(t)$ by the function $\exp(-j2\pi f_c t)$. Let us note that in this operation the correct carrier phase is not applied. A possible frequency difference between the received signal and that used in the conversion of the signal $x_P(t)$ to the baseband is contained in the phase function $\theta(t)$. If the pulse $h_g(t)$ introduces intersymbol interference, then an equalizer is necessary. The equalizer is also complex-valued.[3] As we remember, the equalizer is usually implemented by transversal filters. Similar to most transversal filters, in which the main tap is located somewhere in the middle of the delay line, such an equalizer introduces a significant delay. For that reason the correction of the signal phase is performed on the output of the equalizer. A decision feedback loop is applied in the phase correction circuit. In the other possible configuration of the phase correction we should modify the carrier phase in front of the equalizer on the basis of the decisions made on the equalizer output. As a result, the equalizer would introduce a significant delay in the phase synchronization loop. This delay would have a serious influence on the loop stability, its rate of reaction to the changes in carrier phase or frequency of the received signal. The considered receiver scheme with the decision feedback loop on the output of the equalizer was first proposed by Falconer (1976).

Let us assume for simplicity that the equalizer has already minimized the intersymbol interference to such an extent that the ISI residual values can be incorporated into the noise component. Since data decisions are performed once per modulation period T, it is sufficient to process the complex samples from the equalizer output with the same interval T. Thus, the sample on the equalizer output at the kth moment can be written as

$$y_k = \exp(j\theta_k)d_k p_0 + n_k \tag{8.85}$$

where p_0 is a central sample of the joint channel $h_g(t)$ and equalizer impulse response. We can assume that owing to the appropriately adjusted equalizer we have approximately

---

[3] Compare with our considerations on a complex equalizer in Chapter 6.

$p_i = 0$ for $i \neq 0$, and $p_0 = 1$. The second component of formula (8.85) is a noise sample on the equalizer output in the kth moment. We conclude from expression (8.85) that the equalizer output sample in the kth moment is the data symbol d_k, which is rotated by the angle θ_k and disturbed by an additive noise sample. The carrier synchronization circuit rotates the complex signal by the angle $-\widehat{\theta}$. As a result, we obtain the sample

$$z_k = \exp[j(\theta_k - \widehat{\theta}_k)]d_k + n'_k \tag{8.86}$$

Based on z_k the decision device calculates the estimate $\widehat{d}_k$ of the data symbol d_k. Assume that the signal-to-noise ratio is so high that the symbol error rate $\Pr\{\widehat{d}_k \neq d_k\}$ is very small. Under this assumpion the noise sample is negligible. Then, based on the signal

$$z_k = \exp[j(\theta_k - \widehat{\theta}_k)]d_k \tag{8.87}$$

we can determine the angle $\widehat{\theta}_k - \theta_k$ in the following way. First, we multiply both sides of (8.87) by the expression $\widehat{d}_k^* / |\widehat{d}_k|^2 = d_k^* / |d_k|^2$. As a result we obtain

$$\exp[j(\theta_k - \widehat{\theta}_k)] = \frac{z_k d_k^*}{|d_k|^2} \tag{8.88}$$

Therefore, if we extract the imaginary part of the signal (8.88), we obtain the phase error signal, which is similar to that achievable on the output of the phase detector based on the multiplier. Namely, we have

$$\mathrm{Im}\{\exp[j(\theta_k - \widehat{\theta}_k)]\} = \sin(\theta_k - \widehat{\theta}_k) = \mathrm{Im}\left\{\frac{z_k d_k^*}{|d_k|^2}\right\} \tag{8.89}$$

As a result, the error sample fed to the loop filter is given by the formula

$$e_k = \theta_k - \widehat{\theta}_k = \arcsin\left\{\frac{\mathrm{Im}[z_k d_k^*]}{|d_k|^2}\right\} \tag{8.90}$$

For small angle differences $\theta_k - \widehat{\theta}_k$ for which $\sin(\theta_k - \widehat{\theta}_k) \approx \theta_k - \widehat{\theta}_k$ we obtain

$$e_k \approx \frac{\mathrm{Im}[z_k d_k^*]}{|d_k|^2} \tag{8.91}$$

Normalization of the error signal with respect to the power of the current data symbol is not needed if $|d_k|^2$ is constant, which occurs for all PSK modulations.

    Dynamic properties of the phase loop are determined by the transfer function of the applied loop filter. In order for the loop to be able to compensate for the frequency offset, which shows itself by a monotonically increasing phase offset resulting in gradual rotation of the signal constellation, the loop filter should contain an integrating element.

    Let us note a certain drawback of the decision feedback synchronization loop. Suppose that the QPSK modulation has been applied. The angle between neighboring decision thresholds is equal to $\pi/2$. Consequently, if the phase error is higher than $\pi/4$, a symbol error is committed in the decision process and the error signal given to the loop filter input will indicate the wrong direction of the change in phase. Possibly, then, the decision feedback loop will synchronize to the carrier phase, which generally differs from the real one by $k\pi/2$. The decision device will then generate decisions that differ from the correct data symbol by $k\pi/2$. As we see, phase ambiguity occurs. Fortunately, its influence can be eliminated by the application of differential encoding.

At the end of our considerations on carrier phase recovery let us analyze how the carrier phase recovery block influences the adaptation process of the adaptive equalizer. First, let us recall a simple case of the linear equalizer, whose coefficients c_i ($i = 0, 1, \ldots, N-1$) are updated using the gradient algorithm if there is no carrier phase offset. The gradient algorithm would have the following form for the complex equalizer

$$c_{k+1,i} = c_{k,i} - \alpha\varepsilon_k x_{k-i}^* \quad \text{for } i = 0, 1, \ldots, N-1 \tag{8.92}$$

where $\varepsilon_k = y_k - d_k$ and α is the adaptation step size. However, in the case of carrier phase offset the equalizer output sample y_k is rotated by the angle $-\widehat{\theta}$, therefore the only error that can be calculated is

$$\varepsilon_k = z_k - d_k \tag{8.93}$$

In this context algorithm (8.92) achieves the form

$$c_{k+1,i} = c_{k,i} - \alpha\varepsilon_k \exp(j\widehat{\theta})x_{k-i}^* \quad \text{for } i = 0, 1, \ldots, N-1 \tag{8.94}$$

In the literature we can find many other solutions of the carrier phase synchronization using the decision feedback loop. Examples can be found in Kurzweil (2000) and Proakis (2000). However, modern digital receivers most often use the decision feedback synchronization circuit shown in this section.

8.6 Timing Synchronization

Besides carrier phase and frequency synchronization, timing synchronization plays an important role in digital transmission. Decisions upon transmitted data are made by the receiver for every data symbol. They are performed once every modulation period T, therefore it is extremely important to determine the time instant τ within the modulation period T that ensures maximum decision quality. As for carrier phase synchronization, we can categorize timing recovery circuits as those that use the decision feedback and that do not.

8.6.1 Timing Recovery with Decision Feedback

Let us consider the problem of finding the optimum sample timing according to the maximum likelihood criterion. Let us focus on the baseband transmission model. The basic equation describing the received signal has the form

$$y(t) = x(t, \tau) + n(t) = \sum_{i=-\infty}^{\infty} d_i h(t - iT - \tau) + n(t) \tag{8.95}$$

Let us assume that subsequent data symbols are uncorrelated and equiprobable and $n(t)$ is the additive white Gaussian noise. Let us also assume that in our search for the best timing phase $\hat{\tau}$ in the maximum likelihood sense we consider the finite time interval $(0, NT)$. Thus, we can perform a series of operations that are analogous to those described by formulas (8.55)–(8.65). These operations lead to the final definition of the likelihood function in the form

$$\Lambda(\tau) = \exp\left\{-\frac{1}{2\sigma^2}\int_0^{NT}\left[y(t) - x(t, \tau)\right]^2 dt\right\}$$ (8.96)

which results from the assumed Gaussian noise distribution. This function should be maximized with respect to the timing phase τ. Let us note that this time

$$x(t, \tau) = \sum_i d_i h(t - iT - \tau)$$ (8.97)

where the time index i is related to data symbols from the integration interval $(0, NT)$. By expanding the expression under the integral in (8.96) we get

$$\Lambda(\tau) = \exp\left\{-\frac{1}{2\sigma^2}\int_0^{NT}\left[y^2(t) - 2y(t)x(t, \tau) + x^2(t, \tau)\right] dt\right\}$$

$$= \exp\left[-\frac{1}{2\sigma^2}\int_0^{NT}y^2(t)dt\right]\exp\left[\frac{1}{\sigma^2}\int_0^{NT}y(t)x(t, \tau)dt\right]\exp\left[-\frac{1}{2\sigma^2}\int_0^{NT}x^2(t, \tau)dt\right]$$

$$= C\exp\left[\frac{1}{\sigma^2}\int_0^{NT}y(t)x(t, \tau)dt\right]$$ (8.98)

The reason for the simplification applied in the last part of formula (8.98) is the following. The likelihood function $\Lambda(\tau)$ can be represented as the product of three factors, as shown in the second part of (8.98). The first factor does not depend on the timing phase τ, so it can be treated as a constant. If the integration interval $(0, NT)$ is sufficiently long, the integral in the third factor can be treated as the energy of signal $x(t, \tau)$. Its value does not depend on τ either. Thus, these two factors are excluded from the optimization process and they are futher represented as a constant C. If we apply (8.97) in (8.98), calculate the logarithm of both sides of (8.98) and drop the constant C (which does not have any influence on finding the optimum timing phase $\hat{\tau}$), we obtain the logarithmic likelihood function of the form

$$\Lambda_{\log}(\tau) = C_{\log}\sum_i d_i \int_0^{NT}y(t)h(t - iT - \tau)dt$$

$$= C_{\log}\sum_i d_i z_i(\tau)$$ (8.99)

For a sufficiently long integration interval, the following signal

$$z_i(\tau) = \int_0^{NT} y(t)h(t - iT - \tau)dt \qquad (8.100)$$

can be treated as a signal seen on the output of the filter matched to the pulse $h(t)$, whereas the matched filter input signal is $y(t)$. The matched filter output signal is acquired at the sampling moments $iT + \tau$. The necessary condition for finding the maximum likelihood timing phase is

$$\frac{d\Lambda_{\log}(\tau)}{d\tau} = 0 \qquad (8.101)$$

From this condition we derive the equality that has to be fulfilled for the best timing phase $\widehat{\tau}$

$$\sum_i d_i \frac{d}{d\tau}\left[\int_0^{NT} y(t)h(t - iT - \widehat{\tau})\right] = \sum_i d_i \frac{d}{d\tau}[z_i(\widehat{\tau})] = 0 \qquad (8.102)$$

The timing recovery block diagram shown in Figure 8.12 is a simple consequence of formula (8.102). This solution requires the knowledge of data symbols, so it belongs to the category of decision feedback timing recovery circuits. As we see, the filter matched to the pulse $h(t)$ is placed at the input. Such a filter can often be found in the receiver anyway. The matched filter output is differentiated and sampled with the timing interval T and the sampling phase $\widehat{\tau}$. These samples are subsequently multiplied by known data symbols (the switch in position 1), or by data decisions featuring low symbol error rate (the switch in position 2). The multiplication results are accumulated in the interval of

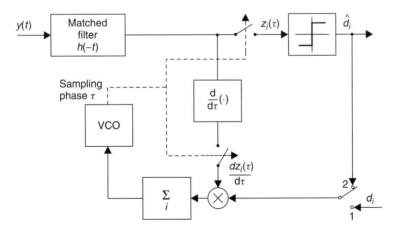

Figure 8.12 Block diagram of the maximum likelihood timing recovery circuit with decision feedback

N modulation periods. The result of this accumulation is a control signal for the VCO, which in turn generates sampling pulses with period T and timing phase $\hat{\tau}$. As we see, the time-discrete PLL is applied in the timing recovery circuit. The role of the loop filter is performed by the accumulator. The averaging and dynamic loop properties depend on the length of the window $(0, NT)$ in which averaging takes place.

Let us consider another timing recovery method that also requires knowledge of data symbols. This method is iterative, so it is realized in the time instants in every modulation period T. The choice of the timing phase τ is based on minimization of the mean square error $\mathcal{E}(\tau)$ on the input of the decision device, i.e.

$$\tau_{opt} = \arg \min_{\tau} \mathcal{E}(\tau) = \arg \min_{\tau} E[|z_i(\tau) - d_i|^2] \qquad (8.103)$$

where $z_i(\tau) = z(iT + \tau)$ is the sample on the input of the decision device in the ith modulation period. Let the receiver have the scheme shown in Figure 8.11. Besides the timing recovery circuit there is also a carrier phase recovery circuit and an adaptive equalizer. The receiver front-end is usually implemented in analog technology, but its remaining part is in most cases implemented digitally. Therefore, the sampling process with adjustable sampling phase usually takes place in front of the filters $g(t)$ and $\tilde{g}(t)$, which extract the in-phase and quadrature signal components, respectively. As we see, the receiver is a rather complex digital structure in which several dynamic processes take place concurrently: carrier phase estimation and correction, compensation of intersymbol interference introduced by the transmission channel and timing phase selection. The minimum mean square error criterion can be applied concurrently in all these processes, which results in their mutual interaction. Usually a training sequence is used in the process of establishing a link. Owing to knowledge of the transmitted data sequence, the receiver gradually acquires carrier phase correction, appropriate timing phase and finally the suboptimal equalizer coefficients. The optimum solution would be joint optimization of all three processes.

Now we will focus on finding the best timing phase $\hat{\tau}$ in the sense of criterion (8.103). As opposed to the case of an adaptive equalizer considered in Chapter 6, criterion (8.103) is not a convex function of the optimized parameter, so adjusting the timing phase in such a way that the derivative of the mean square error $\mathcal{E}(\tau)$ with respect to this phase is equal to zero does not guarantee finding the mean square error minimum. However, using the derivative is sensible if the timing recovery circuit already operates in the tracking mode in which the current timing phase is close to the optimum. In general, the applied algorithm updates the sampling moment with respect to the current moment by subtracting the correction term, whose value is proportional to the derivative of the mean square error with respect to τ, calculated for the error at the current timing phase $\tau = \tau_i$. Thus, the applied algorithm is a gradient algorithm, which in the ideal case would be determined by the formula

$$\tau_{i+1} = \tau_i - \alpha \left. \frac{\partial \mathcal{E}(\tau)}{\partial \tau} \right|_{\tau=\tau_i} \qquad (8.104)$$

Signals shown in Figure 8.11 are complex-valued, therefore the derivative of the mean square error needs more detailed treatment. Denoting $\varepsilon_i(\tau) = z_i(\tau) - d_i$ we have

$$\frac{\partial \mathcal{E}(\tau)}{\partial \tau} = \frac{\partial E[|\varepsilon_i(\tau)|^2]}{\partial \tau} = \frac{\partial E[\varepsilon_i(\tau)\varepsilon_i^*(\tau)]}{\partial \tau}$$

$$= E\left[\varepsilon_i(\tau)\frac{\partial \varepsilon_i^*(\tau)}{\partial \tau} + \varepsilon_i^*(\tau)\frac{\partial \varepsilon_i(\tau)}{\partial \tau}\right] = 2\,\mathrm{Re}\,E\left[\varepsilon_i^*(\tau)\frac{\partial \varepsilon_i(\tau)}{\partial \tau}\right] \qquad (8.105)$$

Since the data symbol applied in the expression for $\varepsilon_i(\tau)$ does not depend on the timing phase τ, algorithm (8.104) achieves the form

$$\tau_{i+1} = \tau_i - \alpha\,\mathrm{Re}\left\{E\left[\varepsilon_i^*(\tau)\frac{\partial z_i(\tau)}{\partial \tau}\right]\right\}\Bigg|_{\tau=\tau_i} \qquad (8.106)$$

where the factor 2 is included in the constant α. However, estimating the ensemble average is difficult and it is a time consuming process, so usually the ensemble average is replaced by its stochastic estimate. Thus, the algorithm is described by the expression

$$\tau_{i+1} = \tau_i - \alpha\,\mathrm{Re}\left\{[z_i(\tau) - d_i]^*\frac{\partial z_i(\tau)}{\partial \tau}\right\}\Bigg|_{\tau=\tau_i} \qquad (8.107)$$

The timing phase recovery circuit shown in Figure 8.13 is a natural consequence of formula (8.107).

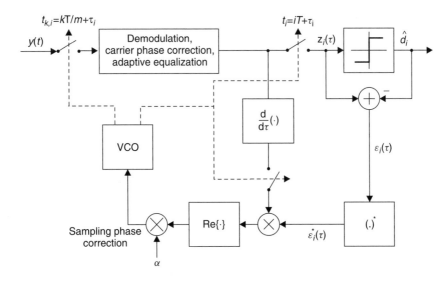

Figure 8.13 Timing recovery using the gradient algorithm that minimizes the mean square error (based on Barry *et al.* (2003))

It is worth mentioning again that most algorithms in the receivers are realized in the digital domain. Therefore, at the input of the scheme shown in Figure 8.13 there is a sampling block that acquires the signal at moments $t_{k,i} = kT/m + \tau_i$. The sampling process is performed with the rate at least equal to the Nyquist frequency, e.g. m times within the modulation period. The index k is the number of samples in one modulation period, whereas the index i denotes the number of modulation periods. Let us note that the decision process is performed once per modulation period, therefore this is the period at which the timing phase correction is performed.

8.6.2 Timing Recovery Circuits without Decision Feedback

Besides the timing phase recovery schemes in which data symbols are applied, there are also other schemes that do not require them. The latter schemes rely on application of the appropriate criterion which is ensemble averaged with respect to data sequences.

Let us come back again to our considerations of the maximum likelihood criterion applied for the time interval $(0, NT)$. As we remember, the likelihood function $\Lambda(\tau)$ is expressed by formula (8.98). Putting (8.100) in (8.98), we obtain

$$\Lambda(\tau) = C \exp\left[\frac{1}{\sigma^2} \sum_i d_i z_i(\tau)\right] \tag{8.108}$$

where $z_i(\tau)$ is the matched filter output signal sampled at the moment $iT + \tau$ and given by formula (8.100). On the basis of (8.108) we obtain

$$\Lambda(\tau) = C \prod_i \exp\left[\frac{1}{\sigma^2} d_i z_i(\tau)\right] \tag{8.109}$$

Assume that binary data symbols are bipolar, equiprobable and statistically independent, i.e. $\Pr\{d_i = 1\} = \Pr\{d_i = -1\} = 1/2$. Then we are able to calculate the ensemble average of the likelihood function with respect to the data symbols in the following way

$$\overline{\Lambda}(\tau) = E\left\{C \prod_i \exp\left[\frac{1}{\sigma^2} d_i z_i(\tau)\right]\right\} = C \prod_i E\left\{\exp\left[\frac{1}{\sigma^2} d_i z_i(\tau)\right]\right\}$$

$$= C \prod_i \left\{\frac{1}{2} \exp\left[\frac{1}{\sigma^2} z_i(\tau)\right] + \frac{1}{2} \exp\left[\frac{-1}{\sigma^2} z_i(\tau)\right]\right\}$$

$$= C \prod_i \cosh \frac{1}{\sigma^2} z_i(\tau) \tag{8.110}$$

Searching for the optimum sampling moment for which the likelihood is maximum can be done after finding the logarithm of the average likelihood function. We get

$$\overline{\Lambda}_{\log}(\tau) = \ln \overline{\Lambda}(\tau) = \sum_i \ln\left[\cosh \frac{1}{\sigma^2} z_i(\tau)\right] \tag{8.111}$$

As before, we obtain the optimum timing phase from the condition resulting from setting the derivative of the function $\overline{\Lambda}_{\log}(\tau)$ to zero. As a result, we achieve the expression

$$\frac{\partial \overline{\Lambda}_{\log}(\tau)}{\partial \tau} = \sum_i \frac{\partial}{\partial \tau} \ln\left[\cosh \frac{1}{\sigma^2} z_i(\tau)\right] = \frac{1}{\sigma^2} \sum_i \frac{\sinh \frac{1}{\sigma^2} z_i(\tau)}{\cosh \frac{1}{\sigma^2} z_i(\tau)} \frac{\partial z_i(\tau)}{\partial \tau}$$

$$= \frac{1}{\sigma^2} \sum_i \tanh\left[\frac{1}{\sigma^2} z_i(\tau)\right] \frac{\partial z_i(\tau)}{\partial \tau} = 0 \qquad (8.112)$$

From this formula we can derive the scheme of timing phase recovery for binary PAM transmission, shown in Figure 8.14.

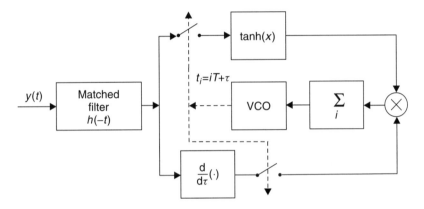

Figure 8.14 Maximum likelihood timing phase recovery scheme without decision feedback

As we see in Figure 8.14, the received signal is fed to the input of the matched filter. The filter output signal is differentiated in time and sampled with the phase τ. In the parallel branch, the signal is also sampled with the same phase τ, and subsequently processed in the nonlinear circuit with the characteristics $\tanh(x)$. The multiplied samples of both branches are accumulated in the interval of NT and the resulting value is a signal that controls the phase of the timing clock.

There also exist many suboptimum timing phase recovery schemes and we will briefly describe two of them.

The first scheme (Figure 8.15) is appropriate for baseband signals and takes advantage of the fact that the data signals are periodically sent every T seconds. Let us assume that the signal-to-noise ratio is so high that we can omit the noise component in the expression for the received baseband-equivalent signal. Then we have

$$y(t) = \sum_{i=-\infty}^{\infty} d_i h(t - iT - \tau) \qquad (8.113)$$

Assuming that data symbols are zero-mean and uncorrelated, we can calculate the ensemble average of the squared magnitude of the signal $y(t)$, which results in the expression

$$E[|y(t)|^2] = E\left[\sum_{i=-\infty}^{\infty} d_i h(t - iT - \tau) \sum_{k=-\infty}^{\infty} d_k h(t - kT - \tau)\right]$$

$$= \sum_{i=-\infty}^{\infty} E[|d_i|^2]|h(t - iT - \tau)|^2 = E[|d_i|^2] \sum_{i=-\infty}^{\infty} |h(t - iT - \tau)|^2 \quad (8.114)$$

As we see, the ensemble average of the squared magnitude of the signal $y(t)$ is a periodical signal with the period equal to T and the phase resulting from the time phase τ, regardless of the pulse shape $h(t)$. The signal of this frequency can be extracted by using the bandpass filter with the center frequency equal to $1/T$. Temporary phase variations on the filter output can be minimized if an additional averaging phase-locked loop is applied. The result of our considerations is shown in Figure 8.15.

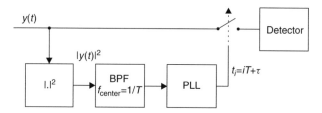

Figure 8.15   Timing phase recovery with the application of a spectral line created by a nonlinear circuit

The last method of timing recovery described in this chapter attempts to establish the sampling moment in the maximum of the pulse observed on the output of the matched filter. If the filter applied in the receiver is really matched to the pulses appearing on its input, then a single pulse seen on the filter output is symmetric with respect to its maximum. Therefore, by taking the signal samples shifted on the time axis by δ in forward and backward directions we should receive samples of the same magnitude, so their difference should be zero. This observation is reflected in the block diagram shown in Figure 8.16. The input signal is given to the matched filter input. Its output is sampled at two phases – the phase delayed by δ and the phase advanced by δ with respect to the current timing clock. The magnitudes of the received samples are determined and their values are compared in the subtracting block. The resulting difference between them is averaged in the loop filter and its output signal controls the timing phase of the VCO.

* * *

In this chapter we have presented examples of methods used in carrier and timing synchronization. The optimal methods are derived from the applied criteria such as minimum mean square error, maximum likelihood, etc. Generally, some methods take advantage of the knowledge of data symbols or decisions related to them, while others find the carrier phase or timing phase in a different way. Specific synchronization methods are also associated with multicarrier (OFDM) transmission. Although they have not been discussed in

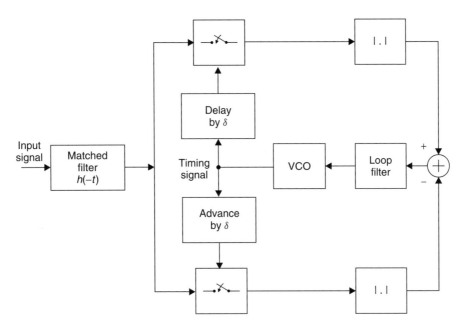

Figure 8.16   Timing phase recovery scheme with the application of early–late gates

this chapter, they were briefly described in the section devoted to OFDM modulation in Chapter 5.

The examples presented in this chapter mostly show the rules of functioning of synchronization recovery systems. Their performance is not analyzed. The examples cover all synchronization issues that appear in many comunication systems and networks. Generally, synchronization is one of the key issues in the design of digital communication systems and networks.

Problems

Problem 8.1   *The one-sided noise equivalent bandwidth of the system characterized by the transfer function $H(f)$ is given by the formula*

$$B_{eq} = \frac{1}{\max_{f} |H(f)|} \int_{0}^{\infty} H(f) df$$

Find the one-sided equivalent bandwidth of the first-order and second-order PLLs whose transfer functions are given by formulas (8.14) and (8.27), respectively.

Problem 8.2   *Consider the second-order PLL with the transfer function of the loop filter given by (8.23), so the closed loop transfer function of the PLL is described by (8.25). Prove the conditions for loop stability expressed by (8.26).*

Problem 8.3 *For the second-order PLL with the proportional-integral filter given by formula (8.29), find the loop transfer function, the stability condition and determine the expression for the one-sided noise equivalent bandwidth.*

Problem 8.4 *Solve Problem 8.3 if the transfer function $H(s)$ of the loop filter is*

$$H(s) = \frac{K_1}{s + K_2}$$

Problem 8.5 *Let us consider the discrete time PLL whose closed loop transfer function is determined by formula (8.44). Simulate its operation using Matlab. Assume that the phase detector has linear characteristics $W(\alpha)$ for $-\pi \le \alpha \le \pi$.*

1. *For the first-order discrete time PLL the loop filter has the transfer function $H(z) = K$. Plot the magnitude of the closed loop transfer function by applying substitution $z = \exp(j2\pi f T_S)$, where T_S is the sampling period and the coefficient K takes values 0.6, 1 and 1.4. Find the one-sided equivalent PLL bandwidth as a function of K_1 and K_2. Calculate and subsequently plot the PLL response to the phase shift θ and frequency shift Δf that occurred at the moment $n = 0$. Perform calculations for $1/T_S = 2400\,Hz$, $\theta = 45°$ and $\Delta f = 50\,Hz$.*
2. *Apply the PLL filter whose transfer function is $H(z) = K_1(1 + \frac{K_2}{z-1})$. Calculate the closed loop transfer function and find the conditions for the PLL to be stable. Find the one-sided equivalent PLL bandwidth as a function of K_1 and K_2. For the selected parameters K_1 and K_2 calculate and subsequently plot the PLL response to the phase shift θ and frequency shift Δf that occurred at the moment $n = 0$. Compare the results with the first-order PLL. Repeat the calculations and plots for a few sets of K_1 and K_2 to obtain the desired dynamic properties of the loop. Perform calculations for $1/T_S = 2400\,Hz$, $\theta = 45°$ and $\Delta f = 50\,Hz$.*

Problem 8.6 *Consider the QPSK transmission and the carrier phase recovery circuit based on a nonlinear quadratic circuit similar to that shown in Figure 8.9. Draw a block diagram of this modified circuit. Assume the ideal QPSK signal arrives at the input of this circuit. Draw the waveforms at the output of each functional block. What are the factors determining the parameters of the bandpass filter with its band centered around $4f_c$, where f_c is the carrier frequency of the QPSK signal? What is the carrier phase ambiguity of the reference signal produced by this circuit?*

Problem 8.7 *Write a computer program (possibly in Matlab) that models the baseband equivalent digital transmission system shown in Figure 8.17a. The system consists of a source of 16-QAM signals distorted by the carrier frequency and phase offset. Additive white Gaussian noise is added to the signal at the receiver input. Since transmit and receive filters with square root raised cosine characteristics are applied and we assume an ideal timing phase once per signaling period T, resulting in no intersymbol interference, we can simplify the system model to the form shown in Figure 8.17b.*

1. *Send the appropriately large number of 16-QAM data symbols (say, 1000 or more) and plot the distorted constellation points at the input of the carrier recovery circuit. The function scatter in Matlab is useful for that purpose. Observe the influence of additive noise on the spread of the constellation points (assume zero phase and frequency shifts*

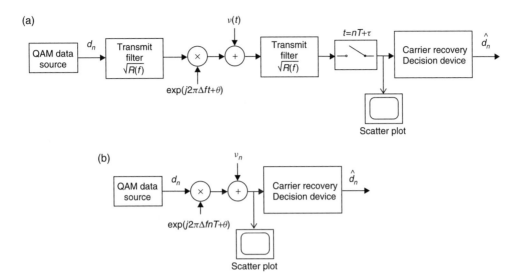

Figure 8.17 Block diagram of the data transmission system with the carrier recovery circuit (a), the simplified block diagram (b)

for a while). Next, observe the influence of the phase shift and later of the frequency shift on the form of the received in-phase and quadrature signal coordinates and their spread around the correct constellation points.

2. *The receiver applies a decision feedback carrier phase and frequency recovery circuit, similar to that drawn in Figure 8.11. Draw a detailed block diagram of the carrier recovery circuit. Apply a discrete proportional-integral loop filter $H(z)$ in it. Select the filter parameters that ensure stability of the digital phase-locked loop and allow the phase θ and frequency offset Δf to be acquired and tracked for the following transmission parameters: $1/T_S = 2400\,Hz$, $\theta = 45°$ and $\Delta f = 50\,Hz$. Plot the PLL error e_k as a function of the timing index k, showing that the carrier recovery circuit is stable and gradually compensates for the phase and constant frequency offset. Estimate the steady-state mean square error of the loop on the loop filter.*

Problem 8.8 *Apply the simulation program of a baseband binary bipolar transmission system that was previously developed in Problem 6.1 to investigate the timing phase recovery circuit based on the PLL with the early–late gates. As previously, apply the oversampling rate $N_s = 16$ (there are N_s samples per signaling period T). Perform the following experiments.*

1. *First, assume that the channel does not introduce amplitude or phase distortion (i.e. in the channel model a = 0 dB, $\beta = 0$). Thus, assuming additive white Gaussian noise as the only signal impairment, notice that the receive filter is strictly matched to the received pulses. Design the timing phase recovery circuit using the early–late gates shown in Figure 8.16 by selecting the timing phase advancement/delay δ (as one of the N_s possible timing phases) and the parameters of the loop filter. What is the main task of this filter? In the simulation run, let the transmitter send an appropriately long*

data sequence to make an observation of how the best timing phase (out of N_s possible phases) is gradually being achieved. How do the parameters of the loop filter influence the dynamic behavior of the timing recovery circuit? Run your simulations starting from different initial timing phases. Run them for $SNR = 15$, 10 and 5 dB. Plot the timing phase as a function of the number of signaling periods.

2. *Repeat similar simulation experiments using the timing recovery circuit and applying the early–late gates for the case when the channel introduces amplitude and phase distortions. Apply the channel parameters similar to those used in Problem 6.1: $R = 2400$ symb/s, $a = 5$ dB, $SNR = 15$ dB and β having such a value that the group delay τ is equal to 1 ms at the frequency $R/2$ Hz.*

Problem 8.9   *One of the practical methods of achieving timing recovery is based on zero crossings of the data pulses. In general, timing positions of zero crossings depend on the pulse-shaping filter properties, the noise level and amplitude and phase distortions of the transmission channel. If the zero crossing detector generates short timing pulses at the moments in which such crossings occur, these pulses can be treated as the excitations for the averaging PLL, which generates a stabilized timing clock. Based on this clock, the appropriate timing (e.g. in the half of the period between subsequent PLL pulses) can be achieved. Apply again the simulation program used in Problem 8.8 and observe the eye pattern at the receive filter output. Supplement the simulated receiver with the timing recovery curcuit based on the above-described principle. Design the whole timing recovery circuit and determine the properties of the applied averaging PLL. Run the simulations as in points 1 and 2 of Problem 8.8 and plot the estimated timing phase that is applied in the decision process. Draw a plot of the squared difference between the signal sample at the decision input and the transmitted data symbol as a function of the number of signaling periods.*

9

Multiple Access Techniques

9.1 Introduction

Resources that are used by communication systems, such as the electromagnetic spectrum and time, are mostly very limited and have to be efficiently shared by many communication links. In typical communication networks the number of links that can be potentially established is much higher than that which is allowed by the amount of network resources. Therefore, a communication system designer has to decide not only about the configuration of communication links but also on the applied method of sharing the resources among simultaneous links. The latter is done by selection of a *multiple access scheme*. Although the problem has to be solved for most of the systems, it is particularly important in the case of radio systems. Owing to the application of the selected multiple access scheme the signals generated by system users can use common system resources such as time and frequency band, and the signals of the particular user can be effectively extracted at the receiver. In radio systems the multiple access scheme largely influences the total system design, including the typically deployed fixed part of the network, and substantially determines the cost and quality of the system operation.

There are three basic multiple access schemes: FDMA – *Frequency Division Multiple Access*; TDMA – *Time Division Multiple Access*; CDMA – *Code Division Multiple Access*. They often appear in a hybrid form as a combination of at least two of them. In radio systems they can be additionally supported by multiple antennas forming antenna arrays that enable SDMA – *Space Division Multiple Access* – to be applied. OFDMA (*Orthogonal Frequency Division Multiple Access*), which can be considered as a special type of FDMA, is another multiple access scheme that is very attractive in certain applications. Basic multiple access schemes have been described in many books on digital communication systems or wireless communications (see Rappaport 1996; Proakis 2000; Sklar 1988; Barry *et al.* 2003; Wesołowski 2002; Pahlavan and Levesque 1995).

With respect to sharing common spectral resources among many users, some authors distinguish the concepts of *multiple access* from *multiplexing* (Sari *et al.* 2000). The latter term refers to the function performed at the communication transmitter (such as a base station) in which the signals available locally are distributed to the receivers located in different places. The former term, i.e. multiple access, refers to the function

Introduction to Digital Communication Systems   Krzysztof Wesołowski
© 2009 John Wiley & Sons, Ltd

performed by transmitters of user terminals communicating with the main station (e.g. a base station). In this case the signals originate from transmitters that are mostly placed in different geographical locations. As a result, their carrier frequencies and/or timing and power have to be adjusted so that the signals arrive at the main station receiver within the assigned frequency band or time frame and at approximately the same power.

Multiple access schemes are mostly accompanied by *duplexing techniques* (Figure 9.1), which allow the terminals to send and receive signals simultaneously or quasi-simultaneously. There are two basic techniques: FDD (*Frequency Division Duplex*) and TDD *Time Division Duplex*). In the first method transmission directions are separated on the frequency axis by having different bands assigned to them, whereas in TDD the direction of transmission is periodically reversed using the same frequency band.

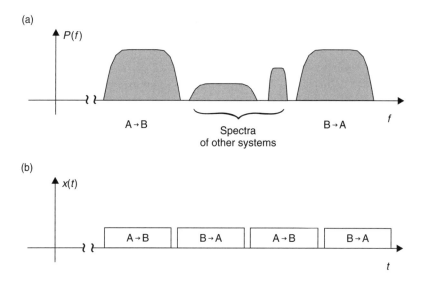

Figure 9.1 Illustration of basic duplexing techniques: (a) FDD and (b) TDD

Multiple access schemes can be applied jointly with several types of modulation. Both single- and multi-carrier transmission techniques are applied in various systems.

Below we provide an overview of basic multiple access schemes. Part of the material presented in this chapter was earlier contained in the public domain deliverable on multiple access schemes prepared within the European Union Sixth Framework project WINNER – World Wireless Initiative New Radio (WINNER Deliverable 2.6 2004).

9.2 Frequency Division Multiple Access

FDMA is historically the first multiple access scheme. Initially, it was applied in analogue telephony transmission (Sklar 1988). For a long time, due to the level of communication technology, it was the only method possible. Introduction of digital communication systems made it possible to use other methods, in particular TDMA and CDMA.

Although FDMA is widely known, there are not so many literature positions in which FDMA is treated in detail compared to other multiple access schemes. In FDMA, individual frequency bands that define transmission channels are assigned to individual users (Figure 9.2). Except for unidirectional systems, such as TV or radio broadcasting, in bidirectional transmission each user is assigned a pair of channels characterized by two different carrier frequencies, so FDD is almost exclusively associated with FDMA.

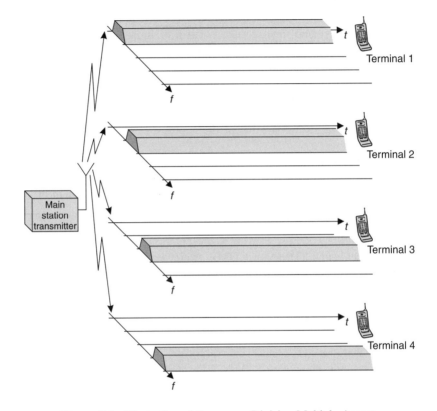

Figure 9.2   Illustration of Frequency Division Multiple Access

In Rappaport (1996) one can find basic features of the FDMA scheme considered from the point of view of wireless systems. The most important features are:

- Each FDMA channel provides only one connection at a time.
- FDMA requires tight RF filtering to separate the user signals and minimize adjacent channel interference. Adjacent channels are separated on the frequency axis by guard bands, which are necessary due to the finite slope between the passband and stopband of the channel filter characteristics. This in turn decreases the FDMA spectral efficiency.

- Due to simultaneous operation of transmitters and receivers when using FDMA in combination with FDD, duplex filters are necessary in terminals and base stations, which increases the cost of the whole system.
- After channel assignment, the base station and terminals transmit simultaneously and continuously.
- The bandwidths of FDMA channels are relatively narrow because each channel is used by only one connection at a time. In this sense, FDMA can be analyzed as a narrowband approach, although the total bandwidth of the whole system may be large.
- In traditional narrowband FDMA systems the channel characteristics are often almost flat. Therefore, inter-symbol interference is small or moderate and in such cases only simple channel equalization, or no equalization at all, is needed.
- FDMA, being a continuous transmission scheme, requires a small number of bits for synchronization and framing.
- A base station power amplifier amplifies the signal, which is the sum of many individual channel signals; thus, the amplifier has to be highly linear because of the high Peak-to-Average Power Ratio (PAPR) of the aggregated signal.

In mobile communications FDMA has been applied in many older systems, such as the first generation cellular systems AMPS (Bell System Technical Journal 1979), NMT (Westin 1993) or others. Frequency division multiplexing is widely applied in analog TV and radio broadcasting.

It is important to stress that if we consider both mobile terminals and base stations in wireless communication systems, FDMA is present as a natural component of virtually all practical schemes using TDMA or CDMA (Baier *et al.* 1996). The reason for this is that the total bandwidth of a typical mobile communication system would be difficult to manage if TDMA or CDMA was applied exclusively. In the case of TDMA this would result in very short burst lengths in order to support an adequate number of simultaneous users. Application of FDMA as a multiple access component relaxes these requirements and allows for higher flexibility of resource management (Baier *et al.* 1996).

In the systems characterized by a large coverage area, such as TV broadcasting systems or cellular systems, the choice of FDMA as a multiple access scheme results in the necessity for sophisticated radio network design in which frequency planning is taken into account. The same carrier frequencies that identify particular FDMA channels can be applied by the stations, which are separated adequately in space to prevent interference among them. A so-called *frequency re-use factor* lower than 1 has to be applied. All these design factors are considered as drawbacks of this multiple access scheme.

9.3 Time Division Multiple Access

TDMA is a well-known access technology· that has been successfully applied in many wired and wireless digital transmission systems. In TDMA, the time axis is typically divided into a sequence of periodically repeating time slots (Figure 9.3). In each slot, only one user is allowed to transmit or receive. Typically, a user has periodical access to the time slot assigned to him. The time slots are organized in frames. Very often higher

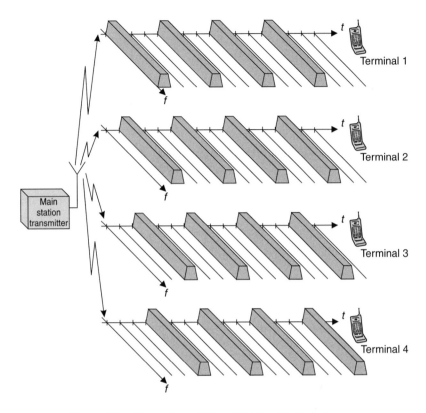

Figure 9.3   Illustration of Time Division Multiple Access

hierarchy time structures are defined, which allow for efficient resource management, signaling and network and frame synchronization. TDMA is accompanied by either TDD or FDD duplex schemes. The hybrid TDMA/FDMA/FDD version is used in GSM mobile telephony.

A basic description of TDMA with emphasis on wireless systems can be found in Rappaport (1996). TDMA applied in mobile communications is characterized by the following main properties:

- A single-carrier frequency is shared by a number of users. Each of them transmits or receives a signal in nonoverlapping time slots.
- Data transmission has a bursty nature, so transmission is exclusively digital. For a certain fraction of time the mobile station can be in an idle state, thus the battery can be saved. Outside the slots in which the mobile station transmits or receives, it can monitor surrounding base stations. This enables and simplifies a mobile-assisted procedure of changing the base station during the call (i.e. the so-called *handover* procedure performed in cellular systems).

- Duplexing filters, i.e. the filters that separate the transmission directions, are not needed in terminals, due to the fact that transmission and reception take place in different time slots, regardless of the duplexing method used. However, TDMA applied jointly with FDD could require duplexing filters.
- Due to the shorter fraction of time assigned to a single user, a larger bandwidth is needed to transmit the same amount of data compared with FDMA. Practically, this results in the necessity of (adaptive) equalization of the transmission channel and sending a training sequence within the data burst.
- Data bursts transmitted in the uplink (i.e. from a user terminal to the base station) have to be separated by guard periods to take into account possible time misalignments as a consequence of synchronization imperfections in the terminals. In the case of TDD, a guard period is also required to take into account the time interval needed to switch from receive to transmit mode, and vice versa.
- A relatively large overhead in the transmitted block is required for frame and slot synchronization.

Application of TDMA allows for flexible time slot assignment, so the number of time slots can be adjusted to the needs of particular users (see data transmission in the GPRS mode of the GSM system; Seurre *et al.* 2003). When TDMA is combined with FDMA, as is common in mobile communication cellular systems, careful frequency planning has to be performed.

9.4 Code Division Multiple Access

There are several types of CDMA schemes. The first category of schemes, which currently is the most popular, is based on transmission using a single carrier (Figure 9.4). This category contains the spread spectrum methods briefly explained in Chapter 7. The second category of CDMA schemes applies multi-carrier transmission as a base. We start our overview with the first category.

9.4.1 Single-Carrier CDMA

CDMA is a multiple access scheme (Prasad 1996) that originates from the direct sequence spread spectrum systems (DS-SS) (Dixon 1984) described in Chapter 7. As we remember, the essential feature of the DS-SS systems used in a CDMA scheme is the robustness of the transmitted signal to jamming. All users transmit and receive signals in the same band, applying unique code sequences assigned to them (Viterbi 1995). The code sequence chip rate is N times higher than the data rate. All applied user code sequences are mutually strictly orthogonal or quasi-orthogonal. Good examples of strictly orthogonal sequences are the Walsh-Hadamard sequences applied as *channelization codes* in the second generation American cellular system IS-95 (Lee and Miller 1998) and the OVSF (*Orthogonal Variable Spreading Factor*) codes applied in UMTS (*Universal Mobile Telecommunication System*) (Holma and Toskala 2004; Springer and Weigel 2002). Quasi-orthogonal sequences are pseudonoise (PN) m-sequences or Gold sequences derived from maximum

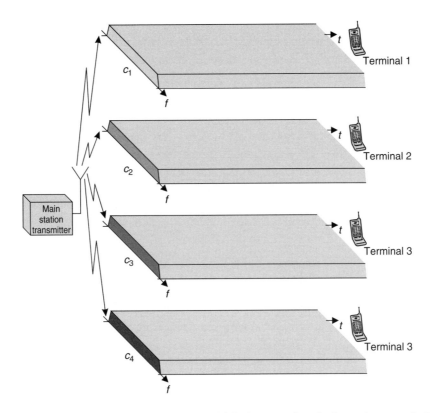

Figure 9.4 Illustration of Code Division Multiple Access using single-carrier transmission

length linear feedback shift registers (LFSR). Particular cells of a cellular CDMA system are typically distinguished among themselves by the application of appropriate *scrambling* codes based on PN or Gold sequences, in addition to the channelization codes.

The Walsh-Hadamard sequences can be derived recurrently from the following formula

$$H_1 = \begin{bmatrix} 1 & 1 \\ 1 & -1 \end{bmatrix}, \qquad H_k = \begin{bmatrix} H_{k-1} & H_{k-1} \\ H_{k-1} & -H_{k-1} \end{bmatrix} \qquad (9.1)$$

Each sequence is formed by a row of the matrix H_k. The 2^k-symbol-long sequences contained in different rows of the matrix H_k are mutually orthogonal. However, their orthogonality is lost if one of these sequences is cyclically shifted with respect to the other one. Figure 9.5 graphically illustrates the Walsh-Hadamard sequences W_i $(i = 0, \dots, 63)$ of length 64 for $k = 5$.

The OVSF codes are defined by the tree shown in Figure 9.6. Considering any node of the tree, we see that new codewords are created by appending the preceding codeword with itself (in the upper branch originating from the node) or with its negation (in

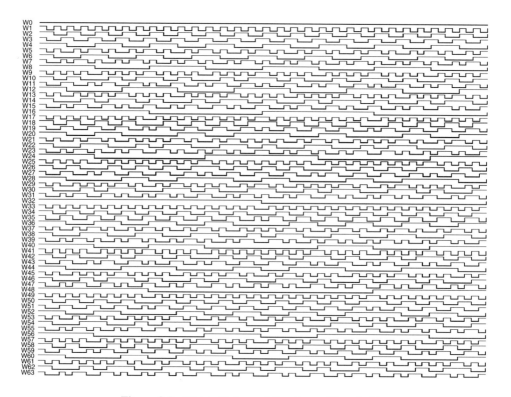

Figure 9.5   Walsh-Hadamard functions of length 64

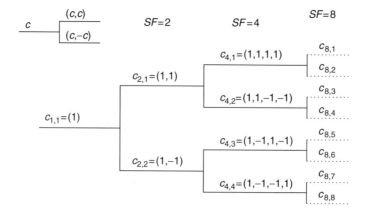

Figure 9.6   OVSF code tree

the lower branch growing from this node). Mutual orthogonality of different codewords is achieved by their appropriate selection from the code tree. One can prove that a codeword c_i is orthogonal to a codeword c_j, if and only if the codeword c_j is not associated with the branch leading from the branch associated with the code c_i to the root of the tree or is not located in the subtree below the codeword c_i. For example, if bits of a particular data stream are spread using the codeword $c_{8,5} = (1, -1, 1, -1, 1, -1, 1, -1)$ with the spreading factor $SF = 8$, then for another data stream requiring the spreading factor $SF = 4$ all the code words $c_{k,i}$ except $c_{4,3}$ can be applied.

Figure 9.7 explains the concept of channelization codes and spreading codes with the example of transmission used in the uplink of the UMTS system. The OVSF codes are uniquely associated with a transmission channel (recall that in FDMA the channel is associated with a carrier frequency, whereas in TDMA the channel is determined by selection of the time slot). Depending on the selected data rate, several spreading factors SF of the assigned OVSF codes can be applied but the chip rate remains constant. The complex scrambling sequence applied in the transmitter has the same chip rate as the channelization codes, which is equal to 3.84 Mbit/s. The reason for application of scrambling codes is the quasi-orthogonalization of signals generated by different CDMA terminals sending their signals to different base stations. This makes it possible to apply the same carrier frequency f_c in transmissions to all base stations.

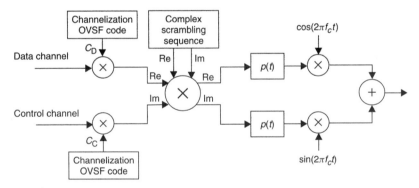

Figure 9.7 The basic scheme of a UMTS uplink transmitter using channelization and scrambling codes

Selection of either a fully orthogonal or a quasi-orthogonal set of spreading sequences is an important factor of a CDMA system design. In the first case, if the sequence period is equal to N, at most N users can transmit simultaneously over flat (nondispersive) and time-invariant channels without mutual interference. In the second case, due to residual cross-correlation between spreading sequences, the number of active users is limited by the tolerated noise level in the receivers and the system performance gradually decreases with increasing number of simultaneous users. The performance is furthermore significantly affected by the presence of multipath propagation. Reception of delayed copies of the codewords will then reduce the performance of single-user receivers.

In Rappaport (1996), the basic features of a classical single-carrier CDMA scheme have been summarized. The most important features are:

- CDMA users share the same frequency band. Both FDD or TDD duplexing methods are applicable.
- In CDMA systems in which pseudo-noise sequences are applied for spreading, the number of users is "soft".
- Influence of multipath fading is potentially reduced due to signal spreading over a large spectrum. This is achieved by applying a RAKE receiver, described in Chapter 7.
- Due to a very high chip rate applied in spreading codes, the receiver is able to extract separate path signals arriving through the multipath channel. Therefore it is possible to combine them efficiently in a RAKE receiver.
- Knowledge of all spreading (channelization) codes of the users in a cell allows for joint (multi-user) detection (Verdù 1998) of user signals, taking into account residual cross-correlation between the spreading codes and loss of orthogonality due to multipath propagation.
- In CDMA cellular systems the carrier frequencies applied by different base stations can be the same, so all surrounding cells use the same frequency band. As a result, the terminal that is located a comparable distance to two different base stations can temporarily transmit to both of them and receive signals from both base stations at the same time. The only requirement is to properly combine the signals arriving from/to both base stations.
- Because of residual cross-correlation of the spreading sequences, or loss of mutual orthogonality caused by a multipath channel, it is desirable that the signals sent by different users arrive with the same mean power, otherwise residual cross-correlation of the stronger signal constitutes a substantial noise in reception of a weaker signal. Thus, the performance of CDMA receivers is very sensitive to the quality of power control. This phenomenon is called the *near–far effect* and is of special concern in CDMA receivers.

So far we have considered CDMA based on direct sequence spreading. However, other spreading methods are also known. The most important among them is frequency hopping (FH), which is the second basic type of spread spectrum transmission. As we remember, in FH systems the available spectrum is divided into contiguous frequency slots (Proakis 2000). During transmission of a data symbol, one or more frequency slots are selected in a pseudo-random manner. Typically, the carrier is FSK modulated. An example of an FH system is included in the wireless local area network IEEE 802.11 standard (IEEE 2007). Two- or four-level Gaussian FSK is applied, resulting in transmission rates of 1 or 2 Mbit/s. Terminals select different hopping patterns resulting in multiple access. Generally, the application of FH methods allows for asynchronous reception and results in less expensive implementation of mobile transceivers.

Example 9.4.1 *Let us consider the example of an FH-SS system such as Bluetooth (Haartsen 2000; Wesołowski 2002). This system applies frequency hopping with the time division duplex (FH/TDD). Time is divided into 625-μs slots. The system band is divided into 79 1-MHz frequency channels. During each time slot the signal occupies one of these channels* *according to the selected hopping pattern. As a result, there are 1600 hops per second. The direction of transmission alternates from slot to slot. In the Bluetooth system, mobile terminals are organized in the so-called piconets, which may consist of up to eight terminals. Different piconets use different hop sequences. The sequences are carefully selected to ensure their statistical properties and immunity to mutual interference. At the same time, their number is so large that they are not fully orthogonal. Lack of orthogonality is compensated by coding and ARQ techniques.*

Frequency hopping systems are typically applied in a difficult environment in which asynchronous reception is preferable and in small-range wireless systems such as Bluetooth, in which system capacity is not a critical issue. In cellular systems, direct sequence spreading is the common technique used to implement CDMA.

As we said earlier, ultra-wideband (UWB) communication has recently become a subject of intensive research and implementation (Siwiak and McKeown 2004). In some proposals for UWB transmission, multiple access can be achieved by application of different patterns of time division spreading sequences (TD-SS). In this arrangement, time is divided into frames, which in turn are divided into narrow slots. The slot in which a signal from the given transmitter will be emitted in the current frame is selected in a pseudorandom fashion. Within the slot a very narrow, properly modulated pulse is transmitted. Due to a large number of slots and short duration of applied pulses, the system can feature an extremely wide bandwidth. Multiple access can be achieved by selection of mutually orthogonal sequences of time slots. The UWB system is a potentially good solution for short-range communications.

9.4.2   Multi-Carrier CDMA

Multi-carrier CDMA (Fazel and Kaiser 2003; Hara and Prasad 1997) is a multiple access technique strictly related to CDMA using direct sequence spreading (DS-SS). This multiple access scheme applies a multi-carrier modulation, mostly OFDM, as a tool for efficient use of the available spectrum. Multi-carrier CDMA was considered for use in third generation cellular radio access networks; however, at the time of technology selection, CDMA based on the single-carrier DS-SS principle was a more mature technology.

Multi-carrier CDMA schemes can be divided into two types. In multiple access schemes of the first type the original data stream is spread using a given spreading code and then a different subcarrier is modulated with each chip (the spreading operation is performed in the frequency domain). As a result, the MC-CDMA can be treated as a serial concatenation of DS spreading and multi-carrier modulation. The transmitter structure of the MC-CDMA scheme is similar to that of a regular OFDM scheme. The main difference is the usage of subcarriers in data transmission. An MC-CDMA scheme transmits the same data symbol in parallel on many subcarriers. Figure 9.8a shows the scheme of the MC-CDMA transmitter used in downlink transmission, so the configuration of transmission is equivalent to a multiplexing scheme.

Data sequences directed to individual terminals (numbered 1 to k) are first spread using different mutually orthogonal PN spreading sequences. The arithmetic sum of the spread sequences is formed into blocks that are subsequently processed in parallel. In the next step block interleaving is performed and the resulting data block constitutes the input

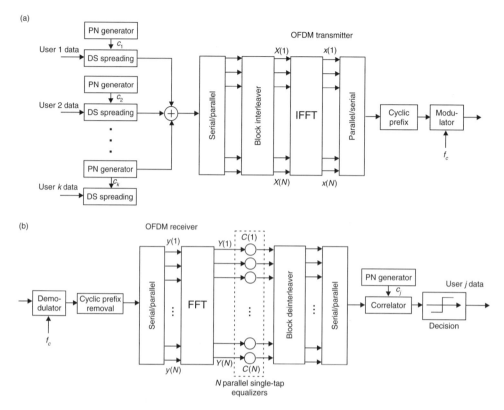

Figure 9.8 Transmitter (a) and receiver (b) of the first type of multi-carrier CDMA scheme in downlink transmission

to the OFDM modulator based on IFFT transformation. A cyclic prefix is added to the resulting time domain signal block.

The receiver performs operations dual to those made in the transmitter. However, only one spread data sequence is intended for a particular receiver, so a single de-spreader is applied, as shown in Figure 9.8b. As we see, the MC-CDMA receiver requires coherent detection to perform the de-spreading operation successfully.

In the second type of MC-CDMA scheme the data stream is initially serial-to-parallel converted into low-rate substreams. The resulting substreams are subsequently spread using a given spreading code and then each spread substream modulates its own subcarrier (the spreading operation is performed in the time domain). Figure 9.9 presents the general scheme of this type of MC-CDMA transmitter and receiver.

9.5   Orthogonal Frequency Division Multiple Access

OFDMA, as with MC-CDMA, is also derived from OFDM. The idea of OFDMA was probably first presented by Sari *et al.* (1996, 1997).

In OFDMA, an individual subcarrier or, more often, a group of subcarriers is assigned to different users. There are several methods for allocating subcarriers to users. Two

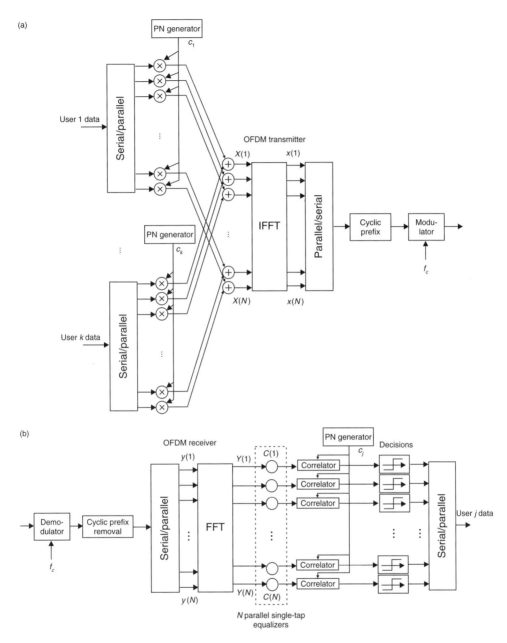

Figure 9.9 Transmitter (a) and receiver (b) of the second type of multi-carrier CDMA scheme in downlink transmission

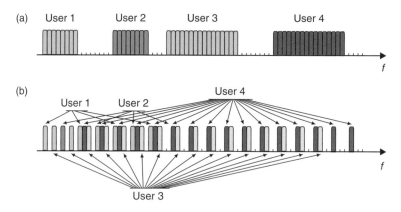

Figure 9.10 Two methods of assigning subcarriers in OFDMA: (a) grouped subcarriers and (b) spread interleaved subcarriers

most common subcarrier configurations are: grouped subcarriers and interleaved spread subcarriers. If the method of grouped subcarriers is used, each user is assigned a group of contiguous subcarriers. In the spread subcarriers method, subcarriers are allocated in a fixed comb pattern, which means that they are spread over the system bandwidth. Figure 9.10 illustrates both types of subcarrier assignment. Let us note that the users can have different numbers of subcarriers at their disposal. They can also apply different modulations and error correction coding with selected coding rates. This makes it possible to assign different radio resources to users, depending on their requirements. The grouped subcarriers method minimizes inter-user interference, but is more sensitive to fading, because a whole group of subcarriers assigned to a given user may suffer from a null in the channel characteristics. The use of spread subcarriers minimizes the sensitivity of the transmission performance to fading, but in the case of transmission from many terminals to a single base station this method is more susceptible to inter-user interference if the users are imperfectly synchronized in frequency and time.

OFDMA can be applied jointly with TDMA, so the hybrid TDMA/OFDMA method can also be used to allocate resources both along time and frequency axes.

The most important advantages of OFDMA and hybrid TDMA/OFDMA are:

- no inter-code interference, since transmission in OFDMA is performed on fully orthogonal subcarriers (as opposed to typical CDMA-based schemes);
- much more flexible radio resource assignment as compared with TDMA;
- a possibility of adaptive radio resource assignment depending on the user's requirements and on channel characteristics;
- high spectral efficiency with respect to FDMA as no guard bands are needed between the user's spectra.

In turn, the disadvantages are:

- sensitivity to frequency offset and high peak-to-average power ratio (PAPR), i.e. the same as for a typical OFDM system;

- difficulty in providing subcarrier synchronization;
- the need for coordinated subcarrier assignments.

Figure 9.11 presents a block diagram of a typical transmitter and receiver for the case of transmission from the base station to the user terminal using the OFDMA method. In the base station, signals are directed to k user terminals. They are placed in a block or spread fashion in the subcarrier mapping block, as shown in Figure 9.10. The OFDM modulator implemented via N-point IDFT (or IFFT) produces the aggregated signal in the time domain. The other blocks are analogous to those in the OFDM transmitter.

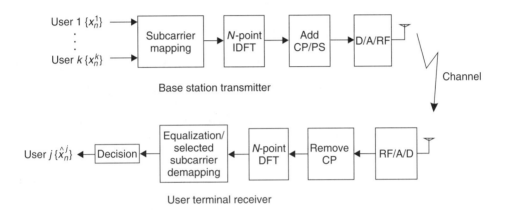

Figure 9.11   Transmitter and receiver for wireless link from a base station to a user terminal

The receiver of the jth user terminal has to detect the data destined for it. Thus, a regular OFDM receiver is applied but it processes the data samples only on those subcarriers that have been assigned to the jth user.

It is worth mentioning that OFDMA is considered to be one of the most promising multiple access schemes for fourth generation cellular radio systems. It has been selected for a broadband wireless access system known as IEEE 802.16 WiMAX (Koffman and Roman 2002).

9.6   Single-Carrier FDMA

Although OFDMA is spectrally very efficient, as we said earlier, it suffers from a high PAPR resulting from inherent properties of the OFDM modulation. Consequently, the transmitter emitting an OFDMA/OFDM signal requires a highly linear amplifier. However, its power efficiency is relatively low. This is a disadvantage that is particularly important for user terminals whose energy source, such as a battery, has a limited capacity. The remedy for this disadvantage is the application of the SC-FDMA (*Single-Carrier Frequency Division Multiple Access*) method (Myung *et al*. 2006) in transmission from user terminals to a base station (Figure 9.12). This method retains many OFDMA features such as spectral efficiency and complexity and uses functional blocks that are typical for OFDMA; however, the value of PAPR results directly from the applied signal constellation

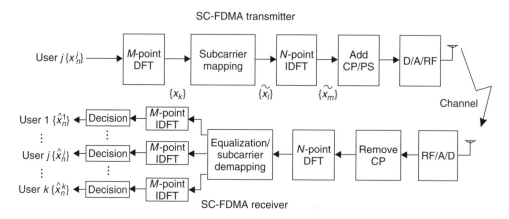

Figure 9.12 Transmitter and receiver for SC-FDMA transmission from a user terminal to the base station

in the time domain and it is significantly lower than in OFDMA systems (Danilo-Lemoine *et al.* 2008).

The transmitter of the user terminal using the SC-FDMA method operates as follows. The jth user's data in the form of the M-element time domain block of QPSK or QAM symbols $\{x_n^j\}$ are the subject of DFT transformation resulting in M samples in the frequency domain. In the subcarrier mapping block the spectral samples are placed in the spectral region assigned to the jth user, which means that M selected subcarriers carry data of the jth user. The remaining $N - M$ subcarriers are unmodulated. After IDFT operation of size N, a time domain sequence of N samples is received. Subsequently a cyclic prefix and (possibly) postfix are appended to the time domain block and the resulting sequence is converted into a continuous signal, amplified and up-converted to the RF frequency range in the RF front-end. Let us note that the transmitted signal is a passband version of the sequence of QPSK or QAM data symbols, so its PAPR value is moderate.

The SC-FDMA receiver is usually located in the base station so data from several user terminal transmitters arrive at its antenna. After down-conversion to the baseband and analog-to-digital conversion the samples are appropriately grouped into blocks. The cyclic prefix is removed and the resulting blocks are the subject of DFT transformation. The received blocks can also be equalized. Sample blocks assigned to different users can be extracted from the received frequency domain block of samples. Thus, after subcarrier de-mapping the extracted user blocks are individually converted into the time domain using M-point IDFT. Finally, decisions on data blocks of each user are made.

Typically, the data traffic from user terminals to base station is lower than in the opposite direction. Thus, if block subcarrier assignment is performed, it is easy to assign subcarrier blocks in a given band in such a way that there are some unoccupied spectral spaces between the user bands. Such a strategy relaxes the stringent requirements on frequency offset compensation and decreases intercarrier interference. Thus, the subcarrier block assignment in the SC-FDMA method can be similar to that shown in Figure 9.10a.

The process of assigning the spectral blocks to users depends on the strategy applied in the considered system. If the wireless channel is assumed to be slowly time-varying with respect to the signaling rate, the base station should estimate the wireless channel properties and assign the spectral resources on the basis of the quality of the channel between a given user and the base station. Several strategies are possible both for OFDMA and SC-FDMA but they are beyond the scope of this introductory section.

9.7 Space Division Multiple Access

SDMA is a multiple access scheme that can enrich the previously described multiple access methods. Besides frequency, time and code, space becomes an additional dimension to be used in the assignment of the system resources. The latter is possible if user terminals are sufficiently separated in angle with respect to the base station antennas. The base station is equipped with the antenna array, which allows it to form highly directional beams that can reach given terminals without interfering with each other. Depending on the advancement of the applied technique, beamforming can be applied in the uplink or in both uplink and downlink. The efficiency of SDMA strongly depends on the angular distribution of terminals located in the area served by a given base station. If the angular separation is sufficiently good, the same channels can be reused by different users. Figure 9.13 shows a simple SDMA arrangment.

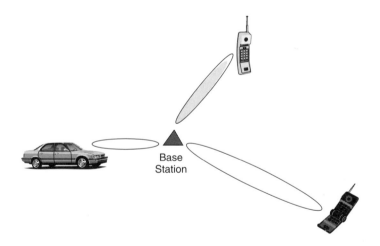

Figure 9.13   The SDMA principle

The antenna array in the base station consists of a certain number of antennas, say M, cooperating with the amplifiers and phase shifters. Their operation can be symbolized by a set of complex gain coefficients. Figure 9.14 illustrates the operation of such an antenna array applied for reception of a single user signal in the base station receiver. The signals received by each antenna are converted into the sampled baseband in-phase and quadrature components, which can be treated as complex samples $x_{i,n}$ $(i = 1, \ldots, M)$. Such samples are weighted by the complex coefficients $w_{i,n}$ $(i = 1, \ldots, M)$ and summed, resulting in the complex signal sample $y_n = \sum_{i=1}^{M} w_{i,n} x_{i,n}$. The block that performs control of

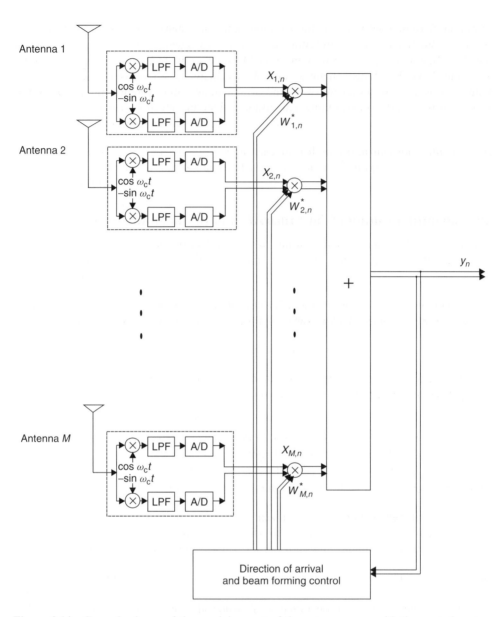

Figure 9.14 General scheme of the receiving part of the antenna array with the control system that can be applied in SDMA (Wesołowski 2002)

direction of arrival and beamforming adaptively tracks the position of a given user by adjusting the set of coefficients $w_{i,n}$ $(i = 1, \ldots, M)$. Such a scheme of functioning has to be applied for all users served simultaneously, unless the blocks can be shared among users if TDMA is applied as an additional multiple access method. Thus, the coefficient settings for each user can be used sequentially.

9.8   Case Study: Multiple Access Scheme in the 3GPP LTE Cellular System

Future wireless communication systems are being discussed on international forums. International cooperation should lead to establishing a universal standard defining the basic rules of operation of the system following UMTS (in Europe and many other countries) or cdma2000 (on the American continent). There is an international organization called 3GPP (*3rd Generation Partnership Project*) that defines improvements and modifications of the UMTS system and whose aim is also to define the path towards the fourth generation wireless communication system following UMTS. A significant step in this direction is the definition of the so-called 3GPP LTE (*Long-Term Evolution of UMTS*). Let us consider basic requirements and parameters of the LTE system in the context of the multiple access method that is to be applied in it.

The assumed peak data rate in the direction from a base station to mobile stations is 100 Mbit/s in a 20-MHz channel. If the channel has a lower bandwidth the maximum rate is proportionally scaled down. The spectral efficiency should be 5 bit/s/Hz in the direction towards mobile stations and 2.5 bit/s/Hz in the opposite direction. Both FDD and TDD duplexing schemes should be available for use. The proposed channel bandwidths are 1.25, 2.5, 5, 10, 15 and 20 MHz.

Such stringent requirements on data rates and spectral efficiency indicate that a DS-SS technique and CDMA are no longer efficient as transmission and multiple access methods. Additionally, a sufficiently dense granulation of data rates has to be supported, which results from the variety of offered services. As a result, OFDM has been selected as a transmission method and OFDMA in conjunction with TDMA as a proposed multiple access scheme. In this way, it is possible to introduce continuous resource management on the frequency and time axis (dynamic assignment of OFDM subcarriers and time slots), called *scheduling*.

Let us consider the time–frequency transmission arrangement in the downlink, i.e. in the direction from a base station to mobile stations. The transmission is organized in 10-ms frames consisting of 20 1/2-ms time slots. Each slot contains six or seven OFDM

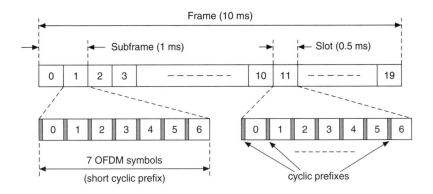

Figure 9.15   Frame structure of the 3GPP LTE system (© 2007. 3GPP™ TSs and TRs are the property of ARIB, ATIS, CCSA, ETSI, TTA and TTC, who jointly own the copyright in them)

symbols, depending on the length of the applied cyclic prefix. Figure 9.15 shows the frame organization.

As we said earlier, a combination of OFDMA and TDMA is applied as a multiple access method. A sufficiently dense granulation of possible data rates is ensured owing to the assignment of the system resources in the form of multiples of resource blocks. Such a resource block is presented in Figure 9.16. After drawing the system resources along time and frequency axes we can interpret a resource block as a rectangle of $N_{RB}^{DL} = 12$ subcarriers and $N_{symb}^{DL} = 6$ or 7 OFDM symbols, depending on the length of the cyclic prefix. Table 9.1 indicates how many resource blocks can be applied in a channel of a given bandwidth. Let us note that the distance between neighboring subcarriers is equal to 15 kHz and it remains constant for any OFDM signal configuration. Thus, 180 kHz is the bandwidth of a single resource block.

Depending on the negotiated service conditions, channel conditions and traffic intensity, a number of resource blocks can be assigned to each link. The amount of radio resources given to the link can be changed dynamically and can be assigned according to temporary needs, thus increasing the overall system throughput.

In the opposite direction, i.e. from mobile terminals to a base station, SC-FDMA is used as the multiple access method. The same rule of division of radio resources among

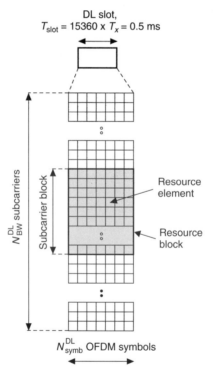

Figure 9.16 Map of the radio resources along time and frequency axes. © 2007. 3GPP™ TSs and TRs are the property of ARIB, ATIS, CCSA, ETSI, TTA and TTC who jointly own the copyright in them. They are subject to further modifications and are therefore provided to you "as is" for information purposes only. Further use is strictly prohibited

Table 9.1 Basic transmission parameters of the 3GPP LTE system

Channel bandwidth [MHz]	1.25	2.5	5	10	15	20
Subframe length				0.5 ms		
Subcarrier separation				15 kHz		
FFT size	128	256	512	1024	1536	2048
Number of OFDM symbols in the frame	7 (short CP) or 6 (long CP)					
Number of short CP samples[a]	9(10)	18(20)	36(40)	72(80)	108(120)	144(160)
Number of long CP samples	32	64	128	256	384	512
Size of the resource block	12 × 15 = 180 kHz					
Number of available resource blocks	6	12	25	50	75	100

[a]Numbers in parentheses describe the cyclic prefix length of the first OFDM symbol in the frame.

the user terminals is valid, i.e. the resource block is 12 subcarriers high and 7 or 6 OFDM symbols long, depending on the number of OFDM symbols applied in a time slot.

9.9 Conclusions

In this chapter we have sketched basic multiple access methods without considering the system capacity and performance evaluation. We concentrated on general rules of operation of these methods, without going into details. Continuous progress in technology has made it possible to apply more and more complicated modulations and multiple access schemes, which leads to more efficient use of the limited resources. This is particularly important in the case of wireless transmission.

Problems

Problem 9.1 *Let us analyze an ideal FDMA system in which the available bandwidth W is equally divided without guard bands between K users, as shown in Figure 9.17. Assume additive white Gaussian noise as a disturbance in the channel. Recalling our considerations on the capacity of a band-limited channel from Chapter 1, calculate the total capacity C_{tot} of the system for two cases:*

1. *Each user is assigned an average power P_u, regardless of the number of users. Find the capacity C_u per user and the total capacity as the sum of the capacities of all system users. Plot the total capacity C_{tot} normalized with respect to the bandwidth as a function of E_b/N_0 for the number of users $K = 1, 2, 5$ and 10.*

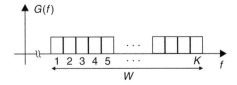

Figure 9.17 FDMA spectrum assignment

2. *Assume that due to administrative or technical limitations the overall power transmitted in the frequency band W assigned to the system is limited to the average power P. Thus, each of K users is able to use the average power P/K. Calculate the capacity of a single user, C_u, and the total capacity C_{tot} as a function of E_b/N_0 for the number of users $K = 1, 2, 5$ and 10. Compare the results obtained with those from Point 1.*

Problem 9.2   *Let us analyze an ideal TDMA system in which the time period T is equally divided without guard intervals between K users, as shown in Figure 9.18. Assume additive white Gaussian noise as a disturbance in the channel. Calculate the total capacity, C_{tot}, for two cases:*

1. *Each user is assigned an average power P_u per time period T, regardless of the number of users, although the transmission takes place in a single time slot only. Find the capacity C_u per user and the total capacity as the sum of the capacities of all system users. Plot the total capacity C_{tot} normalized with respect to the bandwidth as a function of E_b/N_0 for the number of users $K = 1, 2, 5$ and 10.*
2. *Assume that due to administrative or technical limitations the overall power transmitted in the frequency band W assigned to the system is limited to the average power P. Thus, each of K users is able to use the average power P/K. Calculate the capacity of a single user, C_u, and the total capacity C_{tot} as a function of E_b/N_0 for the number of users $K = 1, 2, 5$ and 10. Compare the results obtained with those from Point 1.*
3. *Compare the results obtained for TDMA with those obtained for FDMA in Problem 9.1.*

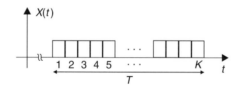

Figure 9.18   TDMA time slot assignment

Problem 9.3   *Let us analyze digital transmission in the GSM system that operates in the TDMA/FDMA multiple access mode. Basic transmission parameters of the GSM system are described in Section 6.10. The spectrum assigned to the GSM system in each direction in the 900 MHz band is divided into 124 FDMA channels of 200 kHz each. Inspect the time structure of the GSM transmission in a single time slot and check how many bits per slot are available for transmission of user data (analyze Figure 6.14).*

1. *Consider a rural area GSM cell in which a single FDMA channel is applied (all connections are performed on the same carrier frequency in the TDMA mode). The TDMA slot #0 is used for control, maintenance and setting of the connections. The remaining slots are used to carry user traffic. Calculate the spectral efficiency in bit/s/Hz, taking into account all bits transmitted in a single frame in all slots.*
2. *Repeat the calculations, taking into account bits transmitted by the users only. How large is the overhead, expressed as a percentage of the overall transmitted bits on a*

single FDMA channel, that is needed to maintain and organize transmission in the GSM system?

3. *Repeat calculations from Points 1 and 2 for a larger cell in which four FDMA channels are applied (transmission takes place on four carriers) and two slots on one of the carriers are used for control, maintenance and setting of the connections.*

4. *In the GSM system a special burst, called an access burst (shown in Figure 9.19), is used by a mobile station to inform the base station about the need for information exchange and connection setting. At the moment of its sending the mobile station does not know the distance to the base station. Assuming that electromagnetic waves propagate at the speed of light, calculate how far the mobile station can be located from the base station so that the access burst sent at the beginning of the slot arrives at the base station still within the same time slot.*

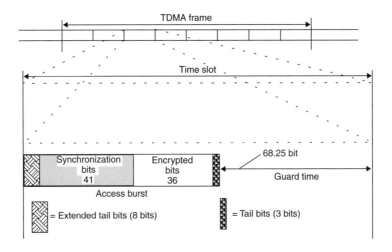

Figure 9.19   GSM access burst

Problem 9.4   *Consider a discrete time synchronous CDMA system. Each user spreads its data symbol $d_{i,n}$ (i is the index of the user, whereas n is a data symbol time index) by the unique bipolar sequence based on the m-sequence of length $N_s = 2^m - 1$. Denote the bipolar spreading sequence of the ith user as $\mathbf{c}_i = [c_{i,1}, c_{i,2}, \ldots, c_{i,N_s}]$. Assume that data symbols of K system users are zero mean and mutually uncorrelated. Their variance is σ_d^2. The signals from all users arrive at the receiver jointly with the additive Gaussian noise, denoted by the vector of samples $v_n = [v_1, v_2, \ldots, v_{N_s}]$. The noise samples are zero mean and mutually uncorrelated with the variance σ_v^2. The receiver extracts the signal of the jth user by applying a simple correlator and using the jth user spreading sequence $\mathbf{c}_j$. Recall the properties of the spreading bipolar sequences resulting from the application of m-sequences of length N_s:*

$$\mathbf{c}_i \mathbf{c}_j^T = \begin{cases} N_s & for \quad i = j \\ -1 & for \quad i \neq j \end{cases}$$

Find the signal-to-noise plus interference power ratio $P/(I + N)$ at the output of the correlator as a function of the spreading length N_s, the number of users K and the data and noise powers.

Problem 9.5 *Generate, by applying Matlab, the set of Walsh-Hadamard spreading sequences of length $N_s = 64$ using formula (9.1). Calculate the value of cross-correlation for two selected sequences $\mathbf{c}_i$ and $\mathbf{c}_j$ ($i, j = 0, \ldots, N_s - 1$). Next calculate the value of the cross-correlation for the sequence $a_n\mathbf{c}_i$ and the sequences $b_{n-1}\mathbf{c}_j$ and $b_n\mathbf{c}_j$ delayed by, say, two or four chips (see Figure 9.20), where $a_n, b_n, b_{n-1} = \pm 1$ are data symbols transmitted using sequences $\mathbf{c}_i$ and $\mathbf{c}_j$, n is the data time index and $\mathbf{c}'_j$ is the cyclically shifted version of $\mathbf{c}_j$. Is orthogonality of the signals further preserved? Perform your calculations for a few pairs of indices i and j.*

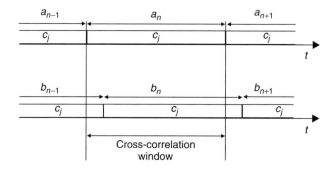

Figure 9.20 Allocation of spreading sequences of the ith and jth user on the time axis

Problem 9.6 *Analyze the orthogonality of several OVSF spreading sequences. Write a program that generates the OVSF tree up to sequences of length 16. Let the spreading code in the ith transmission channel have the spreading factor $N_s = 4$ and the spreading code in the jth transmission channel have the spreading factor $N_s = 16$. Select two spreading sequences having these spreading factors that are allowable due to the OVSF code selection rule. Draw the spread data sequences of the ith user and the jth user if they are equal to $\mathbf{a}_n = (1, -1, 1, 1)$ and $\mathbf{b}_n = 1$, respectively. Check their mutual orthogonality. Check the orthogonality if the signal of the jth user is delayed by one chip and the symbol $\mathbf{b}_{n-1}$ preceding the main data symbol is equal to -1.*

Appendix

Let us consider a normalized Gaussian random variable X of zero mean and unit variance. The probability that a value of the random variable X is greater than x is given by the formula known as the *Q-function*:

$$Q(x) = \frac{1}{\sqrt{2\pi}} \int_x^\infty \exp\left(-\frac{t^2}{2}\right) dt$$

The value of the Q-function is the area under the tail of the normalized Gaussian probability distribution.

Alternatively, the *complementary error function*, the so-called *erfc-function*, is often used in calculations of error probabilities. This is described by the formula

$$\text{erfc}(x) = \frac{2}{\sqrt{\pi}} \int_x^\infty \exp\left(-t^2\right) dt$$

One can easily show using substitution of variables that both functions are related to each other by the equations

$$Q(x) = \frac{1}{2}\text{erfc}\left(\frac{x}{\sqrt{2}}\right)$$

$$\text{erfc}(x) = 2Q\left(\sqrt{2}x\right)$$

Below we show the tables of both functions because they can be useful in problem solutions. They are often contained as special functions in software packages. For example in Matlab they can be invoked as $y = \text{qfunc}(x)$ and $y = \text{erfc}(x)$, respectively.

Introduction to Digital Communication Systems   Krzysztof Wesołowski
© 2009 John Wiley & Sons, Ltd

Table A.1   The Q-function values

x	$Q(x)$	x	$Q(x)$	x	$Q(x)$
0.00	0.50000000	1.35	0.08850799	2.70	0.00346697
0.05	0.48006119	1.40	0.08075666	2.75	0.00297976
0.10	0.46017216	1.45	0.07352926	2.80	0.00255513
0.15	0.44038231	1.50	0.06680720	2.85	0.00218596
0.20	0.42074029	1.55	0.06057076	2.90	0.00186581
0.25	0.40129367	1.60	0.05479929	2.95	0.00158887
0.30	0.38208858	1.65	0.04947147	3.00	0.00134990
0.35	0.36316935	1.70	0.04456546	3.05	0.00114421
0.40	0.34457826	1.75	0.04005916	3.10	0.00096760
0.45	0.32635522	1.80	0.03593032	3.15	0.00081635
0.50	0.30853754	1.85	0.03215677	3.20	0.00068714
0.55	0.29115969	1.90	0.02871656	3.25	0.00057703
0.60	0.27425312	1.95	0.02558806	3.30	0.00048342
0.65	0.25784611	2.00	0.02275013	3.35	0.00040406
0.70	0.24196365	2.05	0.02018222	3.40	0.00033693
0.75	0.22662735	2.10	0.01786442	3.45	0.00028029
0.80	0.21185540	2.15	0.01577761	3.50	0.00023263
0.85	0.19766254	2.20	0.01390345	3.55	0.00019262
0.90	0.18406013	2.25	0.01222447	3.60	0.00015911
0.95	0.17105613	2.30	0.01072411	3.65	0.00013112
1.00	0.15865525	2.35	0.00938671	3.70	0.00010780
1.05	0.14685906	2.40	0.00819754	3.75	0.00008842
1.10	0.13566606	2.45	0.00714281	3.80	0.00007235
1.15	0.12507194	2.50	0.00620967	3.85	0.00005906
1.20	0.11506967	2.55	0.00538615	3.90	0.00004810
1.25	0.10564977	2.60	0.00466119	3.95	0.00003908
1.30	0.09680048	2.65	0.00402459	4.00	0.00003167

Table A.2 The erfc-function values

x	erfc(x)	x	erfc(x)	x	erfc(x)
0.00	1.00000000	1.35	0.05623780	2.70	0.00013433
0.05	0.94362802	1.40	0.04771488	2.75	0.00010062
0.10	0.88753708	1.45	0.04030497	2.80	0.00007501
0.15	0.83200403	1.50	0.03389485	2.85	0.00005566
0.20	0.77729741	1.55	0.02837727	2.90	0.00004110
0.25	0.72367361	1.60	0.02365162	2.95	0.00003020
0.30	0.67137324	1.65	0.01962441	3.00	0.00002209
0.35	0.62061795	1.70	0.01620954	3.05	0.00001608
0.40	0.57160764	1.75	0.01332833	3.10	0.00001165
0.45	0.52451828	1.80	0.01090950	3.15	0.00000840
0.50	0.47950012	1.85	0.00888897	3.20	0.00000603
0.55	0.43667663	1.90	0.00720957	3.25	0.00000430
0.60	0.39614391	1.95	0.00582067	3.30	0.00000306
0.65	0.35797067	2.00	0.00467773	3.35	0.00000216
0.70	0.32219881	2.05	0.00374190	3.40	0.00000152
0.75	0.28884437	2.10	0.00297947	3.45	0.00000107
0.80	0.25789904	2.15	0.00236139	3.50	0.00000074
0.85	0.22933194	2.20	0.00186285	3.55	0.00000052
0.90	0.20309179	2.25	0.00146272	3.60	0.00000036
0.95	0.17910919	2.30	0.00114318	3.65	0.00000024
1.00	0.15729921	2.35	0.00088927	3.70	0.00000017
1.05	0.13756389	2.40	0.00068851	3.75	0.00000011
1.10	0.11979493	2.45	0.00053058	3.80	0.00000008
1.15	0.10387616	2.50	0.00040695	3.85	0.00000005
1.20	0.08968602	2.55	0.00031066	3.90	0.00000003
1.25	0.07709987	2.60	0.00023603	3.95	0.00000002
1.30	0.06599206	2.65	0.00017849	4.00	0.00000002

Bibliography

Abend K and Fritchman BD 1970 Statistical Detection for Communication Channels with Intersymbol Interference. *Proc. IEEE*, **58**, 779–785.

Abramson N 1963 *Information Theory and Coding*. McGraw-Hill.

Aman AK, Cupo RL and Zervos NA 1991 Combined Trellis Coding and DFE through Tomlinson Precoding. *IEEE J. Selected Areas Commun.* **9**, 876–883.

Anderson JB and Mohan S 1984 Sequential Decoding Algorithms: a Survey and Cost Analysis. *IEEE Trans. Commun.* **32**, 169–176.

Austin ME 1967 *Equalization of Dispersive Channels Using Decision Feedback*, QPR 84, pp. 227–243. Research Laboratory of Electronics, MIT, MA.

Bahl LR, Cocke J, Jelinek F and Raviv J 1974 Optimal Decoding of Linear Codes for Minimizing Symbol Error Rate. *IEEE Trans. Inform. Theory* **20**, 284–287.

Baier PW, Jung P, and Klein A 1996 Taking Challenge of Multiple Access for Third-Generation Cellular Mobile Radio Systems – A European View. *IEEE Commun. Mag.* **2**, 82–89.

Baran Z 1982 *Foundations of Data Transmission* (in Polish). WKŁ.

Barry JR, Lee EA and Messerschmitt DG 2003 *Digital Communication*, 3rd edn. Springer.

Belfiore CA and Park Jr. JH 1979 Decision Feedback Equalization. *Proc. IEEE* **67**, 1143–1156.

Bell System Technical Journal 1979. Special issue on AMPS, **58**, No.1.

Bellini S 1988 Bussgang Techniques for Blind Equalization. *Proc. of GLOBECOM'88*, 1634–1640.

Benveniste A, Goursat M and Ruget G 1980 Robust Identification of a Nonminimum Phase System: Blind Adjustment of a Linear Equalizer in Data Communications. *IEEE Trans. Autom. Control* **25**, 385–398.

Benvenuto N and Cherubini G 2002 *Algorithms for Communications Systems and their Applications*. Wiley.

Berlekamp ER 1965 On Decoding Binary Bose-Chaudhuri-Hocquenghem Codes. *IEEE Trans. Inform. Theory* **11**, 577–580.

Berrou C, Glavieux A and Thitimajshima P 1993 Near Shannon Limit Error-Correcting Coding and Decoding: Turbo Codes. *Proc. of 1993 International Conference on Communications*, 1064–1070.

Bing B 2000 *High-Speed Wireless ATM and LANs*. Artech House.

Blahut RE 1983 *Theory and Practice of Error Control Codes*. Addison-Wesley.

Blahut RE 1987 *Principles and Practice of Information Theory*. Addison-Wesley.

Bossert M 1999 *Channel Coding for Telecommunications*. Wiley.

Carey MB, Chen H-T, Desloux A, Ingle JF and Park KI 1984 1982/83 End Office Connection Study: Analog and Voiceband Data Transmission Performance Characterization of the Public Switched Network. *AT&T Bell Lab. Tech. J.* **63**, 2059–2117.

Castoldi P 2002 *Multiuser Detection in CDMA Mobile Terminals*. Artech House.

Chang RW and Hancock JC 1966 On Receiver Structures for Channel Having Memory. *IEEE Trans. Inform. Theory* **12**, 463–468.

Chevillat PR and Elefteriou E 1989 Decoding of Trellis-Encoded Signals in the Presence of Intersymbol Interference and Noise. *IEEE Trans. Commun.* **37**, 669–676.

Chevillat PR, Maiwald D and Ungerboeck G 1987 Rapid Training of a Voiceband Data-Modem Receiver Employing an Equalizer with Fractional-T Spaced Coefficients. *IEEE Trans. Commun.* **35**, 869–876.

Chouly A, Brajal A and Jourdan J 1993 Orthogonal Multi-Carrier Techniques Applied to Direct Sequence Spread Spectrum CDMA Systems. *Proc. of GLOBECOM'93*, 1723–1728.

Cioffi JM and Kailath T 1984 Fast Recursive Least-Squares Transversal Filter for Adaptive Filtering, *IEEE Trans. Acoust., Speech, Signal Process.* **32**, 304–337.

Clark AP 1985 *Equalizers for Digital Modems*. Pentech Press.

Clark GC and Cain JB 1981 *Error-Correction Coding for Digital Communications*. Plenum Press.

Couch LW 1987 *Digital and Analog Communication Systems*. Macmillan Publishing.

Cover TM and Thomas JA 1991, *Elements of Information Theory*. Wiley.

Danilo-Lemoine F, Falconer D, Lam C-T, Sabbaghian M and Wesołowski K 2008 Power Backoff Reduction Techniques for Generalized Multicarrier Waveforms. *EURASIP Journal of Wireless Systems and Networking*, **2008**, Article ID 437801.

Davidson CW 1978 *Transmission Lines for Communications*. McMillan Press.

Dąbrowski A and Dymarski P 1999 *Foundations of Digital Transmission* (in Polish). Warsaw University of Technology.

Ding Z and Ye L. 2001 *Blind Equalization and Identification*. Marcel Dekker.

Dixon RC 1984 *Spread Spectrum Systems*, 2nd edn. Wiley.

Duel-Hallen A 1992 Equalizers for Multiple Input/Multiple Output Channels and PAM Systems with Cyclostationary Input Sequences. *IEEE J. Selected Areas Commun.* **10**, 630–639.

Duel-Hallen A and Heegard C 1989 Delayed Decision-Feedback Sequence Estimation. *IEEE Trans. Commun.* **37**, 428–436.

ETSI 1997 ETS 300 421. *Framing structure channel coding and modulation for 11/12 GHz satellite services*. ETSI Standard.

ETSI 1999 GSM 05.02 (EN 300 908). *Digital cellular telecommunications system (Phase 2+) (GSM). Multiplexing and multiple access on the radio path*. ETSI Standard.

ETSI 2000 GSM 05.05 (EN 300 910). *Digital celllular telecommunications system (Phase 2+); Radio transmission and reception*, V.8.5.1. ETSI Standard.

ETSI 2001a GSM 05.04 (EN 300 959). *Digital cellular telecommunications system (Phase 2+) (GSM); Modulation*. ETSI Standard.

ETSI 2001b TS 101 475. *Broadband Radio Access Networks (BRAN); HIPER LAN Type 2; Physical (PHY) layer*. ETSI Standard.

ETSI 2004 EN 300 744. *Digital Video Broadcasting (DVB); Framing structure, channel coding and modulation for digital terrestrial television*. ETSI Standard.

Eyuboglu MV and Qureshi SUH 1988 Reduced-State Sequence Estimation with Set Partitioning and Decision Feedback. *IEEE Trans. Commun.* **36**, 13–20.

Falconer DD 1976 Jointly Adaptive Equalization and Carrier Recovery in Two-Dimensional Digital Communication Systems. *Bell System Tech. J.* **55**, 317–394.

Fazel K and Kaiser S 2003 *Multi-Carrier and Spread Spectrum Systems*. Wiley.

Forney Jr. GD 1966 *Concatenated Codes*. MIT Press.

Forney Jr. GD 1972 Maximum-Likelihood Sequence Estimation of Digital Sequences in the Presence of Intersymbol Interference. *IEEE Trans. Inform. Theory* **18**, 363–378.

Forney GD, Gallager RG, Lang GR, Longstaff FM and Qureshi SUH 1984 Efficient Modulation on Band-Limited Channels. *IEEE J. Selected Areas Commun.* **2**, 632–647.

Gallager RG 1968 *Information Theory and Reliable Communication*. Wiley.

Garg VK 2000 *IS-95 CDMA and cdma2000*. Prentice Hall.

Gibbs AJ and Addie R 1979 The Covariance of Near-End Crosstalk and Its Application to PCM System Engineering in Multipair Cable. *IEEE Trans. Commun.* **27**, 469–477.

Gitlin RD, Hayes JF and Weinstein SB *Principles of Data Communications*. Plenum Press.

Gitlin RD, Meadors HC and Weinstein SB 1982 The Tap Leakage Algorithm: an Algorithm for the Stable Operation of a Digitally Implemented, Fractionally Spaced Adaptive Equalizer. *Bell Syst. Tech. J.* **61**, 1817–1839.

Godard DN 1974 Channel Equalization Using a Kalman Filter for Fast Data Transmission. *IBM J. Res. Dev.* **18**, 267–273.

Gold R 1966 Maximal Recursive Sequences with 3-Valued Recursive Cross-Correlation Functions. *IEEE Trans. Inform. Theory* **14**, 154–156.

Goldsmith AJ and Varaiya PP 1997 Capacity of Fading Channels with Channel Side Information. *IEEE Trans. Inform. Theory* **43**, 1986–1992.

Golomb SW 1967 *Shift Register Sequences*. Holden-Day.

Haartsen JC 1998 BLUETOOTH – The Universal Radio Interface for ad hoc Wireless Connectivity. *Ericsson Review* **3**, 110–117.

Haartsen JC 2000 The Bluetooth Radio System. *IEEE Pers. Commun.* **7**, 28–36.

Hagenauer J and Hoeher P 1989 A Viterbi Algorithm with Soft-Decision Outputs and Its Applications. *Proc. of IEEE Global Communications Conference*, 47.1.1–47.1.7

Halsall F 1996 *Data Communications, Computer Networks and Open Systems*, 4th edn. Addison-Wesley.

Hara S and Prasad R 1997 Overview of Multi-Carrier CDMA. *IEEE Commun. Mag.* **35**, 126–133.

Hartley RVL 1928 Transmission of Information. *Bell System Tech. J.* **7**, 535–563.

Haykin S 2000 *Communication Systems*, 4th edn. Wiley.

Haykin S 2002 *Adaptive Filter Theory*, 4th edn. Prentice-Hall.

Heegard Ch and Wicker SB 1999 *Turbo Coding*. Kluwer Academic Publishers.

Heise W and Quatrocchi P 1989 *Informations- und Codierungstheorie*. Springer.

Held G 1999 *Data Communications Networking Devices: Operation, Utilization and LAN and WAN Internetworking*, 4th edn. Wiley.

Holma A and Toskala A 2004 *WCDMA for UMTS. Radio Access for Third Generation Mobile Communications*, 3rd edn. Wiley.

Hołubowicz W and Szwabe M 1998 *Spread Spectrum Radio Systems, CDMA. Theory, Standards, Applications* (in Polish). Holkom.

IEEE 1999 Standard 802.11a. Supplement to 802.11. *High-speed Physical Layer in the 5 GHz Band*.

IEEE 2003 Standard 802.11g. Supplement to 802.11. *Amendment 4: Further Higher Data Rate Extension in the 2.4 GHz Band*.

IEEE (2007) Standard 802.11. *Telecommunications and Information Exchange between Systems. Local and Metropolitan Area Networks. Specific Requirements. Part 11: Wireless LAN Medium Access Control (MAC) and Physical Layer (PHY) Specifications*.

ITU-T Recommendations 1997, Series M.

ITU-T Recommendations 1997, Series V.

ITU-T G.961 Recommendation 1993 *Digital Transmission System on Metallic Local Lines for ISDN Basic Rate Access*.

ITU-T G.991.1 Recommendation 1998 *High Bit Rate Digital Subscriber Line (HDSL) Transceivers*.

ITU-T G.991.2 Recommendation 2003 *Single-Pair High-Speed Digital Subscriber Line (SHDSL) Transceivers*.

ITU-T V. 32 Recommendation 1993 *A Family of 2-Wire, Duplex Modems Operating at Data Signalling Rates of up to 9600 bit/s for Use on the General Switched Telephone Network and on Leased Telephone-Type Circuits*.

ITU-T V.34 Recommendation 1997 *A Modem Operating at Data Signalling Rates of up to 33 600 bit/s for Use on the General Switched Telephone Network and on Leased Point-to-Point 2-Wire Telephone-Type Circuits*.

Jakes W 1974 *Microwave Mobile Communications*. Wiley.

Karam G, Maalej K, Paxal V and Sari H 1996 Variable Symbol-Rate Modem Design for Cable and Satellite TV Broadcasting. In *Signal Processing in Telecommunications*. (eds Biglieri E and Luise M), pp. 244–255. Springer.

Kasami T, Lin S and Peterson W 1968 Some Results on Cyclic Codes which Are Invariant under the Affine Group and Their Applications. *Inf. Control* **11**, 475–496.

Kiefer J 2003 Data Compression. In *Wiley Encyclopedia of Telecommunications* (ed. Proakis JG), pp. 631–650. Wiley.

Koffman I and Roman V 2002 Broadband Wireless Access Solutions Based on OFDM Access in IEEE 802.16. *IEEE Commun. Mag.* **40**, 96–103.

Krenz R and Wesołowski K 1997 Comparative Study of Space-Diversity Techniques for MLSE Receivers in Mobile Radio. *IEEE Trans. Veh. Technol.* **46**, 653–663.

Kurzweil J 2000 *An Introduction to Digital Communications*. Wiley.

Lagarde JB, Rouffet D and Cohen M 1995 GLOBALSTAR System: an Overview. *Proc. of First European Workshop on Mobile/Personal Satcoms (EMPS '94)* Springer.

Lathi BP 1998 *Modern Digital and Analog Communication Systems*, 3rd edn. Oxford University Press.

Laurent PA 1986 Exact and Approximate Construction of Digital Phase Modulations by Superposition of Amplitude Modulated Pulses (AMP). *IEEE Trans. Commun.* **34**, 150–160.

Lechleider JW 1986 Loop Transmission Aspects of ISDN Basic Access. *IEEE J. Selected Areas Commun.* **4**, 1294–1301.

Lee JS, and Miller LE 1998 *CDMA Systems Engineering Handbook*. Artech House.

Lee LHCh 2000 *Error-Control Block Codes for Communications Engineers*. Artech House.

Li T 1980 Structures, Parameters, and Transmission Properties of Optical Fibers. *Proc. IEEE* **10**, 1175–1180.

Lin S and Costello DJ 2004 *Error Control Coding*, 2nd edn. Prentice-Hall.

Lindsey WC 1972 *Synchronization Systems in Communication and Control*. Prentice-Hall.

Ling F and Proakis JG 1985 Adaptive Lattice Decision-Feedback Equalizers – their Performance and Application to Time-Variant Multipath Channels. *IEEE Trans. Commun.* **33**, 348–356.

Long G, Ling F and Proakis JG 1989 The LMS Algorithm with Delayed Coefficient Adaptation. *IEEE Trans. Acoust., Speech, Signal Process* **37**, 1397–1405.

Lucky RW, Salz J and Weldon Jr. EJ 1968 *Principles of Data Communication*. McGraw-Hill.

Luise M, Mengali U and Morelli M 2003 Synchronization in Digital Communication Systems. In *Wiley Encyclopedia of Telecommunications* (ed. Proakis JG), pp. 2472–2484. Wiley.

Macchi O 1995 *Adaptive Processing*. Wiley.

Macchi O and Guidoux L 1975 A New Equalizer and Double Sampling Equalizer, *Ann. Telecomm.* **30**, 331–338.

MacKay D 1999 Good Error-Correcting Codes Based on Very Sparsed Matrices. *IEEE Trans. Inform. Theory* **45**, 399–431.

MacKay D 2003 *Information Theory, Inference and Learning Algorithms*. Cambridge University Press.

MacKay D and Neal R 1996 Near Shannon Limit Performance of Low Density Parity Check Codes. *Electron. Lett.* **32**, 1645–1646.

Manning T 1999 *Microwave Radio Transmission Design Guide*. Artech House.

Mansuripur M 1987 *Introduction to Information Theory*. Prentice-Hall.

Martin AL, Coutts RP and Campbell J 1983 Results of a 16-QAM Digital Radio Field Experiment. *Proc. of IEEE International Conference on Communications*, F2.2.1–F2.2.8

Massey JL 1963 *Threshold Decoding*. MIT Press.

Massey JL 1972 Shift Register Synthesis and BCH Decoding. *IEEE Trans. Inform. Theory* **15**, 196–198.

Meggitt JE 1961 Error Correcting Codes and Their Implementation for Data Transmission Systems. *IRE Trans. Inform. Theory* **7**, 234–244.

Mengali U and D'Andrea AN 1997 *Synchronization Techniques for Digital Receivers*. Plenum Press.

Meyr H and Ascheid G 1990 *Synchronization in Digital Communications*. Wiley.

Michelson AM and Levesque AH 2003 BCH Codes – Nonbinary and Reed-Solomon. In *Wiley Encyclopedia of Telecommunications* (ed. Proakis JG), pp. 253–262. Wiley.

Modestino JW, Massey ChS, Bollen RE and Prabhu RP 1986 Modeling and Analysis of Error Probability Performance for Digital Transmission over Two-Wire Plant. *IEEE J. Selected Areas Commun.* **4**, 1317–1330.

Monsen P 1971 Feedback Equalization for Fading Dispersive Channels. *IEEE Trans. Inform. Theory* **17**, 56–64.

Moon TK 2005 *Error Correction Coding: Mathematical Methods and Algorithms*. Wiley.

Myung HG, Lim J and Goodman DJ 2006 Single Carrier FDMA for Uplink Wireless Transmission. *IEEE Veh. Technol. Mag.* **1**, 30–38.

Nowakowski J and Sobczak W 1971 *Information Theory* (in Polish). WNT.

Nyquist H 1924 Certain Factors Affecting Telegraph Speed. *Bell System Tech. J.* **3**, 324.

Okunev Y 1997 *Phase and Phase Difference Modulation in Digital Communication*. Artech House.

Pahlavan K and Levesque AH 1995 *Wireless Information Networks*. Wiley.

Park Jr. JH 1978 On Binary DPSK Detection. *IEEE Trans. Commun.* **26**, 484–486.

Pätzold M 2002 *Mobile Fading Channels*. Wiley.

Peebles PZ 1986 *Digital Communication Systems*. Prentice Hall.

Peterson W and Weldon EJ 1972 *Error-Correcting Codes*. MIT Press.

Prasad R 1996 *CDMA for Wireless Personal Communications*. Artech House.

Price R and Green PE 1958 A Communication Technique for Multipath Channels. *Proc. IEEE* **46**, 555–570.

Proakis JG 2000 *Digital Communications*, 4th edn. McGraw-Hill.

Qureshi SUH 1985 Adaptive Equalization. *Proc. IEEE* **53**, 1349–1387.

Qureshi SUH and Forney Jr. DG 1977 Performance and Properties of a T/2 Equalizer. *Conference Record*, *National Telecommunication Conference*.

Raheli R, Polydoros A and Tzou Ch-K 1991 The Principle of Per-Survivor Processing: A General Approach to Approximate and Adaptive MLSE. *Proc. of GLOBECOM'91*, 1170–1175.

Rappaport TS 1996 *Wireless Communications. Principles and Practice*. Prentice Hall.

Roman S 1992 *Coding and Information Theory*. Springer.

Rummler WD 1979 A New Selective Fading Model: Application to Propagation Data. *Bell System Tech. J.* **58**, 1037–1071.

Rummler WD 1981 More on the Multipath Fading Channel Model. *IEEE Trans. Commun.* **29**, 346–352.

Ryan WE 2004 An Introduction to LDPC Codes. In *Coding and Signal Processing for Magnetic Recording Systems* (eds. Vasic and Kuflas E.). CRC Press.

Sari H, Levy Y and Karam G 1996 Orthogonal Frequency-Division Multiple Access for the Return Channel on CATV Networks. *Proc. of Int. Conference on Telecommunications (ICT'96)*, 602–607.

Sari H, Levy Y and Karam G 1997 An Analysis of Orthogonal Frequency-Division Multiple Access. *Proc. of IEEE GLOBECOM'97*, 1635–1639.

Sari H, Vanhaverbeke F and Moeneclaey M 2000 Extending Capacity of Multiple Access Channels. *IEEE Commun. Mag.* **38**, 74–82.

Satorius EH and Alexander ST 1979 Channel Equalization Using Adaptive Lattice Algorithms. *IEEE Trans. Commun.* **27**, 899–905.

Satorius EH and Pack JD 1981 Application of Least Squares Lattice Algorithms to Adaptive Equalization. *IEEE Trans. Commun.* **29**, 136–142.

Schmid H 1976 *Theorie und Technik der Nachrichtenkabel* (in German). Hüthig.

Seidler J 1983 *Nauka o informacji* (in Polish). WNT.

Seshadri N 1994 Joint Data and Channel Estimation Using Blind Trellis Search Techniques. *IEEE Trans. Commun.* **42**, 1000–1011.

Seurre E, Savelli P and Pietri P-J 2003 *GPRS for Mobile Internet*. Artech House.

SGS-Thomson Microelectronics 1993 *M5913 Combined Single Chip PCM Codec and Filter*.

Shannon CE 1948 A Mathematical Theory of Communication. *Bell System Tech. J.* **27**, 379–423 and 623–656.

Siwiak K and McKeown D 2004 *Ultra-Wideband Radio Technology*. Wiley.

Sklar B 1988 *Digital Communications. Fundamentals and Applications*. Prentice Hall.

Sobczak W 1984 *Statistical Theory of Information-Bearing Systems* (in Polish), WKŁ.

Springer A, and Weigel R 2002 *UMTS. The Physical Layer of Universal Mobile Telecommunications System*. Springer.

Steinbuch K and Rupprecht W 1982 *Nachrichtentechnik*, 2nd edn. Springer.

Stüber GL 2001 *Principles of Mobile Communication*, 2nd edn. Kluwer Academic Publishers.

Taylor DP, Vitetta GM, Hart BD and Mämmelä A 1998 A Wireless Channel Equalisation. *Eur. Trans. Telecomm.* **9**, 117–143.

Togneri R and deSilva ChJS 2002 *Fundamentals of Information Theory and Coding Design*. Chapman & Hall/CRC.

Ungerboeck G 1972 Theory on the Speed of Convergence in Adaptive Equalizers for Digital Communication. *IBM J. Res. Dev.* **16**, 546–555.

Ungerboeck G 1974 Adaptive Maximum-Likelihood Receiver for Carrier-Modulated Data Transmission Systems. *IEEE Trans. Commun.* **22**, 624–636.

Ungerboeck G 1976 Fractional Tap-Spacing Equalizer and Consequence for Clock Recovery in Data Modems. *IEEE Trans. Commun.* **24**, 856–864.

Ungerboeck G 1982 Channel Coding with Multilevel/Phase Signals. *IEEE Trans. Inform. Theory* **28**, 55–67.

Ungerboeck G 1987a Trellis-Coded Modulation with Redundant Signal Sets Part I: Introduction. *IEEE Commun. Mag.* **25**, 5–11.

Ungerboeck G 1987b Trellis-Coded Modulation with Redundant Signal Sets Part II: State of the Art. *IEEE Commun. Mag.* **25**, 12–21.

Verdú S 1998 *Multiuser Detection*. Cambridge University Press.

Viterbi AJ 1967 Error Bounds for Convolutional Codes and an Asymptotically Optimum Decoding Algorithm. *IEEE Trans. Inform. Theory* **13**, 260–269.

Viterbi AJ 1995 *CDMA: Principles of Spread Spectrum Communications*. Addison-Wesley.

Viterbi AJ and Omura JK 1979 *Principles of Digital Communication and Coding*. McGraw-Hill.

Vucetic B and Yuan J 2000 *Turbo-Codes. Principles and Applications*. Kluwer Academic Publishers.

Vucetic B and Yuan J 2003 *Space-Time Coding*. Wiley.

Werner J-J 1990 Impulse Noise in the Loop Plant. *Proc. of IEEE International Conference on Communications*, 1734–1737.

Werner J-J 1991 The HDSL Environment. *IEEE J. Selected Areas Commun.* **9**, 785–800.

Wesołowski K 1987 An Efficient DFE & ML Suboptimum Receiver for Data Transmission Over Dispersive Channels Using Two-Dimensional Signal Constellations. *IEEE Trans. Commun.* **35**, 336–339.

Wesołowski K 2002 *Mobile Communication Systems*. Wiley.

Wesołowski K 2003 Adaptive Equalizers. In *Wiley Encyclopedia of Telecommunications* (ed. Proakis JG), pp. 79–94. Wiley.

Westin D 1993 NMT: The Nordic Solution. In *Cellular Radio Systems* (eds Balston DM and Macario RCV), pp. 73–111. Artech House.

Wicker SB 1995 *Error Control Systems for Digital Communication and Storage*. Prentice Hall.

Wiesner L 1984 *Telegraph and Data Transmission over Shortwave Radio Links*. Siemens AG and Wiley.

WINNER Deliverable 2.6 Assessment of Multiple Access Technologies, 6FP Project IST-2003-507581 WINNER – World Wireless Initiative New Radio.

Xu G, Tong L and Liu H 1994 A New Algorithm for Fast Blind Equalization of Wireless Communication Channels. *Proc. of GLOBECOM'94*, 544–548.

Ziv J and Lempel A 1977 A Universal Algorithm for Data Compression. *IEEE Trans. Inform. Theory* **23**, 337–343.

Ziv J and Lempel A 1978 Compression of Individual Sequences via Variable-Rate Coding. *IEEE Trans. Inform. Theory* **24**, 530–536.

Index